PHASE DIAGRAMS

UNDERSTANDING THE BASICS

Edited
by
F.C. Campbell

The Materials Information Society

ASM International®
Materials Park, Ohio 44073-0002
www.asminternational.org

First printing, March 2012

Prepared under the direction of the ASM International Technical Book Committee (2011–2012), Bradley J. Diak, *Chair.*

ASM International staff who worked on this project include Scott Henry, Senior Manager, Content Development and Publishing; Karen Marken, Senior Managing Editor; Steven L. Lampman, Content Developer; Sue Sellers, Editorial Assistant; Bonnie Sanders, Manager of Production; Madrid Tramble, Senior Production Coordinator; and Diane Whitelaw, Production Coordinator.

Library of Congress Control Number: 2001012345

ISBN-13: 978-1-61503-835-0
ISBN-10: 1-61503-835-3

SAN: 204-7586

ASM International®
Materials Park, OH 44073-0002
www.asminternational.org

Printed in the United States of America

Contents

Preface

Phase diagrams are graphical maps that show the behavior of metal alloys during heating and cooling. In addition, they show the solid phases that are present after an alloy freezes. They are a primary tool in metallurgy because they provide the basis for predicting or interpreting the changes in the internal structure of a material. When an alloy undergoes heating and cooling, it undergoes phase transitions or transformations, in both the liquid and solid state. Therefore, an appreciable amount of this book is devoted to understanding these transformations.

In keeping with ASM's goal in the *Understanding the Basics* series of technical books, many of the mathematical formulations that underlie the theory of phase diagrams have been omitted. There are many excellent texts on chemical thermodynamic that provide this theory. The material in this book can be comprehended by anyone with some degree of a technical background. For those lacking knowledge of basic metallurgical principles, I have included two appendices that are helpful. Appendix A covers the basics of metallic structure, while Appendix B is an introduction to the principles of solidification. This book is helpful to engineers, technicians, production personnel and students with no previous exposure to metallurgy. It is also helpful to metallurgical engineers who have forgotten many of the principles of phase diagrams and need a refresher.

Chapter 1 is a brief introduction to phase diagrams. The next two chapters offer basic information that is helpful in reading subsequent chapters. Chapter 2 is an introduction to solid solutions and phase transformations. While I have attempted to avoid the detailed development of the thermodynamic formulation behind phase diagrams, Chapter 3 gives a somewhat cursory overview of the importance of Gibbs free energy and how it is used to construct and interpret phase diagrams.

The two types of alloy phase diagrams that receive the most emphasis are binary (two metals) and ternary (three metals). While a few quaternary (four metals) diagrams can be found in the literature, computer modeling based on thermodynamic principles (described in Chapter 13) are now almost exclusively used to analyze this degree of complexity.

Important binary phase diagrams involving liquid and solid reactions are covered in Chapter 4 (isomorphous), Chapter 5 (eutectic), Chapter 6 (peritectic), and Chapter 7 (monotectic). Solid-state transformations are covered in Chapter 8 and include the important eutectoid and peritectoid reactions.

Gas-metal reactions are important in metals processing and in-service corrosion; therefore, Chapter 11 covers the basics of these systems where pressure becomes an important variable.

The construction of a phase diagram is a tedious and exacting process. In Chapter 12, some of the important methods that are used in phase diagram determination and construction errors are discussed.

While many of the uses of phase diagrams will become apparent as you progress through the book, Chapter 14 summarizes their usage and gives some real world industrial examples of how phase diagrams were used to solve problems.

While phase diagrams are constructed under as close as possible to equilibrium conditions, many important phase transformations in alloys occur under highly nonequilibrium conditions. There are two chapters that cover these types of transformations. Chapter 15 covers the nonequilibrium cooling ferrous alloys that form the basics for the strengthening of steel by heat treatment. Chapter 16 covers the precipitation hardening process that is important in strengthening ferrous and nonferrous alloys, particularly aluminum and nickel-based alloys.

A first course in materials science would be helpful in understanding the material in this book; however, most of the material is easy to understand and builds as you progress through the book. The purpose of this book is to introduce the basics and not to replace handbooks on engineering alloys.

I would like to acknowledge the help and guidance of Scott Henry and Steve Lampman of ASM, the editorial staff at ASM, and the people that reviewed this manuscript for their valuable contributions.

F.C. Campbell
St. Louis, Missouri

Phase Diagrams—Understanding the Basics
F.C. Campbell, editor

CHAPTER 1

Introduction to Phase Diagrams

IN MATERIALS SCIENCE, the term *phase* refers to the distinct types of atomic bonding and arrangement of elements in a material of a given chemical composition. A phase is a physically homogeneous state of matter with a given chemical composition and structure. The simplest examples are the three states of matter (solid, liquid, or gas) of a pure metal. The three phases of a pure metal have the same chemical composition, but each phase is obviously distinct physically due to differences in the bonding and arrangement of atoms.

The solid phase of a pure element may also refer to physically distinct arrangements of atoms in the solid. *Polymorphism* is the general term for the ability of a solid to exist in more than one form. The bonding of atoms in solids often occurs in a repetitive arrangement—or crystal structure. Most solid metals and alloys have atoms arranged in a specific type of crystalline structure (see Appendix A, "Review of Metallic Structure," in this book). Some pure metals also have different crystalline structure as a function of temperature. For example, solid iron has different crystal structures (and magnetic properties) as a function of temperature. The physical change in the arrangement of iron atoms represents a phase change. *Allotropy* is the general term to describe solid-phase polymorphism of an element. In this regard, iron (along with some other metals) is an allotropic element.

Alloying, or mixing of two or more elements, also affects the occurrence of phase changes. For example, the temperature for complete melting (100% liquid phase) of an alloy depends on the relative concentration of alloying elements. Another example is the effect of alloying on the crystalline phase of a solid. Depending on the mixture of two or more elements, the elements may form different crystalline phases and/or chemical compounds.

Phase diagrams are useful graphical representations of the phase changes in a material. Alloy phase diagrams are useful to metallurgists, materials engineers, and materials scientists in four major areas: (1) development of new alloys for specific applications, (2) fabrication of these alloys into useful configurations, (3) design and control of heat treatment procedures for specific alloys that will produce the required mechanical, physical, and chemical properties, and (4) solving problems that arise with specific alloys in their performance in commercial applications.

1.1 Unary Systems

As noted, pure metals exist in three states of matter, solid, liquid, and vapor, depending on the specific combination of temperature and pressure. Each of these three states is referred to as a phase; that is, the solid, liquid, and vapor phases. Consider a solid block of metal in a container inside a furnace. Assuming atmospheric pressure, if the metal is heated to a high enough temperature, it will melt and form a liquid. On still further heating, it will start to boil and metal atoms will leave the container as a vapor or gas. Because matter can enter or leave the system, this condition is known as an open system. To create a closed system where no matter enters or leaves the system, an airtight cover can be placed on top of the container. If pressure is now applied inside the container, the vapor phase is compressed and the volume becomes smaller. Next, the pressure is released and the temperature is lowered so that the metal cools to the liquid state. On reapplication of pressure, the liquid metal volume remains fairly constant; that is, liquid metals are essentially incompressible. The same situation holds if the liquid metal is cooled so that only a solid remains. Because the metal atoms are almost as closely packed in a liquid as they are in a solid, there is almost no change in volume. Therefore, during the processing of metals, the effects of pressure and volume change can often be ignored. Because most metallic processes are conducted at atmospheric pressure, the pressure in phase diagrams is normally assigned a value of 1 atm and volume changes are ignored. Because the system considered so far contains only one pure metal, it is known as a unary system.

1.2 Binary Systems

Now consider the addition of a second metal, for example, an alloy of nickel and copper. The phase diagram for this system is shown in Fig. 1.1. This is known as an isomorphous phase diagram, because the alloys are completely miscible in each other in the solid state. Nickel and copper both have the face-centered cubic (fcc) crystalline structure and the atoms are of similar size. Alloys of metals containing metals with different crystalline structures will have somewhat more complex diagrams.

The diagram shows the phases present in all possible alloys of the two metals nickel (Ni) and copper (Cu), at all temperatures from 500 to 1500 °C (930 to 2730 °F). Alloy composition is plotted on the horizontal scale,

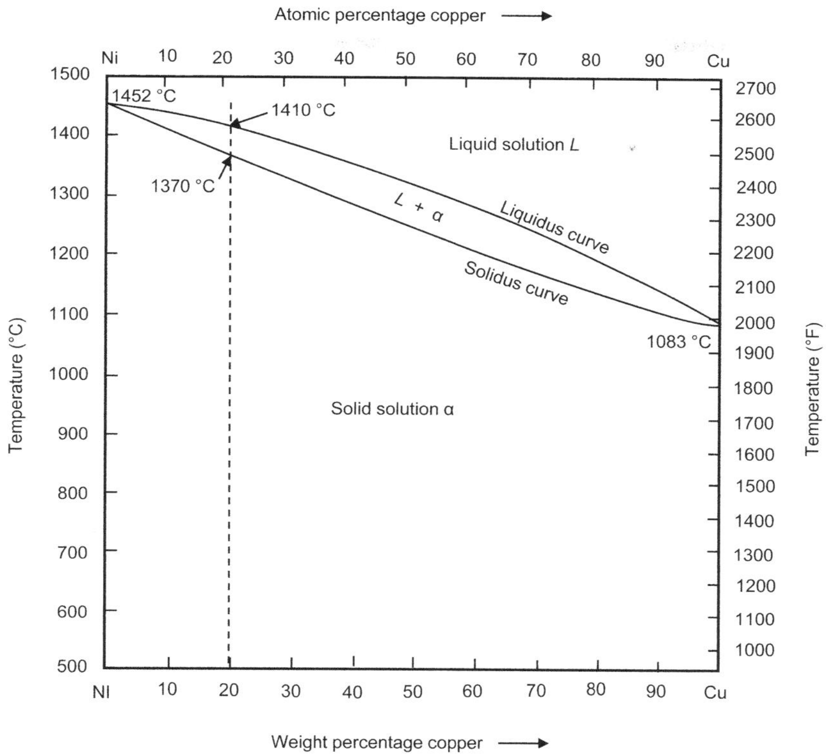

Fig. 1.1 The nickel-copper phase diagram. Adapted from Ref 1.1

along the base of the diagram, where the weight percentage of copper is read directly, the remainder being nickel. Temperature is read vertically, either from the centigrade scale on the left or from the Fahrenheit scale on the right. Two curves cross the diagram from the melting point of nickel at 1452 °C (2646 °F) to the melting point of copper at 1083 °C (1981 °F). Of these, the upper curve, called the liquidus, denotes for each possible alloy composition the temperature at which freezing begins during cooling or, equivalently, at which melting is completed on heating. The lower curve, called the solidus, indicates the temperatures at which melting begins on heating or at which freezing is completed on cooling. Above the liquidus every alloy is molten, and this region of the diagram is, accordingly, labeled L for the liquid phase or liquid solution. Below the solidus all alloys are solid, and this region is labeled α because it is customary to use a Greek letter for the designation of a solid phase that is also a solid solution. At temperatures between the two curves, the liquid and solid phases are present together, as is indicated by the designation $L + \alpha$.

Therefore, the melting range of any desired alloy, for example, an alloy composed of 20% Cu (balance, 80% Ni), may be found by tracing the vertical line originating at 20% Cu on the base of the diagram to its intersections with the solidus and liquidus. In this way, it will be found that the alloy in question begins to melt at 1370 °C (2498 °F) and is completely molten at 1410 °C (2570 °F).

1.3 Temperature and Composition Scales

The centigrade temperature scale is normally used for phase diagrams. However, the Fahrenheit scale may be used as convenience dictates. Alloy composition is normally expressed in weight percentage, but for certain types of scientific work the atomic percentage scale may be preferred. If desired, composition may also be given in terms of the percentage by volume, but this usage is rare in the representation of metal systems.

The conversion from weight percentage (wt %) to atomic percentage (at. %) may be accomplished by the use of the following formulas:

$$\text{wt\% } A = \frac{\text{at.\% } A \times \text{at.wt of } A}{(\text{at.\% } A \times \text{at.wt of } A) + (\text{at.\% } B \times \text{at.wt of } B)} \times 100 \qquad \text{(Eq 1.1)}$$

$$\text{at.\% } A = \frac{\text{wt\% } A / \text{at.wt of } A}{(\text{at.\% } A / \text{at.wt of } A) + (\text{wt\% } B / \text{at.wt of } B)} \times 100 \qquad \text{(Eq 1.2)}$$

The equation for converting from atomic percentages to weight percentages in higher-order systems is similar to that for binary systems, except that an additional term is added to the denominator for each additional component. For example, for ternary systems:

$$\text{at.\% } A = \frac{\text{wt\% } A / \text{at.wt of } A}{(\text{at.\% } A / \text{at.wt of } A) + (\text{wt\% } B / \text{at.wt of } B) + (\text{wt\% } C / \text{at.wt of } C)} \times 100 \qquad \text{(Eq 1.3)}$$

$$\text{wt\% } A = \frac{\text{at.\% } A \times \text{at.wt of } A}{(\text{at.\% } A \times \text{at.wt of } A) + (\text{at.\% } B \times \text{at.wt of } B) + (\text{at.\% } C \times \text{at.wt of } C)} \times 100 \qquad \text{(Eq 1.4)}$$

where *A*, *B*, and *C* represent the metals in the alloy.

1.4 Equilibrium

Phase diagrams record the phase relationships under equilibrium conditions. This is necessary because phase changes observed in practice tend to occur at different temperatures, depending on the rate at which the metal is being heated or cooled. With rapid heating, any phase change, such as melting, occurs at a slightly higher temperature than with slow heating. Conversely, with rapid cooling, the phase change occurs at a lower temperature than with slow cooling. Therefore, transformations observed during heating are at higher temperature than the reverse transformations observed during cooling, except in the hypothetical case wherein the rates of heating and cooling are infinitely slow, where the two observations of temperature would coincide at the equilibrium transformation temperature.

The equilibrium states that are represented on phase diagrams are known as heterogeneous equilibria, because they refer to the coexistence of different states of matter. However, for two or more phases to attain mutual equilibrium, it is necessary that each be internally in a homogeneous state. In general, this means that each phase must be in the lowest free energy state possible under the restrictions imposed by its environment. Thus, the chemical composition must be identical everywhere within the phase, the molecular and atomic species of which the phase is composed (if more than one) must be present in equilibrium proportions, and crystalline phases must be free of internal stresses.

An exception to the rule that only true equilibrium states are recorded on phase diagrams is found in the representation of so-called metastable equilibria. In carbon steels, for example, there is a solid phase, a carbide of iron (Fe_3C) called cementite, that decomposes into graphite and iron under conditions that are favorable to the attainment of true equilibrium. However, the rate of decomposition of the iron carbide is extremely slow under the most favorable conditions and is usually imperceptible under ordinary conditions. Because of its reluctance to decompose, this phase is said to be metastable, and it is represented on the usual (metastable) iron-carbon phase diagram. Evidently, metastability is a concept incapable of definition except by fiat, because there is no fundamental basis for saying that those substances that revert to the stable form at less than a certain rate are metastable, while those that decompose more rapidly are unstable. The recognition of metastable phase diagrams is simply a practical artifice that has been found useful in certain instances, even though in violation of the basic assumptions of the phase rule.

1.5 The Phase Rule

The construction of phase diagrams is greatly facilitated by certain rules that come from thermodynamics. Foremost among these is Gibbs' phase rule. This rule says that the maximum number of phases, P, which can coexist in a chemical system, or alloy, plus the number of degrees of freedom, F, is equal to the sum of the components, C, of the system plus 2:

$$P + F = C + 2 \qquad \text{(Eq 1.5)}$$

The phases, P, are the homogeneous parts of a system that, having definite bounding surfaces, are conceivably separable by mechanical means alone, for example, a gas, liquid, and solid.

The degrees of freedom, F, are those externally controllable conditions of temperature, pressure, and composition, which are independently variable and which must be specified in order to completely define the equilibrium state of the system.

The components, C, are the smallest number of substances of independently variable composition making up the system. In alloy systems, it is usually sufficient to count the number of elements present. In the case of a mixture of stable compounds, such as salt and water, the number of components may be taken as two ($NaCl + H_2O$), unless the mixture is carried to a degree of temperature and pressure where one or both of the compounds decompose, when it becomes necessary to consider four components (Na, Cl, H, and O). Through numerous examples in subsequent chapters the meaning of these definitions becomes clearer. However, for the present, it is sufficient to illustrate the application of the phase rule with a simple example. Suppose that it is desired to ascertain under what conditions a pure metal can exist with the gas, liquid, and solid phases all present in a state of equilibrium. Then, there are three phases. Because only one metal is involved, the number of components is one. The phase rule is used to find the number of degrees of freedom:

$$F = C + 2 - P$$

$$F = 1 + 2 - 3 = 0$$

There are no degrees of freedom, which means that the coexistence of these three phases can occur only at one specific temperature and one specific pressure. Of course, the composition is fixed, because there is only one metal being present. If one such set of specific conditions of temperature and pressure is now found by experiment, it will be unnecessary to look for another set, because the phase rule shows that only one can exist. Moreover, if it is desired to construct a phase diagram in which the coexistence of the three phases is represented, it becomes apparent that the coordinates of the diagram should be temperature and pressure and that the coexistence of the three phases must be indicated by a single point on this diagram.

1.6 Theorem of Le Châtelier and the Clausius-Clapeyron Equation

Although the phase rule tells what lines and fields should be represented on a phase diagram, it does not usually define their shapes or the directions of the lines. Further guidance in the latter respect may be explained by several additional thermodynamic rules.

The theorem of Le Châtelier says that if a system in equilibrium is subjected to a constraint by which the equilibrium is altered, a reaction takes place that opposes the constraint, that is, one by which its effect is partially annulled. Therefore, if an increase in the temperature of an alloy results in a phase change, that phase change will be one that proceeds with heat absorption, or if pressure applied to an alloy system brings about a phase change, this phase change must be one that is accompanied by a contraction in volume.

The usefulness of this rule can again be shown by reference to Fig. 1.2. Consider the line showing freezing, which represents for a typical pure metal the temperature at which melting occurs at various pressures. This line slopes upward away from the pressure axis. The typical metal contracts on freezing. Hence, applying an increased pressure to the liquid can cause the metal to become solid, experiencing at the same time an abrupt contraction in volume. Had the metal bismuth, which expands on freezing, been selected as an example, the theorem of Le Châtelier would dictate that the solid-liquid line be drawn so that the conversion of liquid to solid with pressure change would occur only with a reduction in pressure; that is, the line should slope upward toward the pressure axis.

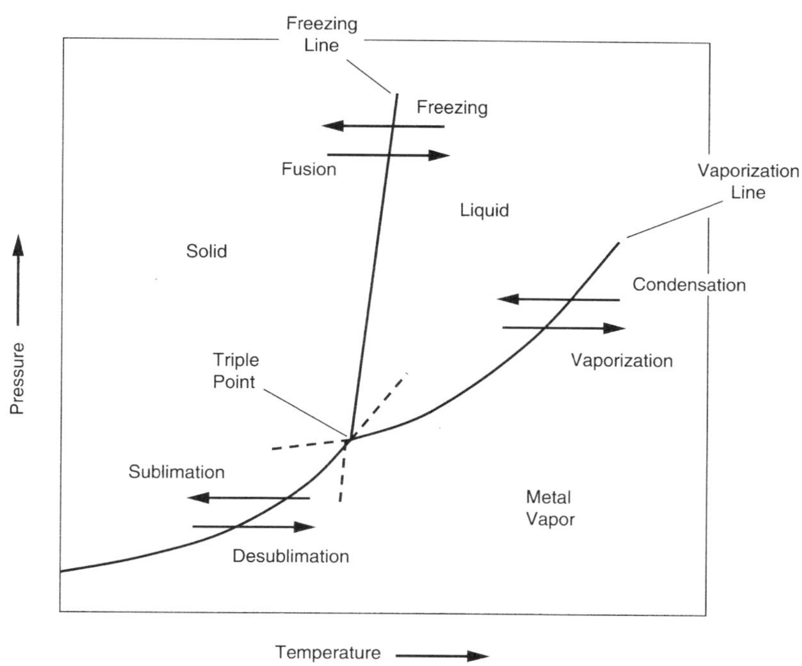

Fig. 1.2 Pressure-temperature diagram for a pure metal. Source: Ref 1.2

A quantitative statement of the theorem of Le Châtelier is found in the Clausius-Clapeyron equation. Referring again to Fig. 1.2, this equation leads to the further conclusion that each of the curves representing two-phase equilibrium must lie at such an angle that on passing through the point of three-phase equilibrium, each would project into the region of the third phase. Thus, the sublimination-desublimination line must project into the liquid field, the vaporization curve into the solid field, and the solid-liquid curve into the vapor field.

1.7 The Lever Rule

The lever rule is one of the cornerstones of understanding and interpreting phase diagrams. A portion of a binary phase diagram is shown in Fig. 1.3. In this diagram, all phases present are solid phases. There are two single-phase fields labeled α and β, separated by a two-phase field labeled α + β. It indicates that, at a temperature such as *b*, pure metal *A* can dissolve metal *B* in any proportion up to the limit of the single-phase α field at composition *a*. At the same temperature, metal *B* can dissolve metal *A* in any proportion up to composition *c*, which, at this temperature, is the boundary of the single-phase β field. Therefore, at temperature *b*, any alloy that contains less than *a*% of metal *B* will exist at equilibrium as the homogeneous α solid solution; and any alloy containing more than *c*% of metal *B* will exist as the β solid solution. However, any alloy whose overall composition is between *a* and *c* (for example, at *d*) will, at the same temperature, contain more metal *B* than can be dissolved by the α and more metal *A* than can be dissolved by the β. It will therefore exist as a mixture of α and β solid solutions. At equilibrium, both solid solutions

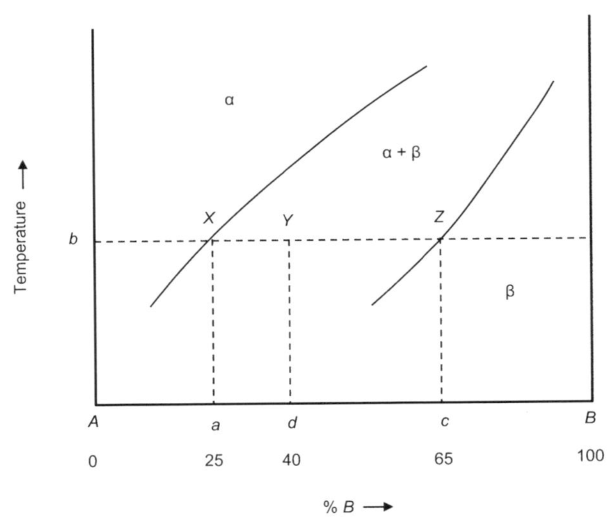

Fig. 1.3 Portion of hypothetical phase diagram

will be saturated. The composition of the α phase is therefore $a\%$ of metal B and that of the α phase is $c\%$ of metal B.

When two phases are present, as at composition Y in Fig. 1.3, their relative amounts are determined by the relation of their chemical compositions to the composition of the alloy. This is true because the total weight of one of the metals, for example metal A, present in the alloy must be divided between the two phases. This division can be represented by:

$$W_0\left(\frac{\%A_0}{100}\right) = W_\alpha\left(\frac{\%A_\alpha}{100}\right) + W_\beta\left(\frac{\%A_\beta}{100}\right) \quad \text{(Eq 1.6)}$$

$$\begin{pmatrix}\text{weight of metal} \\ A \text{ in alloy}\end{pmatrix} = \begin{pmatrix}\text{weight of metal} \\ A \text{ in } \alpha \text{ phase}\end{pmatrix} + \begin{pmatrix}\text{weight of metal} \\ A \text{ in } \beta \text{ phase}\end{pmatrix}$$

where W_0, W_α, and W_β are the weights of the alloy, the α phase, and the β phase, respectively, and $\%A_0$, $\%A_\alpha$, and $\%A_\beta$ are the respective chemical compositions in terms of metal A. Because the weight of the alloy is the sum of the weight of the α phase and the weight of the β phase, the following relationship exists:

$$W_0 = W_\alpha + W_\beta \quad \text{(Eq 1.7)}$$

This equation can be used to eliminate W_α from Eq 1.6, and the resulting equation can be solved for W_β to give the expression:

$$W_\beta = W_0\left(\frac{\%A_0 - \%A_\alpha}{\%A_\beta - \%A_\alpha}\right) \quad \text{(Eq 1.8)}$$

Although a similar expression can be obtained for the weight of the α phase W_α, the weight of the α phase is more easily obtained by means of Eq 1.7.

Because the weight of each phase is determined by chemical composition values according to Eq 1.8, the tie-line ac shown in Fig. 1.3 can be used to obtain the weights of the phases. In terms of the lengths in the tie-line, Eq 1. 8 can be written as:

$$W_\beta = W_0\left(\frac{\text{length of line } d-a}{\text{length of line } c-a}\right) \quad \text{(Eq 1.9)}$$

where the lengths are expressed in terms of the numbers used for the concentration axis of the diagram. *The lever rule, or inverse lever rule can be stated: The relative amount of a given phase is proportional to the length of the tie-line on the opposite side of the alloy point of the tie-line.* Thus, the weights of the two phases are such that they would balance, as shown in Fig. 1.4.

Using Eq 1.9, the *weight* of the β phase at composition *Y* in Fig. 1.3 is:

$$W_\beta = W_\alpha \left(\frac{40-25}{65-25} \right) = W_\alpha (0.375)$$

The *percentage* of β phase can be determined by use of the equation:

$$\text{Percentage of } \beta \text{ phase} = \left(\frac{W_\beta}{W_0} \right) 100 = \left(\frac{\%A_0 - \%A_\alpha}{\%A_\beta - \%A_\alpha} \right) 100 \qquad \text{(Eq 1.10)}$$

At composition *Y* the percentage of β phase is:

$$\%\ \beta \text{ phase} = \left(\frac{40-25}{65-25} \right) 100 = 37.5\%$$

The percentage of α phase is the difference between 100% and 37.5%, or 62.5%.

Other rules and aids of this type are mentioned in subsequent chapters as opportunities for their application arise. Foremost among these is the second law of thermodynamics, which leads to rules governing the con-

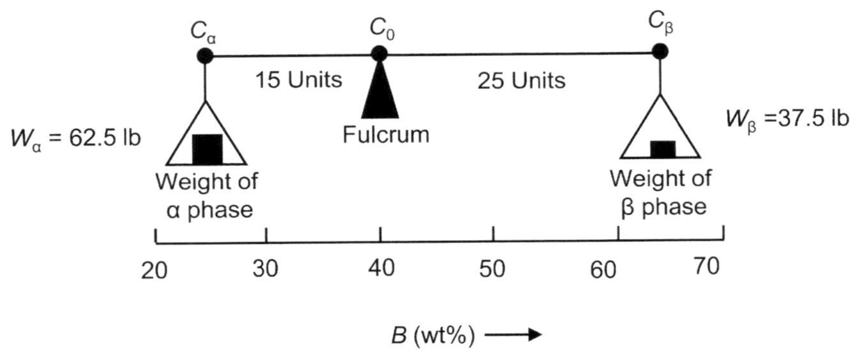

$$\text{Wt\% } \alpha = \frac{C_\beta - C_o}{C_\beta - C_\alpha} = \frac{65-40}{65-25} = 0.625 \text{ or } 62.5\%$$

$$\text{Wt\% } \beta = \frac{C_o - C_\alpha}{C_\beta - C_\alpha} = \frac{40-25}{65-25} = 0.375 \text{ or } 37.5\%$$

Fig. 1.4 Visual representation of lever rule

struction of more complex phase diagrams, such as are encountered in binary and ternary systems.

1.8 Relationships between Alloy Constitution and Physical Properties

One of the more successful correlations is that with the structural alterations that an alloy undergoes during temperature change, as in manufacturing operations. True equilibrium is, of course, rarely attained by metals and alloys in the course of ordinary manufacture and application. Rates of heating and cooling are usually too fast, times of heat treatment too short, and phase changes too sluggish for the ultimate equilibrium state to be reached. However, any change that does occur must constitute an adjustment toward equilibrium. Hence, the direction of change can be ascertained from the phase diagram, and a wealth of experience is available to indicate the probable degree of attainment of equilibrium under various circumstances.

1.9 Phase Diagram Resources

Starting in 1978, ASM International established a cooperative agreement with the National Institute of Standards and Technology (then National Bureau of Standards) to publish and continue to update phase diagrams so that engineers and scientists would have the best thermodynamic tools on which to base new discoveries, explain failures, and so forth. The program that resulted has spanned more than 20 years and involved hundreds of scientific workers, comprehensive scans of the literature, and substantial publications. The binary and ternary books and CDs that have been developed are well-known internationally and are commonly cited as authoritative.

In 1987, the Alloy Phase Diagram International Commission (APDIC) was formed in partnership with other programs operating in Europe and Asia. This group continues to thrive after 20 years and meets once per year to share information regarding projects and alloy systems that are under study. This meeting forms the basis for international collaborations.

Some current phase diagram compilations are given in the Selected References. Another resource is the Alloy Phase Diagram Center published by ASM International (Ref 1.3). This database contains over 30,000 binary and ternary alloy phase diagrams, each with their associated crystal and reaction data.

REFERENCES

1.1 F.N. Rhines, *Phase Diagrams in Metallurgy*, McGraw-Hill, 1956, p 2

1.2 F.C. Campbell, *Elements of Metallurgy and Engineering Alloys*, ASM International, 2008
1.3 *Alloy Phase Diagram Center*, 2011, http://www.asminternational.org/portal/site/www/info/apdc/ (07 November 2011)

SELECTED REFERENCES

- A.T. Dinsdale, SGTE Data for Pure Elements, *CALPHAD*, Vol 15 (No. 4), 1991, p 317–425
- G. Effenberg and S. Ilyenko, Ed., *Landolt-Börnstein Series*, O. Madelung (Editor-in-Chief), Vol IV/11: *Ternary Alloy Systems. Phase Diagrams, Crystallographic and Thermodynamic Data*, Subvolumes A to C, Springer, 2004–2006
- S.P. Garg, M. Venkatraman, N. Krishnamurthy, and R. Krishnan, Ed., *Phase Diagrams of Binary Tantalum Alloys*, Indian Institute of Metals, 1996
- K.P. Gupta, Ed., *Phase Diagrams of Ternary Nickel Alloys*, Parts 1 and 2, Indian Institute of Metals, 1990
- M.E. Kassner and D.E. Peterson, Ed., *Phase Diagrams of Binary Actinide Alloys*, ASM International, 1995
- Yu.V. Levinsky and G. Effenberg, Ed., *Pressure Dependent Phase Diagrams of Binary Alloys*, ASM International, 1997
- F.D. Manchester, Ed., *Phase Diagrams of Binary Hydrogen Alloys*, ASM International, 2000
- T.B. Massalski and H. Okamoto, Ed., *Binary Alloy Phase Diagrams*, 2nd ed., ASM International, 1990
- J.L. Murray, Ed., *Phase Diagrams of Binary Titanium Alloys*, ASM International, 1987
- S.V.N. Naidu and P.R. Rao, Ed., *Phase Diagrams of Binary Tungsten Alloys*, Indian Institute of Metals, 1991
- P. Nash, Ed., *Phase Diagrams of Binary Nickel Alloys*, ASM International, 1991
- H. Okamoto and T.B. Massalski, Ed., *Phase Diagrams of Binary Gold Alloys*, ASM International, 1987
- H. Okamoto, Ed., *Phase Diagrams of Binary Iron Alloys*, ASM International, 1993
- H. Okamoto, *Desk Handbook: Phase Diagrams for Binary Alloys*, ASM International, 2000
- H. Okamoto, *Phase Diagrams of Dilute Binary Alloys*, ASM International, 2002
- G. Petzow and G. Effenberg, Ed., *Ternary Alloys: A Comprehensive Compendium of Evaluated Constitutional Data and Phase Diagrams*, 15 volumes, Wiley-VCH, 1989–1996
- B. Predel, Ed., *Landolt-Börnstein Series*, O. Madelung (Editor-in-Chief), Vol IV/5: *Phase Equilibria, Crystallographic and*

Thermodynamic Data of Binary Alloys, Subvolumes A to J, Springer, 1991–1998, Vol IV/12, 2006

- A.A. Prince, G.V. Raynor, and D.S. Evans, Ed., *Phase Diagrams of Ternary Gold Alloys*, The Institute of Metals, 1990
- V. Raghavan, *Phase Diagrams of Ternary Iron Alloys*, Parts 1–6, ASM International and The Indian Institute of Metals, 1987–1993
- V. Raghavan, S.P. Garg, M. Venkatraman, N. Krishnamurthy, and R. Krishnan, Ed., *Phase Diagrams of Quaternary Iron Alloys*, Indian Institute of Metals, 1996
- P. Rogl and G. Effenberg, Ed., *Phase Diagrams of Ternary Metal-Boron-Carbon Systems*, ASM International, 1998
- J.F. Smith, Ed., *Phase Diagrams of Binary Vanadium Alloys*, ASM International, 1989
- P.R. Subramanian, D.J. Chakrabarti, and D.E. Laughlin, Ed., *Phase Diagrams of Binary Copper Alloys*, ASM International, 1994
- L. Tanner and H. Okamoto, Ed., *Phase Diagrams of Binary Beryllium Alloys*, ASM International, 1987
- E.Yu. Tonkov and E.G. Ponyatovsky, *Phase Transformations of Elements under High Pressure*, CRC Press, 2004
- P. Villars, A. Prince, and H. Okamoto, Ed., *Handbook of Ternary Alloy Phase Diagrams*, 10 volumes, ASM International, 1995
- J.H. Westbrook, Ed., *Moffatt's Handbook of Binary Phase Diagrams*, 5 loose-leaf volumes, Genium Publ., 2006
- C.E.T. White and H. Okamoto, *Phase Diagrams of Indium Alloys and their Engineering Applications*, ASM International, 1991
- D.A. Young, *Phase Diagrams of the Elements*, University of California Press, 1991

CHAPTER 2

Solid Solutions and Phase Transformations

THE CRYSTALLINE STATE is an important mode of aggregation of metal atoms. However, assemblies of metal atoms can also exist in the two other states of matter; that is, the liquid and vapor states. Consider a metal specimen in the solid state that consists of 50% iron atoms and 50% copper atoms (Fig. 2.1a). These atoms do not form a homogeneous solid under equilibrium conditions; rather, they form two different solids, one rich in iron and the other rich in copper. Because it is convenient to distinguish between these two solids, it is conventional to label them with the first two letters of the Greek alphabet, α and β. Each of the distinguishable solid states is then called a phase, and one speaks of the α phase and the β phase.

Two or more different kinds of metal atoms will usually form a homogeneous solution in the liquid state. This behavior is shown for a 50% Fe-50% Cu liquid phase alloy in Fig. 2.1(b). Here also it is conventional to use the word *phase* rather than *state* and to speak of the liquid phase. In a few instances, such as that of the 50% Zn-50% Pb alloy shown in Fig. 2.1(c), two (or more) liquid phases exist at equilibrium. The usual nomenclature in this case is simply to use L_1 (liquid 1) and L_2 (liquid 2). All gases are completely miscible in one another and metal vapors are no exception to this rule. The vapor phase is shown in Fig. 2.1(d) in which the 50% Pb-50% Zn alloy can exist under suitable conditions of temperature and pressure.

2.1 One-Component (Unary) Systems

Before considering in some detail the phases that constitute metal systems, it is helpful to understand the nature of the equilibrium relations among these phases. The relations are relatively simple for a one-component system, that is, a system consisting of a single, pure metal.

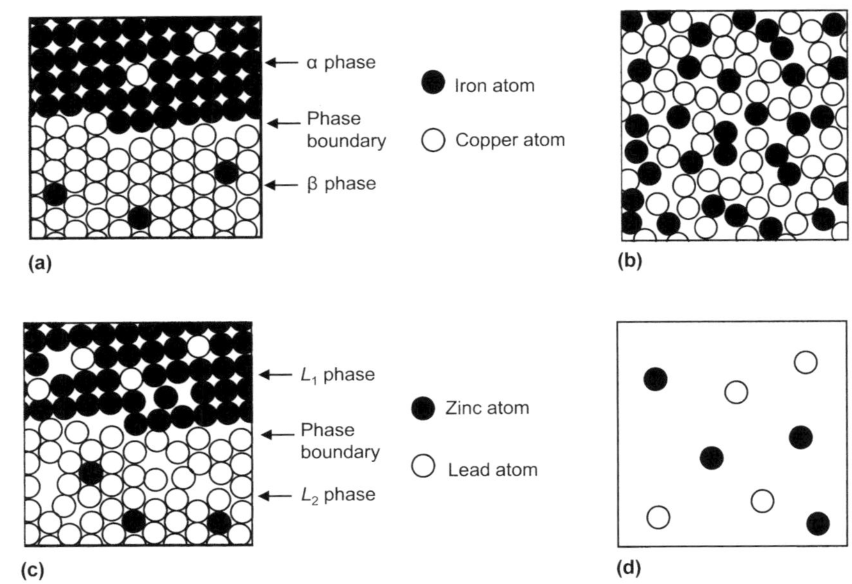

Fig. 2.1 The distribution of atoms in solid, liquid, and vapor phases of alloys. (a) Two solid solutions formed in a 50% Fe-50% Cu alloy. (b) Liquid phase formed by 50% Fe-%50 Cu alloy. (c) Two liquid phases formed in a 50% Zn-50% Pb alloy. (d) Vapor phase formed by a 50% Zn-50% Pb alloy. Adapted from: Ref 2.1

How does one put a metal into the liquid condition for casting, or into the vapor phase for purification? Obviously, one or more external conditions must be altered, with the temperature and pressure being of the greatest importance. Although it is possible to change haphazardly the temperature and pressure of a metal until the desired effect is produced, it is much more satisfactory to choose the optimal conditions from data collected by previous workers. Fortunately, these data are not spread in long columns through many journals, but are concisely presented in a single diagram for each metal.

The one-component, or unary, diagram for magnesium is shown in Fig. 2.2. This diagram shows what phases are present as a function of the temperature and pressure. In this case the three phases that can exist correspond to the three states of matter: vapor, liquid, and solid. The interpretation of equilibrium diagrams is made easy by the following rule: Only points in the diagram have physical significance. Because the variables are temperature and pressure, evidently a point is determined by a temperature and a pressure. In Fig. 2.2 the point A is at a pressure P_A and at temperature T. Because A is in the field of the diagram labeled liquid, magnesium metal held long enough at pressure P_A and temperature T for equilibrium to be established will be entirely liquid. Similarly, magnesium metal at pressure P_B and temperature T will be a mixture of liquid and vapor. The temperature and pressure of the triple point will cause solid, liquid, and

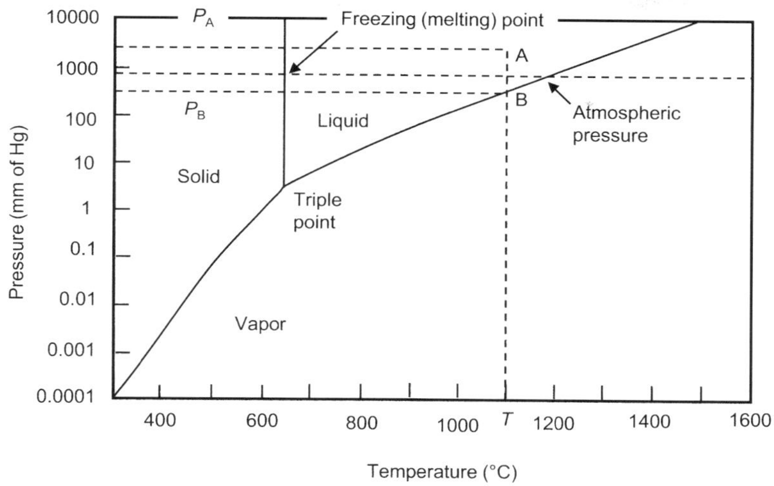

Fig. 2.2 Unary diagram for magnesium. Adapted from: Ref 2.1

vapor to be in equilibrium. When two or more phases coexist, the relative amount of each phase cannot be determined from this diagram.

An elementary principle of chemistry states that every liquid or solid tends to be in equilibrium with a particular pressure of its vapor. How then, it may be asked, can there be a completely liquid region in the one-component diagram? Will not vapor exist in equilibrium with the liquid at point A in Fig. 2.2? The explanation lies in the manner in which the one-component system is determined (Fig. 2.3). Only the metal being investigated is contained in the cylinder that exerts pressure on the system; even air is excluded. If the external pressure is equal to the vapor pressure of the liquid metal at the given temperature, both liquid and vapor exist in equilibrium in the cylinder. This is the condition at point B in Fig. 2.2 and is shown schematically in Fig. 2.3(a). However, if the external pressure is greater than the vapor pressure of the liquid metal, the piston is forced down, the vapor condenses, and only the liquid phase remains. This is the condition at point A in Fig. 2.2 and is shown schematically in Fig. 2.3(b). An entirely different situation exists if magnesium metal is held at temperature *T* in contact with air at one atmosphere pressure (760 mm of mercury) as shown in Fig. 2.3(c). Of the total pressure of one atmosphere, only the partial pressure P_B is supplied by magnesium vapor. In this instance the significant pressure for most purposes is that of the vapor, not that of vapor and air combined.

Just as the line separating the liquid and vapor areas of the diagram in Fig. 2.2 gives the vapor pressure of the liquid metal as a function of temperature, so the corresponding line for the solid gives the vapor pressure of the solid. It is significant that a solid metal need not melt on being heated.

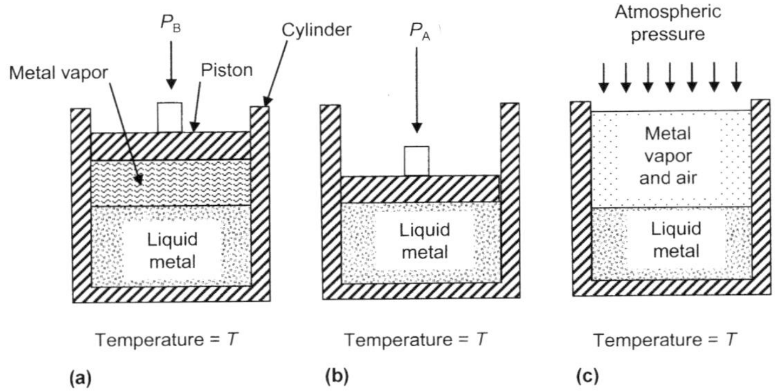

Fig. 2.3 (a) and (b) Difference between the conditions under which one-component diagrams are valid. (c) Usual conditions of heating a metal. Adapted from: Ref 2.1

For example, if magnesium is held in a vacuum of 1 mm of mercury and is heated to slightly over 600 °C (1110 °F), the metal will completely vaporize without the formation of any liquid. Moderate pressures above that of the triple point have little effect on the melting point of a metal, and the line separating the solid and liquid regions is almost vertical.

The application of the phase rule to the one-component diagram for magnesium (Fig. 2.4) will show the significance of its terms. In this case there is only one component, magnesium, and in general the number of components is equal to the number of metals present in the system being considered. At point 1 in Fig. 2.4 there is only one phase present, gaseous magnesium; therefore at this point the phase rule becomes:

$$F = C + 2 - P$$
$$F = 1 + 2 - 1 = 2$$

That is, there are two degrees of freedom. If both the pressure and temperature in a unary system are freely and arbitrarily selected, the situation corresponds to having two degrees of freedom and is called bivariant equilibrium.

The situation at point 2 is significantly different. Because the liquid and vapor phases exist in equilibrium at this point, there are two phases present and the phase rule yields:

$$F = C + 2 - P$$
$$F = 1 + 2 - 2 = 1$$

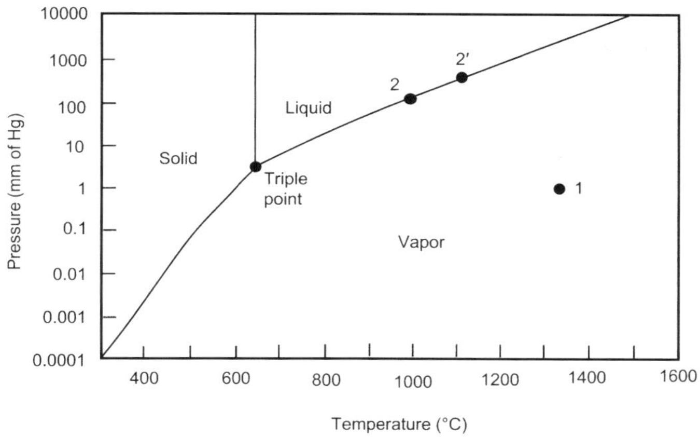

Fig. 2.4 Unary diagram for magnesium with points in one-, two-, and three-phase regions of the diagram. Adapted from: Ref 2.1

This condition is called univariant equilibrium or monovariant equilibrium. Either pressure or temperature may be freely selected, but not both. Once a pressure is selected, there is only one temperature that will satisfy equilibrium conditions, and conversely, once a temperature is selected, there is only one pressure that will satisfy equilibrium conditions.

How can there be only one degree of freedom when it is evident that both temperature and pressure must change in order to proceed from point 2 to point 2′, where the two phases continue to exist? The answer is that only one of the variables can be changed independently. To increase the pressure without losing either of the phases originally present, the temperature must also be raised exactly the proper amount. Similarly, the temperature can be changed arbitrarily but the two phases will continue to exist only if the pressure is also varied by a specific, not an independent, amount. Only one of the variables, therefore, can be independently changed at point 2 if two phases are to remain. A similar analysis shows that there are zero degrees of freedom at point 3, the triple point, in Fig. 2.4. The triple point is also called an invariant point because, at that location on the diagram, all externally controllable factors are fixed (no degrees of freedom). At this point, all three states (phases) are in equilibrium, but any changes in pressure and/or temperature will cause one or two of the states (phases) to disappear.

2.2 Solid Solutions

Pure metals are rarely used for industrial applications unless high conductivity, high ductility, or good corrosion resistance are required. Because pure metals tend to be much weaker than alloys, alloying elements are added to improve strength and hardness. In addition, alloying element additions can often be accomplished without major reductions in the attributes associated with pure metals, that is, conductivity, ductility, and corrosion resistance. When two metals are mixed in the liquid state to produce a solution, the resulting alloy is called a binary alloy. While a great number of alloys are binary, many more alloys contain a number of alloying elements. In the case of high-temperature superalloys, as many as ten alloying elements may used to obtain the desired performance.

The metals nickel and copper completely dissolve in each other in the liquid state and then retain their complete solubility in each other on freezing to form a series of alloys. The improvement in strength properties in these alloys is shown in Fig. 2.5. Note that the yield strength, the tensile strength, and the hardness all are improved as a result of alloying. While the ductility is reduced somewhat with alloying, the reduction is very moderate and these alloys all still have very good ductility.

When a metal is alloyed with another metal, either substitutional or interstitial solid solutions (Fig. 2.6) are usually formed. Substitutional solid solutions are those in which the solute and solvent atoms are nearly the same size, and the solute atoms simply substitute for solvent atoms on the crystalline lattice. Interstitial solid solutions are those in which the solute atoms are much smaller and fit within the spaces between the existing

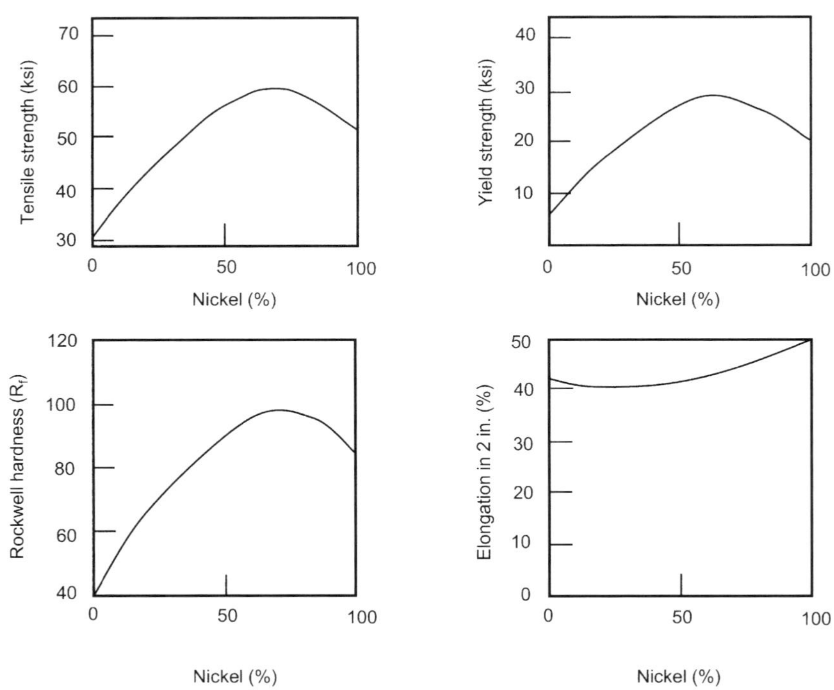

Fig. 2.5 Solid-solution strengthening for copper-nickel alloys. Source: Ref 2.2

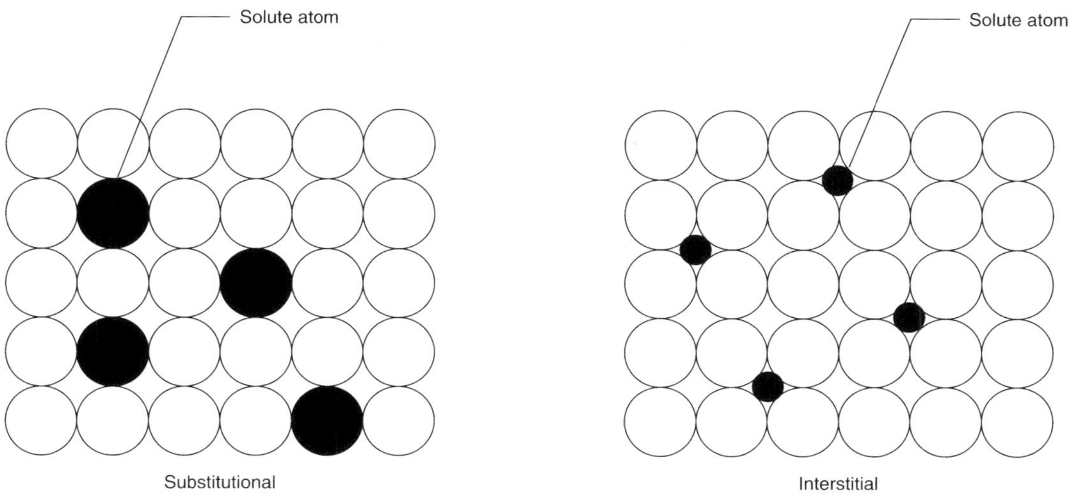

Fig. 2.6 Solid solutions. Source: Ref 2.2

solvent atoms on the crystalline structure. However, the only solute atoms small enough to fit into the interstices of metal crystals are hydrogen, nitrogen, carbon, and boron. The other small-diameter atoms, such as oxygen, tend to form compounds with metals rather than dissolve in them. When both small and large solute atoms are present, the solid solution can be both interstitial and substitutional.

The insertion of substitutional and/or interstitial alloying elements strains the crystalline lattice of the host solvent structure (Fig. 2.7). This increase in distortion, or strain, creates barriers to dislocation movement. The distortion energy causes some hardening and strengthening of the alloy, and is called solid-solution hardening. The solute atoms have a different size than the host atoms, which alters the crystal lattice. As a result, a moving dislocation is either attracted to, or repelled by, the solute; however, both situations result in a strength increase. When the dislocation is attracted to a solute, the additional force required to pull the dislocation away from it is the cause of the added strength. If the dislocation is repelled by the solute, an additional force is required to push the dislocation past the solute atom.

Studies of solid-solution hardening indicate that the hardening depends on the differences in elastic stiffness and atomic size between the solvent and solute. In general, larger differences result in greater strengthening, but at the same time, the greater the difference in sizes between the solute and solvent atoms, the more restricted is their mutual solubilities. The solvent phase becomes saturated with the solute atoms and reaches its limit of homogeneity when the distortion energy reaches a critical value determined by the thermodynamics of the system. The effects of several alloying elements on the yield strength of copper are shown in Fig. 2.8. Nickel and zinc atoms are approximately the same size as copper

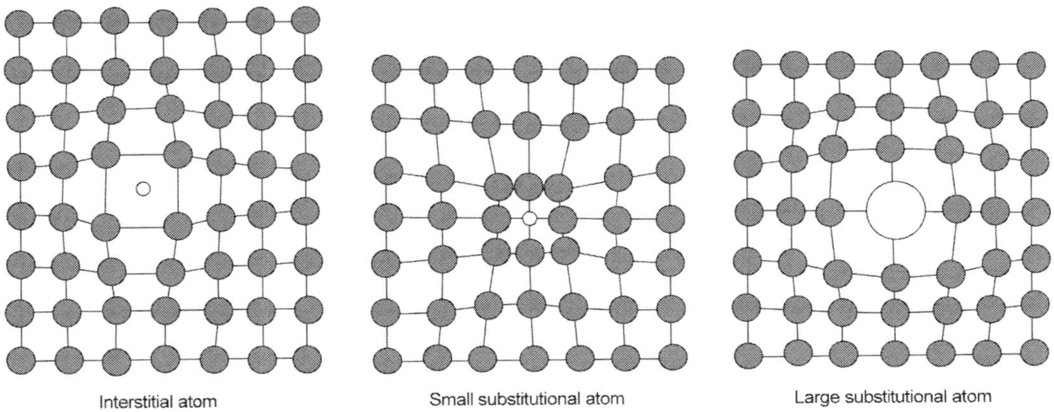

Fig. 2.7 Lattice distortions caused by solute additions. Source: Ref 2.2

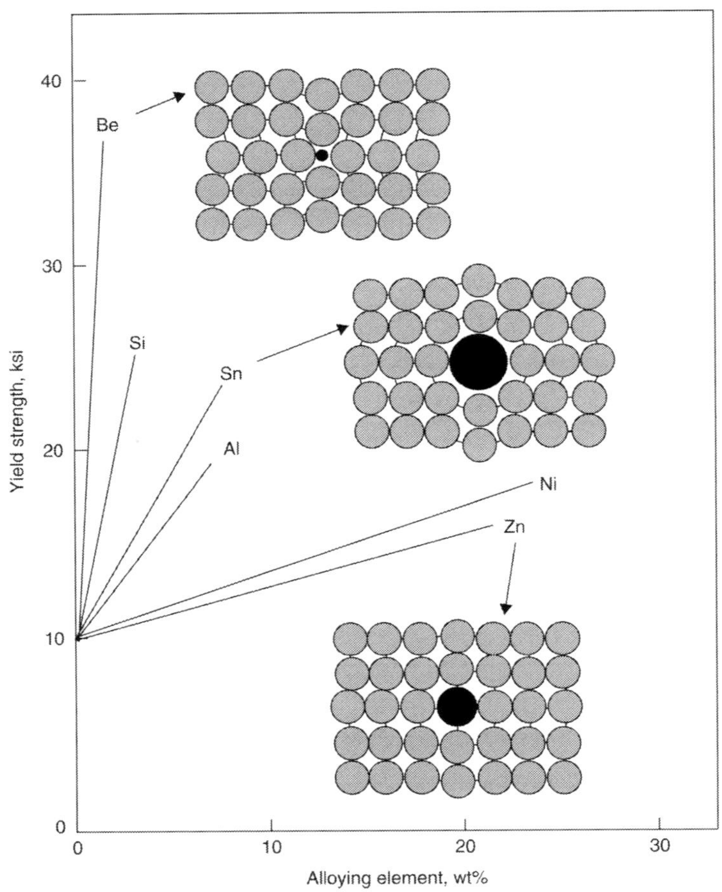

Fig. 2.8 The effects of several alloying elements on the yield strength of copper. Source: Ref 2.3 as published in Ref 2.2

atoms, but beryllium and tin atoms are much different from copper atoms. Increases in the atomic size difference and the amount of alloying both result in increases in solid-solution strengthening.

2.3 Interstitial Solid Solutions

The four elements carbon, nitrogen, hydrogen, and boron have such small diameters that they can form interstitial solid solutions. In general, these interstitial solid solutions have somewhat limited composition ranges. Only the transition metals (e.g., iron, nickel, titanium, and zirconium) have appreciable solubilities for carbon, nitrogen, and boron. Very small atoms, such as carbon, nitrogen, and hydrogen can fit in the spaces between the larger atoms. These sites are called interstitial sites and can be of either the tetrahedral or octahedral variety (Fig. 2.9). Interstitial atoms generally

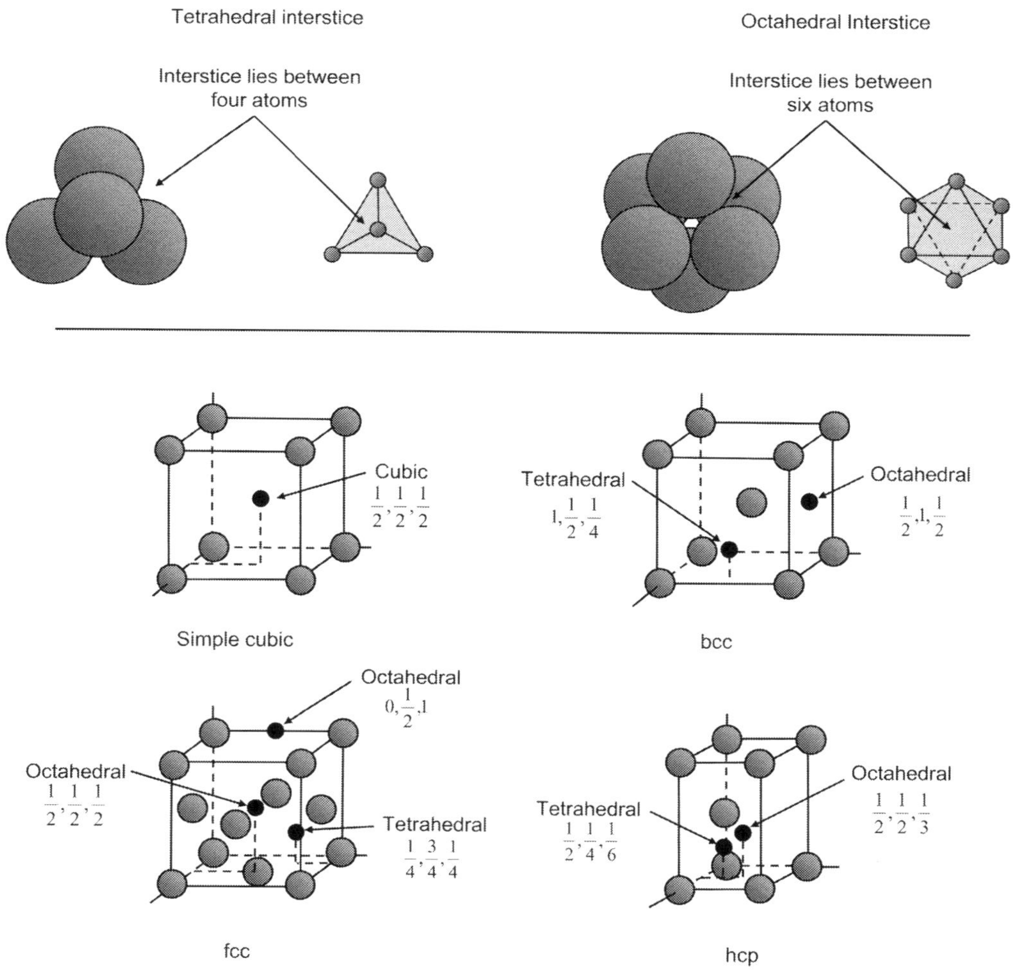

Fig. 2.9 Representative interstitial sites in unit cells. Source: Ref 2.3 as published in Ref 2.2

strengthen a metal more than substitutional atoms do, because the interstitials cause more distortion. Carbon atoms in the body-centered cubic (bcc) form of iron are particularly potent hardeners in this respect.

Carbon, nitrogen, and boron are all important alloying elements in steels. Interstitial carbon in iron forms the basis of steel hardening. Indeed, steels are alloys of iron and small amounts of carbon. Although not as important as carbon, nitrogen and boron are also useful alloying elements in certain steels. In addition, carbon and nitrogen are diffused into the surfaces to provide hardness and wear resistance in processes called carburizing and nitriding. On the other hand, hydrogen is almost never a welcome addition to any metal. It usually results in sharp decreases in ductility and produces brittle fracture modes, a mechanism called hydrogen embrittlement.

2.4 Substitutional Solid Solutions

These four rules giving a qualitative estimate of the ability of two metals to form substitutional solid solutions were developed by Hume-Rothery and are summarized as:

- *Rule 1—Relative Size Factor.* If the sizes of the solute and solvent atoms differ by less than 15%, the metals are said to have a favorable size factor for solid-solution formation. Each of the metals will be able to dissolve an appreciable amount of the other metal, on the order of 10% or more. If the size factor differs by more than 15%, then solid-solution formation tends to be severely restricted.
- *Rule 2—Chemical Affinity Factor.* The greater the chemical affinity of two metals, the more restricted is their solid solubility. When their chemical affinity is great, they tend to form compounds rather than a solid solution.
- *Rule 3—Relative Valency Factor.* If a solute atom has a different valence from that of the solvent metal, the number of electrons per atom, called the electron ratio (*e*/*a*), will be changed by alloying. Crystal structures are more sensitive to a decrease in the electron ratio than to an increase. Therefore, a metal of high valence can dissolve only a small amount of a lower-valence metal, while a lower-valence metal may have good solubility with a higher-valence metal.
- *Rule 4—Lattice Type Factor.* Only metals that have the same type of lattice structure (e.g., face-centered cubic) can form a complete series of solid solutions. Also, for complete solid solubility, the size factor must usually be less than 8%.

There are numerous exceptions to these rules. In general, an unfavorable size factor alone is sufficient to severely limit solid solubility to a minimal level. If the size factor is favorable, then the other three rules should be evaluated to determine the probable degree of solid solubility. Metallic systems that display complete solid solubility are quite rare, with the copper-nickel system being the most important.

2.5 Intermediate Phases

When the electrochemical properties of the alloying element metals are similar, normal substitutional solid solutions will form during solidification. However, when the metals have widely divergent electrochemical properties, they are more likely to form a chemical compound, often with some degree of covalent or ionic bonding present. For example, strongly electropositive magnesium will combine with weakly electropositive tin to form Mg_2Sn, which is often described as being an intermetallic compound.

Between these two extremes of substitutional solid solution on the one hand and intermetallic compound on the other, phases are formed that exhibit a gradation of properties. These phases are collectively termed *intermediate phases*. At one extreme there are true intermetallic compounds while at the other are ordered structures that can be classified more accurately as secondary solid solutions. These intermediate phases are often grouped into categories determined by their structures.

Intermetallic Compounds. Two chemically dissimilar metals tend to form compounds with ordinary chemical valence. These compounds have stoichiometric compositions with limited solubility. These compounds are generally formed when one metal has chemical properties that are strongly metallic, such as magnesium, and the other metal has chemical properties that are only weakly metallic, such as tin. Frequently, such a compound has a melting point that is higher than that of either of the parent metals. For example, the intermetallic compound Mg_2Sn melts at 780 °C (1436 °F), whereas pure magnesium and tin melt at 650 and 230 °C (1202 and 450 °F), respectively. This is an indication of the high strength of the bond in Mg_2Sn. Because they exhibit either covalent or ionic bonding, they exhibit nonmetallic properties such as brittleness and poor electrical conductivity. Examples include the covalent compounds Mn_2Sn, Fe_3Sn, and Cu_6Sn_5, and the ionic compounds Mg_2Si and Mg_2Sn.

Electron Phases. These compounds appear at definite compositions and depend on the ratio of electrons to atoms (*e*/*a*) at those compositions. The most important of these are the intermediate phases of the copper-zinc system. The valence of a metal is defined by the number of electrons in the outer shell of the atom. In electron compounds, the normal valence laws are not obeyed, but in many instances there is a fixed ratio between the total number of valence bonds of all the atoms involved and the total number of atoms in the empirical formula of the compound in question.

There are three such ratios, commonly referred to as Hume-Rothery ratios:

- Ratio 3/2 (21/14): β structures, such as CuZn, Cu_3Al, Cu_5Sn, Ag_3Al
- Ratio 21/13: γ structures, such as Cu_5Zn_8, Cu_9Al_4, $Cu_{31}Sn_8$, Ag_5Zn_8, $Na_{31}Pb_8$
- Ratio 7/4 (21/12): ε structures, such as $CuZn_3$, Cu_3Sn, $AgCd_3$, Ag_5Al_3

For example, in the β structure compound CuZn, copper has a valence of 1 and zinc a valence of 2, giving a total of 3 valences and a ratio of 3 valences to 2 atoms. In the compound $Cu_{31}Sn_8$, copper has a valence of 1 and tin a valence of 4. Therefore 31 valences are donated by the copper atom and 32 (4 × 8) by the tin atoms, making a total of 63 valences. In all, 39 atoms are present and the ratio is:

$$\frac{\text{Total number of valences}}{\text{Total number of atoms}} = \frac{63}{39} = \frac{21}{13}$$

These phases exist over a range of compositions and are metallic in nature. These Hume-Rothery ratios have been valuable in relating structures that appeared unrelated. However, there are many electron compounds that do not fall into any of the three groups.

Interstitial Compounds. These are compounds of transition metals with carbon, nitrogen, hydrogen, or boron; the interstitial atomic radius must be less than 2/3 that of the transition metal atom. These compounds are hard and have very high melting points due to the covalent nature of their bonding. When the solid solubility of an interstitially dissolved element is exceeded, a compound is precipitated from the solid solution. The small nonmetal atoms still occupy interstitial positions but the overall crystal structure of the compound is different from that of the original interstitial solid solution. Compounds of this type have some metallic and some nonmetallic properties and comprise carbides, nitrides, hydrides, and borides. Examples include TiH_2, TiN, TaC, WC, and Fe_3C. All of these compounds are extremely hard and the carbides find application in tool steels and cemented carbide cutting materials. The compound Fe_3C (cementite) is important in steels.

Laves Phases. These compounds have a composition of AB_2 that forms because of the dense packing that can be achieved if the ratio of the B atoms to A atoms (B/A) is approximately 1.2. In these phases, the A atoms have 12 nearest B neighbors and 4 nearest A neighbors. Each B atom has 6 A and 6 B nearest neighbors. This arrangement produces an average coordination number of $(2 \times 12 + 16)/3 = 13.33$. These hard and brittle phases, such as $NbFe_2$, $TiFe_2$, and $TiCo_2$, cause deterioration in ductility and stress-rupture properties. Undesirable Laves phase formation can be a problem in nickel, iron-nickel, and cobalt-base superalloys exposed to elevated temperatures for long times.

2.6 Free Energy and Phase Transformations

Free energy is important because it determines whether or not a phase transformation is thermodynamically possible. Essentially, the change in free energy, either ΔF or ΔG, must be negative for a reaction to occur. If the free-energy change is zero, the system is in a state of equilibrium and no reaction will occur. Likewise, if the free-energy change is positive, the reaction will not occur. To summarize:

- If ΔF or $\Delta G > 0$ or positive, the reaction will not occur.
- If ΔF or $\Delta G = 0$, the system is in equilibrium, and the reaction will not occur.
- If ΔF or $\Delta G < 0$ or negative, the reaction will occur.

The more negative the free-energy change (larger negative number), the greater the driving force for a transformation. In other words, if the free-energy change is a small negative number, the driving force is low, and as it becomes more negative, the driving force increases. The following develops these concepts in a little more detail.

The internal energy, E, of a system (e.g., an alloy) is made up of two parts: the kinetic energy that is due to atomic vibrations of the metallic lattice, and the potential energy that is a function of the bond strengths. The internal energy can also be thought of as the sum of the free energy, F, and the bound energy (TS):

$$E = F + TS \qquad \text{(Eq 2.1)}$$

where E is the internal energy, F is the Hemholtz free energy, T is the absolute temperature in degrees Kelvin (K), and S is entropy, a measure of the randomness of the system.

The Gibbs free energy is defined as:

$$G = H - TS \qquad \text{(Eq 2.2)}$$

Because the enthalpy, H, at constant pressure is the heat content:

$$H = E + PV \qquad \text{(Eq 2.3)}$$

When dealing with liquids and solids, the PV term is usually very small in comparison to E; therefore the enthalpy, H, can be considered as being equal to the internal energy, E ($H \sim E$). Therefore, this allows the Gibbs free energy to be expressed as:

$$G = E - TS \quad \text{(Eq 2.4)}$$

which is the same as the Hemholtz free energy, F:

$$F = E - TS \quad \text{(Eq 2.5)}$$

What is really important in metallurgical processes is not the free energy, G, itself, but the change in free ΔG. It can be shown that, at constant temperature and pressure:

$$\Delta G = \Delta H - T\Delta S \quad \text{(Eq 2.6)}$$

A system is said to be in equilibrium when it attains the state of lowest Gibbs free energy and the change in ΔG is then 0:

$$\Delta G = 0 \text{ at constant } T \text{ and } P \quad \text{(Eq 2.7)}$$

All phase transformations occur to lower the total energy of the system. Any transformation that results in a reduction in free energy, G, is possible; that is:

$$\Delta G = G_f - G_i < 0 \quad \text{(Eq 2.8)}$$

where G_f is the free energy of the final state and G_i is the free energy of the initial state.

At low temperatures, the solid phase is the most stable state because there is strong bonding between atoms and the system has the lowest free energy and entropy. As the temperature increases, the TS term begins to dominate and the solid phase has more freedom of atomic movement due to increasing temperature, until it melts and becomes a liquid phase.

A more detailed review of basic thermodynamics and Gibbs free energy is given in Chapter 3, "Thermodynamics and Phase Diagrams," in this book, and Chapter 13, "Computer Simulation of Phase Diagrams," covers thermodynamic modeling and computer simulation of phase diagrams.

2.7 Kinetics

Thermodynamics allows the calculation of the driving force for a phase transformation; however, it tells nothing about how fast the transformation will occur. Kinetics must be used to calculate the speed at which the transformation will occur. In a large number of metallurgical processes, the reaction rate increases exponentially with increases in temperature and can be described by the Arrhenius equation:

$$\text{rate} = Ce^{-Q/RT} \qquad \text{(Eq 2.9)}$$

where C is a pre-exponential constant that is independent of temperature, R is the universal gas constant, Q is the activation energy, and T is the absolute temperature.

Taking the logarithm of each side of Eq 2.9 allows the equation to be rewritten as:

$$\ln(\text{rate}) = \ln C - \frac{Q}{RT} \qquad \text{(Eq 2.10)}$$

A semilogarithmic plot of ln (rate) versus the reciprocal of absolute temperature ($1/T$) gives a straight line, as shown in Fig. 2.10. The slope of this plot is $-Q/R$ and ln C is obtained by extrapolating the plot to $1/T = 0$ ($T = \infty$). Thus, if the rate at two different temperatures is known, the rate at a third temperature can be determined. Likewise, if the rate and activation energy, Q, is known at one temperature, then the rate at any other temperature can be determined.

The form of the Arrhenius given in Eq 2.10 can also be written as:

$$\text{rate} = Ce^{-q/kT} \qquad \text{(Eq 2.11)}$$

where q is the activation energy per atomic scale unit ($q = Q/N_{AV}$), and k is Boltzmann's constant ($k = R/N_{AV} = 13.8 \times 10^{-24}$ J/K).

The activation energy, q, for an atom to move from one stable position to another is shown in Fig. 2.11. In other words, the activation energy is that energy provided by temperature that must be supplied to overcome

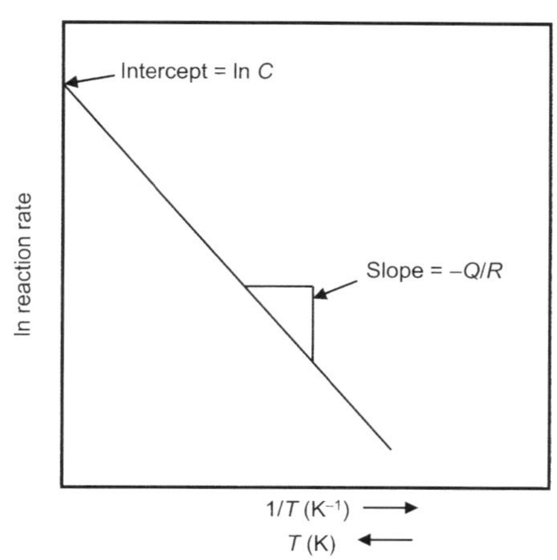

Fig. 2.10 Typical Arrhenius plot. Source: Ref 2.2

the energy barrier. For an Arrhenius type reaction, an increase in temperature of approximately 10 °C (18 °F) nearly doubles the reaction rate of the process.

2.8 Liquid-Solid Phase Transformations

At any temperature, the thermodynamically stable state is the one with the lowest free energy. The equilibrium temperature for transition between two states is the temperature at which both have the same free energy. For the liquid-to-solid transition, this is the melting temperature. Compared to a solid metal, a liquid metal has a higher internal energy (equal to the latent heat of fusion) and a higher entropy due to its more random structure. The result is that as the temperature decreases toward the melting point, T_m, the liquid phase starts to develop more order, the entropy term decreases, and the free energy for the liquid rises at a faster rate than that of the solid, as illustrated in Fig. 2.12. At T_m, the equilibrium melting point, the free energies of both phases are equal. However, solidification does not occur

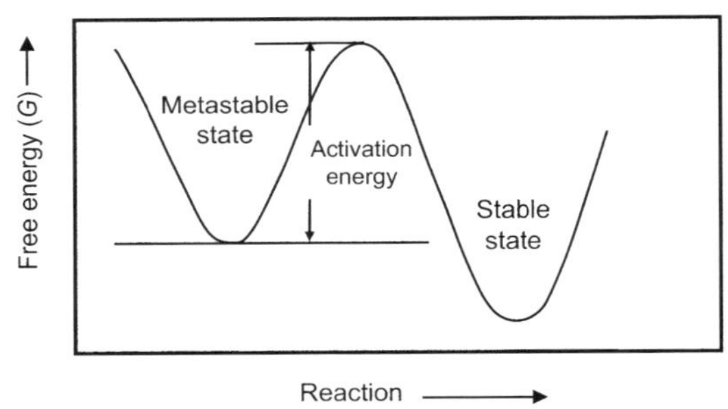

Fig. 2.11 Activation energy, metastable to stable state. Source: Ref 2.2

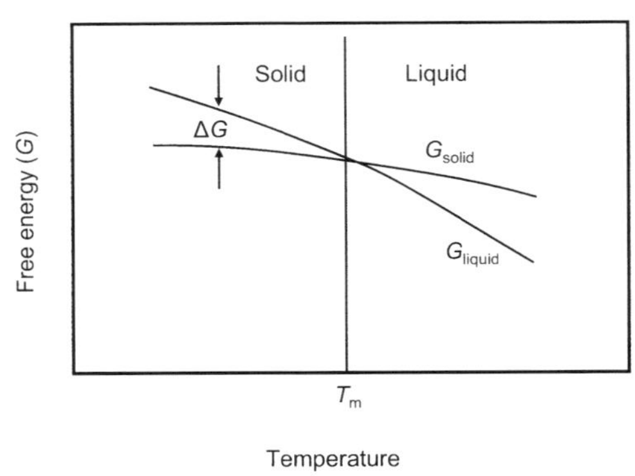

Fig. 2.12 Gibbs free-energy curves during solidification. Source: Ref 2.2

because the free-energy change is zero ($\Delta G = 0$) and it must be negative. Below T_m, the free energy does become negative ($\Delta G < 0$) and the metal solidifies. The free-energy change is thus:

$$\Delta G = \Delta G_S - \Delta G_L < 0 \qquad \text{(Eq 2.12)}$$

Immediately below T_m, the free-energy change is very small, so solidification occurs slowly, but at larger undercooling or supercooling ($T_m - T$), the free-energy change becomes greater and the solidification rate is much more rapid.

As a metal freezes, solidification starts on a small scale with groups of atoms joining together in clusters (Fig. 2.13). As the temperature falls during the solidification process, the thermal agitation of the atoms in the liquid decreases, allowing small random aggregations of atoms to form into small crystalline regions called embryos. An embryo is a small cluster of atoms that has not yet reached a large enough size to become stable and grow. Therefore, embryos are constantly forming and then remelting. Eventually, as the temperature becomes lower, some of the embryos will reach a critical size and become nuclei that are then stable and capable of further growth into crystals. These crystals then continue to grow until they impinge on other growing crystals. Each of these crystals becomes grains in the final solidified structure. The crystalline structure within each grain is uniform but changes abruptly at the interfaces (grain boundaries) with adjacent crystals. This process of forming nuclei in the freezing melt and their subsequent growth is characterized as a nucleation and growth process.

Nucleation can occur by either homogeneous or heterogeneous nucleation. Homogenous nucleation occurs when nucleation takes place throughout the bulk of the liquid without preference for any point; that is, the probability of nucleation is the same everywhere within the volume of the liquid. When a solid nucleates preferentially at certain points in the liquid and the probability of nucleation occurring at certain preferred sites is much more than at other sites, it is called heterogeneous nucleation. Heterogeneous nucleation can occur at small solid inclusions within the melt and at sur-

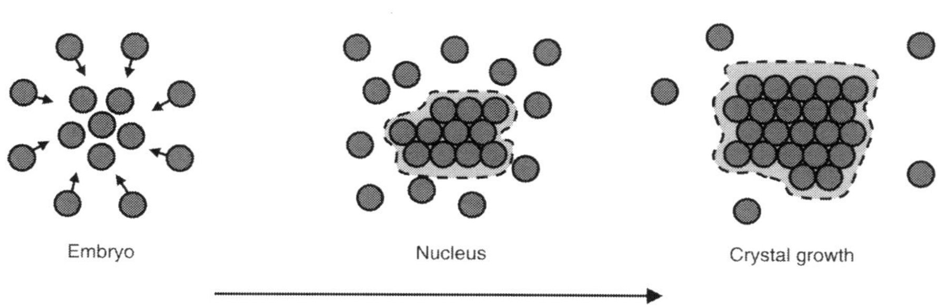

Fig. 2.13 Nucleation and growth during solidification. Source: Ref 2.2

face irregularities on mold walls. While homogeneous nucleation can be made to occur in carefully controlled laboratory conditions, heterogeneous nucleation is the nucleation mechanism that is observed in all commercial casting operations.

Consider homogeneous nucleation in which a spherical nuclei of radius r forms within the bulk of the liquid. Because the formation of a nucleus requires the creation of an interface between the solid and liquid, it creates an increase in the free energy of the system. The surface energy required for a sphere of surface area $4\pi r^2$ is $4\pi r^2\sigma$, where σ is the surface energy per unit area and is shown as increasing the free energy of the system in Fig. 2.14. As was previously illustrated in Fig. 2.12, there is a free-energy reduction when the metal transforms from a liquid to a solid. This free energy is known as the volume free energy G_v, which, for a spherical nuclei, is $4/3\pi r^3\ G_v$, and contributes to a decrease in free energy in Fig. 2.14. Thus, the total change in free energy, ΔG, is the sum of the decrease in volume free energy and the increase in surface free energy.

$$\Delta G = \frac{4}{3}\pi r^3 \Delta G_v + 4\pi r^2 \sigma \qquad \text{(Eq 2.13)}$$

where $^4\!/_3\pi r^3$ is the volume of a spherical embryo of radius r, ΔG_v is the volume free energy, $4\ \pi r^2$ is the surface area of a spherical embryo, and σ is the surface free energy.

The total free-energy, ΔG, curve in Fig. 2.14 shows that there is a critical radius, r^*, that the particle must reach before it becomes a stable nuclei and continued growth is assured. If the embryo is very small, further growth of the embryo would cause the total free energy to increase. Therefore, the

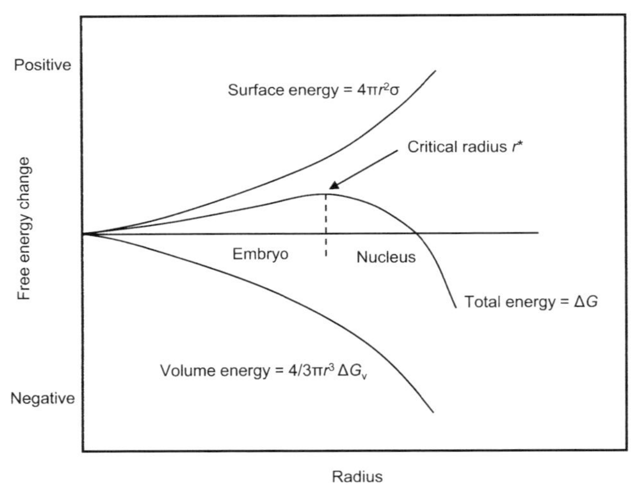

Fig. 2.14 Free-energy curves for homogeneous nucleation. Source: Ref 2.2

embryo remelts and causes a decrease in the free energy. When the particle becomes a nuclei at the critical radius, r^*, growth is assured because the total free-energy curve then decreases continuously as it becomes larger.

Very large undercoolings and highly polished molds are required to cause homogenous nucleation, and it is rarely, if ever, observed in practice. In reality, the degree of undecooling is usually very small, often only a degree or two. Whereas homogeneous nucleation assumes nucleation occurs randomly throughout the liquid, heterogeneous nucleation occurs at preexisting particles in the liquid or at rough areas on the mold. Thus, heterogeneous nucleation does not incur the free-energy penalty of having to form a new surface; rather, it uses a preexisting surface. In actual castings, heterogeneous nucleation usually occurs at defects on the cold mold wall, such as cracks and crevices. However, for an inclusion or a defect on the mold wall to serve as a nucleation site, the liquid metal must be able to wet the surface. In addition, in many castings, a fine grain size is desirable, and nucleation agents are added to the melt to form as many nuclei as possible. Because each of these nuclei eventually grows and forms a grain, a larger number of nuclei results in a finer or smaller grain size. For example, a combination of 0.02 to 0.05% Ti and 0.01 to 0.03% B is added to many aluminum alloys. Solid titanium boride particles form and serve as effective sites for heterogeneous nucleation. Ultrasonic agitation of the mold during casting also promotes finer grain size by breaking up growing particles and helping to distribute them through the melt. Once a melt has been populated with nuclei, the growth of the solid is controlled almost exclusively by the rate of heat removal. Important liquid-solid transformations include the eutectic, peritectic, and monotectic reactions that are covered in Chapters 5, 6, and 7 in this book, respectively.

2.9 Solid-State Phase Transformations

Solid-state phase transformations occur when one or more parent phases, usually on cooling, produces a new phase or phases. The most important mechanisms are nucleation, growth, and diffusion. However, not all transformations rely on diffusion. For example, the important martensitic transformation in steels occurs quite suddenly by a combination of shear mechanisms.

Solid-state transformations differ from liquid-to-solid transformations (solidification) in several important ways. In solids, the atoms are bound much tighter than in liquids and diffuse much more slowly, by a factor of approximately 10^{-5} slower, even at temperatures approaching the melting point. Instead of transforming directly to the equilibrium phase, metastable transition phases can form prior to forming the final equilibrium phase. Like solidification, nucleation in solid-state reactions is almost always heterogeneous. Heterogeneous nucleation occurs at structural defects such as grain boundaries, dislocations, and interstitial atoms.

During a solid-state phase transformation, at least one new phase is normally formed that has different physical/chemical characteristics and/or a different structure than the parent phase. During nucleation, very small particles or nuclei of the new phase, often consisting of only a few hundred atoms, reach a critical size that is capable of further growth. During the growth stage, the nuclei grow in size, resulting in the disappearance of some, or all, of the parent phase. The transformation is complete when the growth of the new phase particles proceeds until the equilibrium fraction is attained.

Particle growth in a solid is controlled by diffusion, in which atoms diffuse from the parent phase across the phase boundary and into the growing second-phase particles. Because this is a diffusion-controlled process, the grow rate, $\dot{G}$, is determined by temperature:

$$\dot{G} = Ce^{-Q/RT} \qquad \text{(Eq 2.14)}$$

where Q is the activation energy and C is the pre-exponential constant.

Both Q and C are independent of temperature. The temperature dependence of the growth rate, $\dot{G}$, and the nucleation rate, $\dot{N}$, is shown in the Fig. 2.15 curves. At a given temperature, the overall transformation rate is given as the product of $\dot{G}$ and $\dot{N}$. For transformations that occur at high temperatures, the nucleation rate will be low and the growth rate will be high. This will result in fewer particles that will grow to large sizes and the resulting transformation product will be coarse. On the other hand, if the transformation occurs at lower temperatures, where there is a much higher driving force for nucleation, many particles will form where the growth rate is lower, and the resulting transformation product will be much finer.

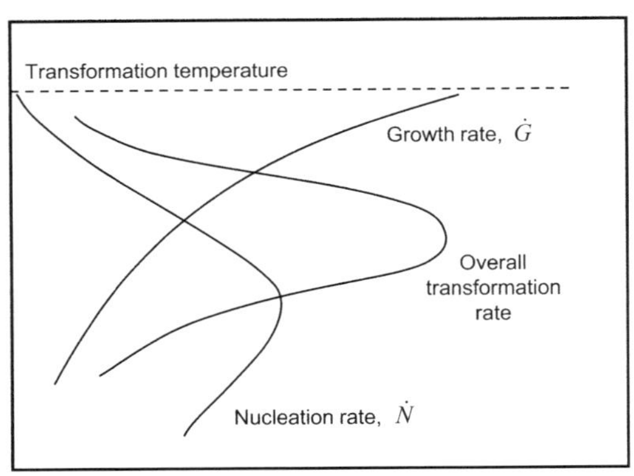

Fig. 2.15 Transformation rate as a function of temperature. Source: Ref 2.4 as published in Ref 2.2

The rate of transformation and the time required for the transformation to proceed to some degree of completion are inversely proportional to one another. The time to reach a 50% degree of completion in the reaction is frequently used. Therefore, the rate of a transformation is taken as the reciprocal of time required for the transformation to proceed halfway to completion, $t_{0.5}$, or:

$$\text{rate} = \frac{1}{t_{0.5}} \qquad \text{(Eq 2.15)}$$

If the logarithm of this transformation time is plotted versus temperature, the result is a C-shaped curve, like the one shown in Fig. 2.16, which is a mirror image of the transformation rate curve previously shown in Fig. 2.15.

When the fraction of material transformed is plotted versus the logarithm of time at a constant temperature, an S-shaped curve similar to that in Fig. 2.17 is obtained, which is typical of the kinetic behavior for most solid-state reactions. For solid-state transformations displaying the kinetic behavior in Fig. 2.17, the fraction of transformation, y, is a function of time, t, and follows the Avrami equation:

$$y = 1 - e^{-kt^n} \qquad \text{(Eq 2.16)}$$

where k and n are time-independent constants for the particular reaction. The value for k depends on the temperature and the properties of the initial phase, while the coefficient n has the values listed in Table 2.1.

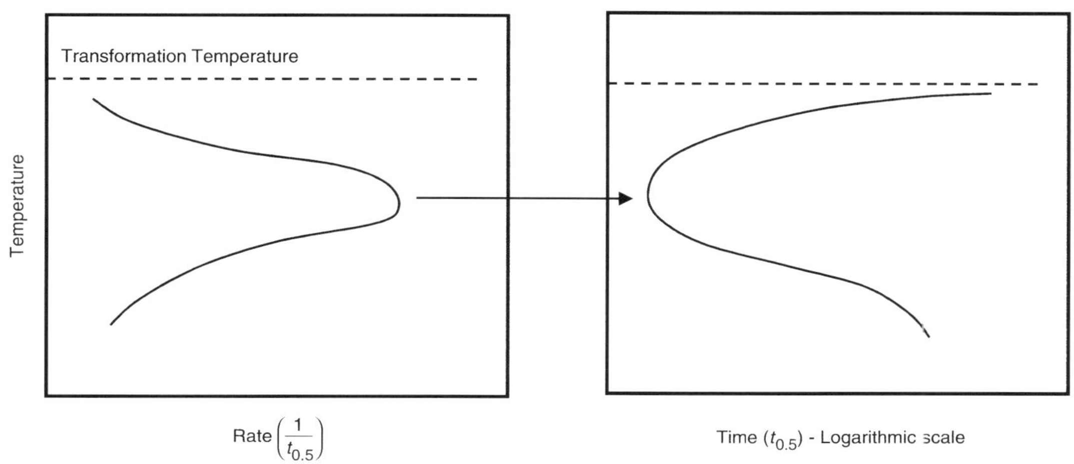

Fig. 2.16 Transformation rate versus temperature. Source: Ref 2.4 as published in Ref 2.2

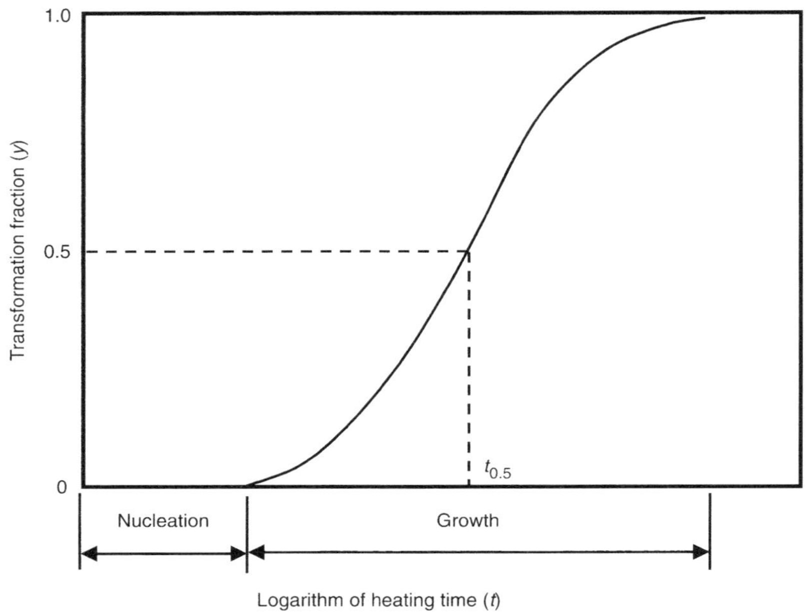

Fig. 2.17 Fraction reacted as a function of time. Source: Ref 2.4 as published in Ref 2.2

Table 2.1 Values of *n* in Avrami equation

Type of transformation		Value of *n*
Cellular	Constant rate of nucleation	4
	Zero rate of nucleation	3
	Nucleation at grain edges	2
	Nucleation at grain boundaries	1
Precipitation	Particle growing from small dimensions	
	1. Constant rate of nucleation	2.5
	2. Zero rate of nucleation	1.5
	Thickening of needles	1
	Thickening of plates	0.5

Source: Ref 2.5 as published in Ref 2.2

The values of *n* in the Avrami equation distinguish between cellular transformations and precipitation transformations. A cellular transformation is one in which an entirely new phase is formed during the transformation, such as a solid-state transformation in which a higher-temperature γ phase transforms in two lower-temperature phases, α and β. In this reaction, γ disappears and is replaced by a mixture of α and β. It is called a cellular reaction because α and β grow as cells or nodules into the γ phase.

2.10 Polymorphism (Allotropy)

The structure of solid elements and compounds under stable equilibrium conditions is crystalline, and the crystal structure of each is unique. Some elements and compounds, however, are polymorphic (multishaped); that is, their structure transforms from one crystal structure to another with changes in temperature and pressure, each unique structure constituting a distinctively separate phase. The term *allotropy* (existing in another form) is usually used to describe polymorphic changes in chemical elements.

The important metal iron undergoes a series of allotropic transformations during heating and cooling, as shown in Fig. 2.18. Note that an allotropic transformation is a solid-state phase transformation, and as such, occurs at a constant temperature during either heating or cooling. Under equilibrium cooling conditions, the solidification of pure iron from the liquid occurs at 1540 °C (2800 °F) and forms what is called delta iron (δ_{Fe}), which has a bcc structure. Delta iron is then stable on further cooling until it reaches 1390 °C (2541 °F), where it undergoes a transformation to a face-centered cubic (fcc) structure called gamma iron (γ_{Fe}). On still further cooling to 900 °C (1648 °F), it undergoes yet another phase transformation, transforming from the fcc structure back to the bcc structure, called ferrite iron (α_{Fe}) to distinguish it from the higher-temperature delta iron. This last transformation, $\gamma_{Fe} \rightarrow \alpha_{Fe}$, is extremely important because it forms the basis for the hardening of steel. Note that the $\gamma_{Fe} \rightarrow \alpha_{Fe}$ transformation occurs at 900

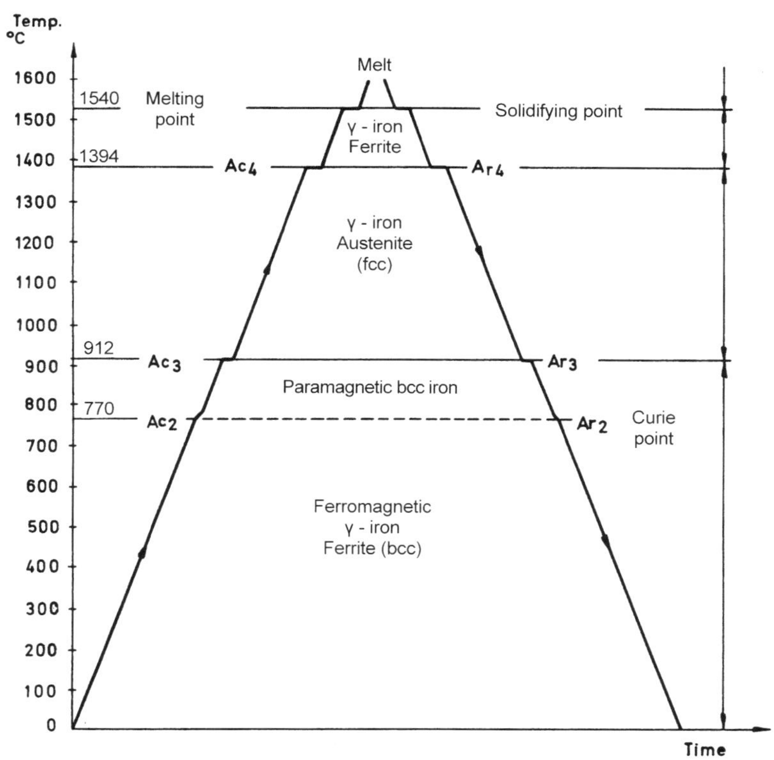

Fig. 2.18 Polymorphic transformations in pure iron. Source: Ref 2.6

°C (1648 °F) on cooling, somewhat lower than the 910 °C (1673 °F) temperature transformation on heating. This temperature differential is known as the temperature hysteresis of allotropic phase transformation and its magnitude increases with increases in the cooling rate. The temperatures (designated *A*) associated with heating contain the subscript "c," which is French for *chauffage,* meaning heating, while the ones for cooling have the subscript "r" for the French *refroidissement,* meaning cooling.

Many other metals, as well as some nonmetals, also exhibit allotropic transformations. For example, titanium, zirconium, and hafnium all exhibit a transition from a hexagonal close-packed (hcp) structure to bcc on heating. Note that in each case, a close-packed structure is stable at room temperature, while a looser packing is stable at elevated temperatures. While this is not always the case, it is a trend experienced with many metals. Allotropic transformations are a type of massive transformation. Massive transformations occur by a transition in crystal structure and are characterized by the chemical invariance between parent and product phases. These transformations can occur both on heating and cooling, although the mechanism requires rapid heating and cooling rates, because the ability of atoms to diffuse the long distances typical of diffusion-controlled transformations is impaired.

Important solid-state transformation structures include eutectoid and peritectoid structures, martensitic and bainitic structures, and precipitation structures (see Chapters 8, "Solid-State Transformations;" 15, "Nonequilibrium Reactions—Martensitic and Bainitic Structures;" and 16 "Nonequilibrium Reactions—Precipitation Hardening;" in this book, respectively).

ACKNOWLEDGMENT

The material in this chapter came from *Elements of Metallurgy and Engineering Alloys* by F.C. Campbell, ASM International, 2008 and from *Elements of Physical Metallurgy* by A.G. Guy, 2nd ed., Addison-Wesley Publishing Company, 1959.

REFERENCES

2.1 A.G. Guy, *Elements of Physical Metallurgy*, 2nd ed., Addison-Wesley Publishing Company, 1959

2.2 F.C. Campbell, *Elements of Metallurgy and Engineering Alloys*, ASM International, 2008

2.3 D.R. Askeland, *The Science and Engineering of Materials,* 2nd ed., PWS-Kent Publishing Co., 1989

2.4 W.D. Callister, *Fundamentals of Materials Science and Engineering,* 6th ed., John Wiley & Sons, Inc., 2003

2.5 J.W. Christian, *The Theory of Transformations in Metals and Alloys,* Pergamon Press, 1965

2.6 *Metallurgy for the Non-Metallurgist,* 2nd ed., A.C. Reardon, Ed., ASM International, 2011, p 42

SELECTED REFERENCES

- M.F. Ashby and D.R.H. Jones, *Engineering Materials 2—An Introduction to Microstructures, Processing, and Design*, 2nd ed., Butterworth Heinemann, 1998
- H. Baker, Introduction to Alloy Phase Diagrams, *Desk Handbook: Phase Diagrams for Binary Alloys*, 2nd ed., H. Okamoto, Ed., 2010 and in *Alloy Phase Diagrams,* Vol 3, *ASM Handbook*, ASM International, 1992
- A.H. Cottrell, *An Introduction to Metallurgy*, 2nd ed., IOM Communications, 1975
- R.A. Higgins, *Engineering Metallurgy—Applied Physical Metallurgy*, 6th ed., Arnold, 1993
- D.A. Porter and K.E. Easterling, *Phase Transformations in Metals and Alloys*, Chapman and Hall, 1981
- R.E. Reed-Hill and R. Abbaschian, *Physical Metallurgy Principles*, 3rd ed., PWS Publishing Company, 1991

CHAPTER 3

Thermodynamics and Phase Diagrams

THERMODYNAMICS is a branch of physics and chemistry that covers a wide field, from the atomic to the macroscopic scale. In materials science, thermodynamics is a powerful tool for understanding and solving problems. Chemical thermodynamics is the part of thermodynamics that concerns the physical change of state of a chemical system following the laws of thermodynamics. The thermodynamic properties of individual phases can be used for evaluating their relative stability and heat evolution during phase transformations or reactions. Traditionally, one of the most common applications of chemical thermodynamics is for the construction and interpretation of phase diagrams.

The thermodynamic quantities that are most frequently used in materials science are the enthalpy, in the form of the heat content of a phase; the heat of formation of a phase, or the latent heat of a phase transformation; the heat capacity, which is the change of heat content with temperature; the Gibbs free energy, which determines whether or not a chemical reaction is possible; and the chemical potential or chemical activity, which describes the effect of compositional change in a solution phase on its energy. All of these thermodynamic quantities are part of the energy content of a system and are governed by the three laws of thermodynamics.

3.1 Three Laws of Thermodynamics

A physical system consists of a substance, or a group of substances, that is isolated from its surroundings, a concept used to facilitate study of the effects of conditions of state. Isolated means that there is no interchange of mass between the substance and its surroundings. The substances in alloy systems, for example, might be two metals, such as copper and zinc; a metal and a nonmetal, such as iron and carbon; a metal and an intermetallic compound, such as iron and cementite; or several metals, such as aluminum, magnesium, and manganese. These substances constitute the components comprising the system. However, a system can also consist of a single component, such as an element or compound.

The sum of the kinetic energy (energy of motion) and potential energy (stored energy) of a system is called its internal energy, E. Internal energy is characterized solely by the state of the system. A thermodynamic system that undergoes no interchange of mass (material) with its surroundings is called a closed system. A closed system, however, can interchange energy with its surroundings.

First Law. The First Law of Thermodynamics, as stated by Julius von Mayer, James Joule, and Hermann von Helmholtz in the 1840s, states that energy can be neither created nor destroyed. Therefore, it is called the Law of Conservation of Energy. This law means that the total energy of an isolated system remains constant throughout any operations that are carried out on it; that is, for any quantity of energy in one form that disappears from the system, an equal quantity of another form (or other forms) will appear. For example, consider a closed gaseous system to which a quantity of heat energy, δQ, is added and a quantity of work, δW, is extracted. The First Law describes the change in internal energy, dE, of the system as:

$$dE = \delta Q - \delta W \qquad \text{(Eq 3.1)}$$

In the vast majority of industrial processes and material applications, the only work done by or on a system is limited to pressure/volume terms. Any energy contributions from electric, magnetic, or gravitational fields are neglected, except for electrowinning and electrorefining processes such as those used in the production of copper, aluminum, magnesium, the alkaline metals, and the alkaline earths. When these field effects are neglected, the work done by a system can be measured by summing the changes in volume, dV, times each pressure, P, causing a change. Therefore, when field effects are neglected, the First Law can be written:

$$dE = \delta Q - PdV \qquad \text{(Eq 3.2)}$$

Second Law. While the First Law establishes the relationship between the heat absorbed and the work performed by a system, it places no restriction on the source of the heat or its flow direction. This restriction, however, is set by the Second Law of Thermodynamics, which was advanced by Rudolf Clausius and William Thomson (Lord Kelvin). The Second Law states that the spontaneous flow of heat always is from the higher-temperature body to the lower-temperature body. In other words, all naturally occurring processes tend to take place spontaneously in the direction that will lead to equilibrium.

The entropy, S, represents the energy (per degree of absolute temperature, T) in a system that is not available for work. In terms of entropy, the Second Law states that all natural processes tend to occur only with an increase in entropy, and the direction of the process is always such as to lead to an increase in entropy. For processes taking place in a system in equilibrium with its surroundings, the change in entropy is defined as:

$$dS \equiv \frac{\delta Q}{T} \equiv \frac{dE + PdV}{T} \qquad \text{(Eq 3.3)}$$

Third Law. A principle advanced by Theodore Richards, Walter Nernst, Max Planck, and others, often called the Third Law of Thermodynamics, states that the entropy of all chemically homogeneous materials can be taken as zero at absolute zero temperature (0 K). This principle allows calculation of the absolute values of entropy of pure substances solely from their heat capacity.

3.2 Gibbs Free Energy

Josiah Willard Gibbs (1839–1903) was an American theoretical physicist, chemist, and mathematician. He devised much of the theoretical foundation for chemical thermodynamics and physical chemistry. Yale University awarded Gibbs the first American Ph.D. in engineering in 1863, and he spent his entire career at Yale. Between 1876 and 1878, Gibbs wrote a series of papers on the graphical analysis of multiphase chemical systems. These were eventually published together in a monograph titled *On the Equilibrium of Heterogeneous Substances*, his most renowned work. It is now deemed one of the greatest scientific achievements of the 19th century and one of the foundations of physical chemistry. In these papers Gibbs applied thermodynamics to interpret physicochemical phenomena, successfully explaining and interrelating what had previously been a mass of isolated facts.

For transformations that occur at constant temperature and pressure, the relative stability of the system is determined by its Gibbs free energy:

$$G \equiv H - TS \qquad \text{(Eq 3.4)}$$

where H is the enthalpy. Enthalpy is a measure of the heat content of the system and is given by:

$$H = E + PV \qquad \text{(Eq 3.5)}$$

The internal energy, E, is equal to the sum of the total kinetic and potential energy of the atoms in the system. Kinetic energy results from the vibration of the atoms in solids or liquids, and the translational and rotational energies of the atoms and molecules within a liquid or gas. Potential energy results from the interactions or bonds between the atoms in the system. If a reaction or transformation occurs, the heat that is absorbed (endothermic) or given off (exothermic) depends on the change of the internal energy of the system. It also depends on the changes in the volume of the system which is accounted for by the term PV. At a constant pressure, the heat absorbed or given off is given by the change in H. When dealing with condensed phases (solids or liquids), the term PV is usually very small in comparison to E, and therefore $H \approx E$. Finally, the entropy, S, is a measure of the randomness of the system.

A system is considered to be in equilibrium when it is in its most stable state and has no desire to change with time. At a constant temperature and pressure, a system with a fixed mass and composition (a closed system) will be in a state of stable equilibrium if it has the lowest possible value of the Gibbs free energy:

$$dG = 0 \qquad \text{(Eq 3.6)}$$

From this definition of free energy, the state with the highest stability will be the one with the lowest enthalpy and the highest entropy. Therefore, at low temperatures, solid phases are the most stable because they have the strongest atomic bonding and therefore the lowest enthalpy (internal energy). However, at high temperatures, the $-TS$ term dominates and the liquid and eventually the vapor phases becomes the most stable. In processes where pressure changes are important, phases with small volumes are most stable at high pressures.

The definition of equilibrium given in Eq 3.6 is illustrated graphically in Fig. 3.1. The various possible atomic configurations are represented by the points along the abscissa. The configuration with the lowest free energy, G, will be the stable equilibrium configuration. Therefore, configuration A would be the stable equilibrium configuration. There are other configurations, such as configuration B, which lie at a local minimum of free energy but do not have the lowest possible value of G. Such configurations are called metastable equilibrium states to distinguish them from the stable equilibrium state. The other configurations that lie between A and B are intermediate states for which $dG \neq 0$ and are unstable and will disappear at the first opportunity; that is, if a change in thermal fluctuations causes the atoms to be arranged into an unstable state, they will rapidly rearrange themselves into one with a free-energy minima. An example of a metastable configuration state is diamond. Given enough time, diamond will convert to graphite, the stable equilibrium configuration. However, as in diamond, metastable equilibrium can, for all practical instances, exist indefinitely.

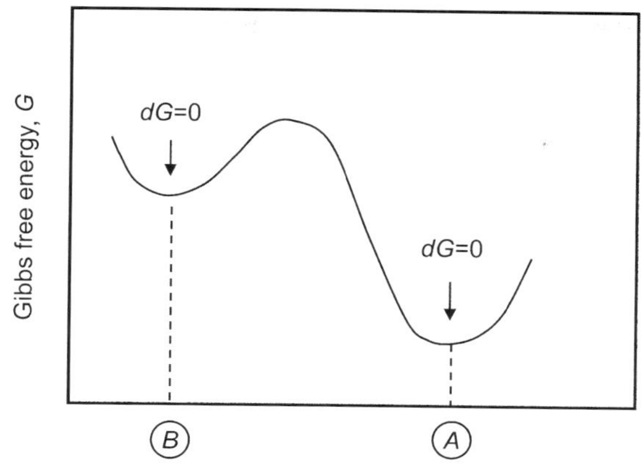

Fig. 3.1 Gibbs free energy for different atomic configurations in a system. Configuration *A* has the lowest free energy and therefore is the arrangement of stable equilibrium. Configuration *B* is in a state of metastable equilibrium. Adapted from Ref 3.1

Any transformation that results in a decrease in Gibbs free energy, *G*, is possible. Any reaction that results in an increase in *G* is impossible and will not occur. Therefore, the necessary criterion for any phase transformation is:

$$\Delta G = G_2 - G_1 < 0 \quad \text{(Eq 3.7)}$$

where G_1 and G_2 are the free energies of the initial and final states, respectively. It is not necessary that the transformation immediately go to the stable equilibrium state. It may go through a whole series of intermediate metastable states.

Sometimes metastable states can be very short-lived, or at other times they can exist almost indefinitely. These are explained by the free-energy hump between the metastable and equilibrium states in Fig. 3.1. In general, higher free-energy humps, or energy barriers, lead to slower transformation rates.

Free Energy of Single-Component Systems

A single-component unary system is one containing a pure metal or one type of molecule that does not disassociate over the temperature range of interest. Consider the phase changes that occur with changes in temperature at a constant pressure of one atmosphere. To predict the phase changes that are stable, or mixtures that are equilibrium at different temperatures, it is necessary to be able to calculate the variation of *G* with *T*.

Specific heat is the quantity of heat required to raise the temperature of the substance by 1 K. At constant pressure, the specific heat, C_p, is given by:

$$C_p = \left(\frac{\partial H}{\partial T}\right)_p \quad \text{(Eq 3.8)}$$

The specific heat of a substance varies with temperature in the manner shown in Fig. 3.2(a). Therefore, the variation of H with T can be obtained from the knowledge of the variation of C_p with T. The enthalpy, H, is usually measured by setting $H = 0$ for a pure element in its most stable state at room temperature (298 K, or 25 °C or 77 °F). The variation of H with T is then calculated by integrating Eq 3.8, that is:

$$H = \int_{298}^{T} C_p dT \quad \text{(Eq 3.9)}$$

The slope of the H-T curve is C_p, as shown in Fig. 3.2(b).

The variation of entropy with temperature can also be derived from the specific heat, C_p. From thermodynamics:

$$\frac{C_p}{T} = \left(\frac{\partial S}{\partial T}\right)_p \quad \text{(Eq 3.10)}$$

Because entropy is zero at 0 K, Eq 3.10 can be integrated to give:

$$S = \int_{0}^{T} \frac{C_p}{T} dT \quad \text{(Eq 3.11)}$$

as shown in Fig. 3.2(c).

By combining Fig. 3.2(a) and (b) and using Eq 3.4, the variation of G with temperature shown in Fig. 3.3 is obtained. When temperature and pressure vary, the change in free energy, G, can be obtained for a system with fixed mass and composition from:

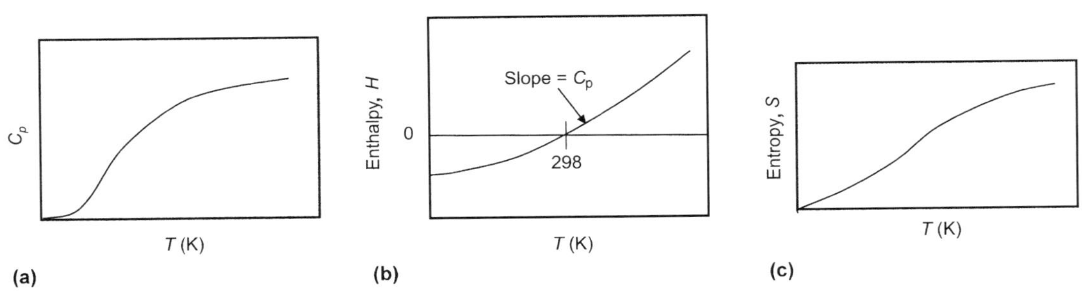

Fig. 3.2 (a) Variation of C_p with absolute temperature, T. (b) Variation of enthalpy, H, with absolute temperature for a pure metal. (c) Variation of entropy, S, with absolute temperature. Adapted from Ref 3.1

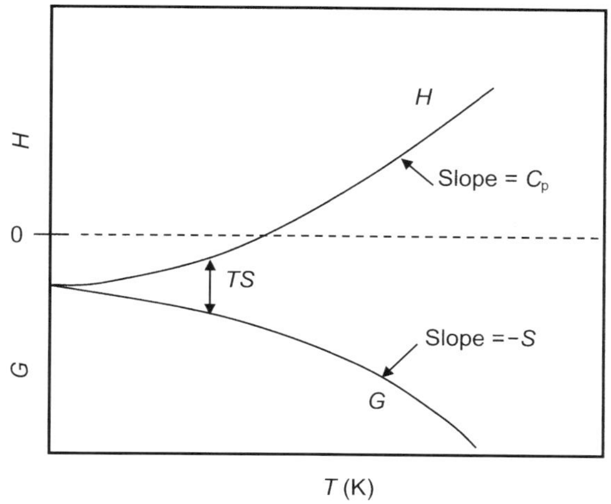

Fig. 3.3 Variation of Gibbs free energy with temperature. Adapted from Ref 3.1

$$dG = -SdT + VdP \qquad \text{(Eq 3.12)}$$

At constant pressure, $dP = 0$ and:

$$\left(\frac{\partial G}{\partial T}\right)_p = -S \qquad \text{(Eq 3.13)}$$

This equation shows that G decreases with increasing T at a rate given by $-S$. The relative positions of the free-energy curves of solid and liquid phases are shown in Fig. 3.4. At all temperatures, the liquid has a higher enthalpy (internal energy) than the solid phase. Therefore, at low temperatures, $G^L > G^S$. However, the liquid phase has a higher entropy than the solid phase and the Gibbs free energy of the liquid therefore decreases more rapidly with increasing temperature than that for the solid. For temperatures up to T_m, the solid phase has the lowest free energy and is therefore the equilibrium state of the system. At T_m, both phases have the same value of G and both the solid and liquid can coexist in equilibrium. Therefore, T_m is the equilibrium melting temperature at the pressure concerned.

If a pure component is heated from absolute zero, the heat supplied will raise the enthalpy at a rate determined by C_p (solid) along line *ab* in Fig. 3.4. Meanwhile, the free energy will decrease along line *ae*. At T_m, the heat supplied to the system will not raise its temperature but will be used to supply the latent heat of melting, L, that is required to convert the solid into a liquid (line *bc* in Fig. 3.4). Note that at T_m the specific heat appears to be infinite because the addition of heat does not appear as an increase in

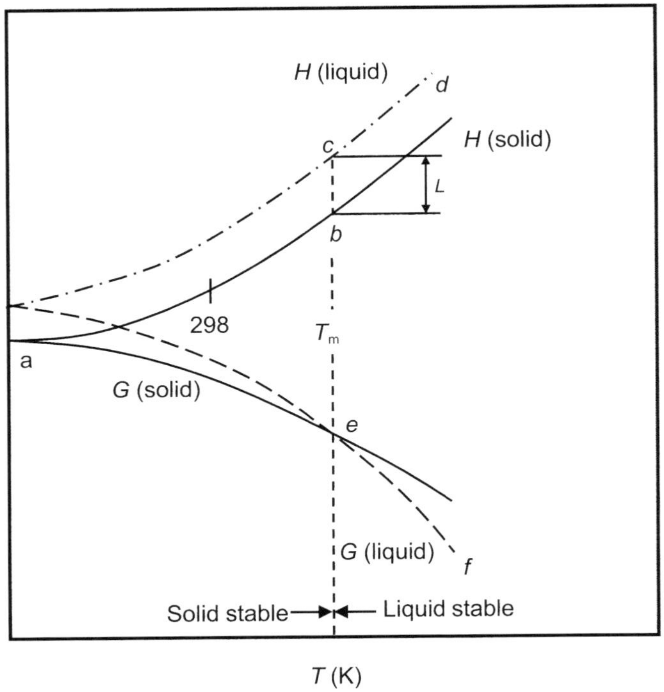

Fig. 3.4 Variation of enthalpy, *H*, and free energy, *G*, with temperature for the solid and liquid phases of a pure metal. *L*, latent heat of melting. *Tm*, equilibrium melting temperature. Adapted from Ref 3.1

temperature. When all the solid has transformed into liquid, the enthalpy of the system follows the line *cd* while the free energy, *G*, decreases along line *ef*. At still higher temperatures than those shown in Fig. 3.4, the free energy of the gas phase at atmospheric pressure becomes lower than the liquid, and the liquid transforms in a gas.

The equilibrium temperatures discussed so far only apply at a specific pressure (1 atm). At other pressures, the equilibrium temperatures will differ. For example, the effect of pressure on the equilibrium temperatures for pure iron is shown in Fig. 3.5. Increasing pressure has the effect of depressing the α–γ equilibrium temperature and raising the equilibrium melting temperature. At very high pressures, hexagonal close-packed (hcp) ε–iron becomes stable. The reason for these changes can be explained by Eq 3.12. At constant temperature, the free energy of a phase increases with pressure such that:

$$\left(\frac{\partial G}{\partial T}\right)_T = V \qquad \text{(Eq 3.14)}$$

If the two phases in equilibrium have different molar volumes, their respective free energies will not increase by the same amount at a given

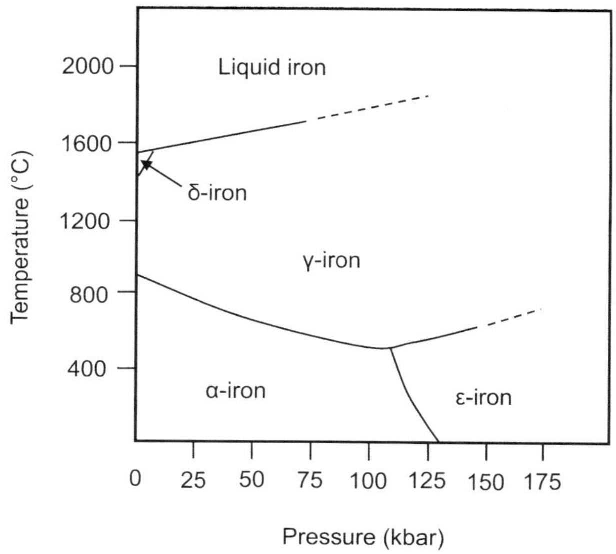

Fig. 3.5 Effect of pressure on the equilibrium phase diagram for pure iron. Adapted from Ref 3.1

temperature and equilibrium will be disturbed by changes in pressure. The only way to maintain equilibrium at different pressures is by varying the temperature.

If the two phases in equilibrium are α and β, the application of Eq 3.12 to 1 mol of both gives:

$$\begin{aligned} dG^{\alpha} &= V_m^{\alpha} dP - S^{\alpha} dT \\ dG^{\beta} &= V_m^{\beta} dP - S^{\beta} dT \end{aligned} \qquad \text{(Eq 3.15)}$$

If α and β are in equilibrium, $G_{\alpha} = G_{\beta}$, and therefore $dG_{\alpha} = dG_{\beta}$, and:

$$\left(\frac{dP}{dT}\right)_{eq} = \frac{S^{\beta} - S^{\alpha}}{V_m^{\beta} - V_m^{\alpha}} = \frac{\Delta S}{\Delta V} \qquad \text{(Eq 3.16)}$$

This equation gives the change in temperature, dT, required to maintain equilibrium between α and β if pressure is increased by dP. The equation can be simplified as follows. From Eq 3.4:

$$G_{\alpha} = H_{\alpha} - TS_{\alpha}$$
$$G_{\beta} = H_{\beta} - TS_{\beta}$$

Putting $\Delta G = G_{\beta} - G_{\alpha}$ gives:

$$\Delta G = \Delta H - T\Delta S$$

But because at equilibrium $G_\beta = G_\alpha$, $\Delta G = 0$ and:

$$\Delta H - T\Delta S = 0$$

As a result, Eq 3.16 becomes:

$$\left(\frac{dP}{dT}\right)_{eq} = \frac{\Delta H}{T\Delta V} \qquad \text{(Eq 3.17)}$$

which is one form of the Clausius-Clapeyron equation. Because close-packed γ-iron has a smaller molar volume than α-iron, $\Delta V = V_{m\beta} - V_{m\alpha} < 0$, while $\Delta H = H_\gamma - H_\alpha < 0$ for the same reason a liquid has a higher enthalpy than a solid, so that dP/dT is negative; that is, an increase in pressure lowers the equilibrium transition temperature. On the other hand, the δ–L equilibrium temperature is raised with increasing pressure due to the larger molar volume of the liquid phase. The effect of increasing pressure is to increase the area of the phase diagram over which the phase with the smallest molar volume is stable (γ-iron in Fig. 3.5).

In dealing with phase transformations, it is important to be concerned with the difference in free energy between two phases at temperatures away from the equilibrium temperature. For example, if a liquid is undercooled by ΔT below the T_m before it solidifies, solidification will be accompanied by a decrease in free energy, ΔG, as shown in Fig. 3.6. This free-energy decrease provides the driving force for solidification. The magnitude of this change can be obtained as follows. The free energies of the liquid and solid at a temperature, T, are given by:

$$G^L = H^L - TS^L$$
$$G^S = H^S - TS^S$$

Therefore, at a temperature, T:

$$\Delta G = \Delta H - T\Delta S \qquad \text{(Eq 3.18)}$$

where:

$$\Delta H = H^L - H^S \text{ and } \Delta S = S^L - S^S$$

At the equilibrium melting temperature, T_m, the free energies of a solid and liquid are equal ($\Delta G = 0$). Therefore:

$$\Delta G = \Delta H - T_m\Delta S = 0$$

and at T_m:

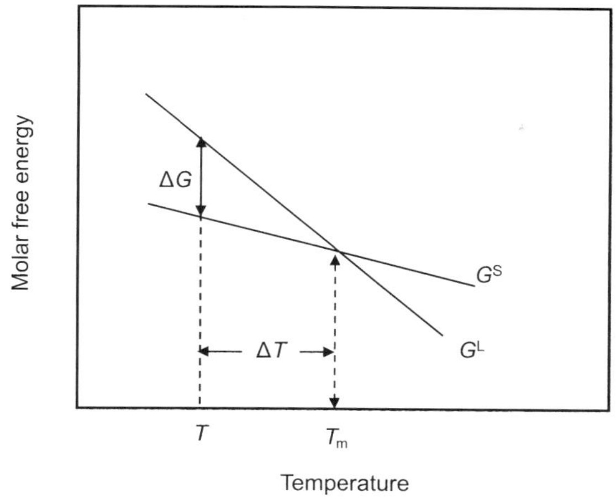

Fig. 3.6 Difference in free energy between liquid and solid close to the melting point. The curvature of G^S and G^L has been ignored. Adapted from Ref 3.1

$$\Delta S = \frac{\Delta H}{T_m} = \frac{L}{T_m} \quad \text{(Eq 3.19)}$$

This is known as the entropy of fusion. It is observed experimentally that the entropy of fusion is a constant ≈ R (8.4 $\text{Jmol}^{-1}\text{K}^{-1}$) for most metals per Richard's rule. This is not unexpected because metals with high bond strengths can be expected to have high values for both L and T_m.

For small undercoolings, ΔT, the difference in specific heats of the liquid and the solid ($C_p^{\,L} - C_p^{\,S}$) can be ignored. Both ΔH and ΔS are therefore independent of temperature. Combining Eq 3.18 and 3.19 gives:

$$\Delta G \cong L = T\frac{L}{T_m}$$

For a small ΔT:

$$\Delta G \cong \frac{L\Delta T}{T_m} \quad \text{(Eq 3.20)}$$

3.3 Binary Solutions

In single-component systems all phases have the same composition, and equilibrium only depends on pressure and temperature as variables. However, in alloys composition is also variable. Therefore, to understand phase changes in alloys requires the knowledge of how the Gibbs free energy of a given phase depends on composition as well as temperature and pressure. Because the important phase transformations in metallurgy mainly occur at a fixed pressure of 1 atm, the focus is on changes in composition and temperature. In order to introduce some of the basic concepts of the thermodynamics of alloys, a simple physical model for binary solid solutions is presented.

3.3.1 Gibbs Free Energy of Binary Solutions

The Gibbs free energy of a binary solution of *A* and *B* atoms can be calculated from the free energies of pure *A* and pure *B*. It is assumed that *A* and *B* have the same crystal structures in their pure states and can be mixed in any proportions to make a solid solution with the same crystal structure. Consider the case where 1 mol of homogeneous solid solution is made by mixing together X_A mol of *A* and X_B mol of *B*. Because there is a total of 1 mol of solution:

$$X_A + X_B = 1 \qquad \text{(Eq 3.21)}$$

and X_A and X_B are the mole fractions of *A* and *B*, respectively, in the alloy. To calculate the free energy of the alloy, mixing can be made in two steps (Fig. 3.7):

1. Bring together X_A mol of pure *A* and X_B mol of pure *B*.
2. Allow the *A* and *B* atoms to mix together to make a homogeneous solid solution.

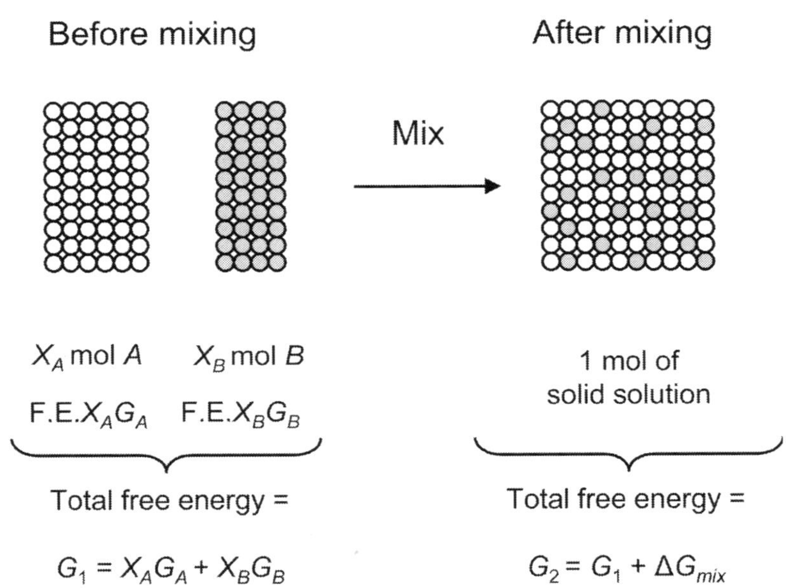

Fig. 3.7 Free energy of mixing. Adapted from Ref 3.1

After step 1, the free energy of the system is given by:

$$G_1 = X_A G_A + X_B G_B \qquad \text{(Eq 3.22)}$$

where G_A and G_B are the molar free energies of pure A and pure B at the temperature and pressure. G_1 can be most conveniently represented on a molar free-energy diagram (Fig. 3.8) in which molar free energy is plotted as a function of X_B or X_A. For all alloy compositions, G_1 lies on the straight line between G_A and G_B.

The free energy of the system will not remain constant during mixing of the A and B atoms, and after step 2, the free energy of the solid solution, G_2, can be expressed as:

$$G_2 = G_1 + \Delta G_{mix} \qquad \text{(Eq 3.23)}$$

where ΔG_{mix} is the change in Gibbs free energy caused by the mixing.

Because:

$$G_1 = H_1 = TS_1$$

and:

$$G_2 = H_2 = TS_2$$

putting:

$$\Delta H_{mix} = H_2 - H_1$$

and:

$$\Delta S_{mix} = S_2 - S_1$$

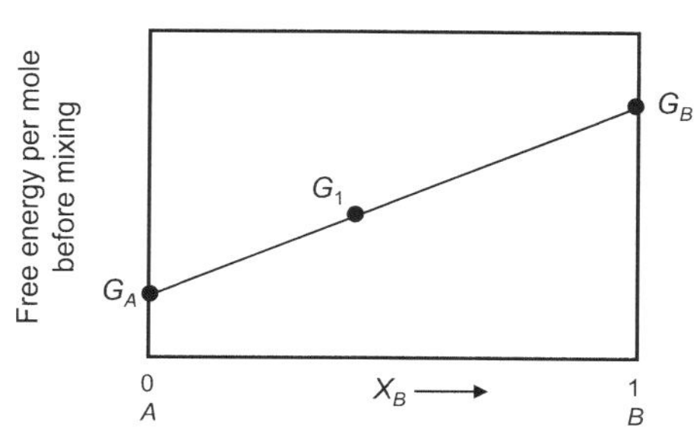

Fig. 3.8 Variation of G_1 (the free energy before mixing) with composition (X_A or X_B). Adapted from Ref 3.1

gives:

$$\Delta G_{mix} = \Delta H_{mix} - T\Delta S_{mix} \quad \text{(Eq 3.24)}$$

where ΔH_{mix} is the heat absorbed or evolved during step 2;that is, it is the heat of solution and, ignoring volume changes during the process, it represents the difference in internal energy, E, before and after mixing. The difference in entropy between the mixed and unmixed states is ΔS_{mix}.

3.3.2 Ideal Solutions

The simplest type of mixing occurs in the ideal solution where $\Delta H_{mix} = 0$, and the free-energy change on mixing is therefore:

$$\Delta G_{mix} = -T\Delta S_{mix} \quad \text{(Eq 3.25)}$$

In statistical thermodynamics, entropy is quantitatively related to randomness by the Boltzmann equation:

$$S = -\,k \ln \omega \quad \text{(Eq 3.26)}$$

where k is Boltzmann's constant and ω is a measure of randomness. There are two contributions to the entropy of a solid solution: (1) a thermal contribution, S_{th}, and (2) a configurational contribution, S_{config}.

The measure of randomness (ω) is the number of ways in which the thermal energy of the solid can be divided among the atoms, that is, the total number of ways in which vibrations can be set up in the solid. In solutions, additional randomness exists due to the different ways in which the atoms can be arranged. This gives extra entropy, S_{config}, for which ω is the number of distinguishable ways of arranging the atoms in the solution.

If there is no volume change or heat change during mixing, then the only contribution to ΔS_{mix} is the change in configurational entropy. Before mixing, the A and B atoms are held separately in the system and there is only one distinguishable way in which the atoms can be arranged. Consequently, $S_1 = k \ln 1 = 0$, and therefore $\Delta S_{mix} = S_2$.

Assuming that A and B mix to form a substitutional solid solution and that all configurations of A and B atoms are equally probable, the number of distinguishable ways of arranging the atoms on the atom sites is:

$$\omega_{config} = \frac{(N_A + N_B)\,!}{N_A!N_B!} \quad \text{(Eq 3.27)}$$

where N_A is the number of A atoms and N_B the number of B atoms.

Because this involves 1 mol of solution, that is, Avogadro's number of N_a atoms:

$$N_A = X_A N_a$$

and:

$$N_B = X_B N_b$$

Substituting into Eq 3.26 and 3.27, and using Stirling's approximation ($\ln N! \approx N \ln - N$) and the relationship $N_a k = R$, gives:

$$\Delta S_{mix} = -R(X_A \ln X_A + X_B \ln X_b) \qquad \text{(Eq 3.28)}$$

Because X_A and X_B are less than unity, ΔS_{mix} is positive; that is, there is an increase in entropy on mixing, as expected. The free energy of mixing, ΔG_{mix}, is obtained from Eq 3.25 as:

$$\Delta G_{mix} = RT(X_A \ln X_A + X_B \ln X_b) \qquad \text{(Eq 3.29)}$$

Figure 3.9 shows ΔG_{mix} as a function of composition and temperature.

The actual free energy of the solution, G, will also depend on G_A and G_B. From Eq 3.22, 3.23, and 3.29:

$$G = G_2 = X_A G_A + X_B G_B + RT(X_A \ln X_A + X_B \ln X_b) \qquad \text{(Eq 3.30)}$$

This is shown schematically in Fig. 3.10. Note that, as the temperature increases, G_A and G_B decrease and the free-energy curves assume a greater

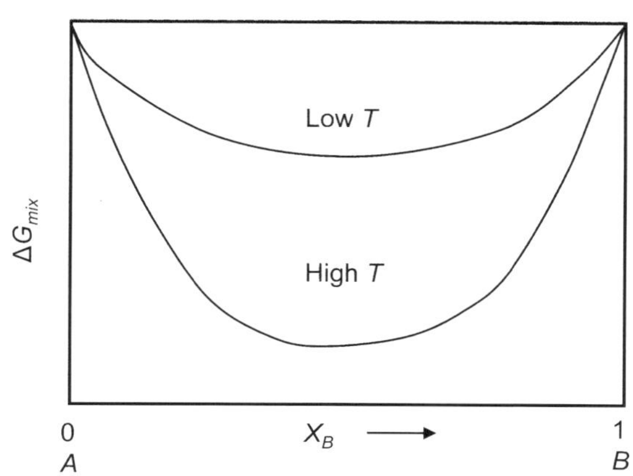

Fig. 3.9 Free energy of mixing for an ideal solution. Adapted from Ref 3.1

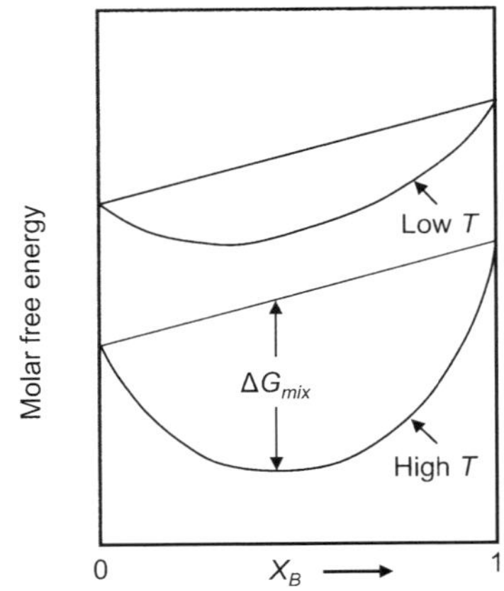

Fig. 3.10 A combination of Fig. 3.8 and 3.9: The molar free energy (free energy per mole of solution) for an ideal solution. Adapted from Ref 3.1

curvature. The decrease in G_A and G_B is due to the thermal entropy of both components and is given by Eq 3.13.

3.4 Chemical Potential

In alloys it is important to know how the free energy of a given phase will change when atoms are added or removed. If a small quantity of A, dn_A mol, is added to a large amount of a phase at constant temperature and pressure, the size of the system will increase by dn_A, and the total free energy of the system will also increase by a small amount, dG'. If dn_A is small enough, dG' will be proportional to the amount of A added. Then it can be written:

$$dG' = \mu_A dn_A \ (T, P, n_B \text{ constant}) \qquad \text{(Eq 3.31)}$$

The proportionality constant, μ_A, is called the partial molar free energy of A, or alternatively, the chemical potential of A in the phase. The chemical potential, μ_A, depends on the composition of the phase, and therefore dn_A must be so small that the composition is not significantly altered. If Eq 3.31 is rewritten, it can be seen that a definition of the chemical potential of A is:

$$\mu_A = \left(\frac{\partial G'}{\partial n_A}\right)_{T,P,n_B} \qquad \text{(Eq 3.32)}$$

The symbol G' has been used for the Gibbs free energy to emphasize the fact that it refers to the whole system. The usual symbol, G, will be used to denote the molar free energy and is therefore independent of the size of the system.

Equations similar to Eq 3.31 and 3.32 can be written for the other components in the solution. For a binary solution at constant temperature and pressure, the separate contributions can be summed:

$$dG' = \mu_A dn_A + \mu_B dn_B \qquad \text{(Eq 3.33)}$$

This equation can be extended by adding further terms for solutions containing more than two components. If T and P changes are also allowed, Eq 3.12 must be added, giving the general equation:

$$dG' = -SdT + VdP + \mu_A dn_A + \mu_B dn_B + \mu_C dn_C = \ldots$$

If 1 mol of the original phase contained X_A mol A and X_B mol B, the size of the system can be increased without altering its composition if A and B are added in the correct proportions; that is, such that $dn_A : dn_B = X_A : X_B$. For example, if the phase contains twice as many A as B atoms ($X_A = 2/3$, $X_B = 1/3$), the composition can be maintained constant by adding two A atoms for every one B atom ($dn_A : dn_B = 2$). In this way, the size of the system can be increased by 1 mol without changing μ_A and μ_B. To do this, X_A mol A and X_B mol B must be added and the free energy of the system will increase by the molar free energy, G. Therefore, from Eq 3.33:

$$G = \mu_A X_A + \mu_B X_B \qquad \text{(Eq 3.34)}$$

where G is a function of X_A and X_B, as in Fig. 3.10. For example, μ_A and μ_B can be obtained by extrapolating the tangent to the G curve to the sides of the molar free-energy diagram (Fig. 3.11). Remembering that $X_A + X_A$

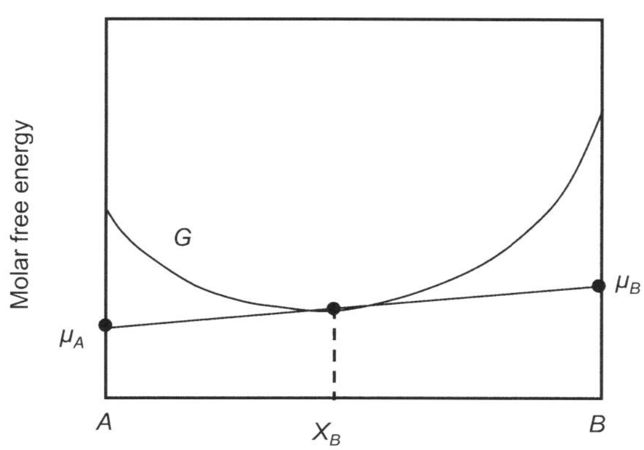

Fig. 3.11 The relationship between the free-energy curve for a solution and the chemical potentials of the components. Adapted from Ref 3.1

= 1; that is, $d\,X_A = -d\,X_B$, this result can be obtained from Eq 3.40 and 3.41. From Fig. 3.11, it is seen that μ_A and μ_A vary systematically with the composition of the phase.

Comparison of Eq 3.30 and 3.34 gives μ_A and μ_A for an ideal solution as:

$$\mu_A = G_A + RT \ln X_A$$
$$\mu_B = G_B + RT \ln X_B \qquad \text{(Eq 3.35)}$$

which is a much simpler way of presenting Eq 3.30. These relationships are shown in Fig. 3.12. The distances *ac* and *bd* are $-RT \ln X_A$ and $-RT \ln X_B$.

3.5 Regular Solutions

Returning to the model of a solid solution, so far it has been assumed that $\Delta H_{mix} = 0$; however, this type of behavior is rare in practice and usually mixing is endothermic (heat absorbed) or exothermic (heat evolved). However, the model used for ideal solution can be extended to include the ΔH_{mix} term by using the so-called quasi-chemical approach.

In the quasi-chemical model, it is assumed that the heat of mixing ΔH_{mix} is only due to the bond energies between adjacent atoms. For this assumption to be valid, it is necessary that the volumes of pure *A* and *B* are equal and do not change during mixing so that the interatomic distances and bond energies are independent of composition.

The structure of a binary solid solution is shown schematically in Fig. 3.13. Three types of interatomic bonds are present:

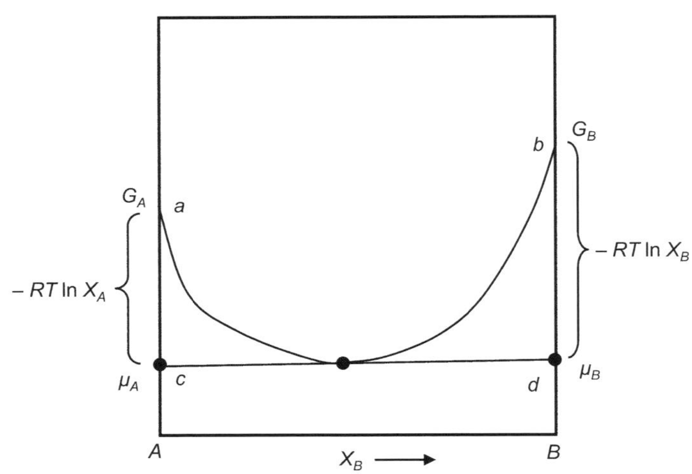

Fig. 3.12 The relationship between free energies and chemical potentials for an ideal solution. Adapted from Ref 3.1

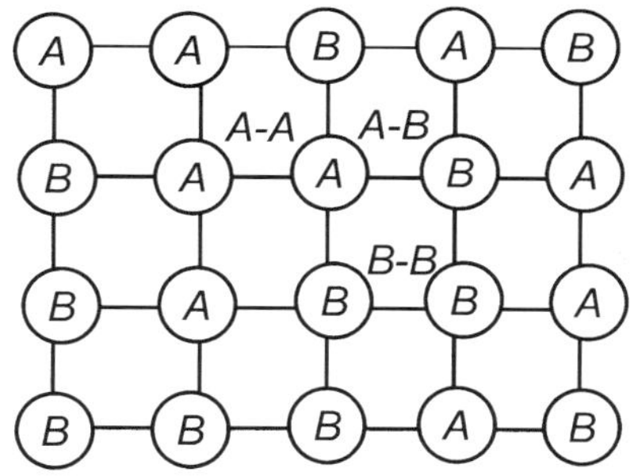

Fig. 3.13 The different types of interatomic bonds in a solid solution. Adapted from Ref 3.1

- A–A bonds, each with an energy ε_{AA}
- B–B bonds, each with an energy ε_{BB}
- A–B bonds, each with an energy ε_{AB}

By considering zero energy to be the state where the atoms are separated to infinity, ε_{AA}, ε_{BB}, and ε_{AB} are negative quantities and become increasingly more negative as the bonds become stronger. The internal energy of the solution, E, will depend on the number of bonds of each type—P_{AA}, P_{BB}, and P_{AB}—such that:

$$\mathrm{E} = P_{AA}\varepsilon_{AA} + P_{BB}\varepsilon_{BB} + P_{AB}\varepsilon_{AB}$$

Before mixing, pure A and B contain only A–A and B–B bonds, respectively, and by considering the relationships between P_{AA}, P_{BB}, and P_{AB} in the solution, it can be shown that the change in internal energy on mixing is given by:

$$\Delta H_{mix} = P_{AB}\varepsilon \qquad \text{(Eq 3.36)}$$

where:

$$\varepsilon = \varepsilon_{AB} - \frac{1}{2}(\varepsilon_{AA} + \varepsilon_{BB}) \qquad \text{(Eq 3.37)}$$

That is, ε is the difference between the A–B bond energy and the average of the A–A and B–B bond energies.

If $\varepsilon = 0$, $\Delta H_{mix} = 0$ and the solution is ideal. In this case the atoms are completely randomly arranged and the entropy of mixing is given by Eq 3.28. In such a solution it can also be shown that:

$$P_{AB} = N_a z X_{AA} X_{BB} \qquad \text{(Eq 3.38)}$$

where N_a is Avogadro's number, and z is the number of bonds per atom.

If $\varepsilon < 0$, the atoms in the solution will prefer to be surrounded by atoms of the opposite type and this will increase P_{AB}, whereas, if $\varepsilon > 0$, P_{AB} will tend to be less than in a random solution. However, provided ε is not too different from zero, Eq 3.38 is still a good approximation, in which case:

$$\Delta H_{mix} = \Omega X_A X_B \qquad \text{(Eq 3.39)}$$

where:

$$\Omega = N_a z \varepsilon \qquad \text{(Eq 3.40)}$$

Real solutions that closely obey Eq 3.39 are known as regular solutions. The variation of ΔH_{mix} with composition is parabolic and is shown in Fig. 3.14 for $\Omega > 0$. Note that the tangents at $X_A = 0$ and 1 are related to Ω as shown.

The free-energy change on mixing a regular solution is given by Eq 3.24, 3.28, and 3.39 as:

$$\Delta G_{mix} = \Omega X_A X_B + RT(X_A \ln X_A + X_B \ln X_b) \qquad \text{(Eq 3.41)}$$

where $\Omega X_A X_B$ is ΔH_{mix}, and $RT(X_A \ln X_A + X_B \ln X_b)$ is $-T\Delta S_{mix}$.

This is shown in Fig. 3.15 for different values of Ω and temperature. For exothermic solutions, $\Delta H_{mix} < 0$, and mixing results in a free-energy decrease at all temperatures (Fig. 3.15a, b). When $\Delta H_{mix} > 0$, however, the situation is more complicated. At high temperatures, $T\Delta S_{mix}$ is greater than ΔH_{mix} for all compositions and the free-energy curve has a positive curvature at all points (Fig. 3.15c). On the other hand, at low temperatures, $T\Delta S_{mix}$ is smaller and ΔG_{mix} develops a negative curvature in the middle (Fig. 3.15d).

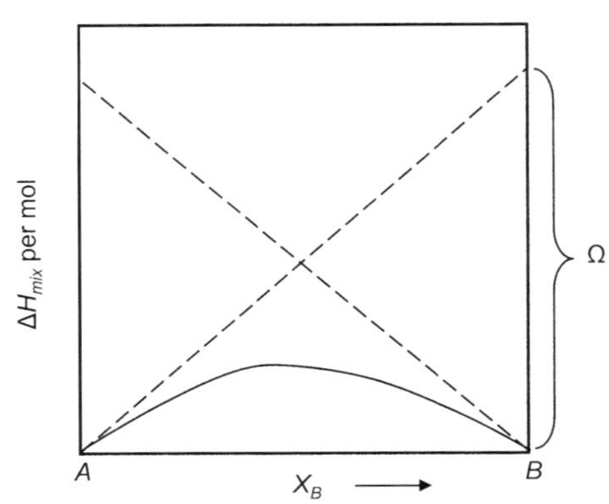

Fig. 3.14 The variation of ΔH_{mix} with composition for a regular solution. Adapted from Ref 3.1

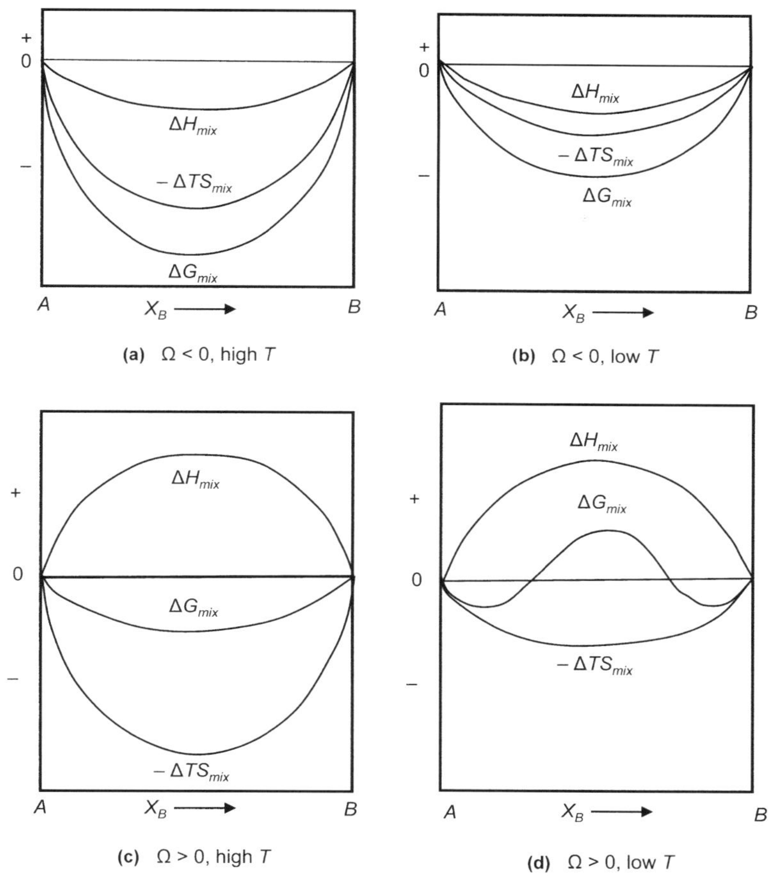

Fig. 3.15 The effect of ΔH_{mix} and T on ΔG_{mix}. Adapted from Ref 3.1

Differentiating Eq 3.28 shows that, as X_A or $X_A \rightarrow 0$, the $-T\Delta S_{mix}$ curve becomes vertical, while the slope of the ΔH_{mix} curve tends to a finite value, Ω (Fig. 3.14). This means that, except at absolute zero, ΔG_{mix} always decreases on the addition of a small amount of solute.

The actual free energy of the alloy depends on the values chosen for G_A and G_B and is given by Eq 3.22, 3.33, and 3.41 as:

$$G = X_A G_A = X_B G_B + \Omega X_A X_B + RT(X_A \ln X_A + X_B \ln X_b) \qquad \text{(Eq 3.42)}$$

This is shown in Fig. 3.16 along with the chemical potentials of A and B in the solution.

Using the relationship $X_A X_B = X_A{}^2 X_B + X_B{}^2 X_A$ and comparing Eq 3.34 and 3.42 shows that for a regular solution:

$$\mu_A = G_A + \Omega(1 - X_A)^2 + RT \ln X_A$$

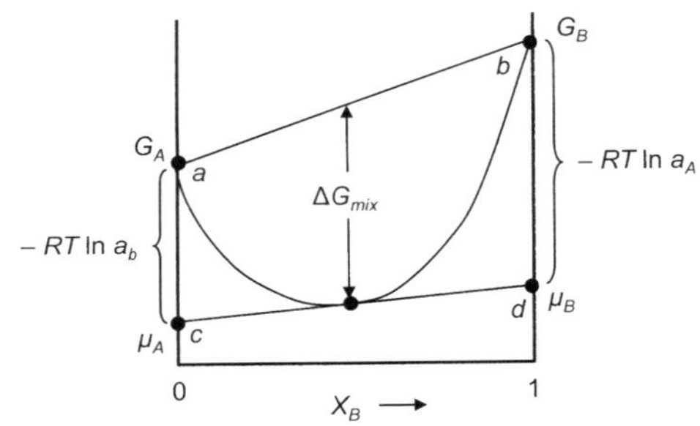

Fig. 3.16 The relationship between molar free energy and activity. Adapted from Ref 3.1

and (Eq 3.43)

$\mu_B = G_B + \Omega(1 - X_B)^2 + RT \ln X_B$

3.5.1 Activity

The chemical potential of an ideal alloy is given by Eq 3.42. A similar expression for any solution is obtained by defining the activity of a component (μ_A) such that the distances *ac* and *bd* in Fig. 3.16 are $-RT \ln a_A$ and $-RT \ln a_B$. In this case:

$\mu_A = G_A + RT \ln a_A$

and (Eq 3.44)

$\mu_B = G_B + RT \ln a_B$

In general, a_A and a_B will be different from X_A and X_B, and the relationship between them will vary with the composition of the solution. For a regular solution, comparison of Eq 3.43 and 3.44 gives:

$$\ln\left(\frac{a_A}{X_A}\right) = \frac{\Omega}{RT}(1 - X_A)^2$$

and (Eq 3.45)

$$\ln\left(\frac{a_B}{X_B}\right) = \frac{\Omega}{RT}(1 - X_B)^2$$

The relationship between *a* and *X* for any solution can be represented graphically as illustrated in Fig. 3.17. Line 1 represents an ideal solution for which $a_A = X_A$ and $a_B = X_B$. If $\Delta H_{mix} < 0$, the activity of the components in solution will be less in an ideal solution (line 2) and vice versa when $\Delta H_{mix} > 0$ (line 3).

The ratio (a_A / X_A) is usually referred to as γ_A, the activity coefficient of *A*, that is:

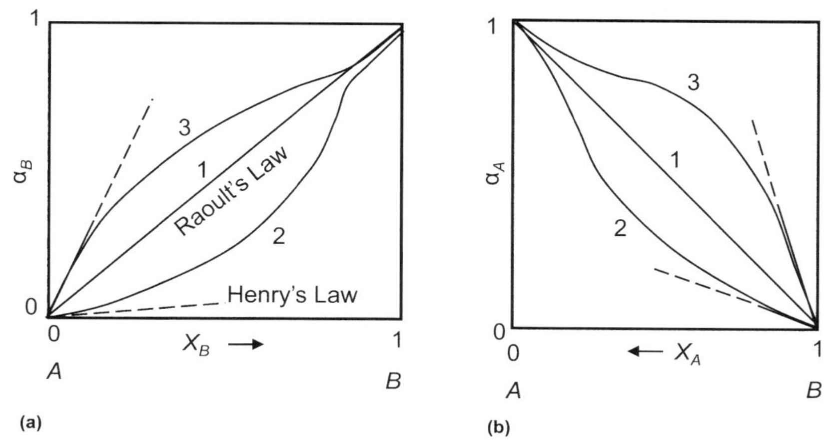

Fig. 3.17 The variation of activity with composition (a) α_B, and (b) α_B. Line 1: ideal solution (Raoult's law). Line 2: $\Delta H_{mix} < 0$. Line 3: $\Delta H_{mix} > 0$. Adapted from Ref 3.1

$$\gamma_A = \frac{a_A}{X_A} \qquad \text{(Eq 3.46)}$$

For a dilute solution of B in A, Eq 3.45 can be simplified by letting $X_B \rightarrow 0$, in which case:

$$\gamma_B = \frac{a_B}{X_B} = \text{constant (Henry's Law)} \qquad \text{(Eq 3.47)}$$

and:

$$\gamma_A = \frac{a_B}{X_B} = 1 \ \text{(Raoult's law)} \qquad \text{(Eq 3.48)}$$

Equation 3.47 is known as Henry's law and Eq 3.46 as Raoult's law. Both apply to all solutions when the solutions are sufficiently dilute.

Because activity is related to chemical potential by Eq 3.44, the activity of a component is just another means of describing the state of the component in a solution. No extra information is supplied, and its use is simply a matter of convenience because it often leads to simpler mathematics.

Activity and chemical potential are a measure of the tendency of an atom to leave a solution. If the activity or chemical potential is low, the atoms are reluctant to leave the solution, which means, for example, that the vapor pressure of the component in equilibrium with the solution will be relatively low. The activity or chemical potential of a component is important when several condensed phases are in equilibrium.

3.6 Real Solutions

While the previous model provides a useful description of the effects of configurational entropy and interatomic bonding on the free energy of binary solutions, its practical use is rather limited. For many systems, the model is an oversimplification of reality and does not predict the correct dependence of ΔG_{mix} on composition and temperature.

As already indicated, in alloys where the enthalpy of mixing is not zero (ε and $\Omega \neq 0$), the assumption that a random arrangement of atoms is the equilibrium, or most stable arrangement, is not true, and the calculated value for ΔG_{mix} will not give the minimum free energy. The actual arrangement of atoms will be a compromise that gives the lowest internal energy consistent with sufficient entropy, or randomness, to achieve the minimum free energy. In systems with $\varepsilon < 0$, the internal energy of the system is reduced by increasing the number of *A–B* bonds, that is, by ordering the atoms as shown in Fig. 3.18(a). If $\varepsilon > 0$, the internal energy can be reduced by increasing the number of *A–A* and *B–B* bonds, that is, by the clustering of the atoms into *A*-rich and *B*-rich groups (Fig. 3.18b). However, the degree of ordering or clustering will decrease as temperature increases due to the increasing importance of entropy. In systems where there is a size difference between the atoms, the quasi-chemical model will underestimate the change in internal energy on mixing, because no account is taken of the elastic strain fields that introduce a strain-energy term into ΔH_{mix}. When the size difference is large, this effect can dominate over the chemical term. When the size difference between the atoms is very large, then interstitial solid solutions are energetically most favorable (Fig. 3.18c). In systems where there is strong chemical bonding between the atoms, there is a tendency for the formation of intermetallic phases. These are distinct from solutions based on the pure components because they have a different crystal structure and may also be highly ordered.

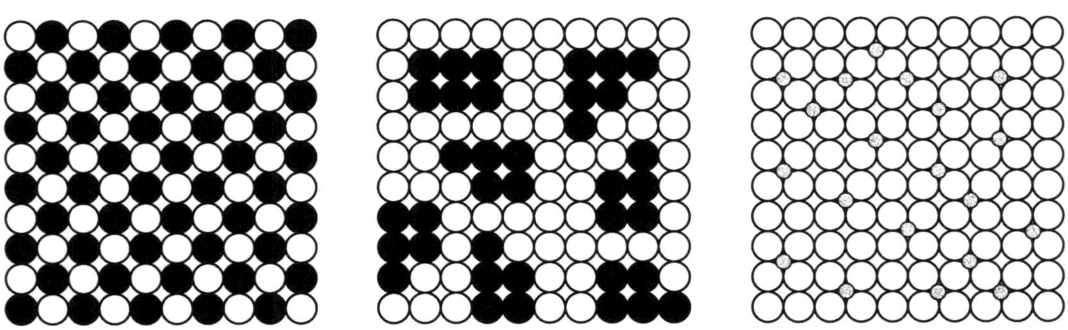

(a) (b) (c)

Fig. 3.18 Schematic representation of solid solutions. (a) Ordered substitutional. (b) Clustering. (c) Random interstitial. Adapted from Ref 3.1

3.6.1 Equilibrium in Heterogeneous Systems

Metals A and B usually do not have the same crystal structure when mixed into a solution. In such cases, two free-energy curves must be drawn, one for each structure. The stable forms of pure A and B at a given temperature (and pressure) can be denoted as α and β, respectively. For the sake of illustration, let α be face-centered cubic (fcc) and β be body-centered cubic (bcc). The molar free energies of fcc A and bcc B are shown in Fig. 3.19(a) as points a and b. The first step in drawing the free-energy curve of the fcc A phase is to convert the stable bcc arrangement of B atoms into an unstable fcc arrangement. This requires an increase in free energy, be. The free-energy curve for the α phase can now be constructed as before by mixing fcc A and fcc B, as shown in the figure. The distance de gives $-\Delta G_{mix}$ for α of composition X. A similar procedure produces the molar free-energy curve for the β phase (Fig. 3.19b). The distance af is now the difference in free energy between bcc A and fcc A.

It is clear from Fig. 3.19(b) that A-rich alloys will have the lowest free energy as a homogeneous α phase and B-rich alloys as β phase. For alloys with compositions near the crossover in the G curves, the situation is not so straightforward. In this case it can be shown that the total free energy can be minimized by the atoms separating into two phases.

It is first necessary to consider a general property of molar free-energy diagrams when phase mixtures are present. Suppose an alloy consists of two phases, α and β, each of which has a molar free energy given by G_α and G_β, respectively (Fig. 3.20). If the overall composition of the phase mixture is $X_B{}^0$, the lever rule gives the relative number of moles of α and β that must be present, and the molar free energy of the phase mixture, G, is given by the point on the straight line between α and β, as shown in the figure. This result can be proven most readily using the geometry of Fig. 3.20. The lengths ad and cf respectively represent the molar free energies of the α and β phases present in the alloy. Point g is obtained by the intersection of be and dc so that bcg and acd, as well as deg and dfc, form

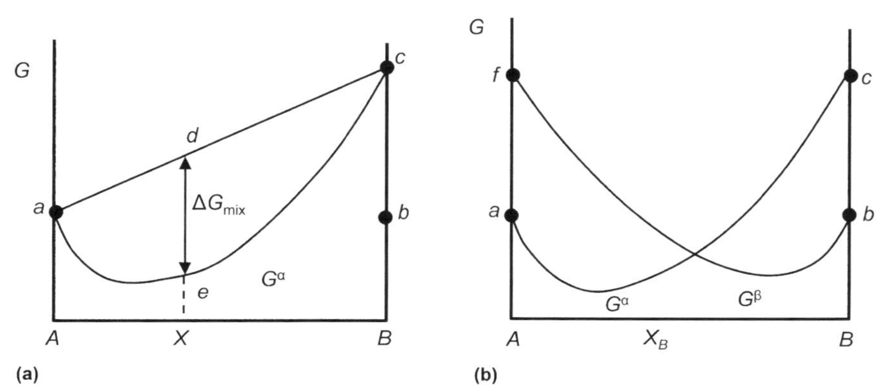

Fig. 3.19 (a) Molar free-energy curve for the α phase. (b) Molar free-energy curves for α and β. Adapted from Ref 3.1

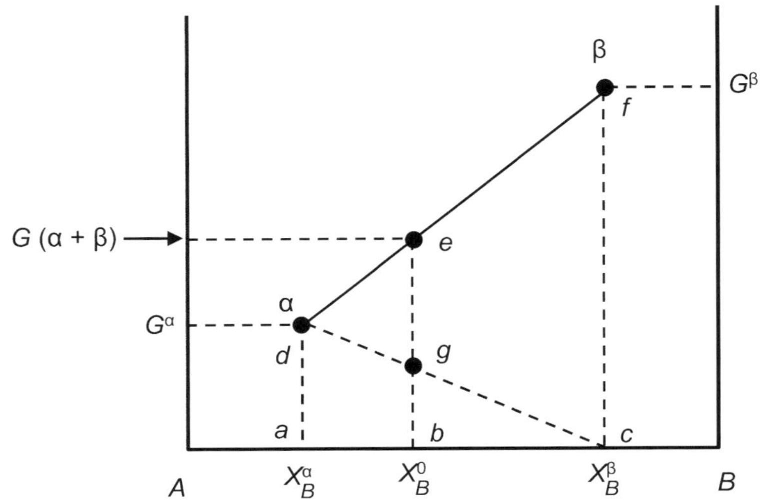

Fig. 3.20 Molar free energy of a two-phase mixture (α + β). Adapted from Ref 3.1

similar triangles. Therefore, $bg/ad = bc/ac$ and $ge/cf = ab/ac$. According to the lever rule, 1 mol of alloy will contain bc/ac mol of α and ab/ac mol of β. It follows that bg and ge represent the separate contributions from the α and β phases to the total free energy of 1 mol of alloy. Therefore, the length be represents the molar free energy of the phase mixture.

Consider now alloy X^o in Fig. 3.21(a). If the atoms are arranged as a homogeneous phase, the free energy will be lowest as α; that is, $G_{0\alpha}$ per mole. However, from the above it is clear that the system can lower its free energy if the atoms separate into two phases, with compositions α_1 and β_1, for example. The free energy of the system will then be reduced to G_1. Further reductions in free energy can be achieved if the A and B atoms interchange between the α and β phases until the compositions α_e and β_e are reached (Fig. 21b). The free energy of the system, G_e, is now a minimum and there is no desire for further change. Consequently, the system is in equilibrium and α_e and β_e are the equilibrium compositions of the α and β phases.

This result is quite general and applies to any alloy with an overall composition between α_e and β_e: Only the relative amounts of the two phases change, as given by the lever rule. When the alloy composition lies outside this range, however, the minimum free energy lies on the G_α or G_β curves and the equilibrium state of the alloy is a homogeneous single phase.

From Fig. 3.21 it can be seen that equilibrium between two phases requires that the tangents to each G curve at the equilibrium compositions lie on a common line. In other words, each component must have the same chemical potential in the two phases, that is, for heterogeneous equilibrium:

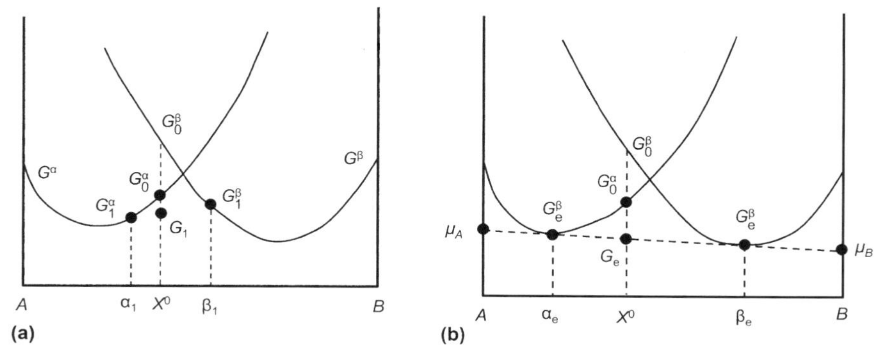

Fig. 3.21 (a) Alloy X^0 has a free energy, G_1, as a mixture of α and β. (b) At equilibrium, alloy X^0 has a minimum free energy, G_e, when it is a mixture αe and βe. Adapted from Ref 3.1

$$\mu^\alpha_A = \mu^\beta_A \text{ and } \mu^\alpha_B = \mu^\beta_B \qquad \text{(Eq 3.49)}$$

From Eq 3.49 it can be seen that the activities of the components must also be equal; that is:

$$a^\alpha_A = a^\beta_A \text{ and } a^\alpha_B = a^\beta_B \qquad \text{(Eq 3.50)}$$

It is easiest to plot the variation of activity with alloy composition, and this is shown schematically in Fig. 3.22. Between A and α_e and β_e and B, where single phases are stable, the activities (or chemical potentials) vary, and, for simplicity, ideal solutions have been assumed, in which case there is a straight line relationship between α_e and X. Between α_e and β_e the phase compositions in equilibrium do not change, and the activities are equal and given by points q and r.

3.6.2 Phase Diagrams

The previous section shows how the equilibrium state of an alloy can be obtained from the free-energy curves at a given temperature. The next step is to see how equilibrium is affected by temperature.

The shapes of liquidus, solidus, and solvus curves (or surfaces) in a phase diagram are determined by the Gibbs free energies of the relevant phases. In this instance, the free energy must include not only the energy of the constituent components, but also the energy of mixing of these components in the phase. Consider, for example, the situation of complete miscibility shown in Fig. 3.23. The two phases, liquid and solid, are in stable equilibrium in the two-phase field between the liquidus and solidus lines. The Gibbs free energies at various temperatures are calculated as a function of composition for ideal liquid solutions and for ideal solid solutions of the two components, A and B. The result is a series of plots similar to those shown in Fig. 3.24(a) to (e):

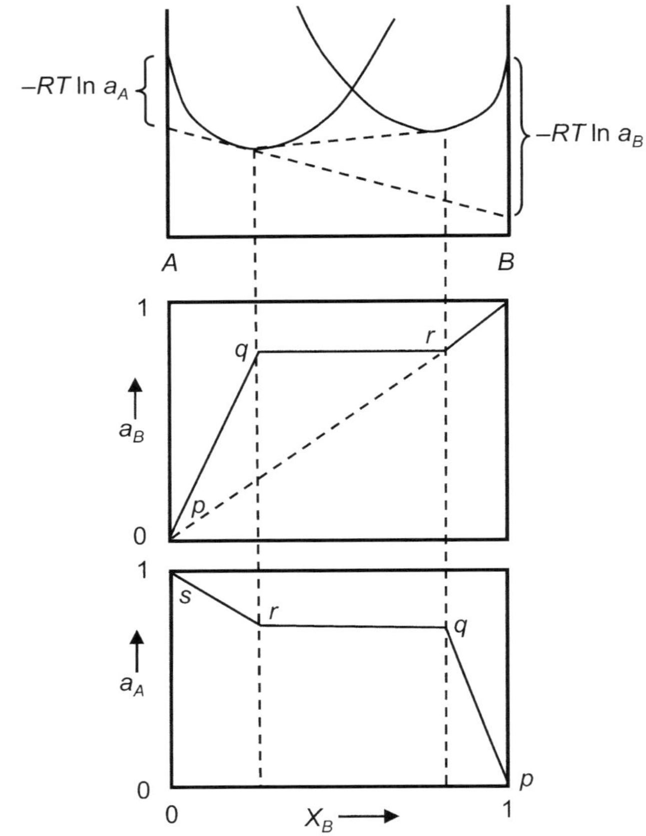

Fig. 3.22 The variation of a_A and a_B, with composition for a binary system containing two ideal solutions, α and β. Adapted from Ref 3.1

- At temperature T_1, the liquid solution has the lower Gibbs free energy and, therefore, is the more stable phase.
- At T_2, the melting temperature of *A*, the liquid and solid are equally stable only at a composition of pure *A*. The remainder of the solution is still liquid.
- At temperature T_3, between the melting temperatures of *A* and *B*, the Gibbs free-energy curves cross. Depending on the composition, there exist fields of liquid, liquid + solid α, and solid α.
- At temperature T_4, the melting temperature of *B*, pure *B* liquid and solid are equally stable. Except for pure *B*, the remainder of the solution is now solid α.
- At temperature T_5 and at all lower temperatures, the free-energy curve for solid α is below the curve for the liquid, and the whole solution is solid α.

Construction of the two-phase liquid-plus-solid field of the phase diagram in Fig. 3.24(f) is as follows. According to thermodynamic principles,

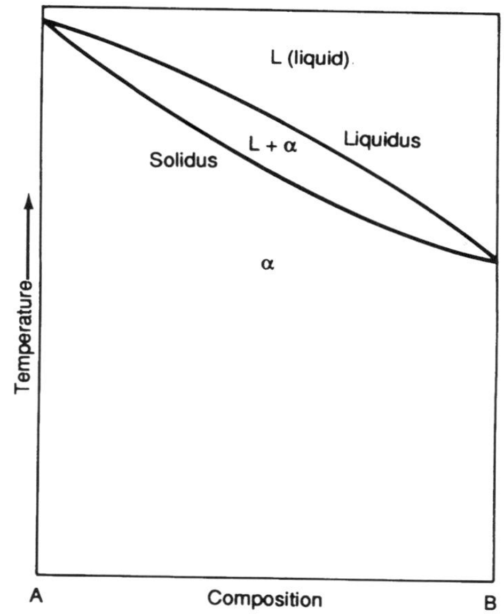

Fig. 3.23 Schematic binary phase diagram showing miscibility in both the liquid and solid states. Source: Ref 3.2

the compositions of the two phases in equilibrium with each other at temperature T_3 can be determined by constructing a straight line that is tangent to both curves in Fig. 3.24(c). The points of tangency, 1 and 2, are then transferred to the phase diagram as points on the solidus and liquidus, respectively. This is repeated at sufficient temperatures to determine the curves accurately.

If, at some temperature, the Gibbs free-energy curves for the liquid and the solid tangentially touch at some point, the resulting phase diagram will be similar to those shown in Fig. 3.25(a) and (b), where a maximum or minimum appears in the liquidus and solidus curves.

The two-phase field in Fig. 3.24(f) consists of a mixture of liquid and solid phases. As stated previously, the compositions of the two phases in equilibrium at temperature T_3 are C_1 and C_2. The horizontal isothermal line connecting points 1 and 2, where these compositions intersect temperature T_3, is the tie line. Similar tie lines connect the coexisting phases throughout all two-phase fields (areas) in binary systems, while tie triangles connect the coexisting phases throughout all three-phase regions (volumes) in ternary systems.

Eutectic phase diagrams, a feature of which is a field where there is a mixture of two solid phases, can also be constructed from Gibbs free-energy curves. Consider the temperatures indicated on the phase diagram in Fig. 3.26(f) and the Gibbs free-energy curves for these temperatures (Fig. 3.26a–e). When the points of tangency on the energy curves are trans-

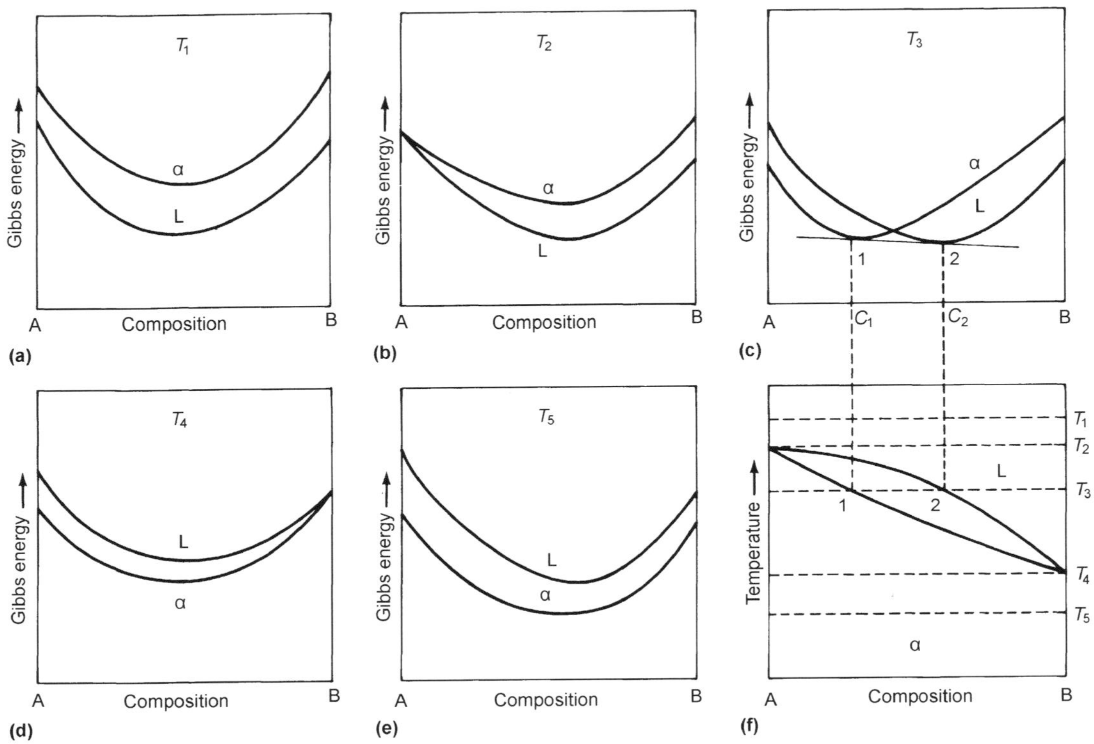

Fig. 3.24 Use of Gibbs energy curves to construct a binary phase diagram that shows miscibility in both the liquid and solid states. Source: Ref 3.3 as published in Ref 3.2

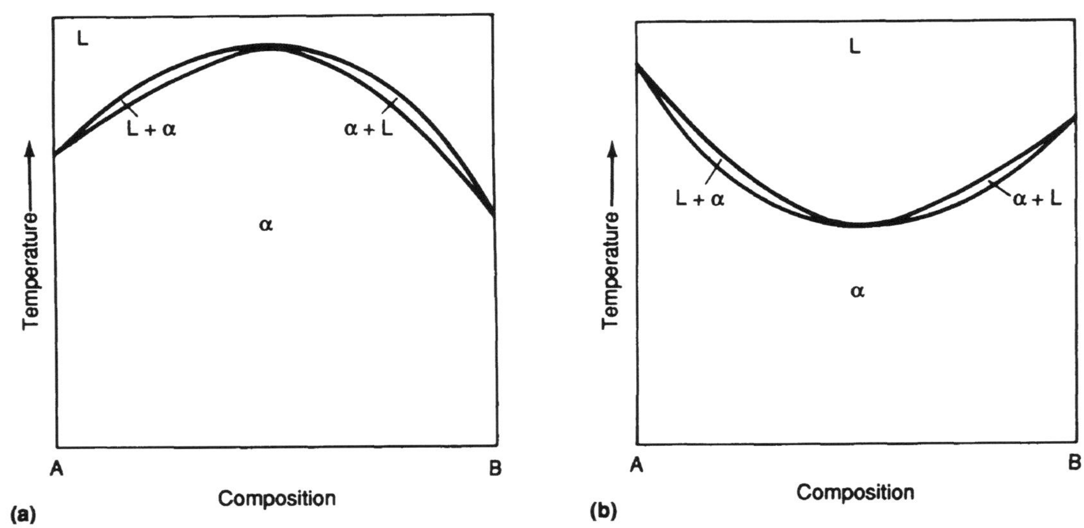

Fig. 3.25 Schematic binary phase diagrams with solid-state miscibility where the liquidus shows (a) a maximum and (b) a minimum. Source: Ref 3.2

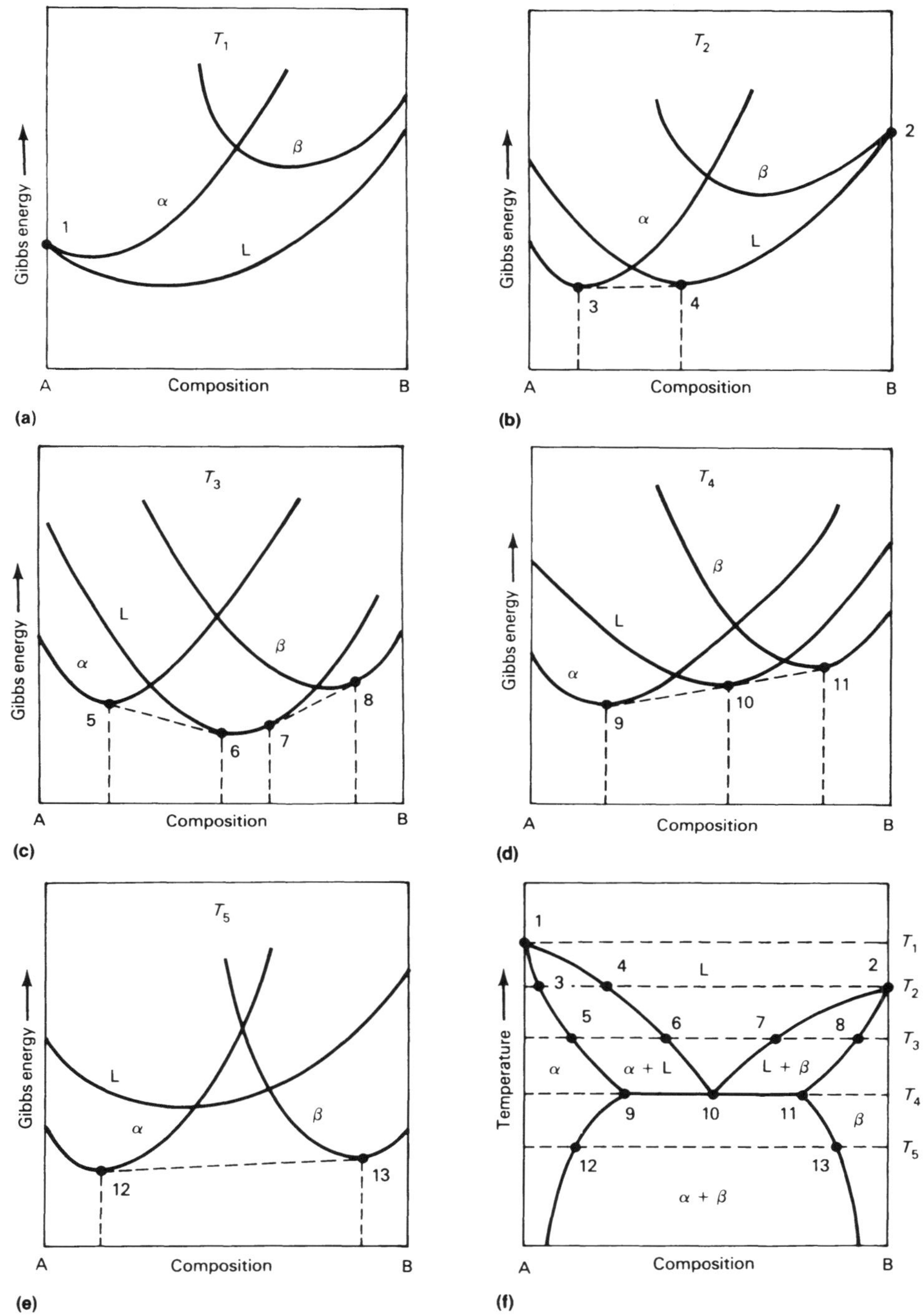

Fig. 3.26 Use of Gibbs energy curves to construct a binary phase diagram of the eutectic type. Source: Ref 3.4 as published in Ref 3.2

ferred to the diagram, the typical shape of a eutectic system results. The mixture of solid α and β that forms on cooling through the eutectic (Point 10 in Fig. 3.26f), has a special microstructure.

Binary phase diagrams that have three-phase reactions other than the eutectic reaction, as well as diagrams with multiple three-phase reactions, also can be constructed from appropriate Gibbs free-energy curves. Likewise, Gibbs free-energy surfaces and tangential planes can be used to construct ternary phase diagrams.

ACKNOWLEDGMENT

The material in this chapter is from "Introduction to Alloy Phase Diagrams," in *Phase Diagrams* by H. Baker, *ASM Handbook* Volume 3, ASM International, 1992, and *Phase Transformations in Metals and Alloys* by D.A. Porter and K.E. Easterling, Chapman and Hall, 1981.

REFERENCES

3.1 D.A. Porter and K.E. Easterling, *Phase Transformations in Metals and Alloys*, Chapman and Hall, 1981
3.2 H. Baker, Introduction to Alloy Phase Diagrams, *Phase Diagrams*, Vol 3, *ASM Handbook*, ASM International, 1992, reprinted in *Desk Handbook: Phase Diagrams for Binary Alloys*, 2nd ed., H. Okamoto, Ed., ASM International, 2010
3.3 A. Prince, *Alloy Phase Equilibria*, Elsevier, 1966
3.4 P. Gordon, *Principles of Phase Diagrams in Materials Systems*, McGraw-Hill, 1968

SELECTED REFERENCES

- B.S. Bokstein, M.I. Mendelev, and D.J. Srolovitz, *Thermodynamics and Kinetics in Materials Science*, Oxford University Press, 2005
- R.T. DeHoff, *Thermodynamics in Materials Science*, McGraw-Hill, 1993
- D.R. Gaskell, *Introduction to the Chemical Thermodynamics of Materials*, 4th ed., Taylor & Francis, 2003
- H.-G. Lee, *Chemical Thermodynamics for Metals and Materials*, Imperial College Press, 1999
- S. Stolen, T. Grande, and N. Allan, *Chemical Thermodynamics of Materials*, John Wiley & Sons, 2004

Phase Diagrams—Understanding the Basics
F.C. Campbell, editor

CHAPTER 4

Isomorphous Alloy Systems

PHASE DIAGRAMS are graphical representations that show the phases present in the material at various compositions, temperatures, and pressures. It should be recognized that phase diagrams represent equilibrium conditions for an alloy, which means that very slow heating and cooling rates are used to generate data for their construction. Because industrial practices almost never approach equilibrium, phase diagrams should be used with some degree of caution. Nevertheless, they are very useful in predicting phase transformations and their resulting microstructures. External pressure also influences the phase structure; however, because pressure is held constant in most applications, phase diagrams are usually constructed at a constant pressure of one atmosphere.

In this chapter, binary isomorphous systems are covered. The term *isomorphous* means the two metals are completely miscible in each other in both the liquid and solid states. Calculation methods are shown that allow the prediction of the phases present, the chemical compositions of the phases present, and the amounts of phases present.

4.1 Binary Systems

Phase diagrams and the systems they describe are often classified and named for the number (in Latin) of components in the system (Table 4.1).

When a system consists of two components, then a three-dimensional diagram is needed to determine how they vary with temperature and pressure. The *P-T-X* (pressure, temperature, composition) diagram for two metals that dissolve in each other is shown in Fig. 4.1. A feature of this diagram is that two-phase and three-phase equilibria extend over a region rather than being limited to a line or point, as in a one-component diagram. This behavior is shown in the *P-T* section at constant composition (Fig.

Table 4.1 Number of components

Number of components	Name of system or diagram
One	Unary
Two	Binary
Three	Ternary
Four	Quarternary
Five	Quinary
Six	Sexinary
Seven	Septenary
Eight	Octanary
Nine	Nonary
Ten	Decinary

Source: Ref 4.1

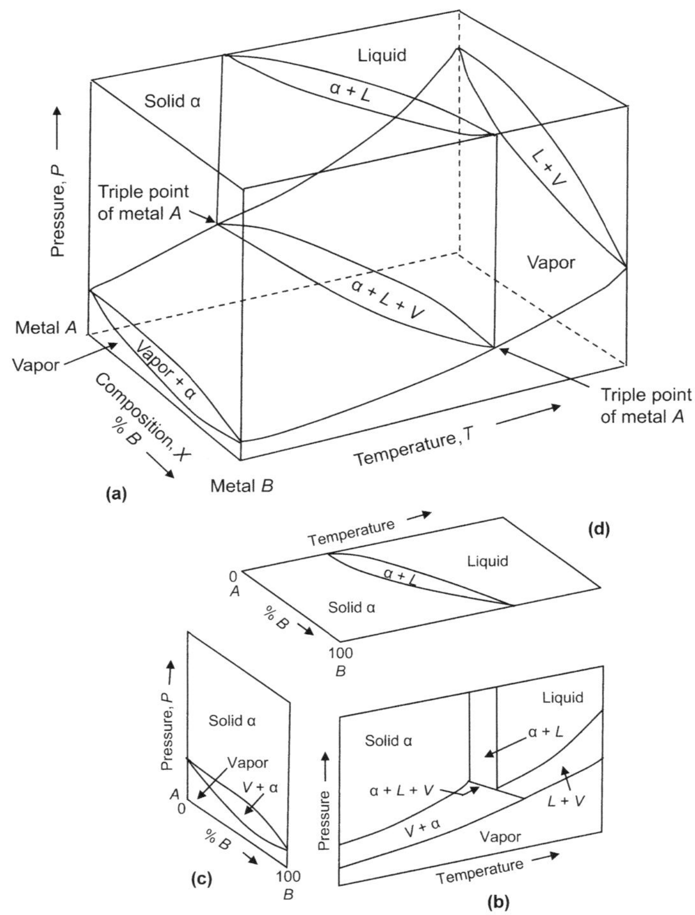

Fig. 4.1 *P-T-X* diagram for two metals, *A* and *B*, that dissolve completely in each other in the liquid and solid states. (a) The *P-T-X* diagram. (b) A *P-T* section at the 50% *B* composition. (c) A *P-X* section below the triple point of both pure metals. (d) A *T-X* section at atmospheric pressure. *P*, pressure. *T*, temperature. *X*, composition. Adapted from Ref 4.2

4.1b). It is often useful to have a section at constant temperature through the *P-T-X* diagram. A *P-X* section of this kind is shown in Fig. 4.1(c) for a temperature below the triple points of both metals. By far the most useful section through the *P-T-X* diagram is the one at constant pressure, at atmospheric pressure in particular. Figure 4.1(d) shows the diagram obtained in this instance. Therefore, the most commonly encountered phase diagram in metallurgy is a *T-X* diagram constructed at a pressure of one atmosphere.

The Gibbs phase rule applies to all states of matter (solid, liquid, and gaseous), but when the effect of pressure is constant, the rule reduces to:

$$F = C - P + 1$$

The stable equilibria for binary systems are summarized in Table 4.2.

The areas (fields) in a phase diagram, and the position and shapes of the points, lines, surfaces, and intersections in it, are controlled by thermodynamic principles and the thermodynamic properties of all of the phases that constitute the system. The phase field rule specifies that at constant temperature and pressure, the number of phases in adjacent fields in a multicomponent diagram must differ by one.

4.1.1 Binary Isomorphous Systems

Some systems are comprised of components having the same crystal structure, and the components of some of these systems are completely miscible (completely soluble in each other) in the solid form, thus forming a continuous solid solution. When this occurs in a binary system, the phase diagram usually has the general appearance of that shown in Fig. 4.2. The diagram consists of two single-phase fields separated by a two-phase field. The boundary between the liquid field and the two-phase field in Fig. 4.2 is called the liquidus; that between the two-phase field and the solid field is the solidus. In general, a liquidus is the locus of points in a phase diagram representing the temperatures at which alloys of the various compositions of the system begin to freeze on cooling or finish melting on heating; a solidus is the locus of points representing the temperatures at which the various alloys finish freezing on cooling or begin melting on heating. The phases in equilibrium across the two-phase field in Fig. 4.2 (the liquid and solid solutions) are called conjugate phases.

Table 4.2 Stable equilibria for binary systems

Number of components	Number of phases	Degrees of freedom	Equilibrium
2	3	0	Invariant
2	2	1	Univariant
2	1	2	Bivariant

Source: Ref 4.1

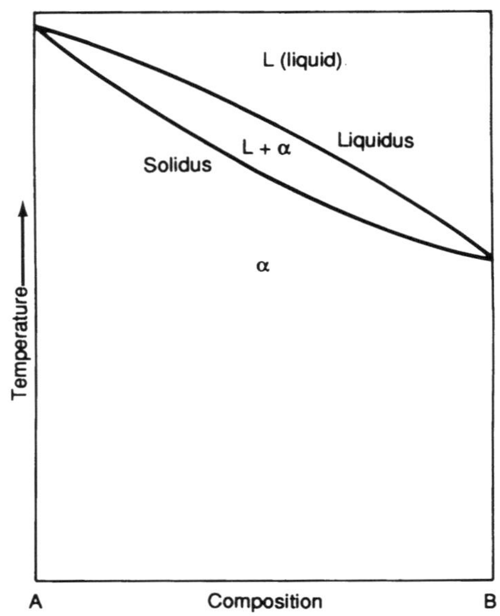

Fig. 4.2 Schematic binary phase diagram showing miscibility in both the liquid and solid states. Source: Ref 4.1

Consider briefly how these diagrams are constructed. A pure metal will solidify at a constant temperature, while an alloy will solidify over a temperature range that depends on the alloy composition. Consider the series of cooling curves for the copper-nickel system (Fig. 4.3). For increasing amounts of nickel in the alloy, freezing begins at increasing temperatures A, A_1, A_2, A_3, up to pure nickel at A_4, and finishes at increasing temperatures B, B_1, B_2, B_3, up to pure nickel at B_4. If the points A, A_1, A_2, A_3, and A_4 are joined, the result is the liquidus line, which indicates the temperature at which any given alloy will begin to solidify. Likewise, by joining the points B, B_1, B_2, B_3, and B_4, the solidus line, the temperature at which any given alloy will become completely solid, is obtained. In other words, at all temperatures above the liquidus, the alloy will be a liquid, and at all temperatures below the solidus, the alloy will be a solid. At temperatures between the liquidus and the solidus, sometimes referred to as the "mushy zone," both liquid and solid coexist in equilibrium.

Very simple phase diagrams of this type can be constructed by using the appropriate points obtained from time-temperature cooling curves, which indicate where freezing began and where it was complete. However, if the alloy system is one in which further structural changes occur after the alloy has solidified, then the metallurgist must resort to other methods of investigation to determine the phase-boundary lines. These techniques are described in Chapter 12, "Phase Diagram Determination," in this book.

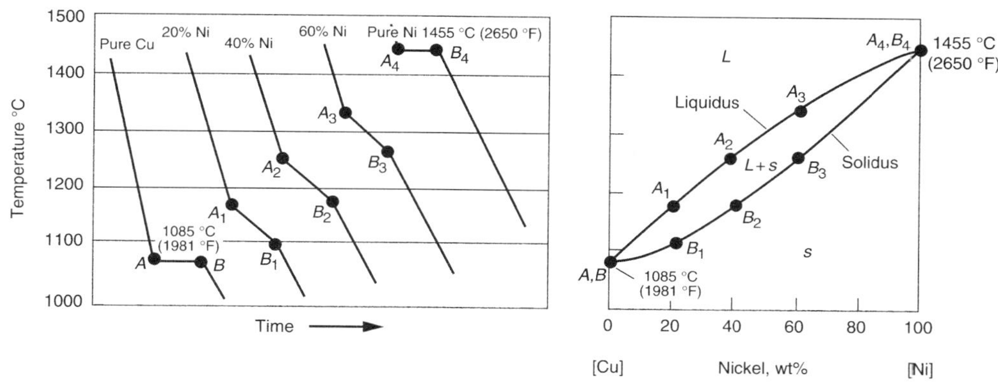

Fig. 4.3 Phase diagram construction from cooling curves. Source: Ref 4.3

A few systems consist of components having the same crystalline structure, and the components of some of these systems are completely soluble, or miscible, in each other in the liquid and solid form, thus forming a continuous series of solid solutions. When this occurs in a binary system, the phase diagram usually has the general appearance of that in the copper-nickel system shown in Fig. 4.4. Temperature is plotted along the ordinate axis, and the alloy composition is shown on the abscissa axis. The composition ranges from 0 wt% Ni (100 wt% Cu) on the extreme left to 100 wt% Ni (0 wt% Cu) on the extreme right. Three different phase regions, or fields, are present on the diagram: a liquid (L) field, a two-phase solid plus liquid field ($\alpha + L$), and a solid-solution alpha (α) field, where α is a solid solution containing both copper and nickel. Each field is defined by the phase or phases that exist over the range of temperatures and compositions bounded by the phase-boundary lines. At high temperatures, the liquid, L, field is a homogenous liquid solution composed of both copper and nickel. The solid solution, α, that exists at lower temperatures is a substitutional solid solution consisting of both copper and nickel atoms with a face-centered cubic (fcc) crystalline structure. When an alloy of any given composition freezes, copper and nickel are mutually soluble in each other and therefore display complete solid solubility. Solid solutions are commonly designated by lowercase Greek letters. The boundaries between the regions are identified as the liquidus and solidus. The upper curve separating the liquid, L, and the two-phase, $L + \alpha$, field is termed the liquidus line. The liquidus is the lowest temperature at which any given composition can be found in an entirely molten state. The lower curve separating the solid solution, α, field and the two-phase, $L + \alpha$, field is known as the solidus line. The solidus is the highest temperature when all atoms of a given composition can be found in an entirely solid state. A solidus is the locus of points representing the temperatures at which the various alloys finish freezing on cooling or begin melting on heating.

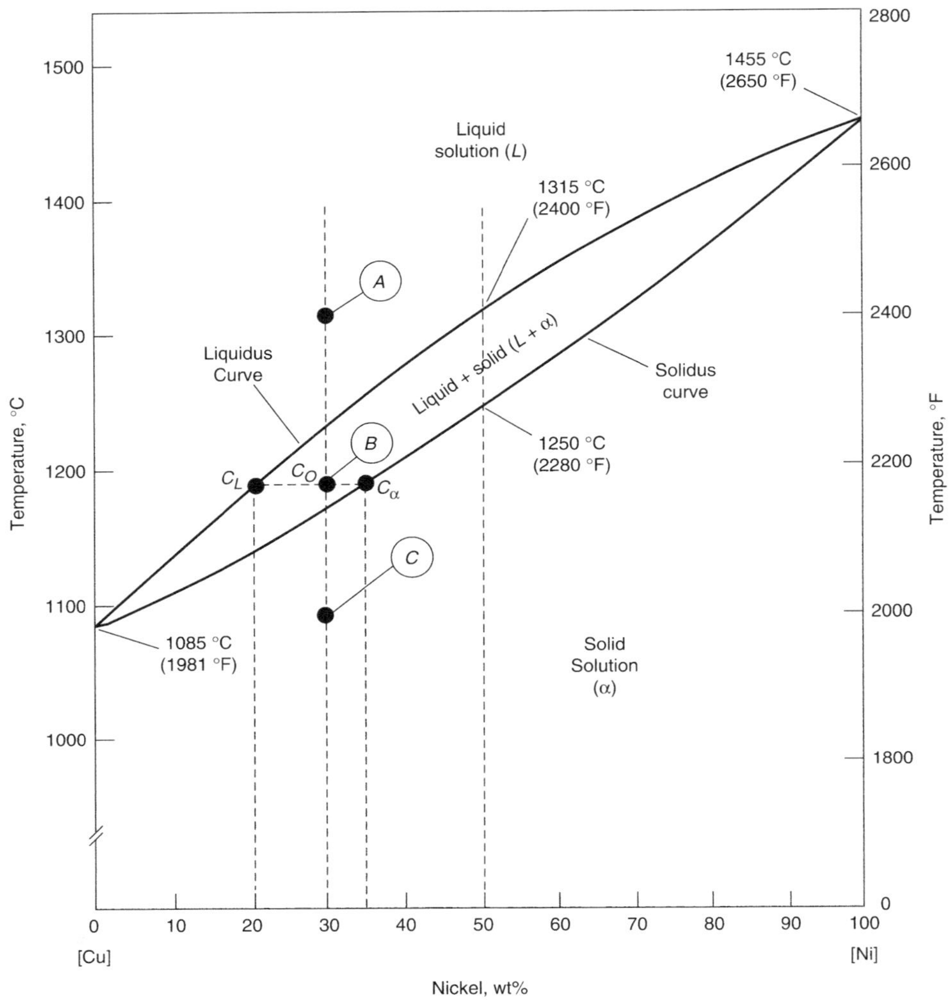

Fig. 4.4 Copper-nickel phase diagram. Source: Ref 4.3

Complete solid solubility is actually the exception rather than the rule. To obtain complete solubility, the system must adhere to the Hume-Rothery rules for solid solutions (see Chapter 2, "Solid Solutions and Phase Transformation," in this book). In this case, both copper and nickel have the fcc crystal structure, have nearly identical atomic radii and electronegativities, and have similar valences. The term *isomorphous* implies complete solubility in both the liquid and solid states. Most alloys do not have such simple phase systems. Typically, alloying elements have significant differences in their atomic size and crystalline structure, and the mismatch forces the formation of a new crystal phase that can more easily accommodate alloying elements in the solid state.

The liquidus and solidus lines intersect at the two composition extremities; that is, at the temperatures corresponding to the melting points of pure copper (1085 °C, or 1981 °F) and pure nickel (1455 °C, or 2644 °F). Because pure metals melt at a constant temperature, pure copper remains a solid until its melting point of 1085 °C (1981 °F) is reached on heating. The solid-to-liquid transformation then occurs and no further heating is possible until the transformation is complete. However, for any composition other than the pure components, melting will occur over a range of temperatures between the solidus and liquidus lines. For example, on heating a composition of 50wt%Cu-50wt%Ni, melting begins at approximately 1250 °C (2280 °F) and the amount of liquid increases until approximately 1315 °C (2400° F) is reached, at which point the alloy is completely liquid.

A binary phase diagram can be used to determine three important types of information: (1) the phases that are present, (2) the composition of the phases, and (3) the percentages or fractions of the phases.

Prediction of Phases. The phases that are present can be determined by locating the temperature-composition point on the diagram and noting the phase(s) present in the corresponding phase field. For example, an alloy of composition 30wt%Ni-70wt%Cu at 1315 °C (2400 °F) would be located at point *a* in Fig. 4.4. Because this point lies totally within the liquid field, the alloy would be a liquid. The same alloy at 1095 °C (2000 °F), designated point *c*, is within the solid solution, α, field, only the single α phase would be present. On the other hand, a 30wt%Ni-70wt%Cu alloy at 1190 °C (2170 °F) (point *b*) would consist of a two-phase mixture of solid solution, α, and liquid, L.

Prediction of Chemical Compositions of Phases. To determine the composition of the phases present, locate the point on the phase diagram. If only one phase is present, the composition of the phase is the overall composition of the alloy. For example, for an alloy of 30wt%Ni-70wt%Cu at 1095 °C (2000 °F) (point *c* in Fig. 4.4), only the α phase is present, and the composition is 30wt%Ni-70wt%Cu. For an alloy with composition and temperature coordinates located in a two-phase region, the compositions of the phases can be determined by drawing a horizontal line, referred to as a tie line, between the two phase boundaries at the temperature of interest. Then, drop perpendicular lines from the intersections of each boundary down to the composition axis and read the compositions. For example, again considering the 30wt%Ni-70wt%Cu alloy at 1190 °C (2170 °F) located at point *b* in Fig. 4.4 and lying with the two-phase, $\alpha + L$, field. The perpendicular line from the liquidus boundary to the composition axis is 20wt%Ni-80wt%Cu, which is the composition, C_L, of the liquid phase. In a similar manner, the composition of the solid-solution phase, C_α, is read from the perpendicular line from the solidus line down to the composition axis, in this case 35wt%Ni-65wt%Cu.

Prediction of Amounts of Phases. The percentages or fractions of the phases present at equilibrium can also be determined with phase diagrams. In a single-phase region, because only one phase is present, the alloy is comprised entirely of that phase; that is, the phase fraction is 1.0 and the percentage is 100%. From the previous example for the 30wt%Ni-70wt%Cu alloy at 1095 °C (2000 °F) (point *c* in Fig. 4.4), only the α phase is present and the alloy is 100% α.

If the composition and temperature position is located within a two-phase field, a horizontal tie line must be used in conjunction with the lever rule. The lever rule is a mathematical expression based on the principle of conservation of matter. First, a tie line is drawn across the two-phase region at the composition and temperature of the alloy. The fraction of one phase is determined by taking the length of the tie line from the overall alloy composition to the phase boundary for the other phase and dividing by the total tie line length. The fraction of the other phase is then determined in the same manner. If phase percentages are desired, each phase fraction is multiplied by 100. When the composition axis is scaled in weight percent, the phase fractions computed using the lever rule are mass fractions—the mass (or weight) of a specific phase divided by the total alloy mass (or weight). The mass of each phase is computed from the product of each phase fraction and the total alloy mass.

Again, consider the 30wt%Ni-70wt%Cu alloy at 1190 °C (2170 °F) located at point *b* in Fig. 4.4 containing both the solid, α, and the liquid, *L*, phases. The same tie line that was used for determination of the phase compositions can again be used for the lever rule calculation. The overall alloy composition located along the tie line is $C_\alpha - C_L$ or 35 – 20 wt%. The weight percentage of liquid present is then:

$$\text{wt\% } L = \frac{C_\alpha - C_o}{C_\alpha - C_L} = \frac{35-30}{35-20} \approx 0.33, \text{ or } 33\%$$

Likewise, the amount of solid present is:

$$\text{wt\% } S = \frac{C_o - C_L}{C_\alpha - C_L} = \frac{30-20}{35-20} \approx 0.67, \text{ or } 67\%$$

Because graphical methods are used to perform these calculations, the results are approximate rather than exact. The lever rule, or more appropriately, the inverse lever rule, can be visualized as a scale, such as the one shown in Fig. 4.5. For the scale to balance, the weight percentage of the liquid, which is less, must have the longer lever arm, in this instance 10 units, compared to the solid phase, α, which has a higher weight percentage and thus a shorter lever arm of 5 units.

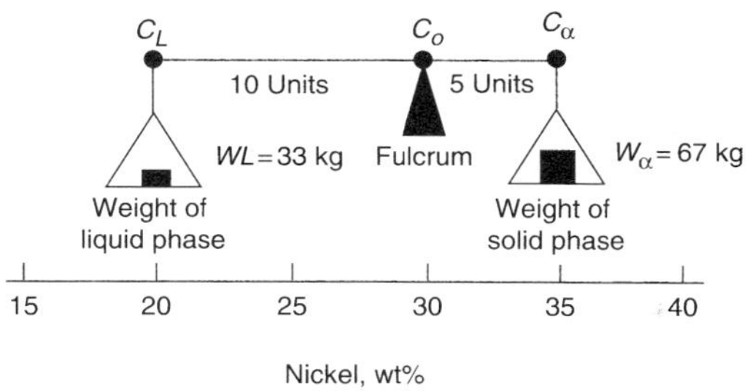

$$\text{wt\% } L = \frac{C_\alpha - C_o}{C_\alpha - C_L} = \frac{35 - 30}{35 - 20} \approx 0.33, \text{ or } 33\%$$

$$\text{wt\% } S = \frac{C_o - C_L}{C_\alpha - C_L} = \frac{30 - 20}{35 - 20} \approx 0.67, \text{ or } 67\%$$

Fig. 4.5 Visual representation of lever rule. Source: Ref 4.2 as published in Ref 4.3

It is important to emphasize that equilibrium phase diagrams identify phase changes under conditions of very slow changes in temperature. In practical situations, where heating and cooling occur more rapidly, the atoms do not have enough time to get into their equilibrium positions, and the transformations may start or end at temperatures different from those shown on the equilibrium phase diagrams. In these practical circumstances, the actual temperature at which the phase transformation occurs will depend on both the rate and direction of temperature change. Nevertheless, phase diagrams provide valuable information in virtually all metal processing operations that involve heating the metal, such as casting, hot working, and all heat treatments.

The copper-nickel system is an example of solid-solution hardening or strengthening. Most of the property changes in a solid-solution system are caused by distortion of the crystalline lattice of the base or solvent metal by additions of the solute metal. The distortion increases with the amount of the solute metal added, and the maximum effect occurs near the center of the diagram, because either metal can be considered as the solvent. As shown in Fig. 4.6, the strength curve passes through a maximum, while the ductility curve, as measured by percent elongation, passes through a minimum. Properties that are almost unaffected by atom interactions vary more linearly with composition. Examples include the lattice constant, thermal expansion, specific heat, and specific volume.

Copper-nickel alloys have a good combination of properties and corrosion resistance. One of the most notable uses is for clad coinage. A Cu-25% Ni alloy is used for the clad coinage of the U.S. dime, quarter, and half-dollar. The coins contain a copper core that is clad on the surfaces with the copper-nickel alloy (Fig. 4.7). The same alloy is used for the U.S. nickel.

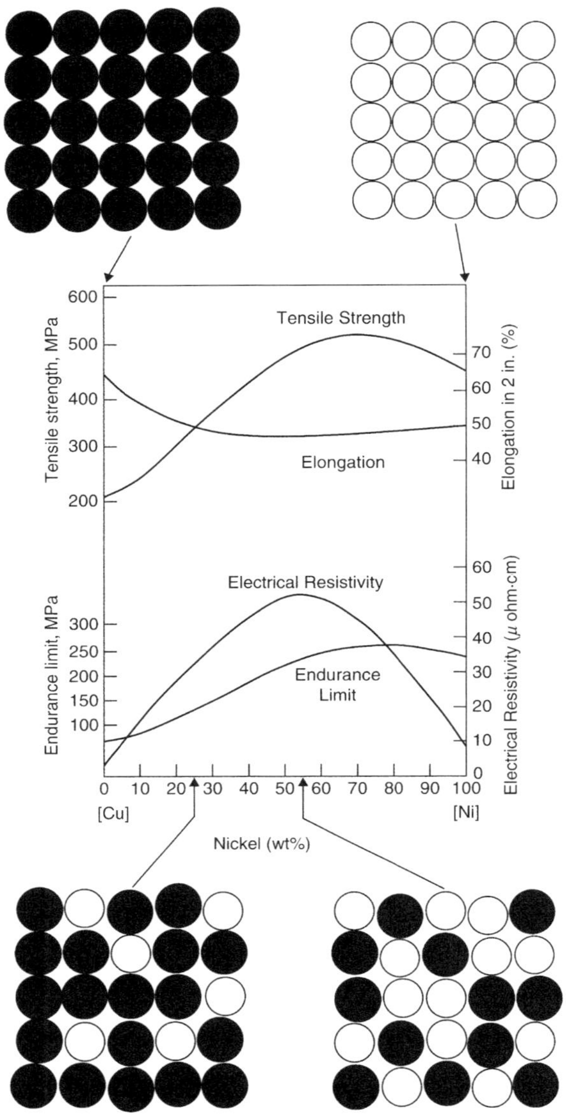

Fig. 4.6 Typical property variations in copper-nickel system. Source: Ref 4.2 as published in Ref 4.3

If the solidus and liquids meet tangentially at some point, a maximum or minimum is produced in the two-phase field, splitting it into two portions (Fig. 4.8). It also is possible to have a gap in miscibility in a single-phase field (Fig. 4.9). Point T_c, above which phases α_1 and α_2 become indistinguishable, is a critical point. Lines a-T_c and b-T_c, called solvus lines, indicate the limits of solubility of component B in A and A in B, respectively.

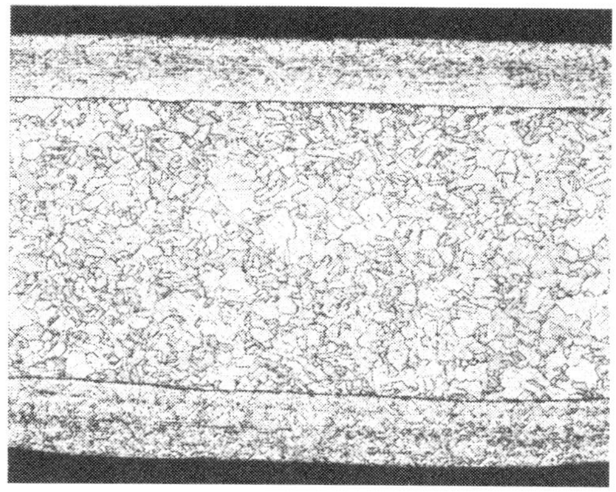

Fig. 4.7 Copper-nickel clad coinage construction. Source: Ref 4.4

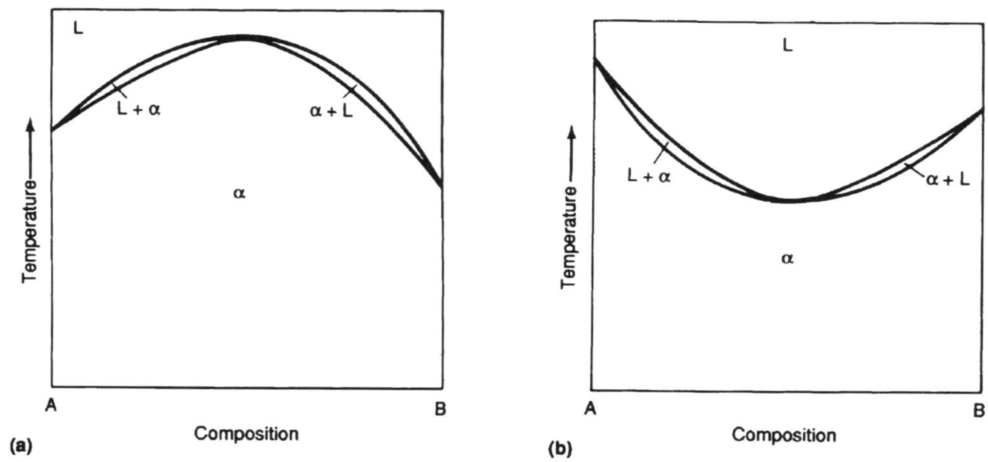

Fig. 4.8 Schematic binary phase diagrams with solid-state miscibility where the liquidus shows (a) a maximum and (b) a minimum. Source: Ref 4.1

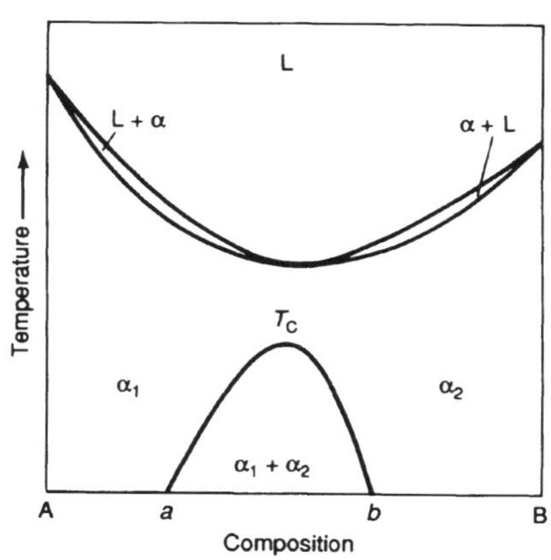

Fig. 4.9 Schematic binary phase diagram with a minimum in the liquidus and a miscibility gap in the solid state. Source: Ref 4.1

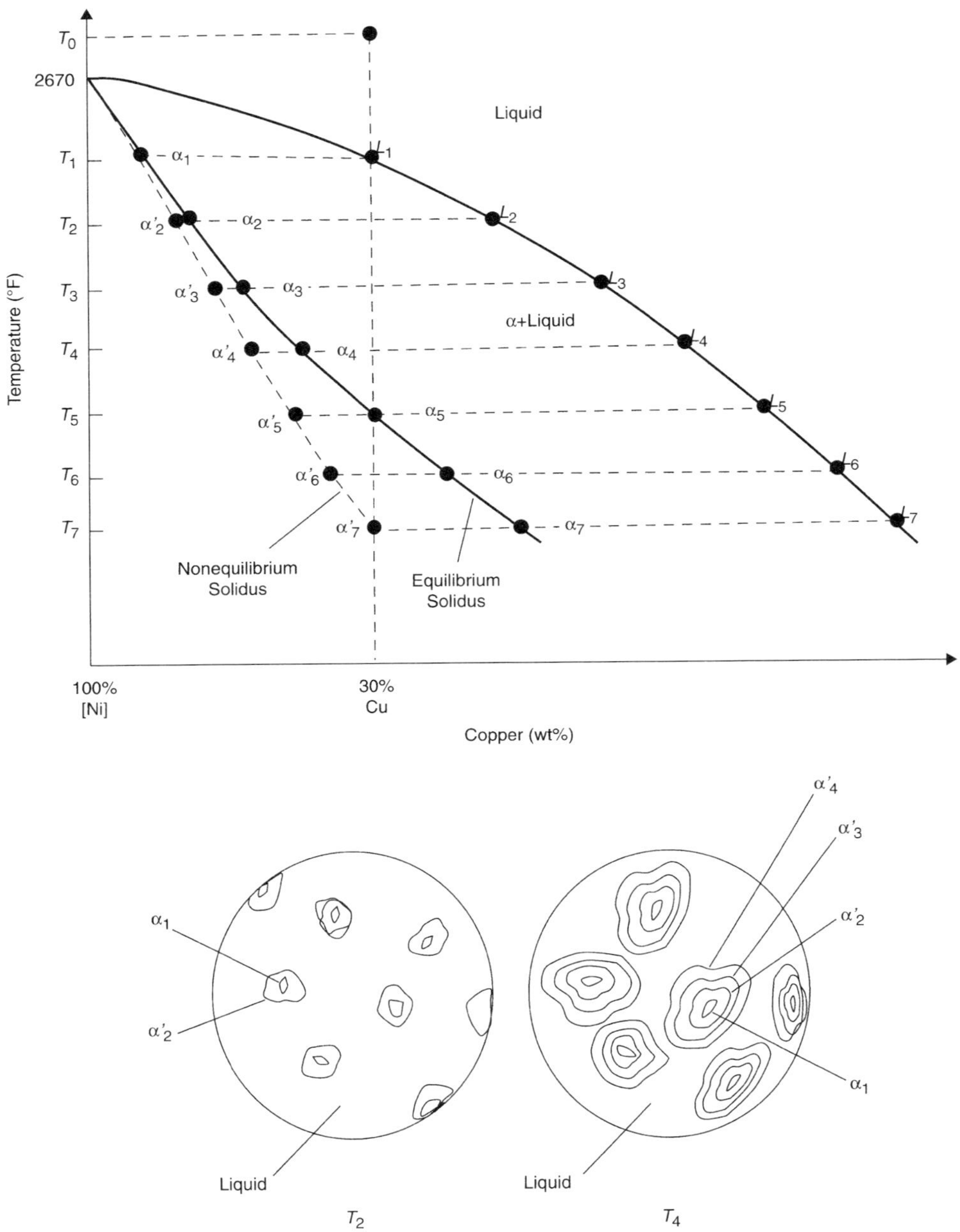

Fig. 4.10 Coring due to nonequilibrium solidification. Source: Ref 4.5 as published in Ref 4.3

4.2 Nonequilibrium Cooling

As previously mentioned, equilibrium phase diagrams are constructed to reflect extremely slow cooling rates that approach equilibrium conditions. This type of cooling is seldom encountered in industrial practice where faster cooling rates can produce segregation in the solidification products. As an example of a type of segregation called coring, produced by non-equilibrium freezing, consider the 70% Ni-30% Cu alloy in Fig. 4.10 that is rapidly cooled from a temperature, T_0. The first solid forms at temperature T_1 with a composition of α_1. On further rapid cooling to T_2, additional layers of composition, α_2, form. The overall composition at T_2 lies somewhere between α_1 and α_2 and is designated as α'_2. Because the tie line at $\alpha'_2\, L_2$ is longer than $\alpha_2\, L_2$, there will be more liquid and less solid in the rapidly cooled alloy than if it had been slowly cooled under equilibrium conditions. As rapid cooling continues through T_3 and T_4, the same processes occur, and the average composition follows the nonequilibrium solidus determined by points α_1, α'_2, α'_3, ... At T_7, freezing is complete and the average composition of the alloy is 30% Cu. The microstructure consists of regions varying from α_1 to α'_7, producing a cored microstructure.

REFERENCES

4.1 H. Baker, Introduction to Alloy Phase Diagrams, *Phase Diagrams*, Vol 3, *ASM Handbook*, ASM International, 1992, reprinted in *Desk Handbook: Phase Diagrams for Binary Alloys*, 2nd ed., H. Okamoto, Ed., ASM International, 2010

4.2 A.G. Guy, *Elements of Physical Metallurgy*, 2nd ed., Addison-Wesley Publishing Company, 1959

4.3 F.C. Campbell, *Elements of Metallurgy and Engineering Alloys*, ASM International, 2008

4.4 Microstructure of Copper and Copper Alloys, *Atlas of Microstructures of Industrial Alloys*, Vol 7, Metals Handbook, 8th ed., American Society for Metals, 1972

4.5 V. Singh, *Physical Metallurgy*, Standard Publishers Distributors, 1999

CHAPTER 5

Eutectic Alloy Systems

IF THE TWO-PHASE FIELD in the solid region of Fig. 5.1 is expanded so that it touches the solidus at some point, as shown in Fig. 5.2(a), complete miscibility of the components is lost. Instead of a single solid phase, the diagram now shows two separate solid terminal phases, which are in three-phase equilibrium with the liquid at point *P*, an invariant point that occurred by coincidence. Then, if this two-phase field in the solid region is even further widened so that the solvus lines no longer touch at the invariant point, the diagram passes through a series of configurations, finally taking on the more familiar shape shown in Fig. 5.2(b). The three-phase reaction that takes place at the invariant point, *E*, where a liquid phase freezes into a mixture of two solid phases, is called a eutectic reaction (from the Greek word for "easily melted"). The alloy that corresponds to the eutectic composition is called a eutectic alloy. An alloy having a composition to the left of the eutectic point is called a hypoeutectic alloy (from

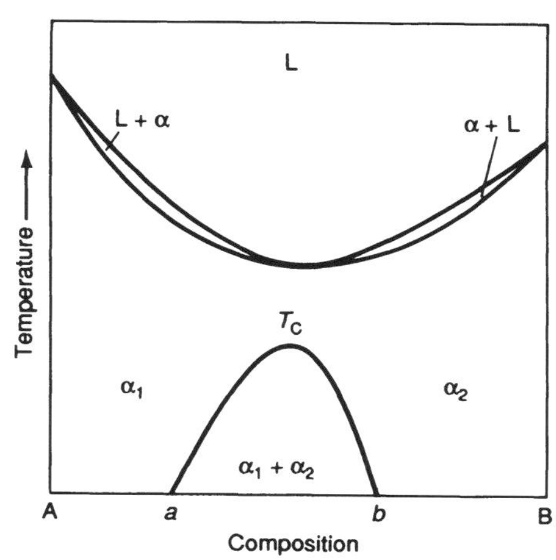

Fig. 5.1 Schematic binary phase diagram with a minimum in the liquidus and a miscibility gap in the solid state. Source: Ref 5.1

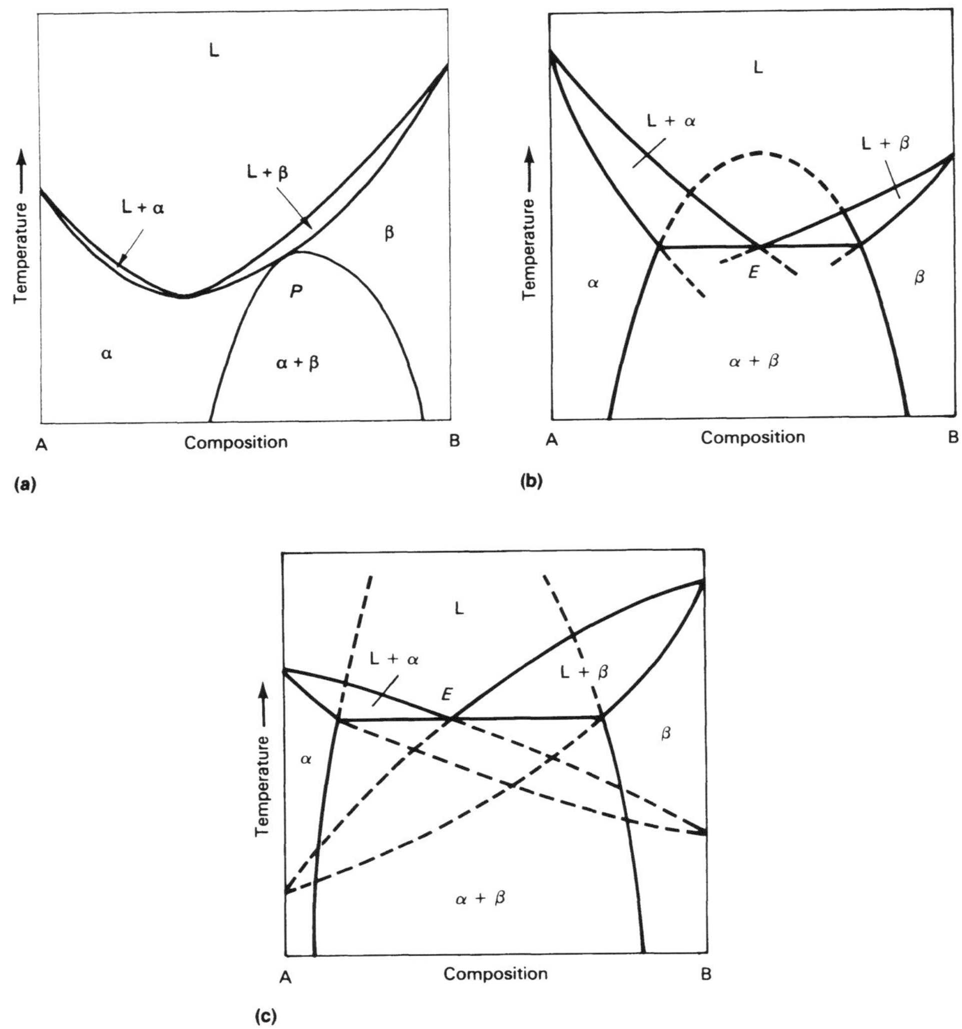

Fig. 5.2 Schematic binary phase diagrams with invariant points. (a) Hypothetical diagram with miscibility gap in the solid that touches the solidus curve at invariant point *P*; an actual diagram of this type probably does not exist. (b) and (c) Typical eutectic diagrams for (b) components having the same crystal structure and (c) components having different crystal structures; the eutectic (invariant) points are labeled *E*. The dashed lines in (b) and (c) are metastable extensions of the stable-equilibria lines. Source: Ref 5.1

the Greek word for "less than"); an alloy to the right is a hypereutectic alloy (meaning "greater than").

In the eutectic system described, the two components of the system have the same crystal structure. This, and other factors, allows complete miscibility between them. Eutectic systems, however, also can be formed by two components having different crystal structures. When this occurs, the liquidus and solidus curves (and their extensions into the two-phase field) for each of the terminal phases (Fig. 5.2c) resemble those for the situation of complete miscibility between system components.

A generic eutectic phase diagram is shown in Fig. 5.3. Eutectic systems form when alloying additions cause a lowering of the liquidus lines from both melting points of the pure elements. At a specific composition, there is a minimum melting point, where the mixed solid-liquid phase regions ($L + \alpha$ and $L + \beta$) vanish. This is the eutectic point, *e*, which defines an alloy composition that has the lowest melting point of the *A*-*B* system. The eutectic composition also solidifies completely at a single temperature that is referred to as an invariant point.

In a eutectic reaction, a liquid freezes to form two solid solutions:

$$\text{Liquid } L \rightarrow \text{Solid } \alpha + \text{Solid } \beta$$

The maximum solid solubility of element *B* is defined by point *a* on the *A*-rich side of the diagram, and the maximum solid solubility of element *A* in the lattice of element *B* is defined by point *b*. The methods of determining the equilibrium temperature ranges for solidification, fractions of phases, and compositions of phases are all similar to those illustrated for the isomorphous systems.

Consider the eutectic phase diagram in Fig. 5.4, which contains five different alloy compositions (*A*-*E*). Alloy *C* has exactly the eutectic composition. On cooling from the liquid state, it converts from a liquid to a solid at the eutectic temperature. The resulting solid consists of α and β phases of the eutectic structure. Next, consider alloys *A* and *E*, which do not cross the eutectic isotherm line. Thus, their final microstructures do not contain any eutectic. Instead, they form solid solutions of α and β for alloys *A* and *B*, respectively. On cooling, alloy *B* forms α crystals in liquid when it passes through the liquidus. When alloy *B* cools through the eutectic

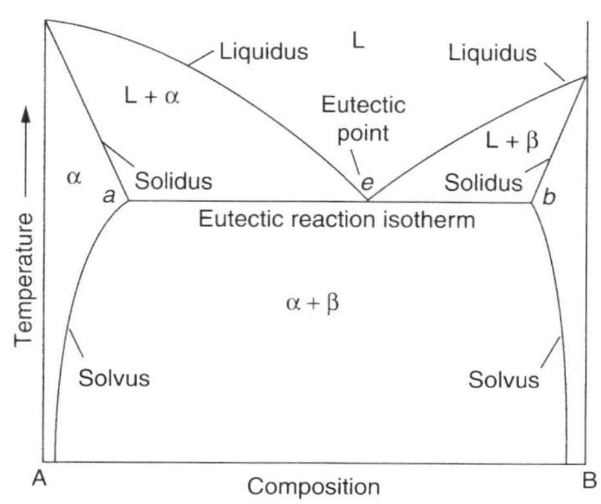

Fig. 5.3 Phase diagram containing a eutectic reaction. Source: Ref 5.2

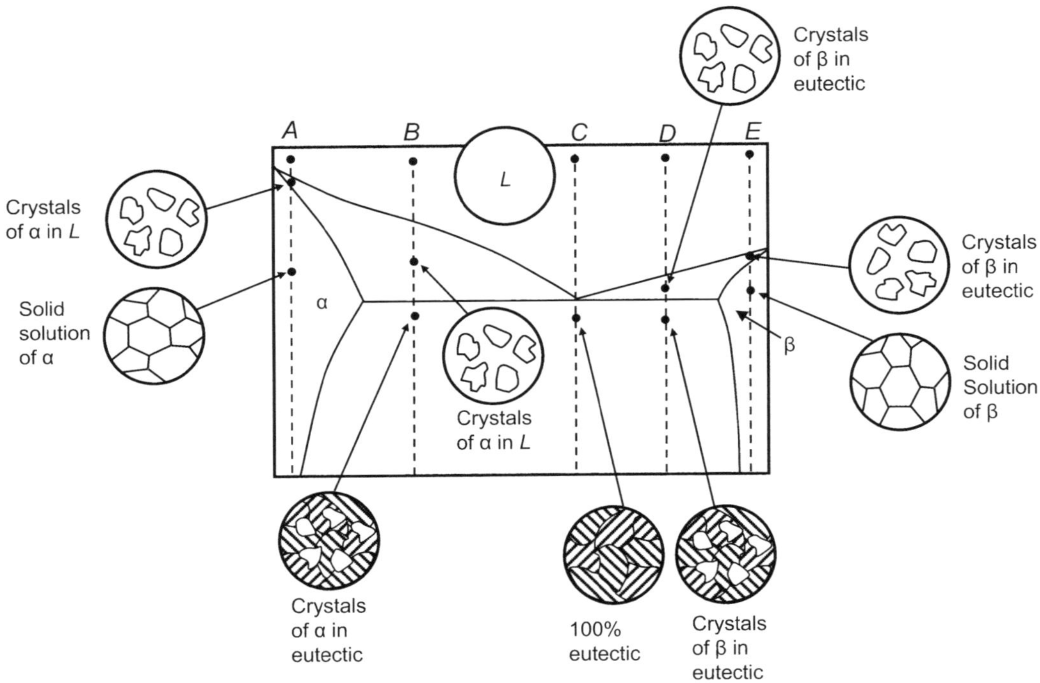

Fig. 5.4 Crystalline structures on cooling in a eutectic phase diagram

temperature, the remaining liquid freezes to form a eutectic mixture of α and β. Therefore, the final microstructure of alloy *B* contains proeutectic α in a matrix of eutectic (α and β). The situation is similar for alloy *D*, except the final microstructure has proeutectic β in a matrix of eutectic.

Because eutectic alloys have a single melting/solidification point, eutectic compositions include important types of commercial alloys. Traditional lead-tin solder alloys are based on their eutectic compositions. Casting alloys are often based on eutectic compositions for various reasons, including the minimization of both energy input and coring, or alloy segregation. For example, the iron-carbon system has a eutectic at composition of 4.3 wt% C, which is the basis of cast irons. Many other binary systems have phase diagrams with multiple eutectic transformation compositions.

Because the eutectic is the lowest melting composition in the alloy, it can often cause problems during hot working and heat-treating operations. During casting of an ingot, due to nonuniform cooling rates, there is often quite a bit of segregation of the alloying elements. The low-melting eutectic composition is normally the last portion of the metal to freeze, normally along the grain boundaries. On reheating the metal for hot working operations, if the lowest melting eutectic temperature is exceeded, melting can occur along the grain boundaries and the part is frequently ruined. It should also be noted that alloying systems with a number of different alloying elements will often form lower-melting-point eutectics than for a simple binary system. To help prevent some of these problems, as-cast ingots are often reheated to temperatures just below the melting point and soaked for long times (called homogenization) to create more uniform structures prior to hot working or heat treating.

5.1 Aluminum-Silicon Eutectic System

Aluminum-silicon alloys are a family of industrially important casting alloys. The phase diagram for the aluminum-silicon system is shown in Fig. 5.5, along with micrographs of representative structures. The micro-

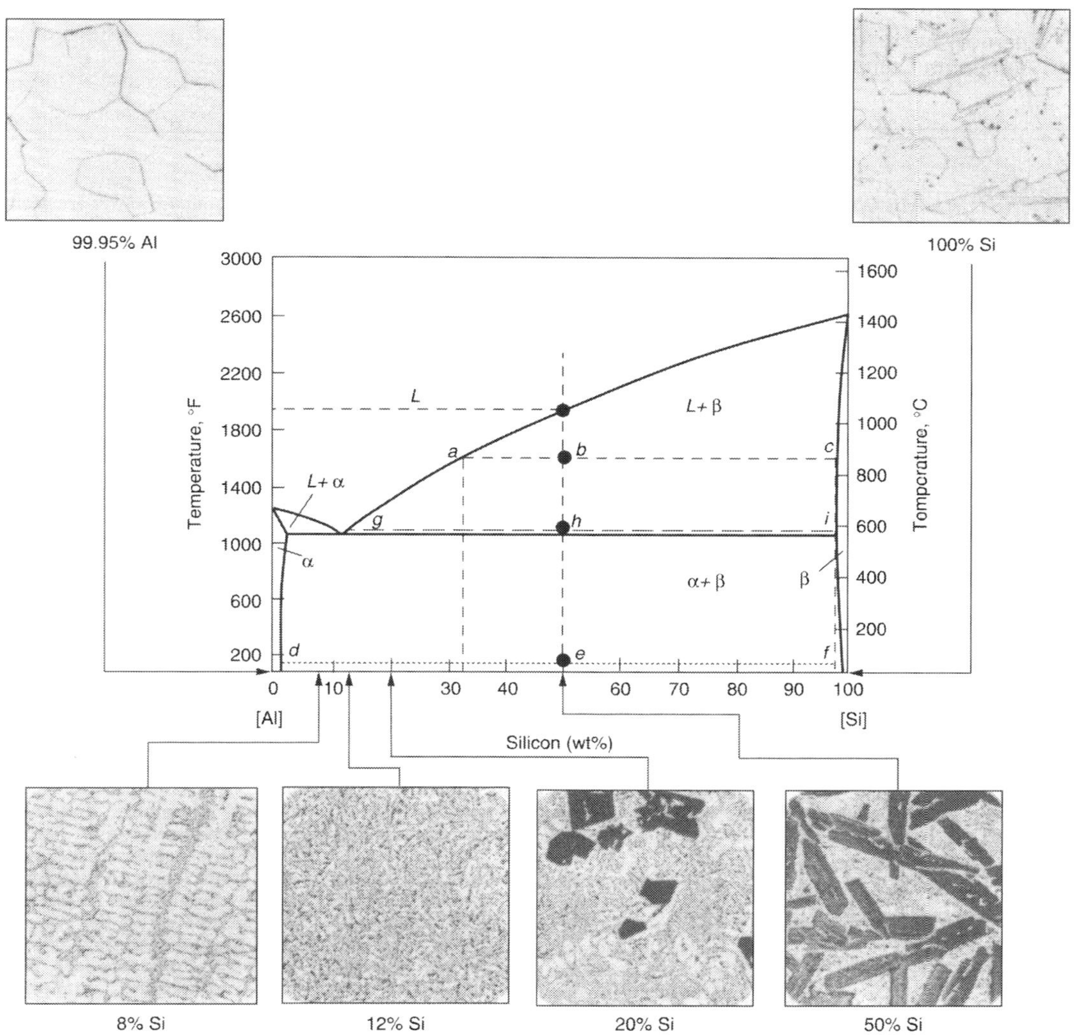

Fig. 5.5 Aluminum-silicon phase diagram. Source: Ref 5.3 as published in Ref 5.4

structure of 99.95 wt% Al has the typical equiaxed structure of a pure metal. The microstructure of the 8 wt% Si alloy shows long dendrites of primary α solid solution surrounded by the eutectic microconstituent. In contrast, when the primary solid phase is the silicon-rich β phase, as in the 20 and 50 wt% silicon alloys, the primary crystals have geometric shapes due to the pronounced difference between aluminum, a metal, and silicon, which has predominately nonmetallic properties.

Consider an alloy containing 50 wt% Si. When the alloy is cooled from the liquid state, it starts to solidify at approximately 1080 °C (1975 °F). Solidification begins with the nucleation and growth of primary crystals of β solid solution. At 870 °C (1600 °F), point *b*, the alloy composition-temperature point lies in the two-phase $L + \beta$ field. The ends of the tie line drawn across the two-phase field determine the chemical composition of the phases: The liquid phase is 34 wt% Si and the β solid-solution phase is 98 wt% Si. The lever rule can be used to determine the amounts of the two phases:

$$\%\ \beta \text{ phase} = \frac{ab}{ac} = \frac{50-34}{98-34} \times 100 \approx 25 \text{ wt\%}$$

Because the remaining portion of the alloy is liquid phase, there must be 75 wt% liquid.

As the 50 wt% Si alloy continues to cool, the amount of primary β phase increases until the eutectic temperature is reached. At this temperature (580 °C, or 1075 °F), the liquid phase reaches the eutectic composition of 12 wt% Si. On cooling through the eutectic temperature, the liquid solidifies at a constant temperature to form the eutectic structure consisting of an intimate mixture of α and β phases. A phase analysis at room temperature is representative of the condition of the alloy in the solid $\alpha + \beta$ region:

$$\%\ \beta \text{ phase} = \frac{de}{df} = \frac{50-1}{99-1} \times 100 \approx 50 \text{ wt\%}$$

This amount of β phase (50 wt%) is the sum of the weight fractions of primary and secondary β. To determine the amount of primary β that formed prior to the eutectic reaction (in the $L + \beta$ region of the phase diagram), a phase analysis can be conducted at a temperature immediately above the eutectic temperature. At this temperature, solidification of the primary β is complete, and all the remaining liquid will solidify as part of the eutectic microconstituent.

$$\%\ \text{primary } \beta \text{ phase} = \frac{gh}{gi} = \frac{50-12}{98-11} \times 100 \approx 44 \text{ wt\%}$$

As previously mentioned, aluminum-silicon alloys are extensively used for castings. The primary reason for their widespread use is that silicon greatly increases the fluidity, or lowers the viscosity, of aluminum during casting, allowing for better mold filling and fewer casting defects. As shown in Fig. 5.6, tensile strength increases with silicon additions, while ductility (elongation) decreases. Slowly cooled aluminum-silicon castings form a eutectic structure in which the eutectic grows as thin, flat plates that appear needlelike. The silicon plates are stress risers that reduce ductility and toughness. In commercial castings, modifiers, such as 0.2% Na or 0.01% Sr are added to the melt and modify the eutectic structure producing more rounded particles. If the alloy is rapidly cooled, such as in a die casting, modification is not necessary because rapid cooling produces a fine structure. In hypereutectic alloys that are used for engine blocks because of their high silicon contents and thus high wear resistance, the primary silicon that forms is normally very coarse, causing poor machinability, poor castability, and gravity segregation where the lighter silicon particles float to the top of the casting. In this case, the addition of 0.05% P encourages the nucleation of primary silicon, refining its size and minimizing its deleterious properties.

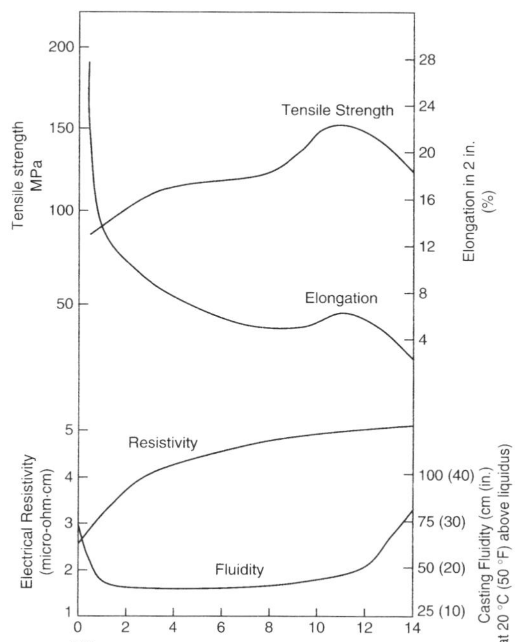

Fig. 5.6 Properties of aluminum-silicon alloys. Source: Ref 5.3 as published in Ref 5.2

5.2 Lead-Tin Eutectic System

Although lead-tin alloys are too weak for use as structural materials, they are widely used as solders for everything from joining copper plumbing to soldering electrical circuits. However, there is an international effort to develop lead-free solders because of the health concerns associated with lead. Nevertheless, lead-tin solders are still widely used. Consider alloy 1 in the lead-tin phase diagram (Fig. 5.7). At 183 °C (361 °F), which is the eutectic temperature, all of the liquid solidifies by the eutectic reaction and forms a eutectic mixture of solid solutions of α (19.2% Sn) and β (97.5% Sn) according to the reaction:

$$\text{Liquid (61.9\% Sn)} \rightarrow \alpha \text{ (19.2\% Sn)} + \beta \text{ (97.5\% Sn)}$$

Next, consider alloy 2 in Fig. 5.7. At point *a*, the alloy is 100% liquid. On cooling to the liquidus at point *b*, proeutectic α crystals will start forming in the liquid. Within the α + *L* region at 40% Sn and 230 °C (446 °F), which is point *c*, a phase analysis yields:

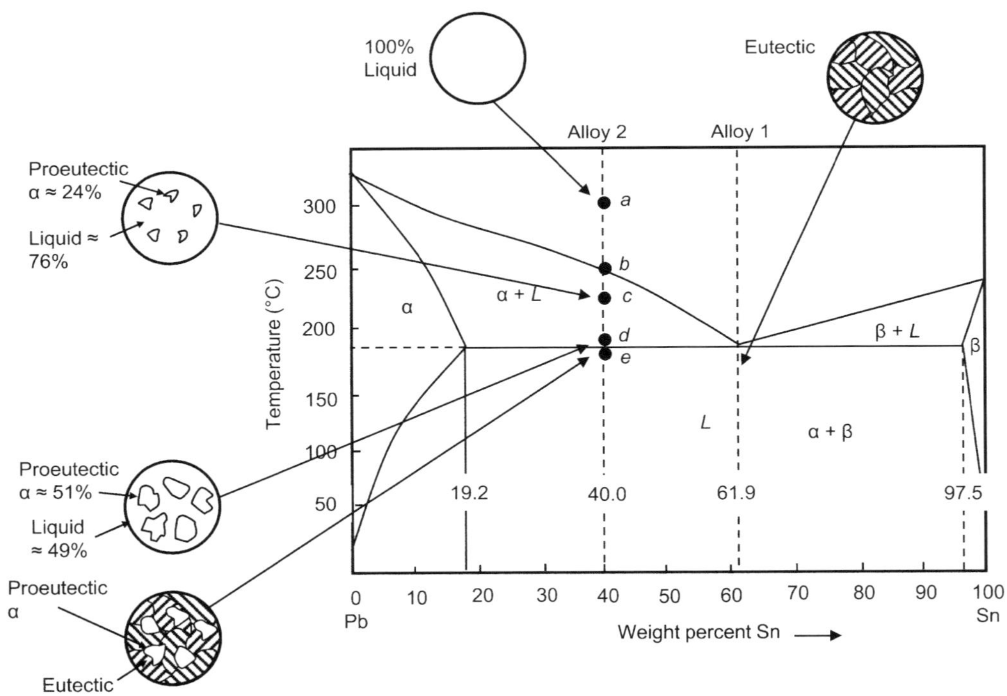

Fig. 5.7 Lead-tin phase diagram. Adapted from Ref 5.5

Phases present:	Liquid	α
Compositions of phases:	48% Sn in L	15% Sn in α
Amounts of phases:	wt% in L phase	wt% in α phase
	$\left(\frac{40-15}{48-15}\right)(100\%) \approx 76\%$	$\left(\frac{48-40}{48-15}\right)(100\%) \approx 24\%$

On further cooling to just above the eutectic temperature (183 °C, or 361 °F), the 40% Sn alloy at point *d* yields:

Phases present:	Liquid	α
Compositions of phases:	61.9% Sn in L	19.2% Sn in α
Amounts of phases:	wt% in L phase	wt% in α phase
	$\left(\frac{40-19.2}{61.9-19.2}\right)(100\%) \approx 49\%$	$\left(\frac{61.9-40}{61.9-19.2}\right)(100\%) \approx 51\%$

When the alloy cools to just below the eutectic temperature, the 40% Sn alloy at point *e* now yields:

Phases present:	α	β
Compositions of phases:	19.2% Sn in α	19.2% Sn in β
Amounts of phases:	wt% in α phase	wt% in β phase
	$\left(\frac{40-19.2}{61.9-19.2}\right)(100\%) \approx 49\%$	$\left(\frac{61.9-40}{61.9-19.2}\right)(100\%) \approx 51\%$

Eutectic alloys frequently form a lamellar, or platelike, structure on freezing (Fig. 5.8). This structure permits the lead and tin atoms to diffuse short distances through the liquid to form the lamellar-type structure. It should be pointed out that although eutectics frequently form a lamellar-type structure on freezing, this not always the case.

The lead-tin phase diagram is useful for selecting the best solder for a particular application. For soldering delicate electronic components where it is desirable to minimize heat, a low-melting-point solder such as the eutectic solder in Fig. 5.9 would be selected. On the other hand, when soldering copper plumbing connections, the plumber's solder would be a logical choice because it stays molten longer and allows the worker to wipe off excess flow from the joint. Finally, for joints that will be exposed to moderately high temperatures, the high-temperature solder may be the best choice.

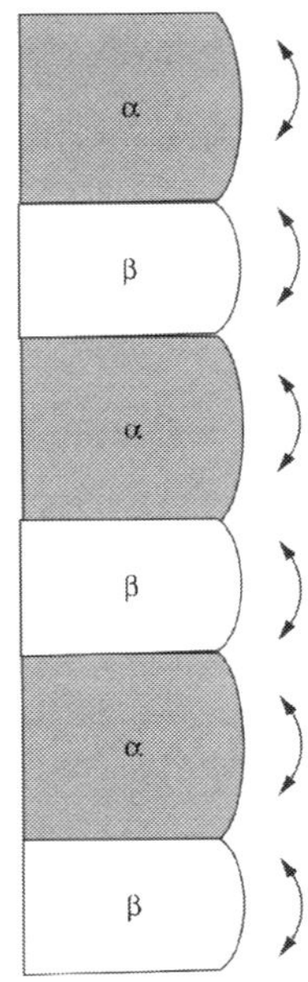

Fig. 5.8 Lamellar eutectic growth. Source: Ref 5.2

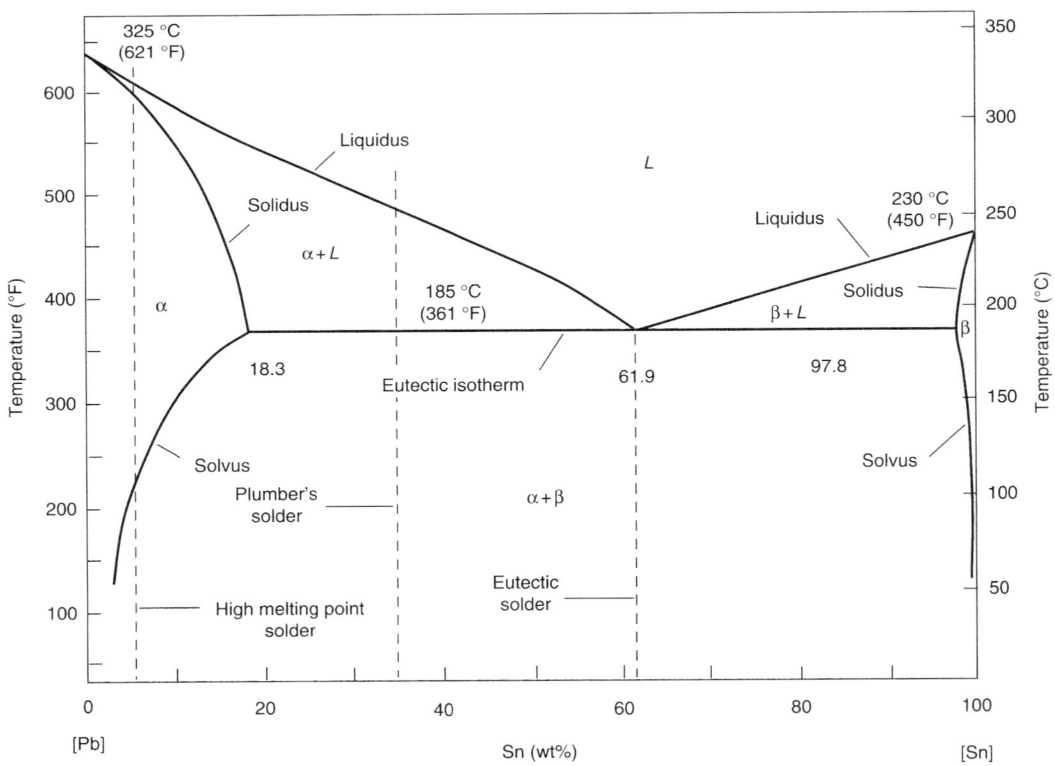

Fig. 5.9 Selection of lead-tin solders. Source: Ref 5.2

5.3 Eutectic Morphologies

Eutectic structures are characterized by the simultaneous growth of two or more phases from the liquid. Three or even four phases are sometimes observed growing simultaneously from the melt. However, because most technologically useful eutectic alloys are composed of two phases, only this type is discussed here. Eutectic alloys exhibit a wide variety of microstructures, which can be classified according to two criteria:

- Lamellar or fibrous morphology of the phases
- Regular or irregular growth

Lamellar and Fibrous Eutectics. When there are approximately equal volume fractions of the phases (nearly symmetrical phase diagram), eutectic alloys generally have a lamellar structure, for example, Al-Al_2Cu (Fig. 5.10). On the other hand, if one phase is present in a small volume fraction, this phase will in most cases tend to form fibers, for example, molybdenum in NiAl-Mo (Fig. 5.11).

In general, the microstructure obtained will usually be fibrous when the volume fraction of the minor phase is lower than 0.25; otherwise, it will be lamellar. This is because of the small separation of the eutectic phases (typically several microns) and the resulting large interfacial area (of the order of 1 m^2/cm^3) that exists between the two solid phases. The system will therefore tend to minimize its interfacial energy by choosing the morphology that leads to the lowest total interface area. For a given spacing (imposed by growth conditions), the interface area is smaller for fibers than for lamellae at volume fractions below 0.25. However, when the minor phase is faceted, a lamellar structure may be formed even at a very low

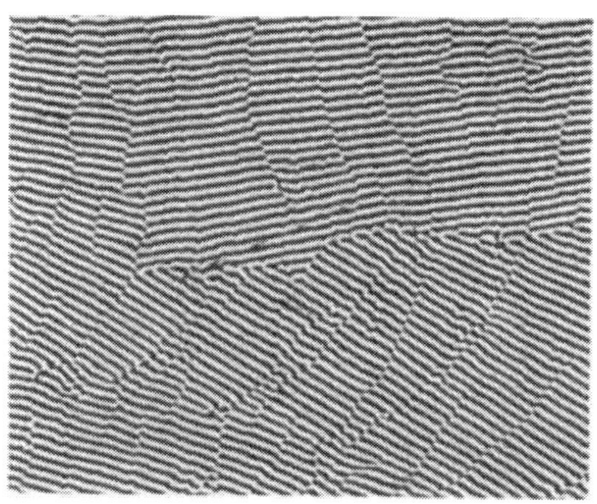

Fig. 5.10 Example of a lamellar eutectic microstructure (Al-Al_2Cu) with approximately equal volume fractions of the phases. Transverse section of a directionally-solidified (DS) sample. As-polished. Source: Ref 5.6

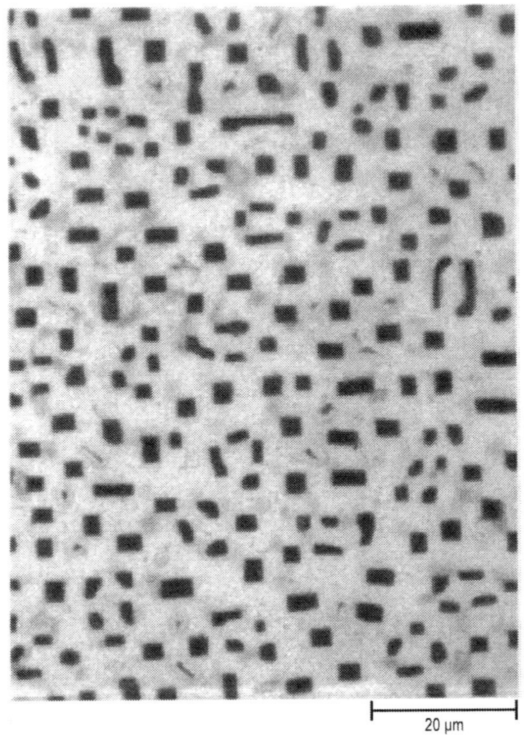

Fig. 5.11 Example of a fibrous eutectic microstructure with a small volume fraction of one phase (molybdenum fibers in NiAl matrix). Transverse section of a directionally-solidified (DS) sample. As-polished. Courtesy of E. Blank. Source: Ref 5.6

volume fraction, because the interfacial energy is then considerably lower along specific planes, along which the lamellae can be aligned. Growth of faceted phases occurs on well-defined atomic planes, thus creating planar, angular surfaces (facets). Typical examples of faceted phases are graphite, silicon, and intermetallic compounds. In the case of gray cast iron (Fig. 5.12), the volume fraction of the graphite lamellae is 7.4%.

Many eutectic microstructures can be classified as lamellar or fibrous, but there is an important exception, namely, spheroidal graphite cast iron (Fig. 5.13). In this case, there is no cooperative eutectic-like growth of both phases; instead, there is separate growth of spheroidal graphite particles as a primary phase (at least during the initial stages), together with austenite dendrites, known as divorced growth. Cast iron often exhibits intermediate microstructural forms, such as vermicular or chunky graphite.

Regular and Irregular Eutectics. If both phases are nonfaceted (usually when both are metallic), the eutectic will exhibit a regular morphology. The microstructure is then made up of lamellae or fibers having a high

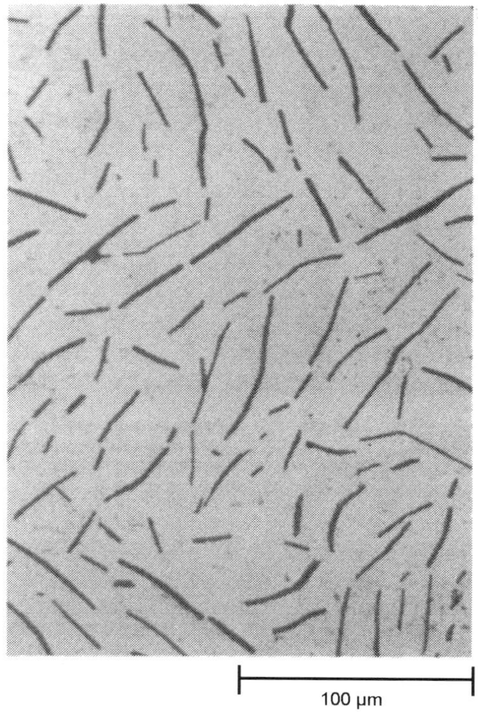

Fig. 5.12 Microstructure of a gray cast iron showing flake graphite. Transverse section etched with nital. Source: Ref 5.6

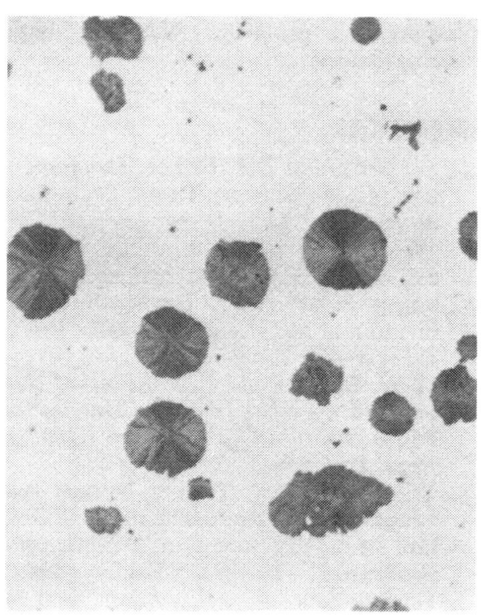

Fig. 5.13 Graphite in spheroidal cast iron, which results from the divorced growth of the phases. Etched with nital. Original magnification: 130×. Source: Ref 5.6

degree of regularity and periodicity, particularly in unidirectionally-solidified specimens (Fig. 5.10).

On the other hand, if one phase is faceted, the eutectic morphology often becomes irregular (Fig. 5.12 and 5.14). This is because the faceted phase grows preferentially in a direction determined by specific atomic planes. Because the various faceted lamellae have no common crystal orientation, their growth directions are not parallel, and the formation of a regular microstructure becomes impossible. The two eutectic alloys of greatest practical importance, iron-carbon (cast iron) and aluminum-silicon, belong to this category.

Although the examination of metallographic sections of irregular eutectics seems to reveal many dispersed lamellae of the minor phase, these lamellae are generally interconnected in a complex three-dimensional arrangement. In the foundry literature, such eutectic grains are often referred to as eutectic cells. In the solidification literature, the term *cell* defines a certain interface morphology; therefore, the term *eutectic grain* is used throughout this section.

The regularity of some eutectic microstructures can be used to make in situ composites. By using a controlled heat flux to achieve slow directional solidification, it is possible to obtain an aligned microstructure throughout the entire casting. When one of the phases is particularly strong, as in the case of TaC fibers in the Ni-TaC eutectic, the mechanical properties of the alloy can be enhanced in the growth direction. In contrast, an equiaxed microstructure can be formed by inoculation, and there is no long-range orientation.

Interpretation of Eutectic Microstructures. Eutectic microstructures, as seen in metallographic section, are two-dimensional images of a three-dimensional arrangement of two (or more) phases. One must therefore be

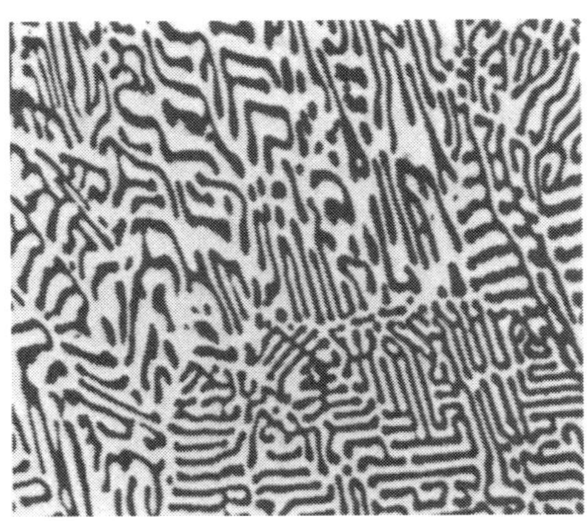

Fig. 5.14 Irregular "Chinese script" eutectic consisting of faceted Mg_2Sn phase (dark) in a magnesium matrix. Etched with glycol. Original magnification: 250×. Source: Ref 5.6

very careful in interpreting these metallographic sections. For example, Fig. 5.15 shows a longitudinal section of a directionally-solidified (DS) lamellar eutectic (white cast iron) covering two different grains. Despite their different appearances, the two grains have the same lamellar spacing. However, the sectioning plane is perpendicular to the lamellae of one grain, but is at a small angle with respect to the lamellae of the other grain. Therefore, in DS samples, the lamellar spacing of eutectic microstructures must always be measured perpendicular to the growth direction. In a casting containing equiaxed grains, only a mean spacing can be measured. Eutectic grains are often difficult to identify, as can be seen in Fig. 5.10.

5.4 Solidification and Scale of Eutectic Structures

A schematic eutectic phase diagram is shown in Fig. 5.16. When a liquid, L, of eutectic composition C_E is frozen, the α and β solid phases solidify simultaneously when the temperature of the melt is below the eutectic temperature, T_E. A variety of geometrical arrangements can be produced. For simplicity, the case of a lamellar microstructure is considered in this discussion; the solidification of fibers can be described in terms of similar mechanisms. Because eutectic growth is essentially solute diffusion controlled, there is no fundamental difference between equiaxed

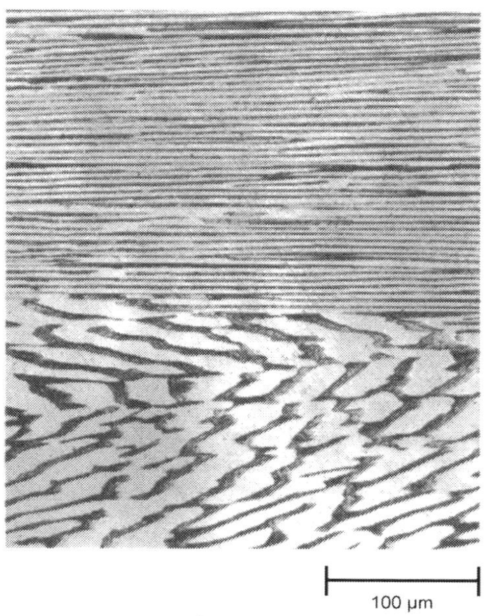

Fig. 5.15 Longitudinal section of directionally-solidified (DS) white cast iron. The two grains in the micrograph have the same lamellar spacing but are oriented differently with regard to the plane of polish. Etched with nital. Source: Ref 5.6

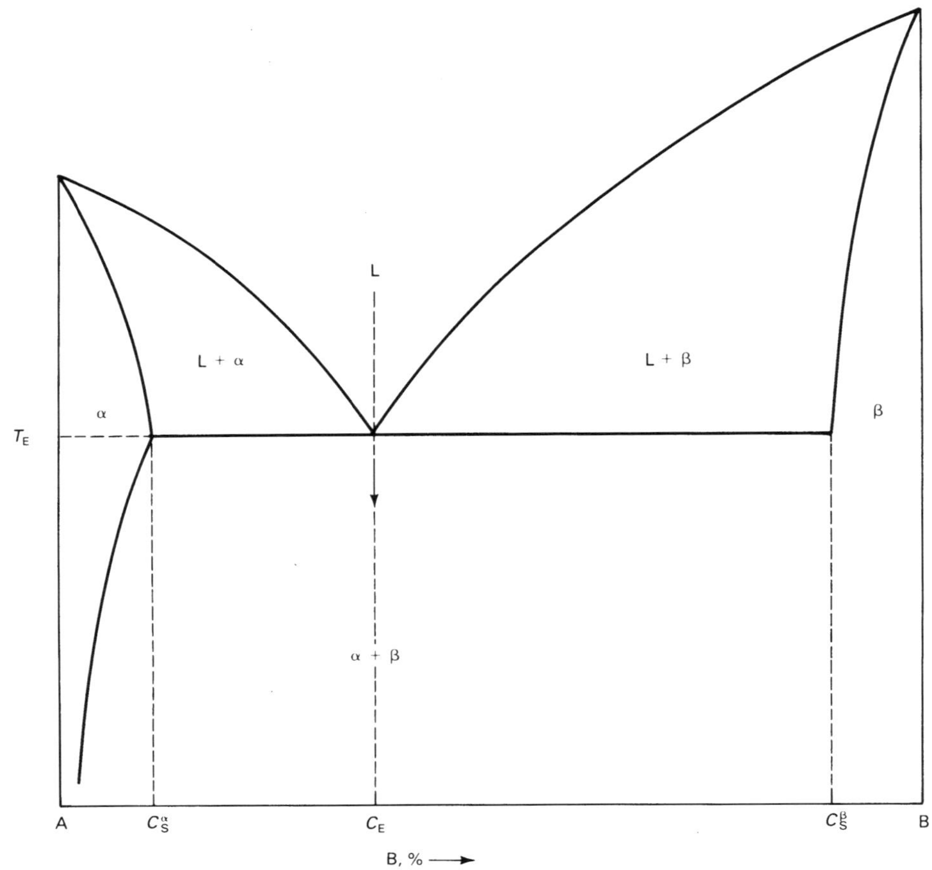

Fig. 5.16 Schematic eutectic phase diagram. See text for explanation. Source: Ref 5.6

and directional solidification. Therefore, the mechanisms described are valid for both cases.

Regular Eutectic Growth. During eutectic solidification, the growing α phase rejects *B* atoms into the liquid because of their lower solubility with respect to the liquid concentration. Conversely, the β phase rejects *A* atoms. If the α and β phases grow separately, solute rejection would occur only in the growth direction. This involves long-range diffusion. Therefore, a very large boundary layer would be created in the liquid ahead of the solid-liquid interface, as shown in Fig. 5.17(a).

However, during eutectic solidification, the α and β phases grow side by side in a cooperative manner; the *B* atoms rejected by the α phase are needed for the growth of the β phase, and vice versa. The solute then needs only to diffuse along the solid-liquid interface from one phase to the other (Fig. 5.17b). The solute buildup in the liquid ahead of the growing solid-liquid interface is considerably lowered by this sidewise diffusion (diffusion coupling), thus being thermodynamically favorable. This is the fundamental reason for the occurrence of eutectic growth. As can be seen in Fig. 5.17(b), the smaller the lamellar spacing, λ, the smaller the solute buildup, if the driving force for diffusion provided by the concentration gradient remains constant.

On the other hand, at the three-phase junction α-β-*L*, the surface tensions must be balanced to ensure mechanical equilibrium (Fig. 5.18). This imposes fixed contact angles, leading to a curvature of the solid-liquid interface. This curvature is thermodynamically disadvantageous. Because

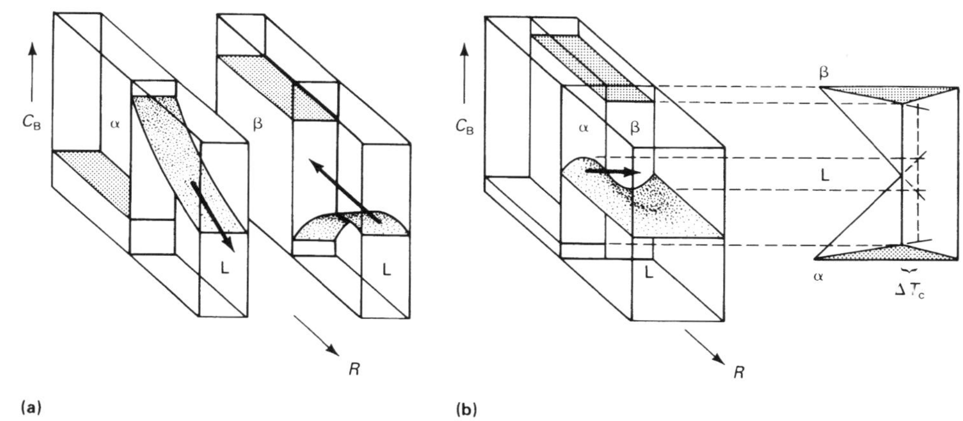

Fig. 5.17 Diffusion fields ahead of the growing α and β phases in (a) isolated and (b) coupled eutectic growth. The dark arrow represents the flux of *B* atoms. Source: Ref 5.7 as published in Ref 5.6

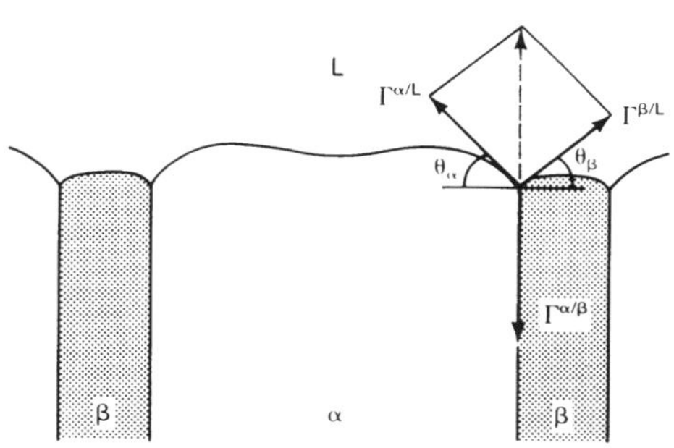

Fig. 5.18 Surface tension balance at the three-phase (α-β-*L*) junction and the resulting curvature of the solid-liquid interface. Source: Ref 5.6

the contact angles are material constants, this curvature is higher when the lamellar spacing is small.

The scale of the eutectic structure is therefore determined by a compromise between two opposing factors:

- Solute diffusion, which tends to reduce the spacing
- Surface energy (interface curvature), which tends to increase the spacing

The lamellar spacing, λ, and the growth undercooling, ΔT (defined as the difference between the eutectic temperature, T_E, and the actual interface temperature during growth), are given by:

$$\lambda = \frac{\varphi k_1}{\sqrt{R}} \quad \text{(Eq 5.1)}$$

$$\Delta T = \frac{(\varphi + 1/\varphi)}{2} k_2 \sqrt{R} \quad \text{(Eq 5.2)}$$

where R is the solidification rate (velocity at which the solid/liquid interface advances), k_1 and k_2 are constants related to the material properties, and ϕ is a regularity constant whose value is close to unity for regular eutectics. Figure 5.19 shows typical values for the $\lambda(R)$ relationship (Eq 5.1). It can be seen that regular eutectics have spacings between the coarse ones of irregular eutectics and the fine ones of eutectoids. In the latter, the effect of diffusion on spacing is more marked because it occurs in a solid phase.

The scale of the eutectic microstructure depends on the solidification rate, not directly on the cooling rate. The reason is that the thermal gra-

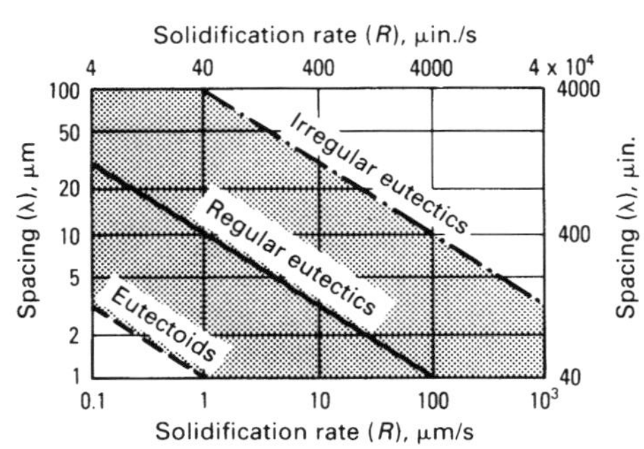

Fig. 5.19 Typical spacings of eutectics and eutectoids as a function of growth rate. Source: Ref 5.7 as published in Ref 5.6

dient has a negligible effect on the size of the eutectic microstructure. Because the cooling rate is the product of the solidification rate and the thermal gradient, two growth conditions characterized by the same cooling rate but with different thermal gradients lead to different solidification rates and therefore to different spacings. An important characteristic of regular eutectic growth is that the lamellae (or fibers) are parallel to the heat flow direction during solidification and perpendicular to the solid-liquid interface.

Irregular Eutectic Growth. Irregular eutectics grown under given growth conditions exhibit an entire range of spacings because the growth direction of the faceted phase (for example, graphite in cast iron or silicon in aluminum-silicon) is determined by specific atomic orientations and is not necessarily parallel to the heat flux. In this case, growth involves the following mechanism: When two lamellae converge, the growth of one simply ceases when λ becomes smaller than a critical spacing, λ_{min}, because the local interface energy becomes too large. Thus, the spacing is increased. This mechanism is illustrated in Fig. 5.20. Conversely, diverging lamellae can grow until another critical spacing, λ_{br}, is reached. When this occurs, one of the lamellae branches into two diverging lamellae, thus reducing the spacing. Growth of an irregular eutectic thus occurs within the range of interlamellar spacings between λ_{min} and λ_{br}.

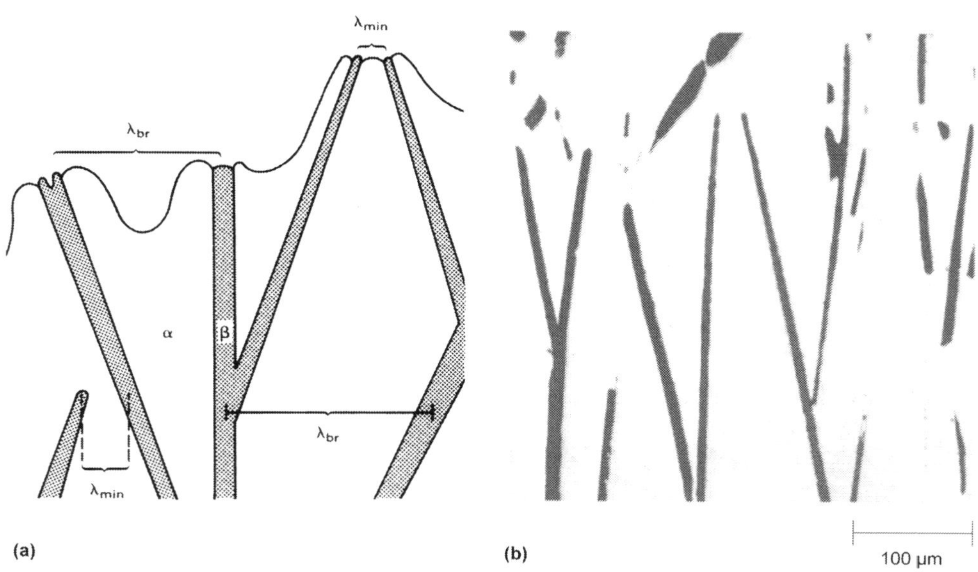

Fig. 5.20 Growth of irregular eutectics. (a) Schematic of branching of the faceted phase at λ_{br}, termination at λ_{min}, and the corresponding shape of the solid/liquid interface. (b) Iron-carbon eutectic alloy directionally solidified at $R = 0.017$ μm/s. Branching was induced by a rapid tenfold increase in R. Longitudinal section. As-polished. Source: Ref 5.8 as published in Ref 5.6

It can be shown that the growth temperature of the region of small λ is higher than that in the large λ zones. The solid-liquid interface is therefore nonisothermal; that is, its shape is irregular (Fig. 5.20a) and is the opposite of the isothermal planar solid-liquid interface that characterizes regular eutectic growth (Fig. 5.18 and 5.21).

A mean spacing, $<\lambda>$, and a mean undercooling, $<\Delta T>$, can be defined and are still given by Eq 5.1 and 5.2. In this case, ϕ (the ratio of the mean spacing, $<\lambda>$, to the minimum undercooling spacing, which is close to λ_{min}) is greater than unity. Therefore, the spacings and undercoolings obtained are higher than those observed in regular eutectics (Fig. 5.19).

5.5 Competitive Growth of Dendrites and Eutectics

The solidification of a binary alloy of exactly eutectic composition was examined earlier in this chapter. In this case, provided the growth is regular, the solid-liquid interface is planar. However, when alloy composition departs from eutectic, or when a third alloying element is present, the interface can become unstable for the same reason as in the case of a simple solid-liquid interface. As shown in Fig. 5.22, two types of morphological instability can develop: single-phase and two-phase.

A single-phase instability (Fig. 5.22a) leads to the solidification of one of the phases in the form of primary dendrites plus interdendritic eutectic. This situation is primarily observed in off-eutectic alloys because one phase becomes much more constitutionally undercooled than the other. For example, during the solidification of a hypoeutectic alloy, the α phase is heavily undercooled because the liquidus temperature at that composition

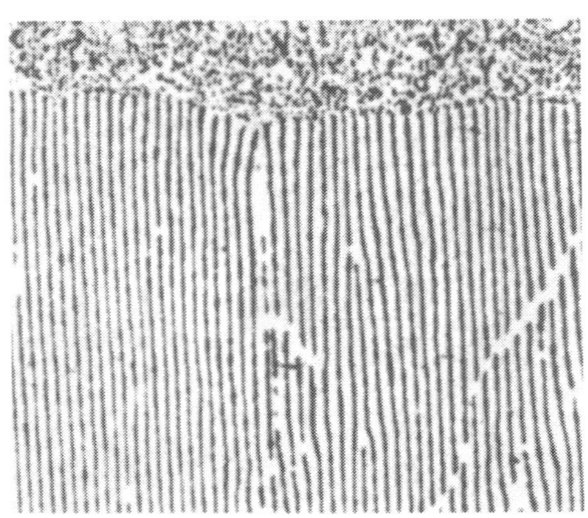

Fig. 5.21 Nearly planar solid-liquid interface of a regular cadmium-tin eutectic as revealed by quenching. Etched with ferric chloride. Original magnification: 210×. Source: Ref 5.6

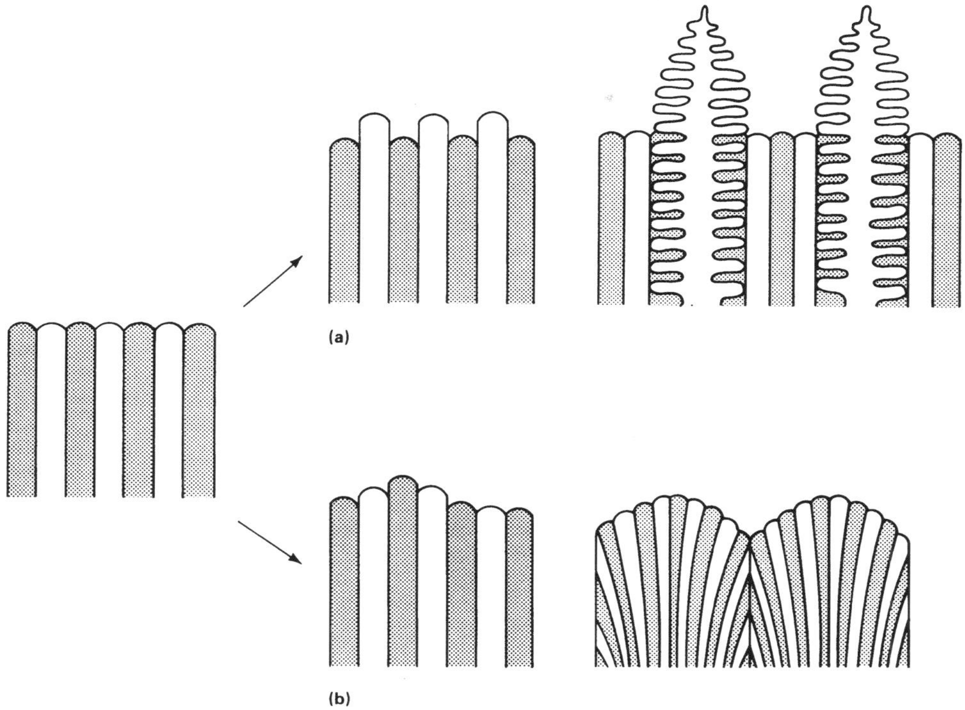

Fig. 5.22 Types of instability of a planar solid-liquid eutectic interface. (a) Single-phase instability leading to the appearance of dendrites of one phase. (b) Two-phase instability leading to the appearance of eutectic cells or colonies in the presence of a third alloying element. Source: Ref 5.6

is much higher than T_E (Fig. 5.16). The α phase can therefore grow faster (or at higher temperature) than the eutectic.

A two-phase instability (Fig. 5.22b) is characterized by cellular-like growth and leads to the appearance of eutectic colonies. This situation is observed when a third alloying element that partitions similarly at both the α-L and β-L interfaces produces a long-range diffusion boundary layer ahead of the solid-liquid interface, thus making the growing eutectic interface constitutionally undercooled with respect to this element.

Coupled Zone of Eutectics. The eutectic-type phase diagram appears to indicate that microstructures consisting entirely of eutectic can be obtained only at the exact eutectic composition. In fact, experimental observations show that purely eutectic microstructures can be obtained from off-eutectic alloys over a range of growth conditions. On the other hand, dendrites can sometimes be found in alloys with the exact eutectic composition if the growth rate is high. This is of considerable practical importance because the properties of a casting can be significantly changed when single-phase dendrites appear.

To explain these observations, one must consider the growth mechanisms of the competing phases. Because of the differing growth characteristics of eutectics and dendrites, the solidification of eutectic (high-efficiency diffusion coupling) can be faster than the isolated growth of one phase (primary dendrites), even for off-eutectic alloys. In this case, the dendrites are overgrown, and a purely eutectic microstructure is obtained over a range of off-eutectic compositions (the volume fraction of both phases in this case is determined by alloy composition and is therefore different from that obtained in the eutectic alloy). Conversely, if one of the phases (for example, β) is faceted, the growth of this phase (and consequently of the eutectic) is slower at a given undercooling. Dendrites of α phase may then grow more rapidly than the eutectic at the eutectic composition; purely eutectic microstructures are obtained only in hypereutectic alloys.

The temperature of a growing eutectic solid-liquid interface is a function of the growth rate. This relationship is used, together with the dendrite tip temperatures of α and β primary crystals, to establish the coupled zone. In the diagrams shown in Fig. 5.23, each point below the eutectic temperature is associated with a solidification rate through Eq 5.2 (that is, the lower the temperature, the higher the solidification rate). The coupled zone (a shaded region) then represents the solidification rate dependent composition region in which the eutectic grows more rapidly (or at a lower undercooling) than α- or β-phase dendrites. This zone corresponds to an entirely eutectic microstructure. Outside the coupled zone, the microstructure consists of primary dendrites and interdendritic eutectic.

The coupled zone of a regular eutectic system is shown in Fig. 5.23(a). In this case, a purely eutectic microstructure is obtained at the eutectic composition for all growth conditions. However, in the case of the skewed coupled zone of an irregular eutectic (Fig. 5.23b, where β is the faceted phase), the alloy composition must be carefully chosen as a function of the growth rate imposed by the casting process if a completely eutectic microstructure is required. For example, the composition of cast iron or aluminum-silicon alloys must often be hypereutectic if one wants to eliminate metal dendrites, especially when using high-solidification-rate casting techniques.

5.6 Terminal Solid Solutions

The terminal solid-solution Al-4% Cu alloy shown in Fig. 5.24 is completely a solid solution of α above 480 °C (895 °F). If the alloy is brought to equilibrium by subjecting it to a long heat treatment at a temperature of approximately 510 °C (950 °F), it will be composed of the single α phase. During this homogenization heat treatment, equilibrium is approached in two ways simultaneously: (1) coring segregation is eliminated by a redistri-

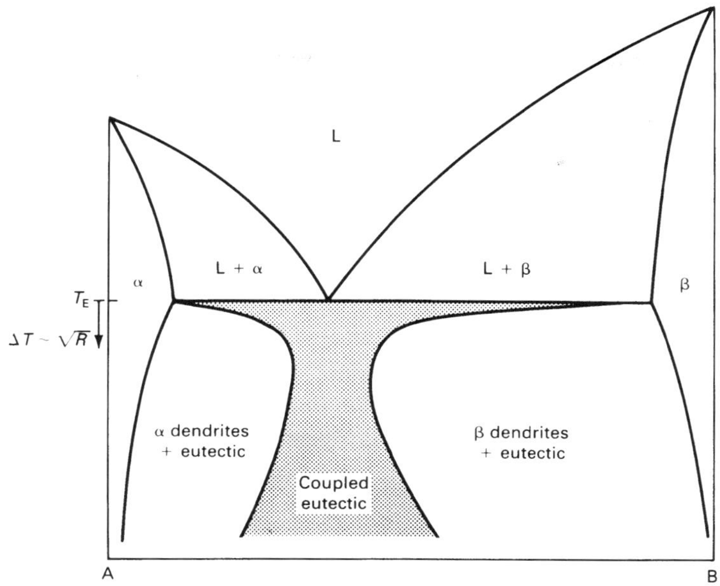

(a)

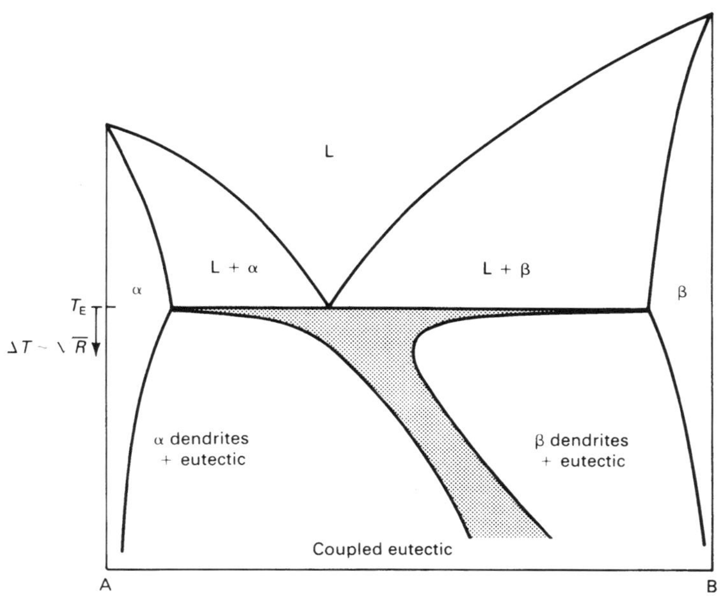

(b)

Fig. 5.23 Coupled zones (shaded regions) on eutectic phase diagrams. The coupled zones represent the interface temperature (solidification rate) dependent composition region in which a completely eutectic structure is obtained. (a) Nearly symmetrical coupled zone in regular eutectic. (b) Skewed coupled zone in an irregular eutectic. In both cases, the widening of the coupled zone near the eutectic temperature is observed only in directional solidification (positive thermal gradient). Source: Ref 5.6

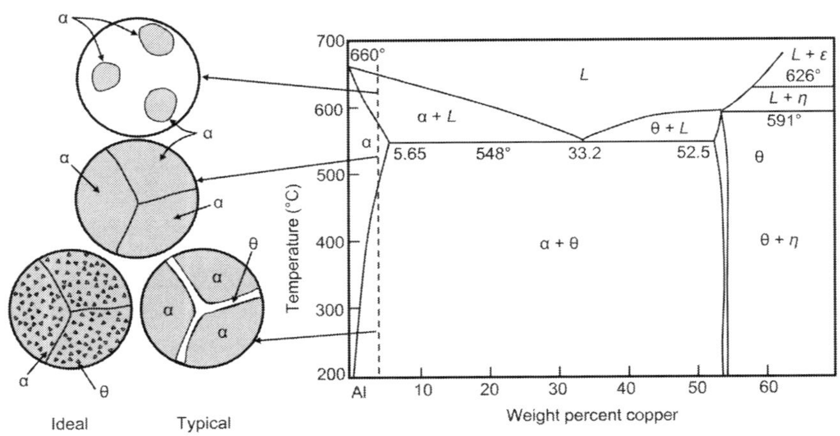

Fig. 5.24 Aluminum-copper phase diagram and the microstructures that may develop during cooling of an Al-4% Cu alloy. Adapted from Ref 5.9

bution of the components, and (2) the θ phase is dissolved in the α phase. When the alloy is slowly cooled from 510 °C (950 °F), it starts rejecting or precipitating θ when it reaches the solvus temperature at 480 °C (895 °F).

However, if the alloy is quenched to room temperature, there is not enough time for the θ phase to precipitate, and the alloy will contain a supersaturated solid solution containing only the α phase. On aging at room temperature or moderately elevated temperatures, such as 150 °C (300 °F), a series of transition phases form that strains the matrix and forms the basis for precipitation hardening, as discussed in Chapter 16, "Nonequilibrium Reactions—Precipitation Hardening," in this book.

The remainder of this section covers precipitation structures formed during slower cooling rates from temperatures in the single-phase region.

General or continuous precipitation refers to the uniform appearance of second-phase particles throughout the grains of the matrix. General precipitation does not imply homogeneous nucleation, rather, the nonlocalized precipitation of the second phase. Discontinuous precipitation can occur at regions such as grain boundaries, or cellular precipitation where precipitation begins at grain-boundary allotriomorphs but does not continue through the entire grain. General precipitation in aluminum alloys is shown in Fig. 5.25. The difference between discontinuous and continuous precipitation in AZ91 (a Mg-Al-Zn alloy) is shown in Fig. 5.26.

Widmanstätten Structures. The continuous precipitation of plate or lathlike structures is referred to as Widmanstätten morphology. This distinctive plate or lathlike morphology is characterized by the presence of both high- and low-angle boundaries. Widmanstätten morphologies form in many alloy systems, as illustrated in Fig. 5.27 for a Ti-6Al-4V alloy. As evident by the microstructure, there are specific orientation relations between the precipitate habit plane and matrix. The long broad faces of

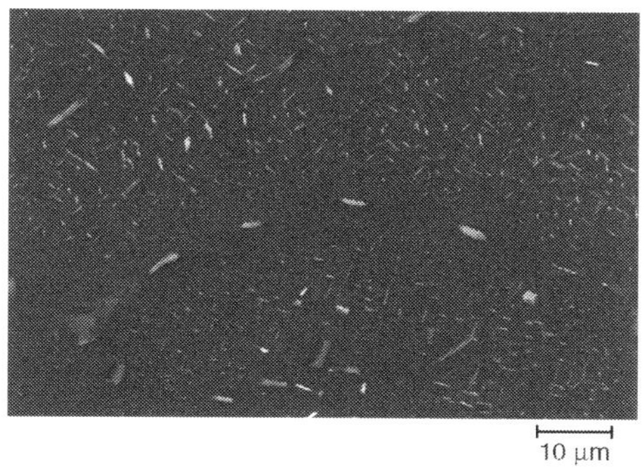

Fig. 5.25 Scanning electron micrograph of continuous precipitation in 6061 aluminum alloy, where the smaller precipitates are Mg2Si, and the larger particles are AlFeSi intermetallics at the grain boundary. Note the precipitate-free zone near the AlFeSi intermetallics. Source: Ref 5.10 as published in Ref 5.11

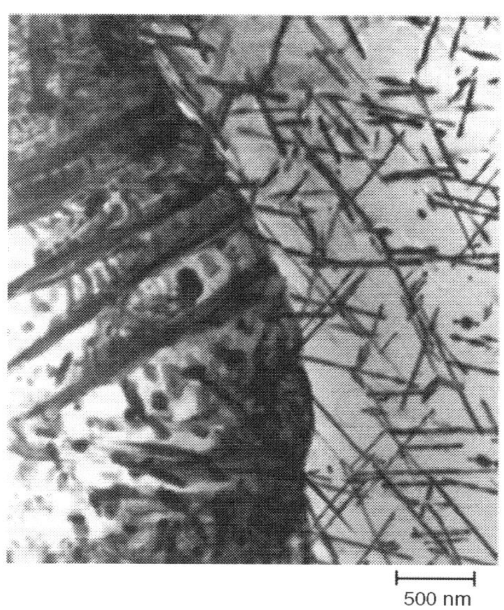

Fig. 5.26 Transmission electron micrograph showing a region of discontinuous (left) and continuous (right) precipitation in a specimen of AZ91 aged at 200 °C (390 °F) for 4 h. Source: Ref 5.12 as published in Ref 5.11

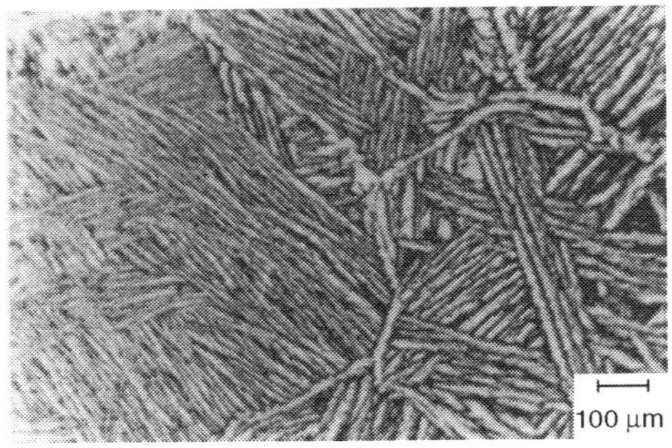

Fig. 5.27 Widmanstätten structure in Ti-6Al-4V alloy cooled at 3.40 °C/s (6.12 °F/s). Source: Ref 5.13 as published in Ref 5.11

the precipitates are the coherent, low-energy interfaces. During growth, small ledges form on these faces, allowing for diffusional thickening while maintaining coherency. Typically, this morphology forms during slow cooling rates, but the Widmanstätten morphology can occur if a sufficient driving force for growth is provided by either a fast cooling rate or large undercooling.

The development of the side-plate morphology normally starts from a grain-boundary allotriomorph (as in the case of steels) and growth occurs into the grain. As cooling rates increase, the diffusion of atoms to the high-angle boundaries (where no special orientation exists between austenite and ferrite) is slow, but if diffusion distances are minimized by phase morphology and growth occurs in preferred crystallographic directions, growth rates can be increased.

Cellular or Discontinuous Precipitation. Grain-boundary precipitation may result in cellular or discontinuous precipitation (DP). Figures 5.28 and 5.29 are examples of the alternating lamellar structure that is common to many cellular precipitation transformations. During discontinuous precipitation, the second phase nucleates at the grain boundary, which then moves with the advancing precipitation reaction. It typically starts at a high-angle incoherent boundary, which is the most likely point to support the process of heterogeneous nucleation and boundary migration. Misfit or atomic mismatch strain are factors, although neither misfit nor atomic mismatch strain appear to be necessary conditions in several instances (e.g., Al-Li, Ni-Al, Ni-Ti, Al-Ag, and Cu-Co). The general conditions and criteria of discontinuous precipitation are not completely understood due to the complex interrelationships among boundary structure, energy, mobility, and diffusivity.

Following nucleation, the grain boundary moves with the advancing precipitation reaction. This is illustrated in Fig. 5.29b, where the grain

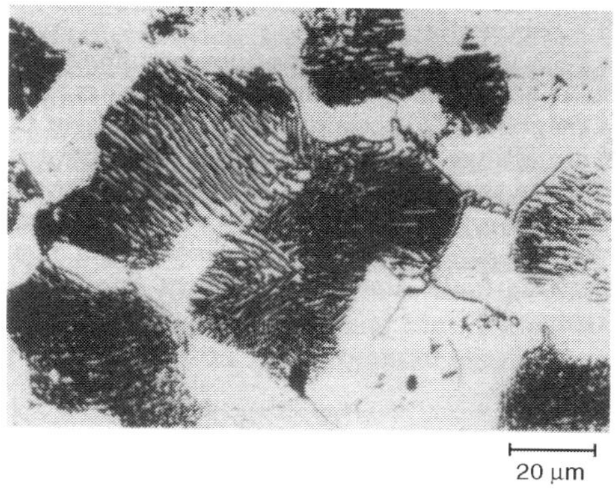

Fig. 5.28 Discontinuous precipitation of β phase ($Mg_{17}Al_{12}$) in cast AZ80 zirconium-free magnesium casting alloy. Source: Ref 5.14 as published in Ref 5.11

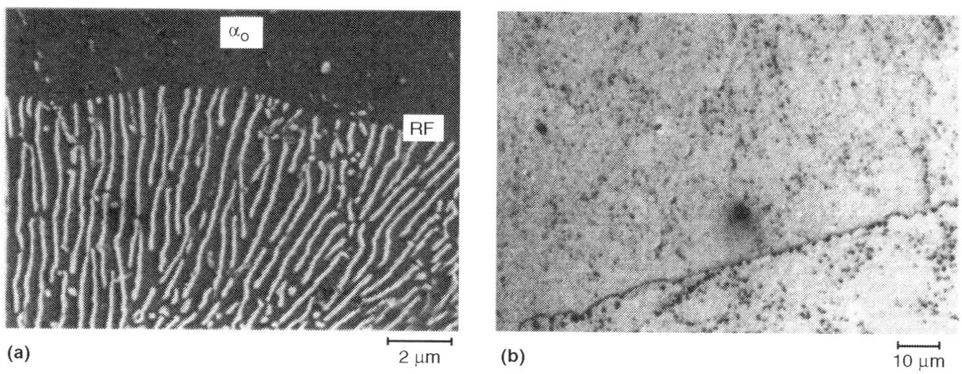

Fig. 5.29 Discontinuous precipitation (DP). (a) Scanning electron micrograph of lamellar structure within a DP cell in Mg-10Al (wt%) annealed at 500 K for 40 min. RF, reaction front. α_0, supersaturated solid solution. (b) Light optical micrograph of early stage of the DP reaction in Mg-10Al (wt%) annealed at 500 K for 20 min. The bowing out of the grain boundary between two allotriomorphs is clearly seen. Source: Ref 5.15 as published in Ref 5.11

boundary is seen bowing out at the beginning of the reaction. Morphologically, it resembles eutectoid decomposition. Discontinuous precipitation results in reaction products that are often lamellar, fibrous, or rodlike, but rarely globular. Discontinuous precipitation is also often referred to as a cellular, grain-boundary, recrystallization, or autocatalytic reaction.

Lamellar spacing of the precipitates is dependent on aging temperature, with wider spacing occurring at higher temperatures. Free energy is limited for creating of interfaces due to the lower driving force (at higher temperature). It is termed discontinuous because the matrix composition changes discontinuously as the cell advances. Cellular precipitation is

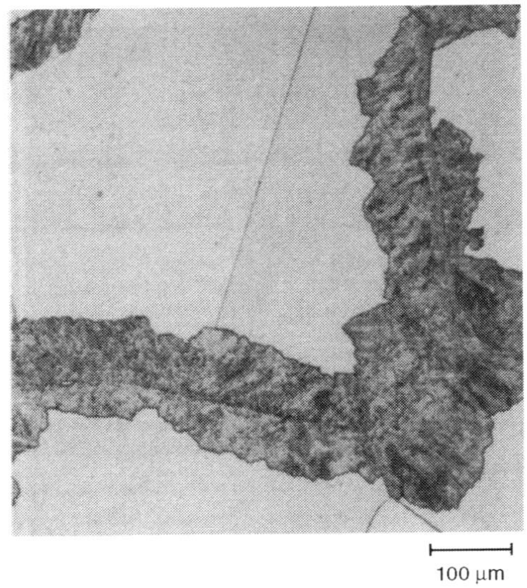

Fig. 5.30 Cellular colonies growing out from grain boundaries in Au-30Ni alloy aged 50 min at 425 °C (795 °F). Etchant: 50 mL 5% ammonium persulfate and 50 mL 5% potassium cyanide. Original magnification: 100×. Source: Ref 5.16 as published in Ref 5.11

observed in Fig. 5.30, clearly showing the relationship between the cells and the grain boundaries.

ACKNOWLEDGMENT

The material for this chapter came from "Solidification of Eutectics," by P. Magnin and W. Kurz in *Casting*, Vol 15, *ASM Handbook*, ASM International, 1988; "Structures by Precipitation from Solid Solution," by M. Epler, in *Metallography and Microstructures*, Vol 9, *ASM Handbook*, ASM International, 2004; and *Elements of Metallurgy and Engineering Alloys* by F.C. Campbell, ASM International, 2008.

REFERENCES

5.1 H. Baker, Introduction to Alloy Phase Diagrams, *Phase Diagrams*, Vol 3, *ASM Handbook*, ASM International, 1992, reprinted in *Desk Handbook: Phase Diagrams for Binary Alloys*, 2nd ed., H. Okamoto, Ed., ASM International, 2010

5.2 F.C. Campbell, *Elements of Metallurgy and Engineering Alloys*, ASM International, 2008

5.3 A.G. Guy, *Elements of Physical Metallurgy*, 2nd ed., Addison-Wesley Publishing Company, 1959

5.4 E.L. Rooy and A. Kearney, Aluminum Foundry Products, *Properties and Selection: Nonferrous Alloys and Special-Purpose Materials*, Vol 2, *ASM Handbook*, ASM International, 1990

5.5 W.F. Smith, *Principles of Materials Science and Engineering*, McGraw-Hill, 1986, p 408
5.6 P. Magnin and W. Kurz, Solidification of Eutectics, *Casting*, Vol 15, *ASM Handbook*, ASM International, 1988
5.7 W. Kurz and D.J. Fisher, *Fundamentals of Solidification*, Trans Tech Publications, 1984
5.8 P. Magnin and W. Kurz, *Acta Metall.*, Vol 35, 1987, p 1119
5.9 D.R. Askeland, *The Science and Engineering of Materials*, 2nd ed., PWS-KENT Publishing, 1989, p 318
5.10 S.R. Claves, D.L. Elias, and W.Z. Misiolek, Analysis of the Intermetallic Phase Transformation Occurring During Homogenization of 6xxx Aluminum Alloys, *Proc. International Conference of Aluminium Alloys 8,* Trans Tech Publications, 2002
5.11 M. Epler, Structures by Precipitation from Solid Solution, *Metallography and Microstructures,* Vol 9, *ASM Handbook*, ASM International, 2004, p 134–139
5.12 S. Celotto and T.J. Bastow, Study of Precipitation in Aged Binary Mg-Al and Ternary Mg-Al-Zn Alloys Using 72AlNMR Spectroscopy, *Acta Mater.,* Vol 49, 2001, p 41–51
5.13 F.J. Gil, M.P. Ginebra, J.M. Manero, and J.A. Planell, Formation of Widmanstätten Structure: Effects of Grain Size and Cooling Rate on the Widmanstätten Morphologies and on the Mechanical Properties in Ti6Al4V Alloy, *J. Alloy. Compd.,* Vol 329 (No. 1–2), 2001, p 142–152
5.14 I.J. Polmear, *Light Alloys—Metallurgy of the Light Metals,* 3rd ed., Arnold, 1995
5.15 D. Bradai, P. Zieba, E. Bischoff, and W. Gust, Correlation between Grain Boundary Misorientation and the Discontinuous Precipitation Reaction in Mg-10 wt.% Al Alloy, *Mater. Chem. Phys.,* Vol 78 (No. 1), 2003, p 222–226
5.16 W.A. Soffa, Structures Resulting from Precipitation from Solid Solution, *Metallography and Microstructures,* Vol 9, *ASM Handbook,* ASM International, 1985, p 646–650

CHAPTER 6

Peritectic Alloy Systems

SIMILAR TO THE EUTECTIC group of invariant transformations is a group of peritectic reactions, in which a liquid and solid phase decomposes into a new solid phase on cooling through the peritectic isotherm. The generalized form of this peritectic reaction is:

$$L + \alpha \rightarrow \beta \text{ (peritectic reaction)}$$

Because the β phase surrounds the solid α particles, as shown in Fig. 6.1, α atoms must diffuse through the β crust to reach the liquid for the reaction to continue. However, diffusion through the solid phase is much slower than diffusion through a liquid. As the peritectic reaction continues, the α layer gets thicker and the reaction slows down even more. Unless the cooling rate is very slow, a cored segregated structure will result. Thus, equilibrium peritectic reactions are almost never observed in practice.

Actually, the term *peritectic*, in the science of heterogeneous equilibria, has a broader meaning. The term may be used to define all reactions in which two or more phases (gas, liquid, solid) react at a defined temperature (T_p) to form a new phase that is stable below T_p. Usually, the term *peritectic*

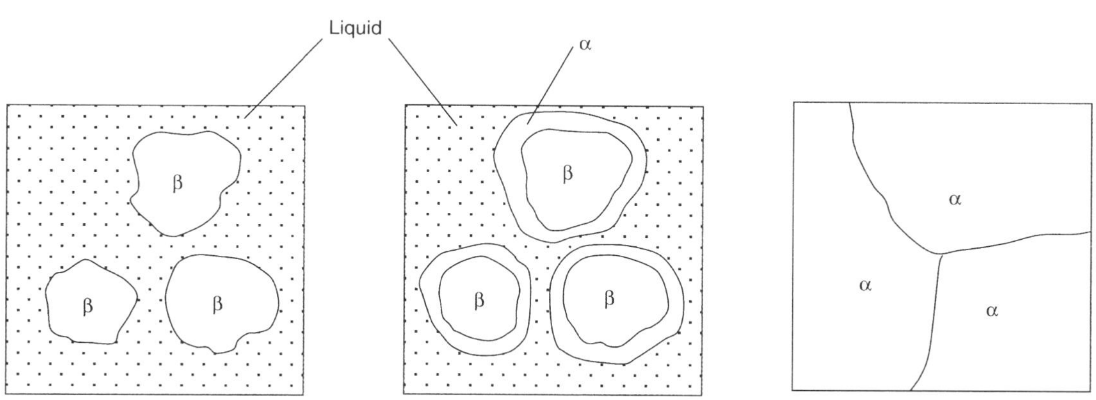

Fig. 6.1 Schematic of peritectic freezing

refers to reactions in which a liquid phase reacts with at least one solid phase to form one new solid phase. This reaction can be written as $\alpha + L \rightarrow \beta$. Furthermore, the term *peritectoid* denotes the special case of an equilibrium phase in which two or more solid phases (which are stable above the temperature T_p) react at T_p to form a new solid phase. This reaction can be written as $\alpha + \beta \rightarrow \gamma$. The phases formed during a peritectic or peritectoid reaction are a solid solution of one of the components, an allotropic phase of one of the components, or an intermetallic compound. Schematics of different types of peritectic phase diagrams are shown in Fig. 6.2.

Phase diagrams are very instructive when describing peritectic phase transformations.

Figure 6.3 shows a phase diagram with a peritectic reaction. This diagram shows that, under equilibrium conditions, all alloys to the left of I will solidify to α crystals. Similarly, all alloys to the right of III will

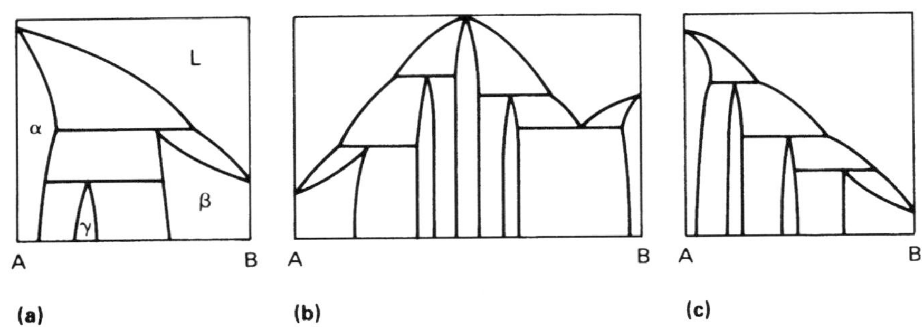

Fig. 6.2 Typical peritectic phase diagrams. (a) Peritectic reaction α + liquid $\rightarrow \beta$ and peritectoid reaction $\alpha + \beta \rightarrow \gamma$. (b) Peritectic formation of intermetallic phases from a high-melting intermetallic. (c) Peritectic cascade between high- and low-melting components. Adapted from Ref 6.1

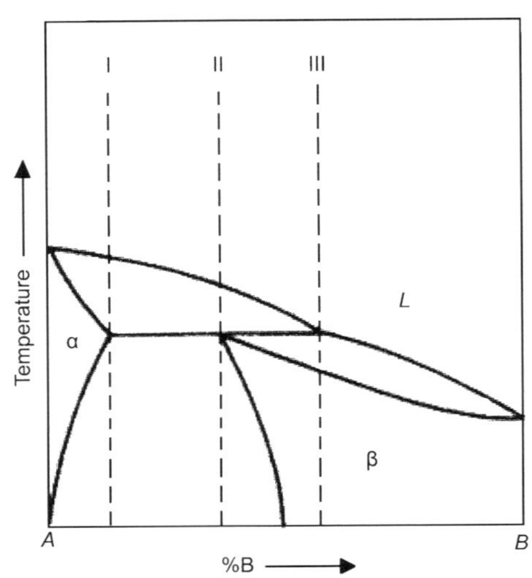

Fig. 6.3 Phase diagram with a peritectic reaction. Source: Ref 6.2

solidify to β crystals. Alloys between II and III first solidify to α crystals and then transform to stable β crystals. Alloys between I and II also solidify to α crystals, but they are partially transformed to β crystals later. The volume fraction of each phase is determined with the lever rule if the alloy solidifies under equilibrium conditions. In practice, the lever rule usually will not give the volume fraction of the different phases, because the nucleation and growth kinetics, determined by the diffusion rate, in the solid phases determine the time required to reach equilibrium.

6.1 Freezing of Peritectic Alloys

Equilibrium Freezing of Peritectic Alloys. Consider the course of freezing of the peritectic Alloy 1 in Fig. 6.4. On cooling from the melt, the liquidus is reached at temperature T_2, where crystals of the α phase (α_2) begin to form. The liquid composition is displaced to the right as the temperature falls, and more solid is deposited. The composition of the α phase changing along the solidus reaches the end of the peritectic line, α_4, when the liquid composition has reached the opposite end, L_4:

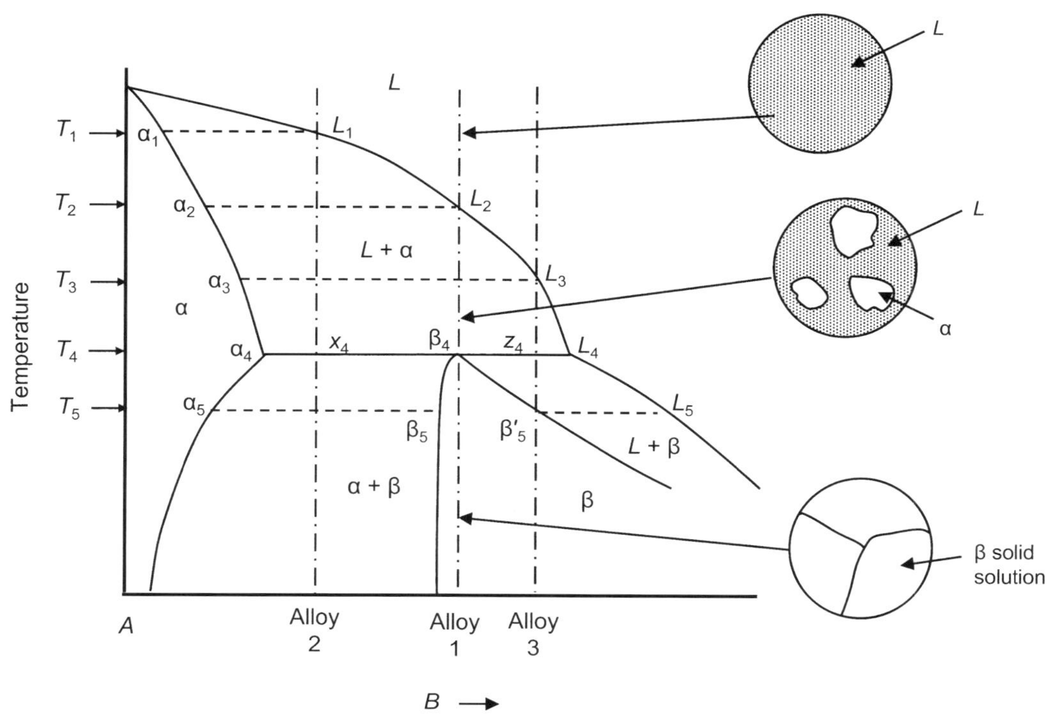

Fig. 6.4 Hypothetical peritectic phase diagram. Adapted from Ref 6.3

$$\% \, \alpha \text{ (Alloy 1 above peritectic)} = \left(\frac{\beta_4 L_4}{\alpha_4 L_4}\right)(100) \approx 35\%$$

$$\% \, L \text{ (Alloy 1 above peritectic)} = \left(\frac{\alpha_4 \beta_4}{\alpha_4 L_4}\right)(100) \approx 65\%$$

The peritectic reaction now occurs; the liquid and α react together to form β Under equilibrium conditions, freezing must be completed isothermally by this process, and all the previously formed α as well as the liquid must be consumed.

If the alloy composition was at the left of the peritectic point, as Alloy 2 in Fig. 6.4, the α phase would not have been totally consumed in forming β. In this alloy, freezing begins at T_1, where α_1 first appears. Just above the peritectic temperature the quantities of α and liquid will be:

$$\% \, \alpha \text{ (Alloy 2 above peritectic)} = \left(\frac{x_4 L_4}{\alpha_4 L_4}\right)(100) \approx 80\%$$

$$\% \, L \text{ (Alloy 2 above peritectic)} = \left(\frac{\alpha_4 x_4}{\alpha_4 L_4}\right)(100) \approx 20\%$$

Just below the peritectic temperature, the liquid will have disappeared and there will be only the solid phases α and β:

$$\% \, \alpha \text{ (Alloy 2 below peritectic)} = \left(\frac{x_4 \beta_4}{\alpha_4 \beta_4}\right)(100) \approx 70\%$$

$$\% \, \beta \text{ (Alloy 2 below peritectic)} = \left(\frac{\alpha_4 x_4}{\alpha_4 \beta_4}\right)(100) \approx 30\%$$

Notice that the quantity of the α phase has decreased during peritectic reaction from approximately 80% to approximately 70%.

If the composition of the alloy is to the right of the peritectic point, Alloy 3 in Fig. 6.4, then an excess of liquid would have survived the peritectic reaction. Freezing in this case begins at T_3, where α_3 is first rejected. There is a small quantity of the α phase present by the time the peritectic temperature is reached:

$$\% \, \alpha \text{ (Alloy 3 above peritectic)} = \left(\frac{z_4 L_4}{\alpha_4 L_4}\right)(100) \approx 10\%$$

$$\% \, L \, (\text{Alloy 3 above peritectic}) = \left(\frac{\alpha_4 z_4}{\alpha_4 L_4} \right) (100) \approx 90\%$$

But this is entirely consumed in forming the β phase:

$$\% \, \beta \, (\text{Alloy 3 below peritectic}) = \left(\frac{z_4 L_4}{\beta_4 L_4} \right) (100) \approx 30\%$$

$$\% \, L \, (\text{Alloy 3 below peritectic}) = \left(\frac{\beta_4 z_4}{\beta_4 L_4} \right) (100) \approx 70\%$$

Observe that the quantity of liquid has been reduced to approximately 70% and that the quantity of β formed is much greater than that of the α originally present.

Nonequilibrium Freezing of Peritectic Alloys. As pointed out earlier, the departure from equilibrium during the natural freezing of peritectic alloys is usually very large. Referring back to Fig. 6.1, the β phase, in forming from the liquid and α phases, surrounds or encases the α-phase particles. This surrounding or encasement shields the α phase from further reaction with the liquid, and diffusion in the solid phases is usually insufficient to allow equilibrium to be established during cooling. If the solidified alloy is subjected to a homogenization heat treatment followed by hot working, the diffusion process is accelerated and equilibrium may be established.

6.2 Mechanisms of Peritectic Formation

Peritectic reactions or transformations are very common in the solidification of metals. Many interesting alloys undergo these types of reactions—for example, iron-carbon and iron-nickel-base alloys as well as copper-tin and copper-zinc alloys. The formation of peritectic structures can occur by at least three mechanisms:

- *Peritectic reaction,* where all three phases (α, β, and liquid) are in contact with each other
- *Peritectic transformation,* where the liquid and the α solid-solution phase are isolated by the β phase. The transformation takes place by long-range diffusion through the secondary β phase.
- *Direct precipitation of* β *from the melt,* when there is enough undercoating below the peritectic temperature, T_p. Direct precipitation of β from the melt can occur when a peritectic reaction or peritectic transformation is sluggish, as is often the case.

The three classes of peritectic phase diagram based on the shape of the peritectic (β) solid-solution region are shown in Fig. 6.5.

Peritectic reactions can proceed only as long as α and liquid are in contact. The β-phase solid nucleates at the α-liquid interface and readily forms a layer isolating α from the liquid. This mechanism occurs from short-range diffusion (Fig. 6.6a). In contrast, the term *peritectic transformation* is used to describe a mechanism of long-range diffusion, where A atoms and B atoms migrate through the α layer to then form the β-phase solid at the α-β and the β-liquid interfaces, respectively (Fig. 6.6b).

From the final microstructure, it is not apparent by which mechanism β has formed. In any case, all three of the mechanisms require some undercooling, because the driving force is zero at the peritectic temperature. The time dependence is pronounced for the peritectic transformation. Therefore, the amount of β phase formed will depend on the cooling rate or on holding time if isothermal conditions are established.

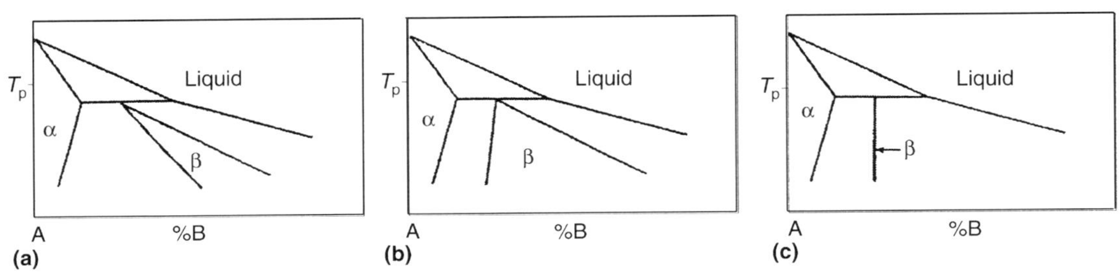

Fig. 6.5 Types of peritectic systems. (a) Type A system where the β/α + β solvus and the β solidus have slopes of the same sign. (b) Type B system where the slopes have opposite signs. (c) Type C system where the β phase has a limited composition. Source: Ref 6.1 as published in Ref 6.4

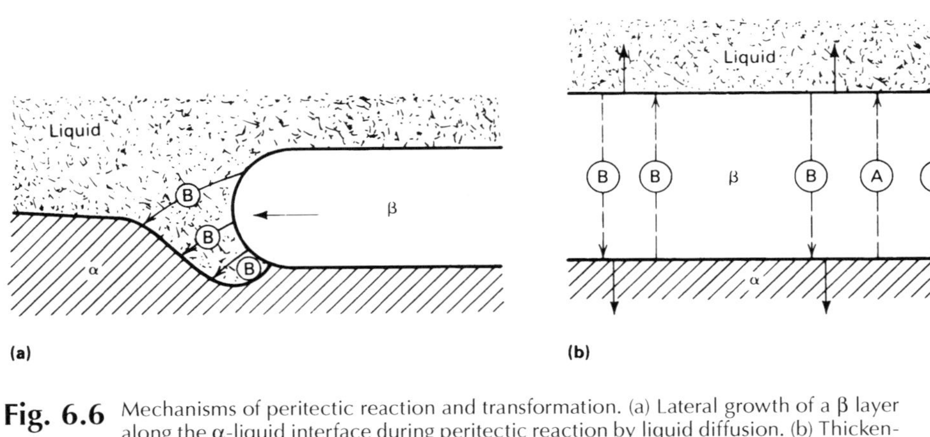

Fig. 6.6 Mechanisms of peritectic reaction and transformation. (a) Lateral growth of a β layer along the α-liquid interface during peritectic reaction by liquid diffusion. (b) Thickening of a β layer by solid-state diffusion during peritectic transformation. The solid arrows indicate growth direction of β; dashed arrows show the diffusion direction of the atomic species. Ref 6.5 as published in Ref 6.4

Localized nucleation of β and shape changes of β by solution reprecipitation is driven by surface energy through diffusion in the liquid and influences the microstructure of peritectic alloys. Typical and specific microstructures are discussed in the following sections. For demonstration purposes, results are included from experimental alloys that were cooled rapidly from above the liquidus to a temperature above the peritectic equilibrium temperature, T_p, held for some time to achieve large homogeneous primary α crystals, then cooled to a temperature below T_p and held for extended times, inducing the formation of β by peritectic reaction and transformation, rather than by direct precipitation from the melt.

Nucleation-Controlled Peritectic Structures. The classical description of peritectic reactions postulates heterogeneous nucleation of β at the α-liquid interface at the peritectic equilibrium temperature, T_p. Undercoolings of up to 4% of T_p are required for the systems investigated. If nucleation is limited to a few locations and lateral growth of the β nuclei does not readily occur, no continuous layer of the peritectic phase is formed, as in nickel-zinc and aluminum-uranium systems. A typical microstructure is shown in Fig. 6.7. Small crystals of the peritectic phase nucleate at the interface and grow into the primary crystals. After extended annealing below T_p, the reaction goes to completion (Fig. 6.8). The original shape of

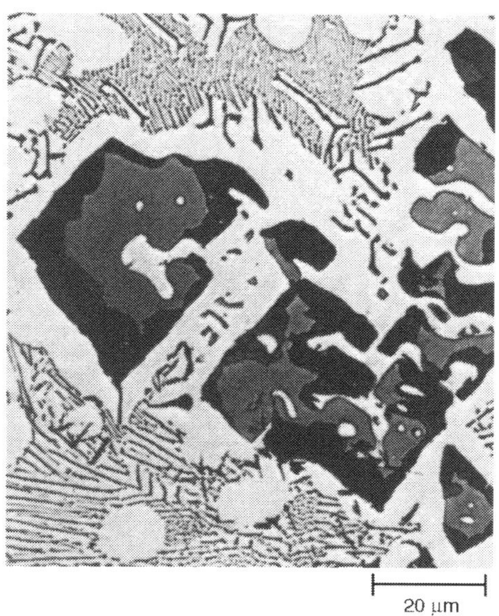

Fig. 6.7 Primary UAl_3 (gray) partially surrounded by peritectically-formed UAl_4 (dark) in an Al-6U alloy that was cooled slowly from above liquidus to 760 °C (1400 °F) and held 10 min, then cooled to 670 °C (1240 °F) and held 15 min (peritectic temperature: 732 °C, or 1350 °F). The matrix is aluminum (white) with UAl_4 (dark) eutectic. This $UAl_3 + Al \rightarrow UAl_4$ reaction leads to unfavorable rolling behavior. Original magnification: 700×. Source: Ref 6.4

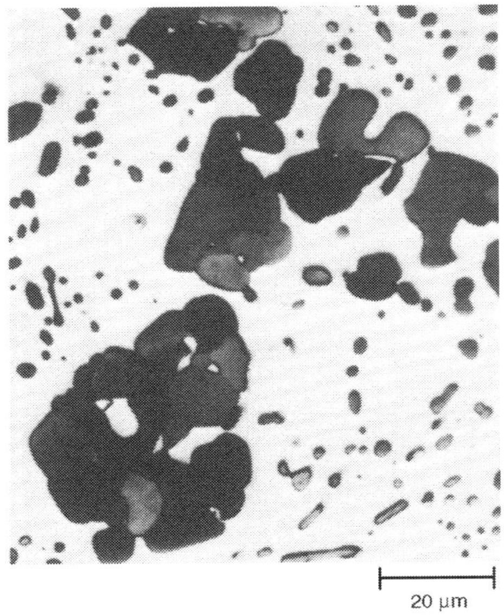

Fig. 6.8 Peritectically-formed UAl_4 in an Al-6U alloy that was cooled from above liquidus to 760 °C (1400 °F) and held 10 min, then cooled to 600 °C (1110 °F) and held 7 days (peritectic temperature: 732 °C, or 1350 °F; eutectic temperature: 640 °C, or 1184 °F). Note the rounded crystals and the necking between crystals of different orientation. The matrix is aluminum (white) with coarsened eutectic UAl_4 (dark). Original magnification: 700×. Source: Ref 6.1 as published in Ref 6.4

the primary crystals is lost by decay into individual β crystals that coarsen by Ostwald ripening during further annealing.

The peritectic reaction in the aluminum-uranium system was of special interest in the production of fuel elements for early nuclear reactors. The reaction UAl_3 + Al (liquid or solid) → UAl_4 is sluggish. In Fig. 6.7, this reaction could not be completely suppressed during the cooling cycles used, resulting in unfavorable rolling behavior. Additions of silicon or zirconium stabilize UAl_3, extending the UAl_3-Al equilibrium region below room temperature.

Peritectic Reactions. Depending on surface tension conditions, two different types of the peritectic reactions can occur:

- Nucleation and growth of the β crystals in the liquid without contact with the α crystals
- Nucleation and growth of the β crystals in contact with the primary α phase

In the first case, the secondary phase is nucleated in the liquid and does not contact the primary phase. This occurs because of the surface tension conditions. Following nucleation, the secondary phase grows freely in the liquid. At the same time, the primary phase will dissolve. The secondary phase will not develop a morphology similar to a precipitated primary phase. This type of peritectic reaction has been observed for the reaction $\gamma + L \rightarrow \beta$ in the aluminum-manganese system. There has also been a tendency for the secondary phase to grow around the primary phase at increasing cooling rates. Similar reactions have been observed in nickel-zinc, and aluminum-uranium systems. In the second type of reaction, which is the most common, nucleation of the secondary β phase occurs at the interface between the primary α phase and the liquid. A lateral growth of the β phase around the α phase then takes place.

In an ideal peritectic reaction, undercooling is rather low (up to a few degrees Kelvin), and a plateau is observed in the cooling curve, as in the aluminum-titanium system. Envelopes of the peritectic phase around the primary phase form by direct reaction in some systems—copper-tin and silver-tin, for example—through interrupted directional solidification experiments. Figure 6.9 shows the microstructures of a Cu-20Sn alloy that demonstrates the onset of the peritectic reactions α + liquid $\rightarrow \beta$ and ε + liquid $\rightarrow \eta$, respectively. Figure 6.10 shows, at a higher magnification, the homogeneous thickness of the β layer around the α dendrites. The peritectic reaction can proceed very rapidly by liquid diffusion over a very short distance in the lateral direction, as shown in Fig. 6.6(a). The thickness of the layers has been calculated with fairly good agreement to experimental results for copper-tin and silver-tin alloys on the basis of maximum growth rate or minimum undercooling from the laws derived for solidification at low undercoolings. The thickness depends to some extent on the cooling rate, but more strongly on the interfacial energies, σ, with σ (liquid-β) + σ (α-β) – σ (liquid-α) as the determining factor.

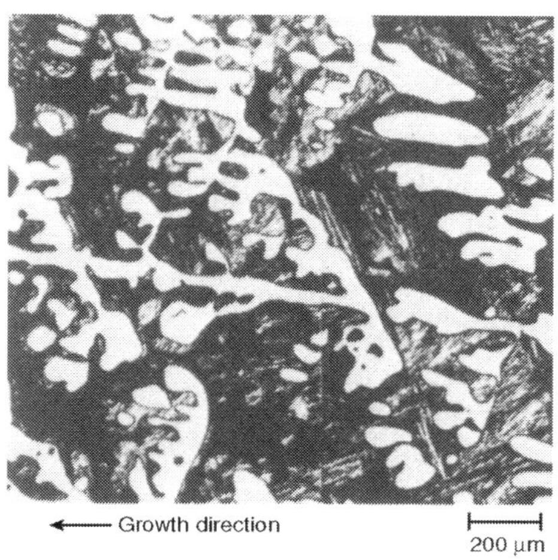

Fig. 6.9 Start of the peritectic reaction in a directionally-solidified Cu-20Sn alloy. Primary α dendrites (white) are covered by peritectically-formed β layer (gray) shortly after the temperature reaches T_p. Matrix (dark) is a mixture of tin-rich phases. Original magnification: 40×. Source: Ref 6.6 as published in Ref 6.4

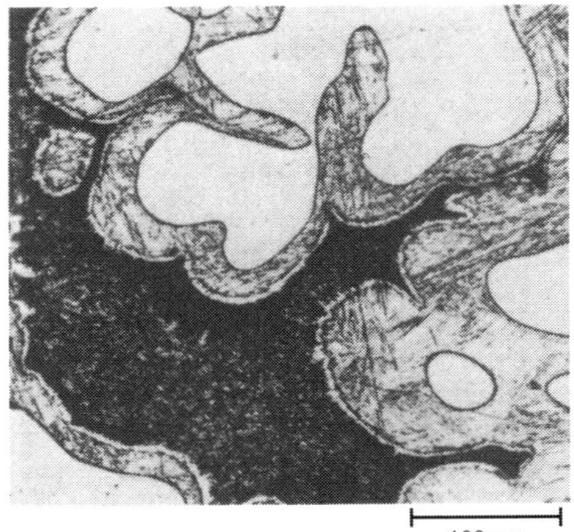

Fig. 6.10 Start of the peritectic transformation in the same directionally-solidified Cu-20Sn alloy shown in Fig. 6.9, but at higher magnification. Note the homogeneous thickness of the β layers (gray) around the primary α (white). The matrix (dark) is a mixture of tin-rich phases. Original magnification: 160×. Source: Ref 6.6 as published in Ref 6.4

Peritectic Transformations. The precipitation of β directly from the liquid and the solid depends on the shape of the phase diagram and the cooling rate. After isolation of primary α from the liquid by the β layer, the direct peritectic reaction can no longer take place. The diffusion process through the β layer depends on the diffusion rate, the shape of the phase diagram, and the cooling rate. The thickness of the β layer will also normally increase during subsequent cooling. There are three reasons for this:

- Diffusion through the β layer
- Precipitation of β directly from the liquid
- Precipitation of β directly from the α phase

The thickness of the β phase envelope surrounding the α phase is determined by the peritectic reaction followed by an increase in thickness due to a precipitation directly from the liquid. The rate of the peritectic transformation is influenced by the diffusion rate and the extension of the β phase region in the phase diagram. If the diffusion rate is small, the peritectic transformation will be negligible compared to the peritectic reaction. During continuous cooling, this diffusional growth is affected by precipitation from the liquid and from the primary α or by dissolution of β, according to the slopes of the solubility limits in the phase diagram. Under simplifying conditions, the growth rates have been calculated numerically and found to be in reasonable agreement with the experimental findings in the copper-tin and silver-tin systems. The kinetics of peritectic transformations can be studied under isothermal conditions. Then, β is formed exclusively by diffusion of the two atomic species in the β layer at the α-β and the β-liquid interfaces.

A typical microstructure from a peritectic transformation is shown in Fig. 6.11. Similar results have been obtained in the copper-tin system for the transformation ε + liquid → η and for peritectic transformations in the cobalt-tin, gold-bismuth, and chromium-antimony systems. In contrast, two maxima of the thickness of the peritectically-formed $CuCd_3$ were observed in the copper-cadmium system, as shown in Fig. 6.12. The first maximum is attributed to the contribution of grain-boundary diffusion, which is small for the faceted, large grains formed at high temperatures (Fig. 6.13) and large for the fine-grained, smooth layer formed below the eutectic temperature (Fig. 6.14).

Peritectic Cascades. In many systems, one peritectic reaction at a high temperature is followed by one or more peritectic reactions at lower temperatures. If the diffusion rate is low in the initially formed peritectic layer, a second peritectic layer can be formed when the second peritectic temperature is reached. This type of series of peritectic reactions is referred to as a cascade. The individual thicknesses depend on the growth rate and the rate of consumption by other growing phases, that is, on the diffusivities and the molar volumes of the individual phases.

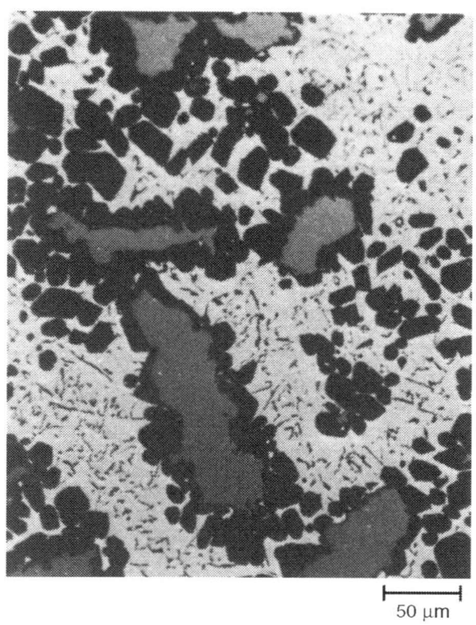

Fig. 6.11 Peritectic transformation of an Sb-14Ni alloy that was slowly cooled to 650 °C (1200 °F) and held 1 h, then cooled to 615 °C (1140 °F) and held 10 min (peritectic temperature: 626 °C, or 1159 °F). An irregular layer of $NiSb_2$ crystals (dark) is formed around the coarse primary NiSb crystals. The matrix is the coarsened $NiSb_2$-Sb eutectic. Original magnification: 200×. Source: Ref 6.1 as published in Ref 6.4

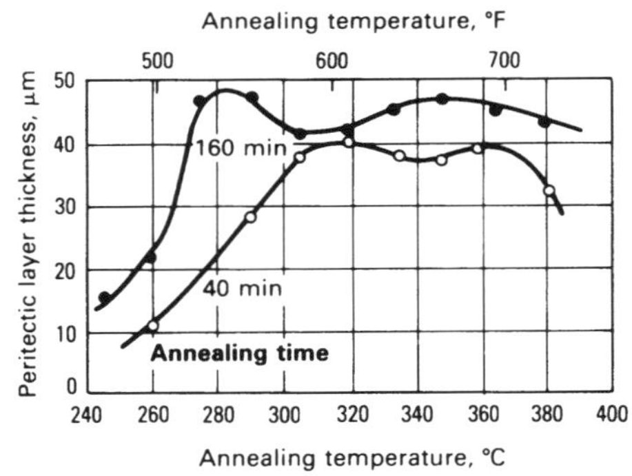

Fig. 6.12 Temperature dependence of the peritectic transformation Cu_5Cd_8 + liquid → $CuCd_3$ in a Cd-10Cu alloy at 40 and 160 min isothermal annealing. Source: Ref 6.7 as published in Ref 6.4

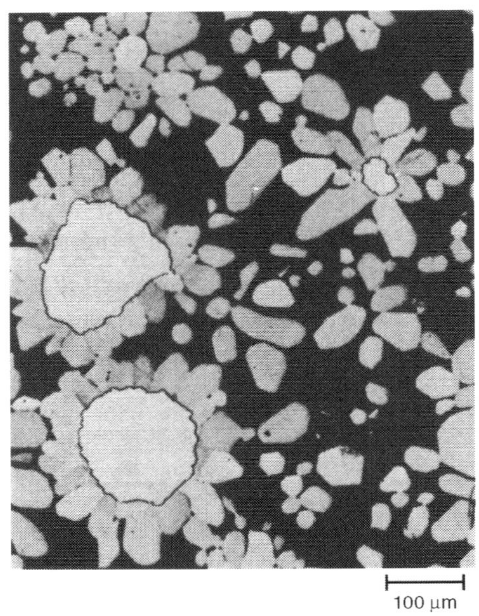

Fig. 6.13 Microstructure of a Cd-10Cu alloy that was cooled to 410 °C (770 °F) and held 20 h, then cooled to 305 °C (580 °F) and held 160 min (peritectic temperature: 397 °C, or 747 °F). Note the faceted coarse crystals of the peritectically-formed $CuCd_3$ envelopes (gray). The primary Cu_5Cd_8 crystals are white; the dark matrix is cadmium. Original magnification: 100×. Source: Ref 6.1 as published in Ref 6.4

For example, the binary tin-antimony systems exhibit a peritectic cascade. In this case, it is possible to get layers of each phase around the initial pro-peritectic phase, as shown in Fig. 6.15. The cascade in the phase diagram for this alloy includes a peritectic and a peritectoid transformation. As shown in a detailed study of the zirconium-aluminum system, peritectoid reactions and transformations follow the same principles as

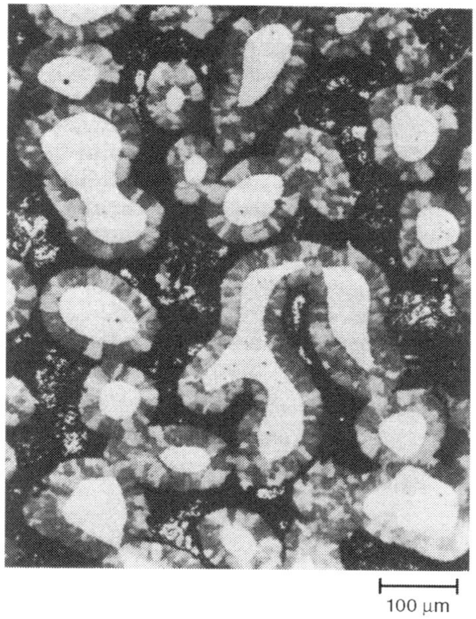

Fig. 6.14 Same as Fig. 6.13, except alloy was cooled to 410 °C (770 °F) and held 20 h, then cooled to 275 °C (525 °F) and held 160 min (peritectic temperature: 397 °C, or 747 °F; eutectic temperature: 314 °C, or 597 °F). Note large number of grain boundaries in the peritectic $CuCd_3$ phase (gray) and its smooth interfaces with the primary Cu_5Cd_8 crystals (white) and the matrix Cd (dark). Original magnification: 100×. Source: Ref 6.1 as published in Ref 6.4

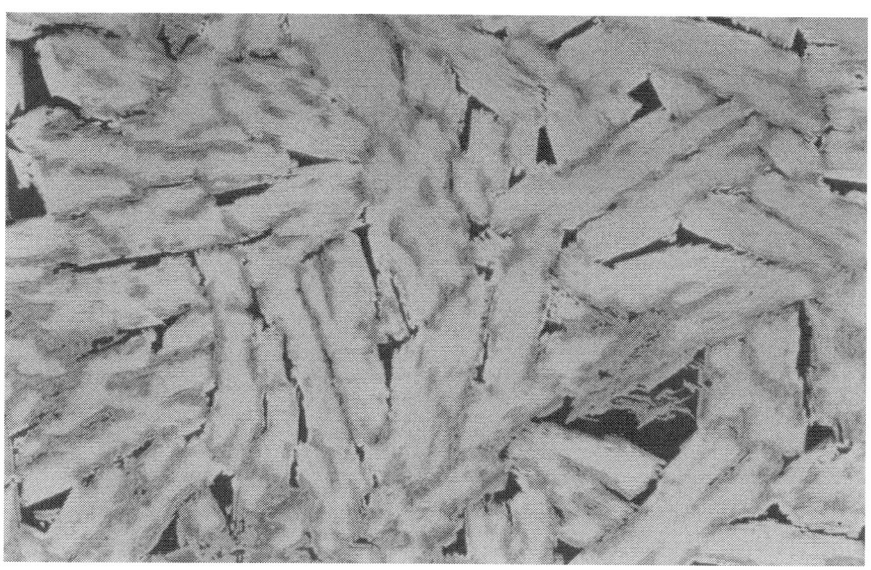

Fig. 6.15 Microstructure of a Sn-50wt%Sb alloy. The primary β phase is light, surrounded by a dark gray structure that was originally Sb_2Sn_3 but has decomposed to β + Sn. Original magnification: 50×. Source: Ref 6.4

the peritectic ones. The theoretical analysis is confirmed by experimental results of the reaction $Zr + Zr_2Al \rightarrow Zr_3Al$, with a parabolic growth dependence and a maximum growth rate approximately 100 K below the peritectoid temperature.

6.3 Peritectic Structures in Iron-Base Alloys

From a technical viewpoint, the peritectic formation of austenite, γ, from primary ferrite, δ, is the most important peritectic reaction. Thermal analysis indicates that the reaction $\delta + L \rightarrow \gamma$ proceeds to a great extent during continuous cooling, and δ-ferrite usually disappears completely on cooling into the austenite region if not stabilized by alloying additions. Phase diagrams of iron with an austenite-stabilizing element (such as carbon, nitrogen, nickel, and manganese) always show a peritectic reaction. In most steels, austenite- and ferrite-stabilizing elements are present, as in stainless and high-speed steels for which peritectic formation of γ has been studied in detail.

A directionally-solidified high-speed steel with peritectic γ envelopes around the highly branched δ dendrites is shown in Fig. 6.16. The varying

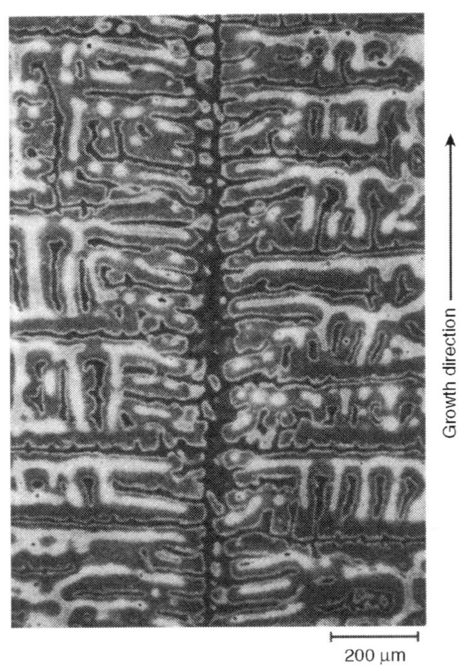

Fig. 6.16 Longitudinal section through directionally-solidified high-speed steel (AISI T1) that was cooled at 0.23 K/s from above liquidus. The peritectic envelopes of austenite (gray) around the highly branched dendrites of δ-ferrite (discontinuously transformed to austenite and carbide, dark) are clearly distinguishable. The matrix is fine ledeburite (white). Original magnification: 60×. Source: Ref 6.8 as published in Ref 6.4

thickness of the layers on the front and the back of the secondary dendrite arms are shown in Fig. 6.17. This is attributed to the wandering of the arms toward the tip of the dendrite, that is, in the solidification direction, during directional solidification due to temperature gradient zone melting. Under some cooling conditions, the layers appear to be partially missing on the back. This and other mechanisms related to the peritectic transformation complicate the interpretation of the microstructures found in high-speed steels and their weldments.

6.4 Multicomponent Systems

Alloys often consist of more than two alloying elements. However, very little information is given in the literature about the peritectic reaction in multicomponent alloys. Recent investigations of iron-base alloys have shown that peritectic reactions are very common in stainless steels. The peritectic reaction in these alloys gives the type of distribution shown in Fig. 6.18. In stainless steels, the peritectic reaction will transfer to a eutectic reaction if the chromium content is increased to 20% or more. This transition is also influenced by the molybdenum content (Fig. 6.19).

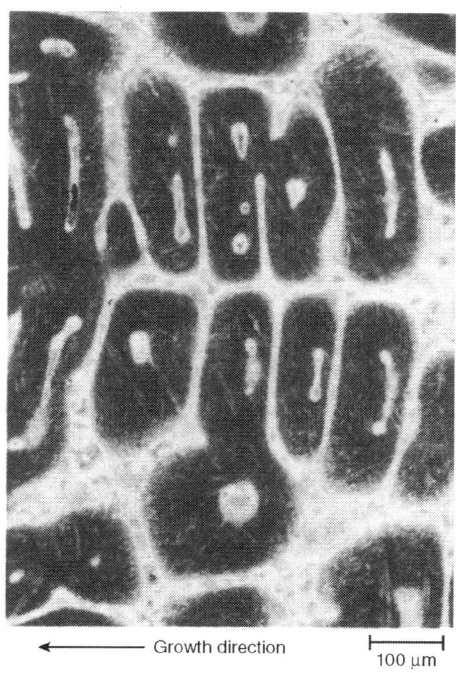

Fig. 6.17 Longitudinal section through directionally-solidified high-speed steel (AISI M2 with 1.12% C and 1% Nb) that was cooled at 0.1 K/s to approximately 1320 °C (2410 °F), that is, 20 K below the onset of the peritectic transformation. Note the thicker layers of peritectic austenite on the front faces of the secondary dendrites compared to the back. Original magnification: 100×. Source: Ref 6.1 as published in Ref 6.4

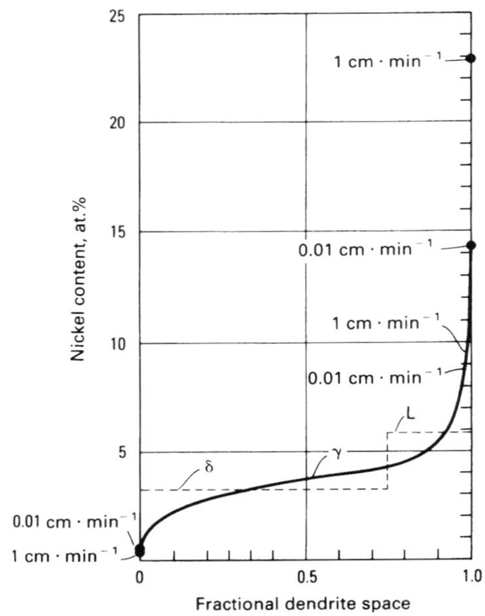

Fig. 6.18 Nickel distribution after peritectic reaction in a steel containing 4 wt% Ni. The temperature gradient was 60 K/cm. Calculations were made at different solidification rates. The dotted line shows the nickel distribution at the start of the peritectic reaction. δ, primary ferrite; γ, austenite. Source: Ref 6.9 as published in Ref 6.4

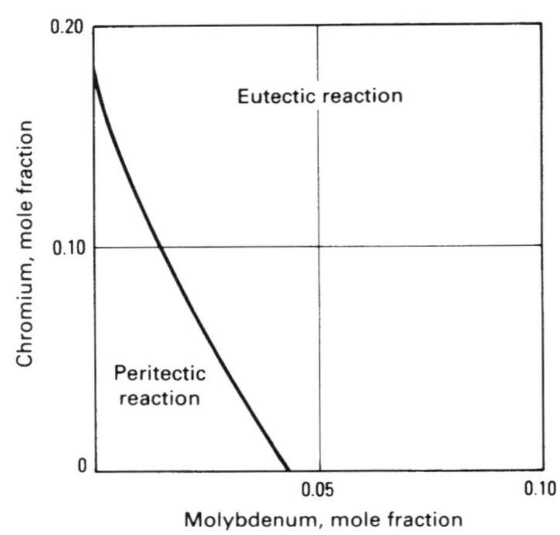

Fig. 6.19 The transition from a peritectic to a eutectic reaction as a function of chromium and molybdenum content in a stainless steel containing 11.9% Ni. Source: Ref 6.10 as published in Ref 6.4

Iron-base alloys often consist of carbon with some other elements. Both chromium and nickel are substitutional alloying elements, while carbon is interstitially dissolved and has a very high diffusion rate. The other substitutional alloying elements have much lower diffusion rates. This gives

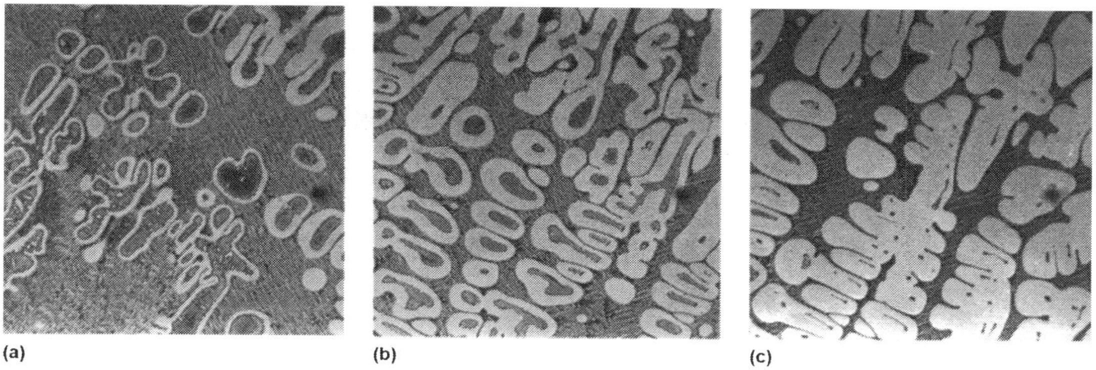

Fig. 6.20 Three stages of a peritectic reaction in a unidirectionally-solidified high-speed steel. (a) First-stage structure. Dark gray is austenite; white is ferrite. The mottled structure is quenched liquid. (b) Subsequent peritectic transformation of (a). (c) Further peritectic transformation of (a) and (b). Dark gray in the middle of the white ferrite is newly formed liquid. Source: Ref 6.11 as published in Ref 6.4

rise to transformations that are determined by the movement of the substitutional elements, and carbon is distributed according to equilibrium conditions. As a result, a normal peritectic transformation does not occur. To fulfill the criterion that carbon should follow the equilibrium conditions, liquid must be formed at the border between ferrite and austenite. This reaction is shown in Fig. 6.20.

ACKNOWLEDGMENT

Portions of this chapter came from Invariant Transformation Structures in *Metallography and Microstructures,* Vol 9, *ASM Handbook*, ASM International, 2004.

REFERENCES

6.1 H.E. Exner and G. Petzow, Peritectic Structures, *Metallography and Microstructures,* Vol 9, *Metals Handbook,* 9th ed., American Society for Metals, 1985, p 675–680

6.2 Peritectic Solidification, *Casting*, Vol 15A, *ASM Handbook*, ASM International, 2008

6.3 F.N. Rhines, *Phase Diagrams in Metallurgy*, McGraw-Hill, 1956, p 84

6.4 Invariant Transformation Structures, *Metallography and Microstructures,* Vol 9, *ASM Handbook*, ASM International, 2004, p 152–164

6.5 M. Hillert, Keynote Address: Eutectic and Peritectic Solidification, *Solidification and Castings of Metals,* The Metals Society, London, 1979, p 81–87

6.6 H. Frederiksson and T. Nylén, Mechanism of Peritectic Reactions and Transformations, *Met. Sci.,* Vol 16, 1982, p 283–294

6.7 G. Petzow and H.E. Exner, Zur Kenntnis peritektischer Unwandlungen, *Radex-Rundsch.,* Issue 3/4, 1967, p 534–539
6.8 R. Riedl, "Erstarrungsverlauf von Schnellarbeitsstrahlen," Ph.D. thesis, University of Leoben, Austria, 1984
6.9 H. Fredriksson, The Solidification Sequence in an 18-8 Stainless Steel, *Metall. Trans.,* Vol 3, 1972, p 2989–2997
6.10 H. Fredriksson, Solidification of Peritectics, *Casting,* Vol 15, *ASM Handbook,* ASM International, 1988, p 125–129
6.11 H. Frederiksson, The Mechanism of the Peritectic Reaction in Iron-Base Alloys, *Met. Sci.,* Vol 10, 1976, p 77–86

Phase Diagrams—Understanding the Basics
F.C. Campbell, editor

CHAPTER 7

Monotectic Alloy Systems

ANOTHER THREE-PHASE reaction of the eutectic class is the monotectic, in which one liquid phase decomposes with decreasing temperature into a solid phase and a new liquid phase:

$$L_1 \rightarrow \alpha + L_2 \text{ (monotectic reaction)}$$

Over a certain composition range, the two liquids are mutually immiscible, such as oil and water, and so constitute individual phases. The phase diagram shown in Fig. 7.1 gives the terminology for this type of system: monotectic point, monotectic reaction isotherm, hypomonotectic, and hypermonotectic. The phase diagram shows a dome-shaped region within which the two liquids mix and coexist. The maximum temperature of this dome, T_c, is called the critical (or consolute) temperature. It should be noted that the liquidus and solidus curves are differently located and that these have been designated as "upper" and "lower" to distinguish them. There is no special name for the boundary of the $L_1 + L_2$ field; it is simply called the limit of liquid immiscibility. The eutectic reaction, depicted by dashed lines in this example, is included merely to carry the diagram into the temperature range where all phases are solid.

Liquid copper and liquid lead are completely soluble in each other at high temperatures. However, as shown in the Fig. 7.2 phase diagram, alloys containing between 36 and 87 wt% Pb separate into two liquids on further cooling. The two liquids coexist in the miscibility gap, or dome, that is typical of all alloys that undergo a monotectic reaction. During solidification of a copper-lead alloy containing 20 wt% Pb, the copper-rich α phase forms first. The liquid composition shifts toward the monotectic composition of 36 wt% Pb. Then, the liquid transforms to more solid α and a second liquid containing 87 wt% Pb. The lever rule shows that only a very small amount of the second liquid is present. On further cooling, the

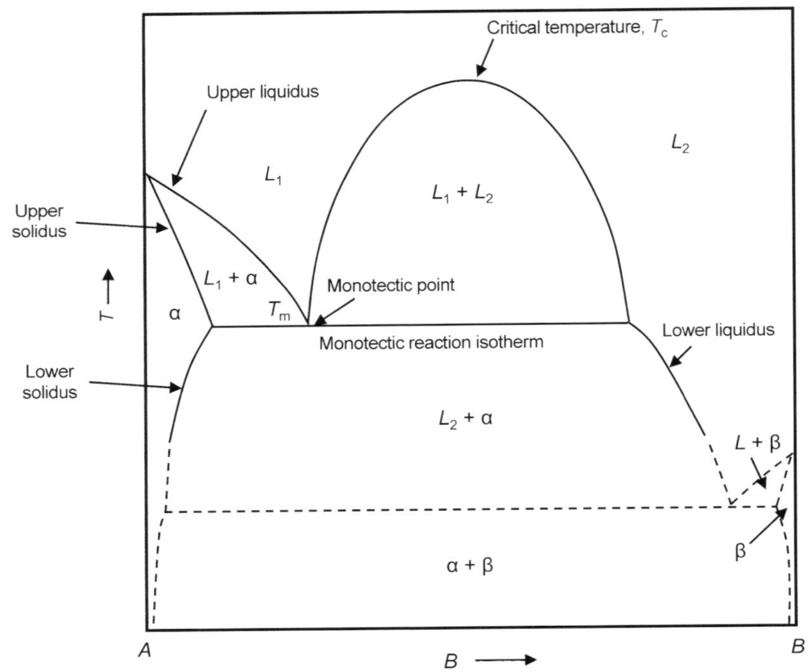

Fig. 7.1 Monotectic phase diagram showing the invariant point, *m*, where liquid 1 (L_1) transforms to another liquid (L_2) and solid solution (α). Adapted from Ref 7.1

second liquid eventually undergoes a eutectic reaction, producing α and β, where the β is almost pure lead. The microstructure of this alloy contains spherical β particles randomly distributed in a matrix of copper-rich α.

Because of the continuity of the copper-rich a phase in the monotectic structure, the physical properties of this alloy more nearly resemble those of copper than those of lead. Lead is often added to alloys because it makes the machining of the metal easier by reducing the ductility just enough to cause chips to break away, without seriously decreasing hardness and strength. Leaded alloys are also used for bearings, where the continuous phase of the high-melting metal gives strength to the member, while the lead, occurring in pockets at the running surface, serves to reduce the friction between bearing and axle.

7.1 Solidification Structures of Monotectics

Monotectic alloys can be classified based on the difference between the critical temperature, T_c, and the monotectic temperature, T_m; that is, the difference $T_c - T_m$. High-dome alloys have a large difference in $T_c - T_m$

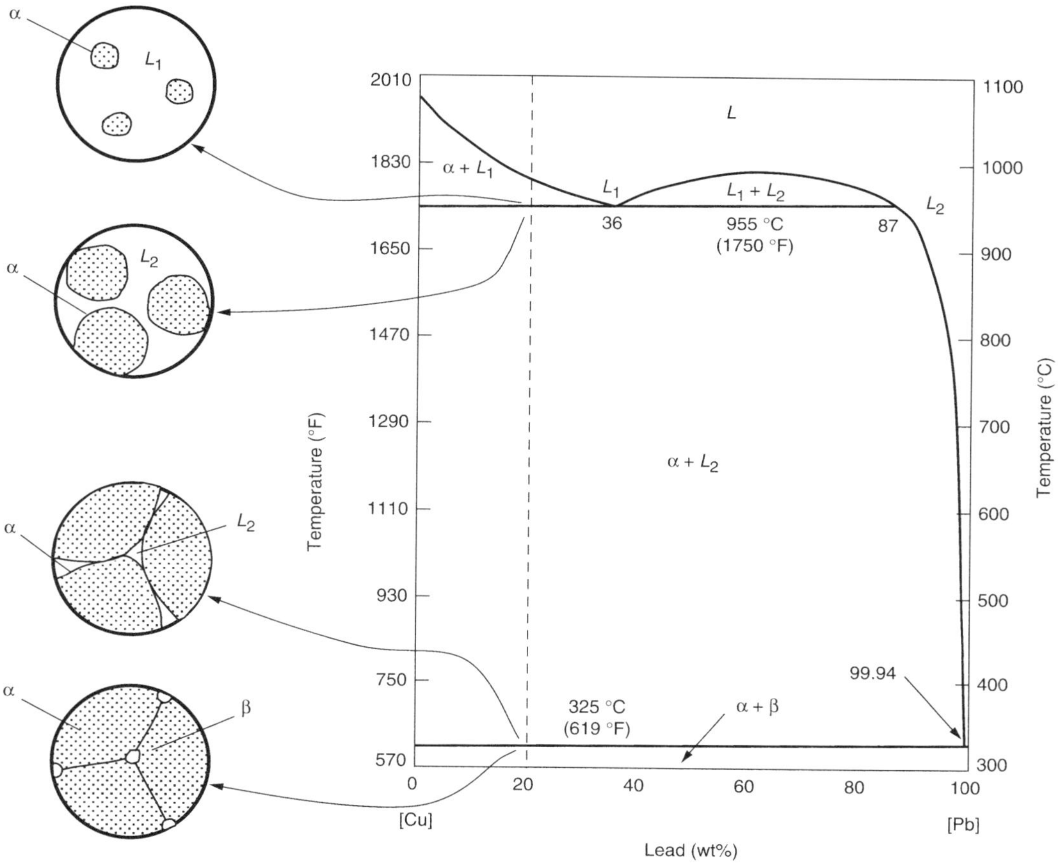

Fig. 7.2 Monotectic reaction in copper-lead system. Source: Ref 7.2 as published in Ref 7.3

(hundreds of °C), while low-dome alloys have a small difference in $T_c - T_m$ (tens of °C). They can also be classified based on the T_m/T_c ratio. High-dome alloys have $T_m/T_c < 0.9$, while low-dome alloys have $T_m/T_c > 0.9$.

The morphology of the microstructure produced during directional solidification is a function of the density difference between the two liquids, and of the wetting between L_2 and α. For low-dome alloys, the phases α and L_2 are separated by L_1, as shown in Fig. 7.3. At low growth rate, L_2 particles are pushed by the solid-liquid interface (Fig. 7.3a). If the solidification velocity increases above a critical velocity (V_{cr}), L_2 is incorporated with formation of an irregular fibrous composite (Fig. 7.3b). Particle engulfment and pushing by solidifying interfaces has been observed experimentally.

Fibrous growth in the succinonitrile-20% ethanol system, in which the velocity is above the critical velocity for particle engulfment, is shown in Fig. 7.4. Similar microstructures are observed in metallic systems such as

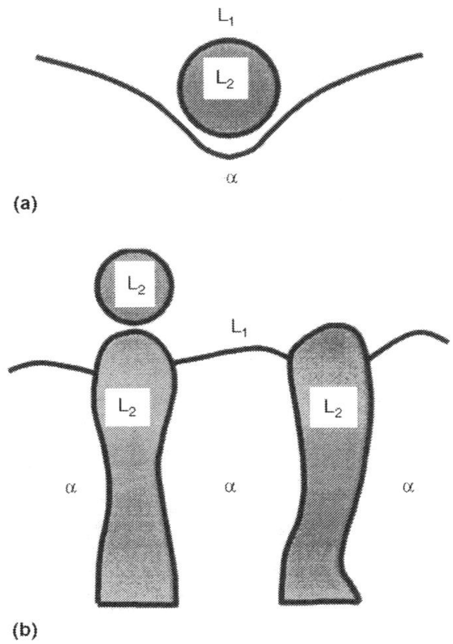

Fig. 7.3 Monotectic solidification for low-dome alloys. (a) Low growth velocity. (b) High growth velocity. Source: Ref 7.4 as published in Ref 7.5

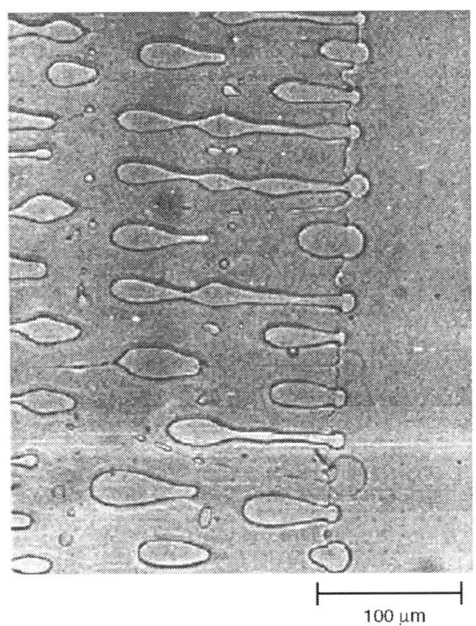

Fig. 7.4 Growth front in succinonitrile-20 wt% ethanol, showing incorporation of ethanol droplets. $V = 0.27$ μm/s. Source: Ref 7.6 as published in Ref 7.5

Cu-60%Pb and Bi-50%Ga alloys, as shown in Fig. 7.5 for a copper-lead alloy. Note that the velocity in this microstructure is very high.

The range of existence of the fibrous composite is limited by the constitutional undercooling on one side and by the critical velocity of pushing-to-engulfment transition (V_{cr}) on the other (Fig. 7.6). When the solidification velocity is smaller than V_{cr}, a banded structure may result. An example of such a structure is provided in Fig. 7.7. It is suggested that the L_2 phase, which precipitates at the solid-liquid interface, piles up and covers the solid-liquid interface. This produces a lead-rich layer and increases the undercooling of the L_1-L_2 interface with respect to the

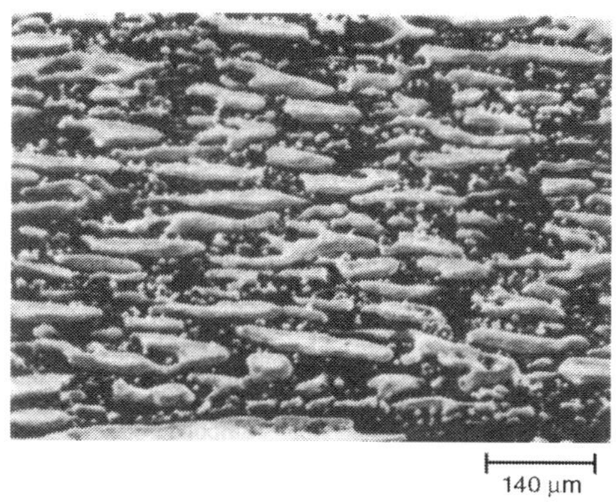

Fig. 7.5 Microstructure of a Cu-70Pb alloy solidified at $V = 778$ μm/s. Solidification direction is right-to-left. Source: Ref 7.7 as published in Ref 7.5

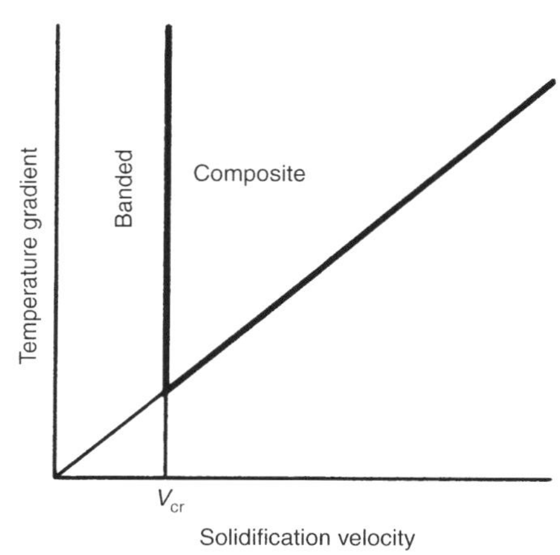

Fig. 7.6 Restriction on composite growth imposed by the critical velocity for the pushing engulfment transition and by constitutional undercooling. Source: Ref 7.4 as published in Ref 7.5

Fig. 7.7 Microstructure of upward directional solidification of a Cu-37.7Pb alloy in longitudinal section. $V = 4.4$ μm/s. Source: Ref 7.8 as published in Ref 7.5

monotectic temperature. Then, nucleation of the α-Cu phase occurs on the lead-rich layer. The temperature at the growth front is also returned to the monotectic temperature. The repetition of this process will result in the banded structure as shown in Fig. 7.8. It should be noted that some fibrous structure might form even in the case of banding (Fig. 7.9).

For high-dome alloys, the lowest energy exists when an α-L_2 interface exists. Consequently, α and L_2 will grow together (L_2 wets α), resulting in a regular (uniform) fibrous composite. The λ^2-V relationship is approximately two orders of magnitude larger for irregular than for regular monotectic composites (with the exception of the aluminum-bismuth alloy) and approximately one order of magnitude higher for regular monotectic composites than for regular eutectics. The differences come from the controlling mechanism. For irregular fibrous eutectics, the controlling mechanism is the pushing-engulfment transition, which is a function of solidification velocity and surface energy. For regular fibrous monotectics, the spacing is controlled by surface energy. For eutectics, the spacing is controlled by diffusion.

ACKNOWLEDGMENT

Portions of this chapter came from "Fundamentals of Solidification," by D.M. Stefanescu and R. Ruxanda in *Metallography and Microstructures*, Vol 9, *ASM Handbook*, ASM International, 2004.

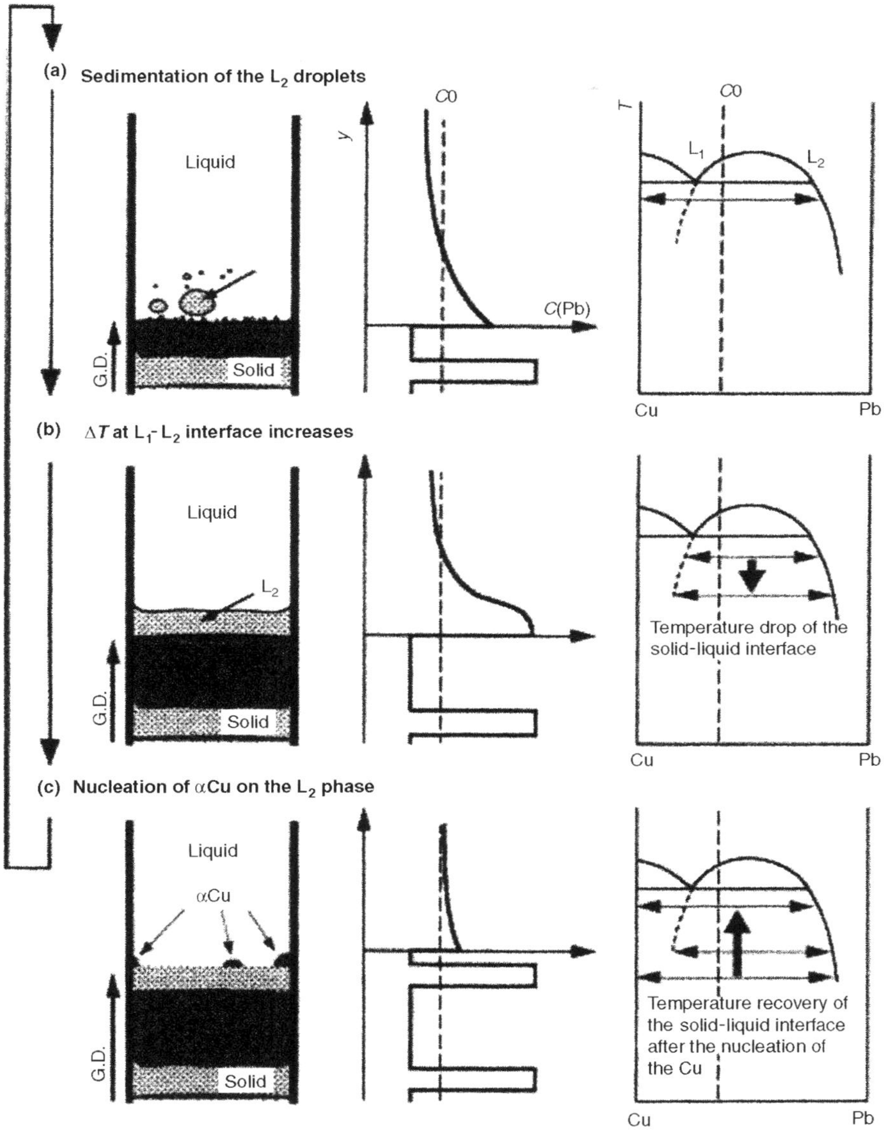

Fig. 7.8 Forming mechanism of the banded structure of copper-lead alloy in upward directional solidification. G.D., growth direction. Source: Ref 7.8 as published in Ref 7.5

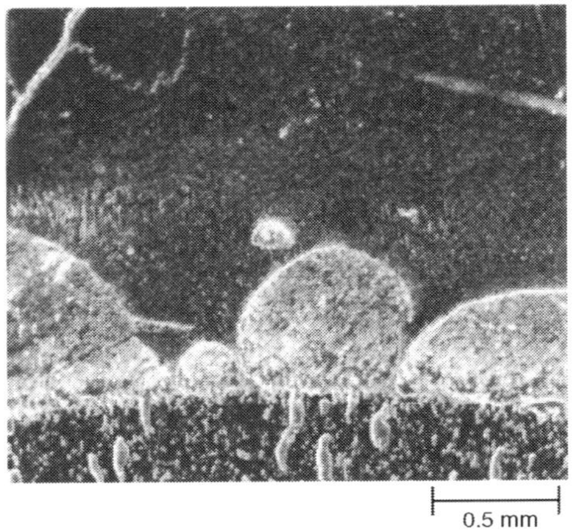

Fig. 7.9 The solid-liquid interface covered with coalesced L_2 phase. Cu-35.4Pb alloy, upward directional solidification, $V = 2.2$ μm/s. Source: Ref 7.8 as published in Ref 7.5

REFERENCES

7.1 F.N. Rhines, *Phase Diagrams in Metallurgy*, McGraw-Hill, 1956, p 72

7.2 D.R. Askeland, *The Science and Engineering of Materials*, 2nd ed., PWS-KENT Publishing Co., 1989

7.3 F.C. Campbell, *Elements of Metallurgy and Engineering Alloys*, ASM International, 2008

7.4 D.M. Stefanescu, *Science and Engineering of Casting Solidification*, Kluwer Academic, 2002

7.5 D.M. Stefanescu and R. Ruxanda, Fundamentals of Solidification, *Metallography and Microstructures*, Vol 9, *ASM Handbook*, ASM International, 2004, p 71–92

7.6 R.N. Grugel, T.A. Lagrasso, and A. Hellawell, *Metall. Trans. A*, Vol 15A, 1984, p 1003–1010

7.7 B.K. Dhindaw, D.M. Stefanescu, A.K. Singh, and P.A. Curreri, *Metall. Trans. A*, Vol 19A, 1988, p 2839

7.8 I. Aoi, M. Ishino, M. Yoshida, H. Fukunaga, and H. Nakae, *J. Cryst. Growth*, Vol 222, 2001, p 806–815

SELECTED REFERENCE

- D.M. Stefanescu and R. Ruxanda, Fundamentals of Solidification, *Metallography and Microstructures*, Vol 9, *ASM Handbook*, ASM International, 2004, p 71–92

CHAPTER 8

Solid-State Transformations

EUTECTOID AND PERITECTOID transformation are classified as solid-state invariant transformations. Invariant transformations are isothermal reversible reactions that occur at an invariant point on the phase diagram of an alloy, where the initial (parent) phase may be either a liquid or a crystalline solid. Solid-state invariant reactions are a category of heterogeneous phase transformations that involve moving reaction boundaries and phase separation. Invariant transformations differ from precipitation reactions in that all reaction products from an invariant transformation have a different crystal structure than that of the parent phase. In contrast, solid-state transformation from a discontinuous precipitation involves the generation of new second phase within a matrix that has the same crystal structure as the parent phase in precipitation structures.

As in the case of invariant reactions during solidification, important solid-state transformations from invariant reactions are of three types:

- *Eutectoid transformation,* where a solid solution converts into two or more intimately mixed solids with different crystal structures than that of the parent phase. The number of solid phases formed equals the number of components in the system. The reaction is:

$$\text{Solid } \gamma \rightarrow \text{Solid } \alpha + \text{Solid } \beta$$

- *Peritectoid transformation,* where two solid phases of a binary alloy transform into one phase on cooling. Peritectoid reactions are similar to peritectic reactions, except that one of the initial phases is liquid in a peritectic reaction. As in all invariant reactions, peritectoid reactions are reversible; that is, the $\alpha + \beta$ are recovered on heating the reaction product, γ. The reaction is:

Solid α + Solid β → Solid γ

- *Monotectoid transformations,* where cooling or heating of a solid solution completely converts it into a solid solution with a different crystal structure (Fig. 8.1). A monotectoid reaction differs from a eutectoid reaction in that only one reaction phase is produced. The reaction is:

Solid α → Solid γ

The monotectoid reaction is not prevalent in commercial alloys and is not discussed further in this chapter. This chapter focuses primarily on structures from eutectoid transformations with emphasis on the classic iron-carbon system of steel. Peritectoid phase equilibria also are very common in several binary systems but are only briefly reviewed.

Solid-state reactions differ in two important aspects from liquid reactions in the manner in which they attain the equilibrium conditions predicted by the phase diagram. Solid-state reactions occur much more slowly, are subject to greater undercooling, and rarely attain true equilibrium conditions. Solid phases consist of atoms arranged in certain crystal structures, and new solid phases forming out of an existing solid phase tend to take definite positions with respect to the existing crystal structure. In other words, the crystal structure of the new phase has a definite orientation relationship to the crystal structure of the phase from which it formed.

8.1 Iron-Carbon Eutectoid Reaction

Most carbon steels contain up to 1.5% C, while cast irons normally contain 2 to 4% C. As discussed in Chapter 2, "Solid Solutions and Phase Transformation," in this book, iron is allotropic and changes its crystalline

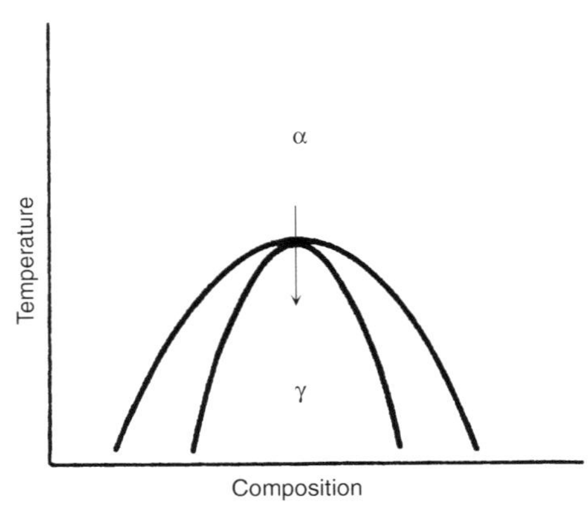

Fig. 8.1 Schematic phase diagram of a monotectoid reaction. Source: Ref 8.1

structure on heating or cooling. When pure iron solidifies on cooling from its melting point of 1540 °C (2800 °F), it assumes a body-centered cubic (bcc) structure designated as δ-ferrite. On further cooling between 1395 and 910 °C (2541 and 1673 °F), it has a face-centered cubic (fcc) structure called austenite, designated as γ. Below 910 °C (1673 °F), it again has a bcc crystal structure called ferrite, designated as α.

The fact that steel can be hardened is a direct result of the eutectoid reaction in iron-carbon alloys. Just as a eutectic reaction involves the decomposition of a liquid solution, a eutectoid reaction involves the decomposition of solid solution into two other solid phases. The iron-carbide phase diagram is shown in Fig. 8.2. The composition only extends to 6.70% C; at this composition the intermediate compound iron carbide, or cementite (Fe_3C), is formed, represented by the vertical line on the right hand side of the diagram. Because all steels and cast irons have carbon contents less than 6.70%, the remainder of the diagram with 6.7 to 100% C is of no engineering interest. Cementite is a hard and brittle intermetallic compound with an orthorhombic crystalline structure.

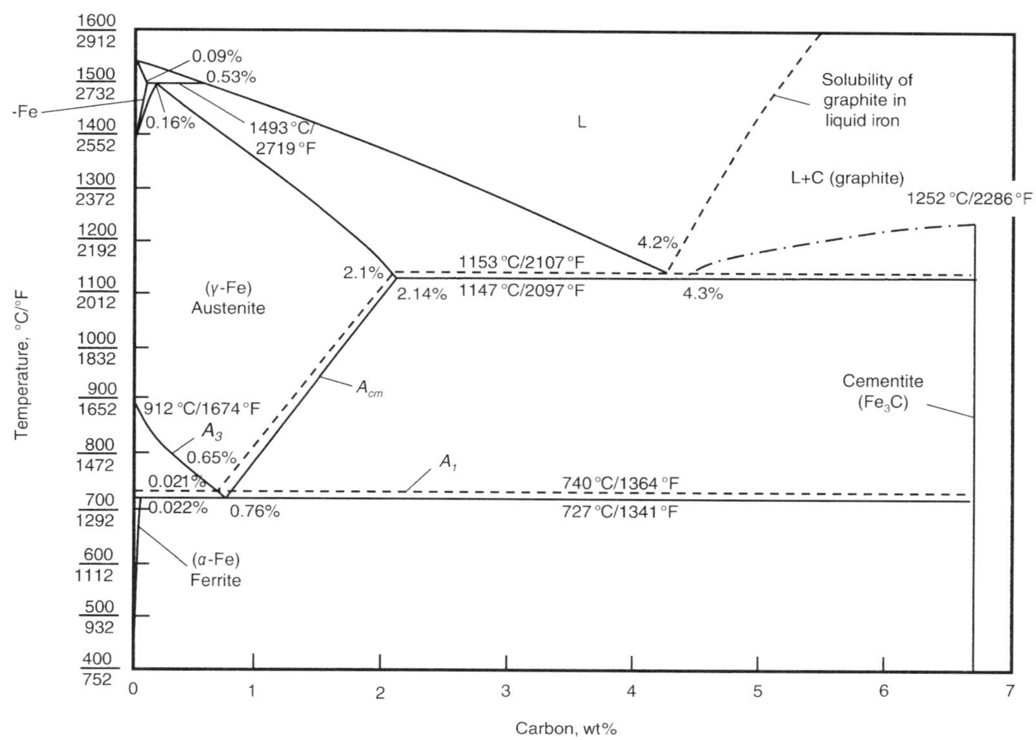

Solid lines indicate $Fe-Fe_3C$ diagram; dashed lines indicate iron-graphite diagram.

Fig. 8.2 The $Fe-Fe_3C$ diagram. Solid lines indicate $Fe-Fe_3C$ diagram; dashed lines indicate iron-graphite diagram. Source: Ref 8.2 as published in Ref 8.3

Carbon forms a series of solid solutions with iron, as indicated by the δ, γ, and α fields. However, carbon has very limited solubility in iron. The maximum solubility of carbon in ferrite is only approximately 0.022 wt% at 171 °C (340 °F) that increases to 2.14 wt% C at 1150 °C (2098 °F). At room temperature it is even less, with the solubility of carbon in iron only 0.005%. Thus, the solubility of carbon in fcc iron is approximately 100 times greater than in bcc ferrite. This is a result of the fcc interstitial positions being larger, and therefore the strains imposed on the surrounding iron lattice are much lower. Even though present in very small amounts, carbon significantly influences the mechanical properties of ferrite.

Steels contain carbon in the form of cementite, while cast irons can contain either cementite or carbon in the form of graphite. Graphite is a more stable carbon-rich phase than cementite. Graphite formation is promoted by a higher carbon content and the presence of large amounts of certain alloying additions, in particular silicon. Therefore, graphite is an important phase in cast irons but is rarely found in steels. Cementite is metastable and will remain a compound at room temperature indefinitely. However, if it is heated to 650 to 705 °C (1200 to 1300 °F) for several years, it will gradually transform into iron and graphite. When graphite does form, the solubility limits and temperature ranges of phase stability are changed slightly, as indicated by the dashed lines in Fig. 8.2.

In ferrous alloys, iron is the prime alloying element, but carbon and other alloying elements are frequently used. Based on carbon content, there are three types of ferrous alloys: irons, steels, and cast irons. Commercially pure iron contains less than 0.008 wt% C and is composed almost exclusively of ferrite when it cools to room temperature. In steels, which have carbon contents in the range of 0.008 to 2 wt%, the microstructure consists of ferrite and cementite when they are slowly cooled to room temperature. Although a steel may contain as much as 2 wt% C, carbon contents are usually restricted to 1.5 wt% or less because of excessive brittleness. Generally, the carbon content is kept low in steels that require high ductility, high toughness, and good weldability, but is used at higher levels in steels that require high strength, high hardness, fatigue resistance, and wear resistance. Cast irons are classified as those alloys that contain between 2 to 6.70 wt% C; however, most cast irons contain 2 to 4 wt% C.

8.1.1 Ferrite

Ferrite is usually present in steels as a solid solution of iron containing carbon or one or more alloying elements such as silicon, chromium, manganese, and nickel. Carbon is an interstitial element, while the larger elements are substitutional. Carbon occupies specific interstitial sites in the bcc iron crystalline lattice, while the larger substitutional elements replace or substitute for iron atoms. The two types of solid solutions impart different characteristics. For example, interstitial carbon can easily diffuse through the open bcc lattice, whereas substitutional elements diffuse much more slowly. Therefore, carbon responds quickly during heat treatment, whereas substitutional alloying elements behave more sluggishly. Interstitial alloying elements have a much stronger influence on the properties of iron than substitutional alloying elements. Even a small addition of carbon, less than its solubility limit, into pure iron substantially increases the room-temperature yield strength of the material (Fig. 8.3).

The influence of alloying elements on the yield strength of ferrite is shown in Fig. 8.4. The strong effects of the interstitial elements carbon and nitrogen, as well as phosphorous, are clearly evident. On the other hand, the substitutional solid-solution elements silicon, copper, manganese, molybdenum, nickel, aluminum, and chromium are far less effective as

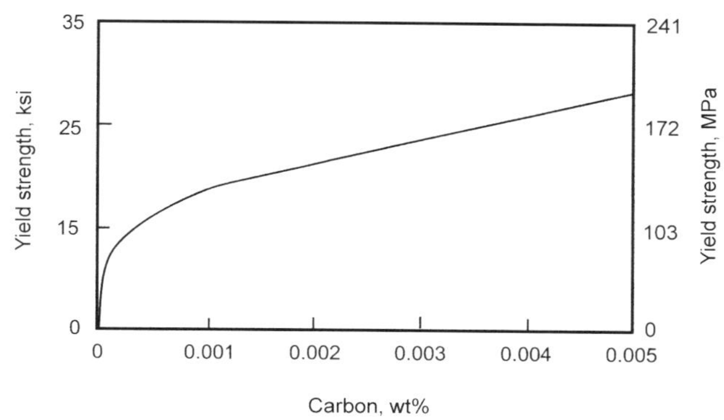

Fig. 8.3 Effect of small carbon additions on strength of iron. Source: Ref 8.4 as published in Ref 8.3

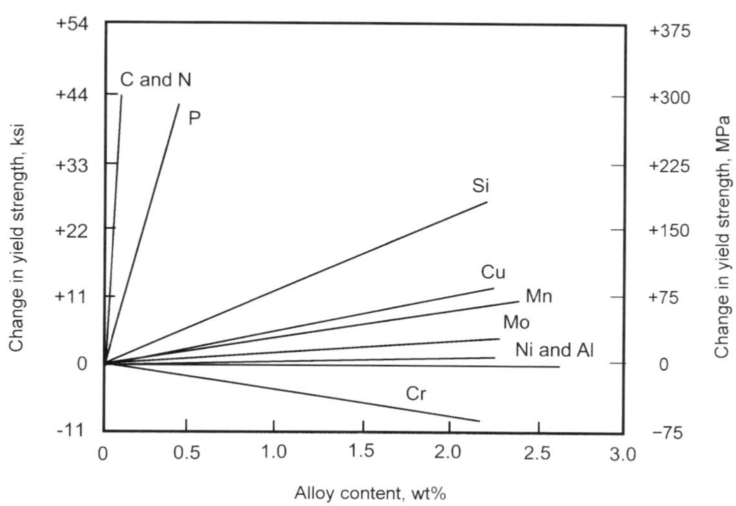

Fig. 8.4 Effect of alloying elements on yield strength. Source: Ref 8.4 as published in Ref 8.3

ferrite strengtheners. The strength of a ferritic steel is also determined by its grain size according to the Hall-Petch relationship, as shown in Fig. 8.5; that is, a finer grain size results in a higher yield strength. Control of the grain size through thermomechanical treatments, heat treatments, and/or alloying is vital to the control of strength and toughness of most steels. While a wide variety of steels contain ferrite, only a few commercial steels are totally ferritic. An example of the microstructure of a fully ferritic ultralow-carbon steel is shown in Fig. 8.6.

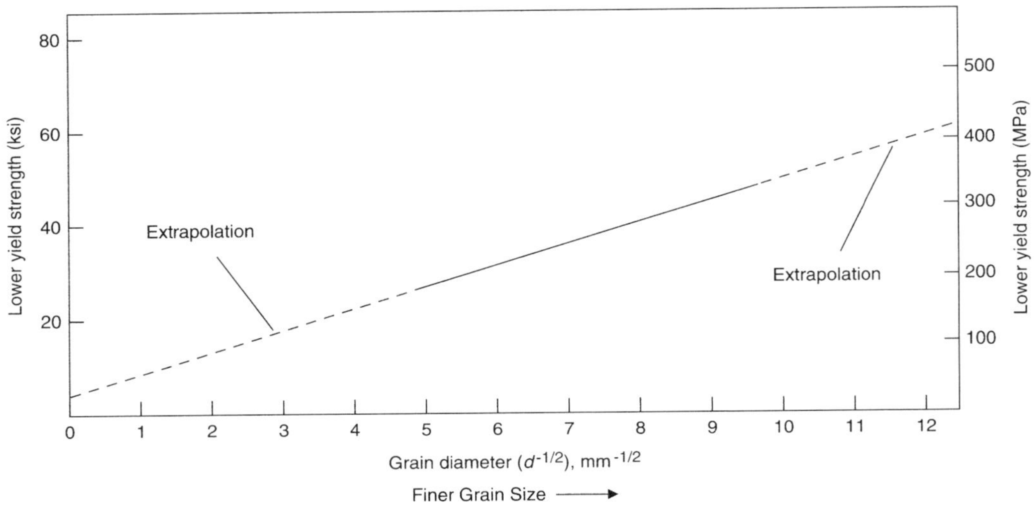

Fig. 8.5 Hall-Petch relationship for low-carbon ferritic steel. Source: Ref 8.3

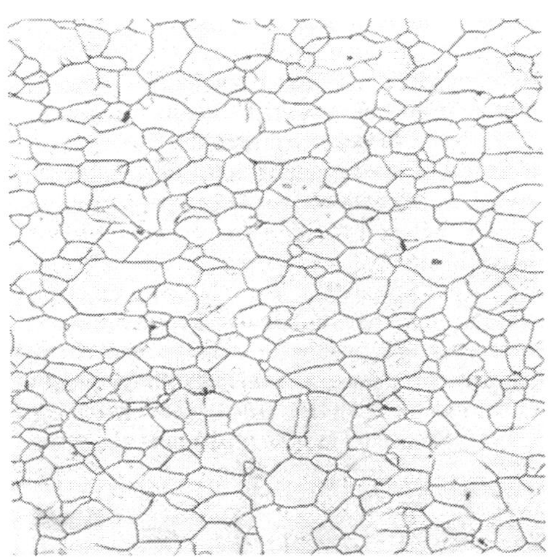

Fig. 8.6 Microstructure of ultralow-carbon ferritic steel. Source: Ref 8.4 as published in Ref 8.3

8.1.2 Eutectoid Structures

The eutectoid reaction is defined as a single-parent phase decomposing into two different product phases through a diffusional mechanism. When a steel with the eutectoid composition (0.76 wt% C) is slowly cooled through the eutectoid temperature of 725 °C (1341 °F), austenite decomposes into ferrite and cementite, in the manner shown in Fig. 8.7:

γ-austenite (0.76 wt% C) → α-ferrite (0.22 wt% C) + Fe_3C-cementite (6.7 wt% C)

At a temperature of 999 °C (1830 °F), a phase analysis of an alloy containing the eutectoid composition (0.8% C) gives:

Point: 0.8% C alloy at 1000 °C (1830 °F)	
Phase	Austenite (γ)
Composition	0.8% C
Amount	100%

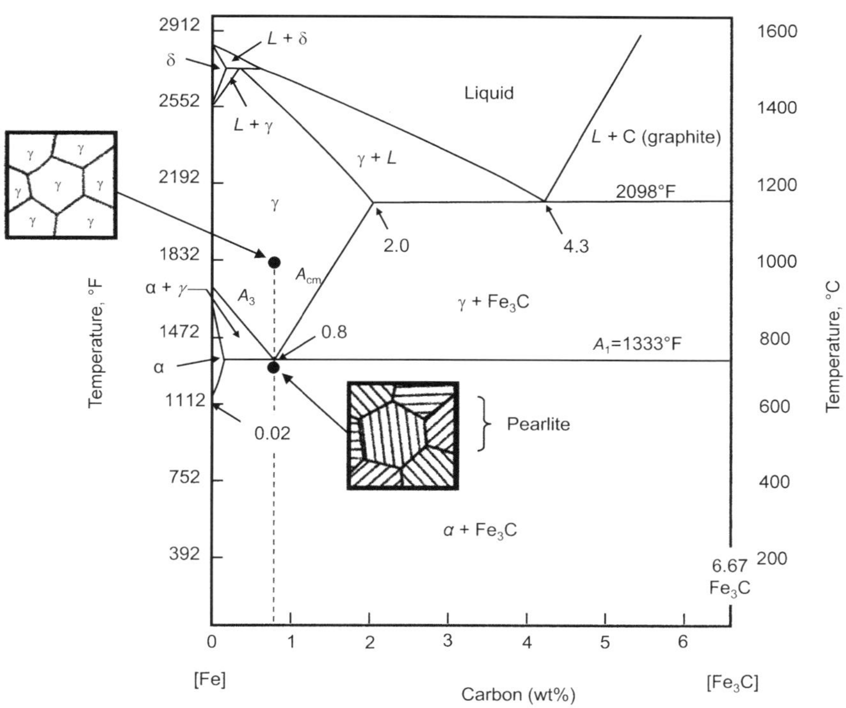

Fig. 8.7 Equilibrium cooling of a eutectoid steel. δ-ferrite, bcc; γ-austenite, fcc; α-ferrite, bcc; Fe3C, cementite. Source: Ref 8.3

When the alloy is cooled past the eutectoid temperature (720 °C, or 1333 °F), ferrite and cementite form side by side within the austenite to create a nodule of pearlite, the eutectoid microconstituent. A phase analysis just below the eutectoid temperature reveals:

Point: 0.8% C at 700 °C (1290 °F)		
Phases	Ferrite (α)	Cementite (Fe_3C)
Compositions	0.3% C	6.7% Fe_3C
Amounts	100% – 12% ≈ 88%	$\text{wt\%Fe}_3\text{C} = \frac{0.80 - 0.03}{6.7 - 0.03} \approx 12\%$

Because the corresponding microstructure is 100% pearlite, it follows that pearlite consists of 88% ferrite and 12% cementite. Pearlite is a lamellar construction consisting of alternating layers of ferrite and cementite. As shown in Fig. 8.8, either ferrite or cementite nuclei form at austenite grain boundaries and then grow as colonies into the austenite grains, with the layers oriented in essentially the same direction within the colony. The concurrent growth of several colonies leads to the formation of a larger nodule. The interface between the growing pearlite and austenite is incoherent. Provided that the temperature is constant, pearlite advances at a constant velocity and tends to produce a hemispherical nodule. The mechanism of pearlite growth involves the rejection of carbon by the growing ferrite plates and its incorporation into the cementite. This process requires diffusion of carbon by one or more of three paths: through the austenite, through the ferrite, or along the interface. The lower the transformation

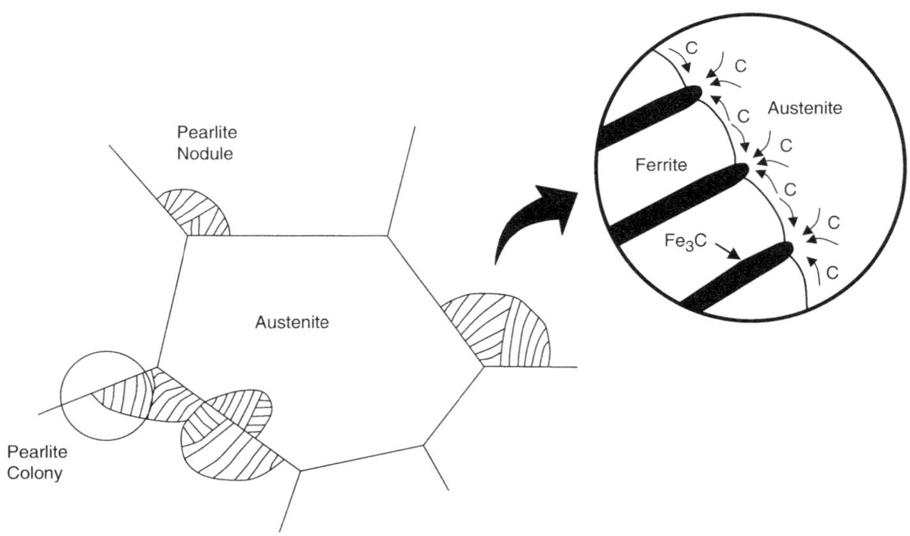

Fig. 8.8 Formation of pearlite from austenite. Source: Ref 8.5 as published in Ref 8.3

temperature, the greater is the chemical energy driving the reaction, resulting in a finer pearlite size. Pearlite gets its name because its appearance at low magnification under a light microscope resembles mother-of-pearl (Fig. 8.9).

The alternating α-ferrite and Fe_3C layers in pearlite form because the carbon content of the austenite parent phase (0.76 wt% C) is different from that of both ferrite (0.022 wt% C) and cementite (6.7 wt% C), and the phase transformation requires that there be a redistribution of the carbon by diffusion. Carbon atoms diffuse away from the ferrite regions to the cementite layers as the pearlite extends from the grain boundary into the austenite grains. The relative layer thickness of the eutectoid ferrite-to-cementite layers is approximately 8 to 1, meaning that the ferrite layers are a lot thicker than the cementite layers. When the rate of cooling is slow enough for the transformation to begin and end just below the eutectoid isotherm, the product will be coarse pearlite. As the cooling rate increases, pearlite with increasingly fine structure will develop.

Although energy minimization would normally predict a structure of dispersed spheroids, the alternating platelet structure forms because the carbon diffuses a shorter distance than is required to form a dispersed spheroid microstructure. Diffusion is possible because the small interstitial carbon atoms can diffuse quite rapidly through the iron lattice at elevated

Fig. 8.9 Pearlitic structure in eutectoid steel. Source: Ref 8.6 as published in Ref 8.3

temperatures. In addition, diffusion is faster in the more loosely packed bcc ferrite structure than the closer packed fcc austenite structure. However, diffusion in solids during a eutectoid reaction is much slower than in liquids during the eutectic reaction. Therefore, nonequilibrium transformations are even more important in eutectoid solid-state reactions than in the liquid-to-solid eutectic reactions. Subsequent cooling of pearlite from the eutectoid to room temperature produces only insignificant changes in the microstructure.

The morphology of a pearlite nodule is illustrated Fig. 8.10. Pearlite nodules nucleate at prior austenite grain boundaries and triple points to minimize the free energy needed for the transformation. Each nodule contains sub-units (colonies) of cementite and ferrite lamellae; each colony has a specific orientation relation with the parent austenite grain. Pearlite nucleation occurs when either ferrite or cementite nucleates on a heterogeneity in the parent structure, such as the austenite (or parent phase) grain boundaries or grain-boundary triple points. The selection of which phase nucleates is determined by the orientation and local composition. For the case of cementite, which is orthorhombic, the relationship between the newly formed cementite and the parent austenite is:

$$(100)_c \| (1\bar{1}1)_\gamma, (0\bar{1}0)_c \| (110)_\gamma, (001)_c \| (\bar{1}12)_\gamma$$

This creates a low-mobility semicoherent interface with the grain, γ_1, with which the orientation relation was developed, and a high-mobility incoherent interface with austenite grain, γ_2, in Fig. 8.11(a). Thus, the newly formed pearlite colony grows by the high-mobility incoherent interface expanding

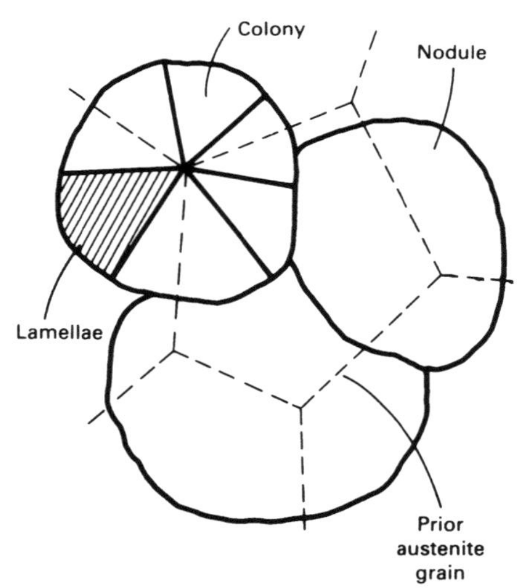

Fig. 8.10 Relationship of pearlite lamellae, colonies, and nodules to prior austenite grains. Source: Ref 8.7 as published in Ref 8.1

into the austenite grain with which the pearlite does not have an orientation relationship. The nucleation of this cementite creates a "carbon-free" region around it, promoting the nucleation of ferrite on both sides of the cementite. The interface between the newly formed cementite and ferrite is semicoherent as well, which promotes lamellae lengthening through the higher-mobility incoherent interface between the lamellae and the parent austenite, rather than lamellae spacing coarsening. This causes the colony shape to become radial from the point of original nucleation as existing lamellae lengthen into the austenite and new lamellae nucleate at the edges (Fig. 8.11b–c and 8.12). This semicoherent interface in a commercial steel is illustrated in Fig. 8.13 and 8.14. Thus, the pearlite colony grows by the incoherent interface expanding into grain γ_2 (Fig. 8.11a, part 4). This growth occurs by a ledge-growth mechanism (Fig. 8.15). This cooperative nucleation and growth is key to the development of the pearlite morphology. This cooperative growth is dependent on both diffusion of carbon

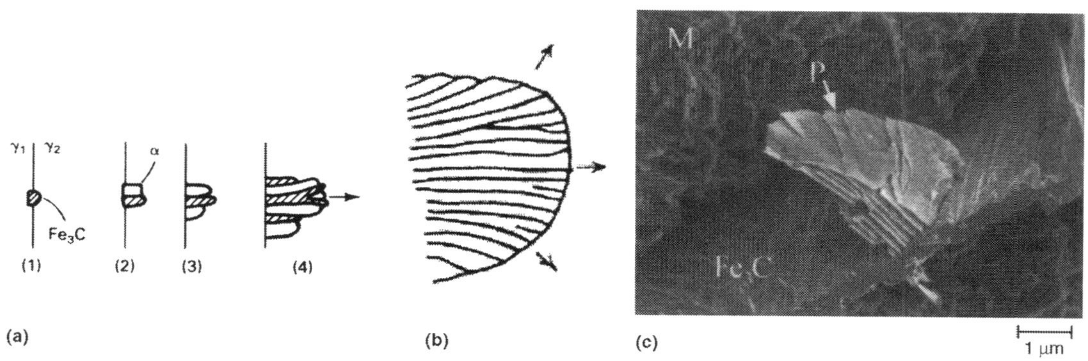

Fig. 8.11 (a) Pearlite nucleation. (b) Colony growth. (c) Deep-etched steel sample showing pearlite colony growth from a proeutectoid cementite plate. Source: Ref 8.8 as published in Ref 8.1

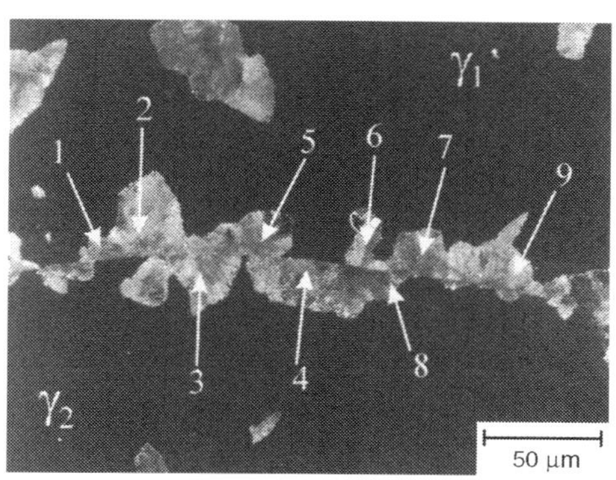

Fig. 8.12 Growth of intergranular pearlite nodules (numbered light regions) into the austenite matrix (dark). Source: Ref 8.9 as published in Ref 8.1

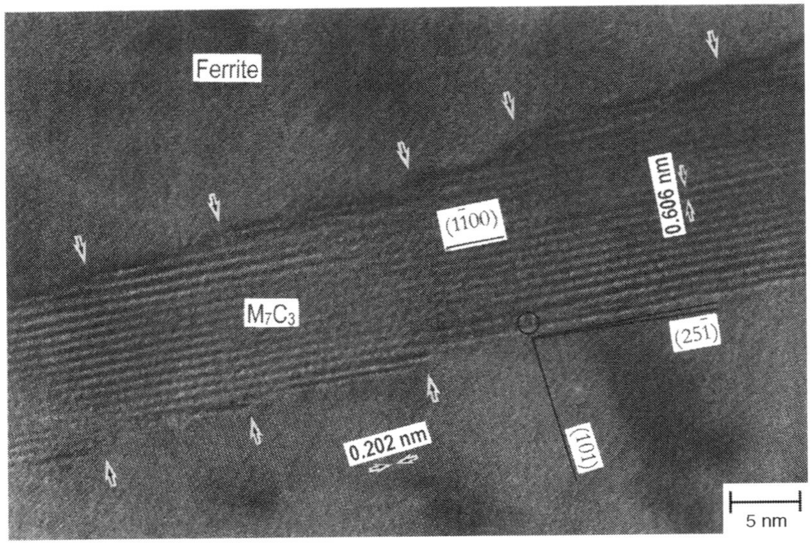

Fig. 8.13 High-resolution electron micrograph of two ferrite regions split by a carbon-rich M_7C_3 lamellae in an Fe-8.2Cr-0.92C alloy. Source: Ref 8.10 as published in Ref 8.1

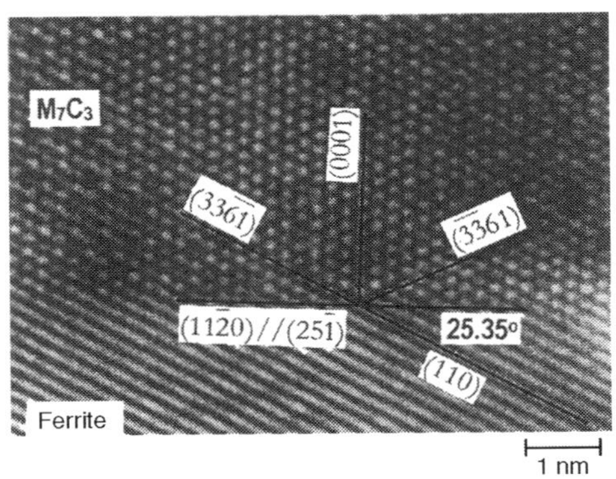

Fig. 8.14 Interface between M_7C_3 and ferrite seen in Fig. 8.13. The interface is semicoherent, pinning any movement. Source: Ref 8.10 as published in Ref 8.1

as well as structural sharing of growth ledges between the cementite and ferrite (Fig. 8.16).

If the ferrite nucleates first, it forms an orientation relation with γ_1 that is close to the Kurdjumov-Sachs relation:

$$\{110\}_f \| \{111\}_\gamma \ \langle 1\bar{1}1\rangle_f \| \langle 1\bar{1}0\rangle_\gamma$$

Carbon is rejected into the austenite, promoting the nucleation of cementite. The pearlite colony grows into the austenite grain with which it does

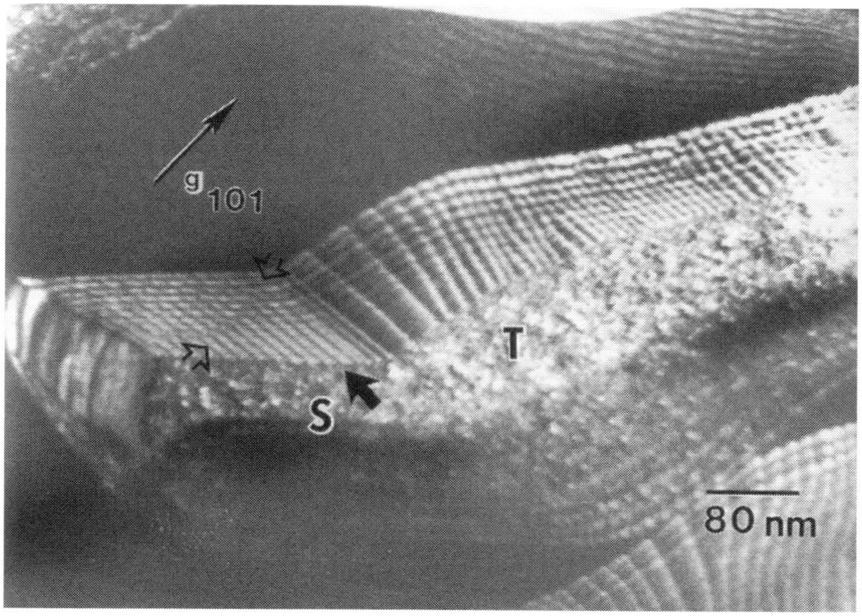

Fig. 8.15 Dark-field ferrite growth front illustrating ledge growth. Source: Ref 8.1

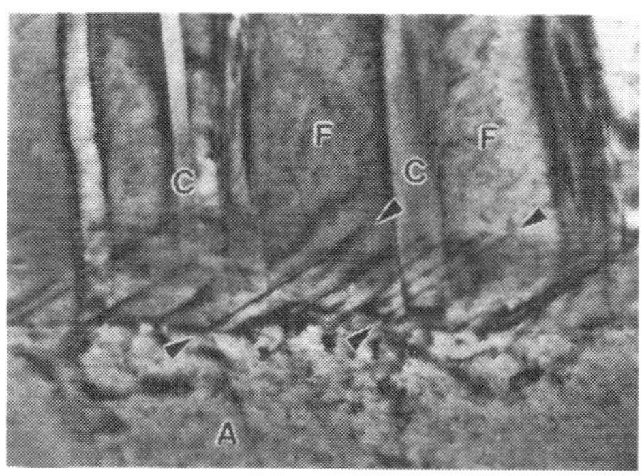

Fig. 8.16 Growth front of pearlite indicating that ledges span both cementite (C) and ferrite (F) as they grow into the austenite (A). Source: Ref 8.11 as published in Ref 8.1

not have the orientation relationship due to the relative ease of moving the high-energy incoherent interface.

The growth rate of pearlite is best represented by the Johnson-Mehl-Avrami equation:

$$f = 1 - \exp(-kt^n)$$

where f is fraction transformed, k accounts for the growth and nucleation rates, t is time, and n is the Avrami exponent. The exponent n varies from

1 to 4, where 1 represents a needle-shaped precipitate and 4 represents a spherical precipitate. The k factor can be rewritten as:

$$k = k_0 \exp\left[\frac{-Q}{RT}\right]$$

where k_0 is a constant, Q is the activation energy, R is the universal gas constant, and T is the temperature. Thus, the growth rate is very dependent on temperature, particle shape, and time. Figure 8.17 illustrates this with a time-temperature-transformation (TTT) diagram showing the time for transformation at two different temperatures. It is seen that the fraction transformed curve is sigmoidal. The initial rate is slow due to relatively few nodules existing. The rate increases with the nucleation rate until nodule impingement, when the transformation rate slows to completion. The C-curve nature of the TTT diagram in Fig. 8.17(a) shows that transformation times are slow at both temperatures close to (small undercooling) and far from (large undercooling) the reaction temperature. This can be explained by the small amount of driving force for nucleation at small undercoolings and very slow diffusion rates at large undercoolings. The intermediate temperatures (approximately temperature T_2 in Fig. 8.17a, corresponding to the "nose" of the C curve), provide an optimal combination of nucleation driving force and diffusion rates for the fastest transformation rate.

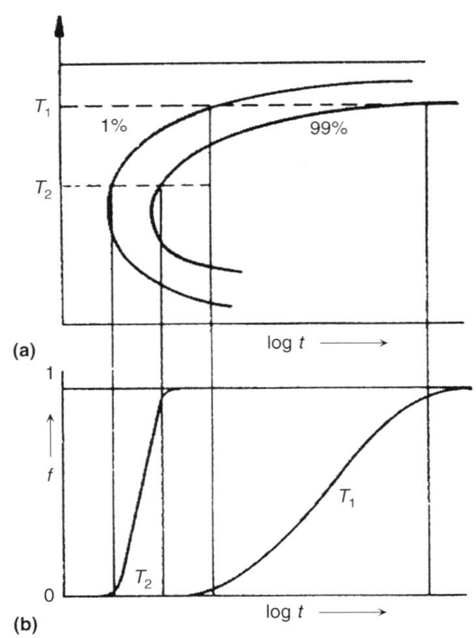

Fig. 8.17 (a) Time-temperature-transformation diagram indicating two temperatures. (b) Time required for transformation as a function of temperature. Source: Ref 8.12 as published in Ref 8.1

The lamellar spacing of the pearlite structure gives an indication of the transformation temperature. As the transformation temperature decreases, the diffusivity of carbon in austenite decreases, which acts to limit the interlamellar spacing. The variation of pearlite spacing versus transformation temperature for steel is shown in Fig. 8.18. The properties of fully pearlitic steels are determined by the interlamellar spacing. A relationship similar to the Hall-Petch relationship for grain size also exists for the interlaminar spacing in pearlite, with a smaller interlaminar spacing producing a higher yield strength. Thus, strength is related to interlamellar spacing, pearlite colony size, and prior austenite grain size, with finer sizes giving better properties. The thickness of the cementite lamellae can also influence the properties of pearlite. Fine cementite lamellae can be deformed, as compared with coarse lamellae which tend to crack during deformation.

Although fully pearlitic steels with a lamellar microstructure have high strength, high hardness, and good wear resistance, they also have rather poor ductility and toughness. For example, a low-carbon, fully ferritic steel will typically have a total elongation of more than 50%, whereas a fully pearlitic steel will typically have a total elongation of only approximately 10%. A low-carbon fully ferritic steel will have a room temperature Charpy V-notch impact energy of approximately 200 J (150 ft · lbf), whereas a fully pearlitic steel will have room-temperature impact energy of under

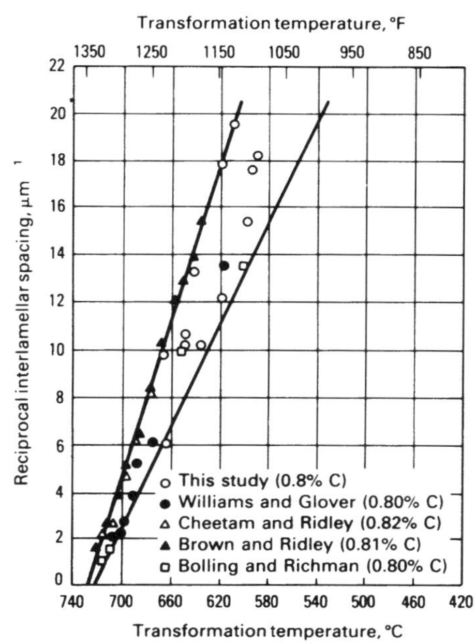

Fig. 8.18 Pearlite interlamellar spacing versus transformation temperature. Source: Ref 8.13 as published in Ref 8.1

10 J (7 ft · lbf). In addition, the transition temperature of a fully pearlitic steel is always well above room temperature. This means that at room temperature, the general fracture mode is the brittle fracture mode cleavage. Therefore, fully pearlitic steels with lamellar microstructure should not be used in applications where toughness is important. Also, pearlitic steels with carbon contents slightly or moderately higher than the eutectoid composition (i.e., hypereutectoid steels) have even lower toughness. Because of their low ductility/toughness, there are only a few applications for fully pearlitic steels, examples being railroad rails and wheels and high-strength wire. A fully pearlitic rail steel provides excellent wear resistance during railroad wheel-rail contact.

8.1.3 Hypoeutectoid and Hypereutectoid Structures

Hypoeutectoid, meaning compositions with carbon content less than the eutectoid, alloys contain 0.022 to 0.76 wt% C. Cooling of a hypoeutectoid alloy is represented by the vertical line in the Fig. 8.19 phase diagram. After solidification, at high temperatures, such as point *a*, the alloy is in the single-phase austenite field and is entirely austenite. When the alloy is cooled to approximately 800 °C (1475 °F) (point *b*), the alloy enters the two-phase $\alpha + \gamma$ phase field. Small particles of ferrite start forming along the austenite grain boundaries. This ferrite is called proeutectoid ferrite because it forms prior to cooling through the eutectoid temperature. This is in contrast to the eutectoid ferrite that forms at the eutectoid temperature as part of the constituent pearlite. While cooling the alloy through the $\alpha + \gamma$ phase region, the composition of the ferrite phase changes with temperature along the α-($\alpha + \gamma$) phase boundary, line *pr*, and becomes slightly richer in carbon. The change in composition of the austenite is more drastic, proceeding along the ($\alpha + \gamma$)-γ boundary, line *pq*, as the temperature is reduced. As the alloy further cools within the $\alpha + \gamma$ field (point *c*), the ferrite particles become larger in size and thicken along the austenite grain boundaries. When the alloy is finally cooled through the eutectoid temperature (point *d*), all of the remaining austenite transforms to pearlite. Because this reaction does not really affect the proeutectoid ferrite that has already formed at the austenite grain boundaries, the final microstructure is one of pearlite within the grain interiors surrounded by proeutectoid ferrite at the prior austenite grain boundaries.

The microstructure of a hypoeutectoid steel depends on how rapidly it is cooled from the austenitic phase region. Very slow cooling, such as furnace cooling, will result in the formation of proeutectoid ferrite at the austenite grain boundaries. The remaining austenite grains will then convert to pearlite as the steel cools through the eutectoid temperature. As the temperature decreases, the morphology of proeutectoid ferrite changes. Growth along the grain boundaries becomes markedly easier than growth normal to them. Therefore, films of proeutectoid ferrite form at the sites

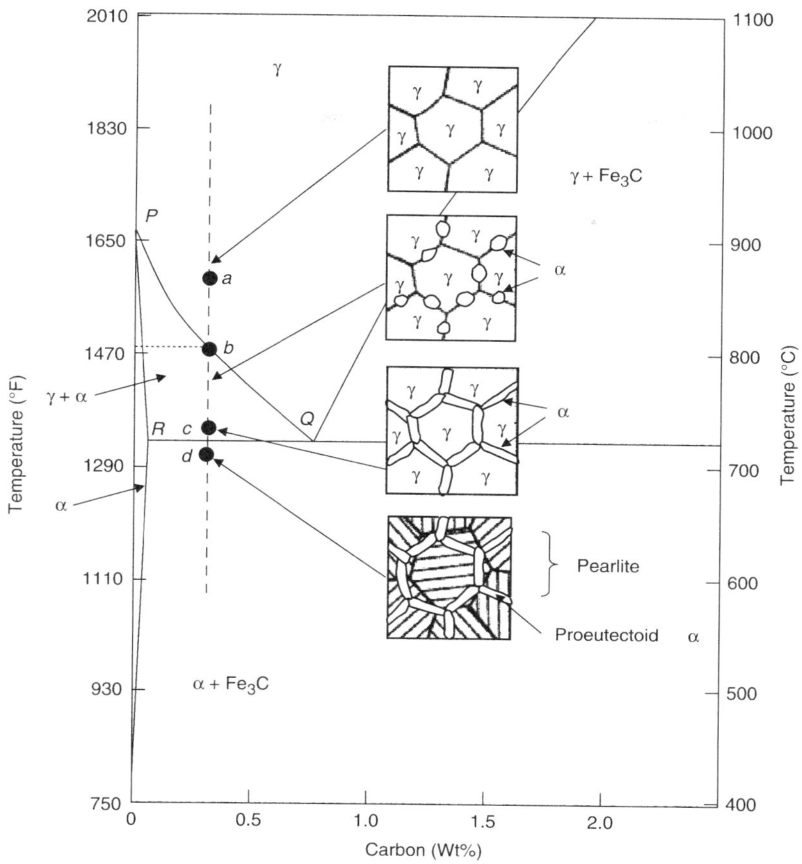

Fig. 8.19 Equilibrium cooling of a proeutectoid steel. Source: Ref 8.3

of the former austenite boundaries. If the cooling rate is fast, ferrite side-plates protrude from the proeutectoid ferrite and project into the austenite grains. These side-plates, called Widmanstätten plates, have a wedge-like shape, and their broad faces comprise flat areas separated by ledges (Fig. 8.20). The growth process now involves the nucleation and sideways migration of the ledges that separate the flats. Thus, the microstructure of hypoeutectoid steel that is cooled rapidly will contain Widmanstätten ferrite and fine pearlite.

The transformations of hypereutectoid steels, those containing 0.077 to 2 wt% C, is exactly analogous to hypoeutectoid steels, except that in this instance, proeutectoid cementite forms on the grain boundaries in the manner shown in the Fig. 8.21 phase diagram. This phase diagram also illustrates why hypereutectoid steels are only used where extreme hardness is required, such as in cutting tools. Note that the cementite forms a continuous network along the prior austenite grain boundaries. Because

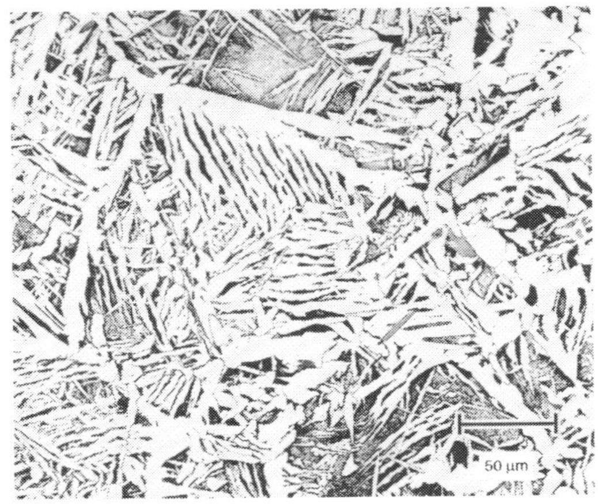

Fig. 8.20 Widmanstätten ferrite. Source: Ref 8.14 as published in Ref 8.3

cementite is a hard and brittle compound, these steels have very little ductility and can be prone to sudden brittle failures. When hypereutectoid steels are fractured, it is common to find a region of the fracture that appears to be intergranular where the crack propagated along the proeutectoid cementite grain-boundary films.

Compositions below 0.77 wt% C in iron-carbon systems will nucleate proeutectoid ferrite before reaching the eutectoid temperature and formation of pearlite. This ferrite nucleates at the grain boundaries to reduce energy and follows the Kurdjumov-Sachs relationship stated previously. A pearlite colony will then grow from the proeutectoid ferrite once the eutectoid temperature has been reached.

For compositions above the eutectoid point, proeutectoid cementite nucleates at the grain boundaries. This tends to form as a layer having the Bagaryatski orientation relationship with the γ_1 austenite grain ($[100]_c \parallel [01\bar{1}1]_\gamma$, $[010]_c \parallel [1\bar{1}\bar{1}]_\gamma$, $(001)_c \parallel (211)_\gamma$). A pearlite colony will then grow from the proeutectoid cementite once the eutectoid temperature has been reached (Fig. 8.22). In this case it is seen that pearlite nucleation is not limited predominantly to the grain boundaries.

In most ferrite-pearlite steels, the carbon content and the grain size determine the microstructure and resulting properties. For example, the effect of carbon on tensile and impact properties is shown in Fig. 8.23. The ultimate tensile strength steadily increases with increasing carbon content. This is caused by the increase in the volume fraction of pearlite in the microstructure, which has much higher strength than that of ferrite. Thus, increasing the volume fraction of pearlite has a profound effect on increasing tensile strength (Fig. 8.24). However, the yield strength is relatively unaffected by carbon content, rising from approximately 280 to 420 MPa (40 to 60 ksi) over the range of carbon content shown. This occurs because yielding in a ferrite-pearlite steel is controlled by the ferrite matrix, which

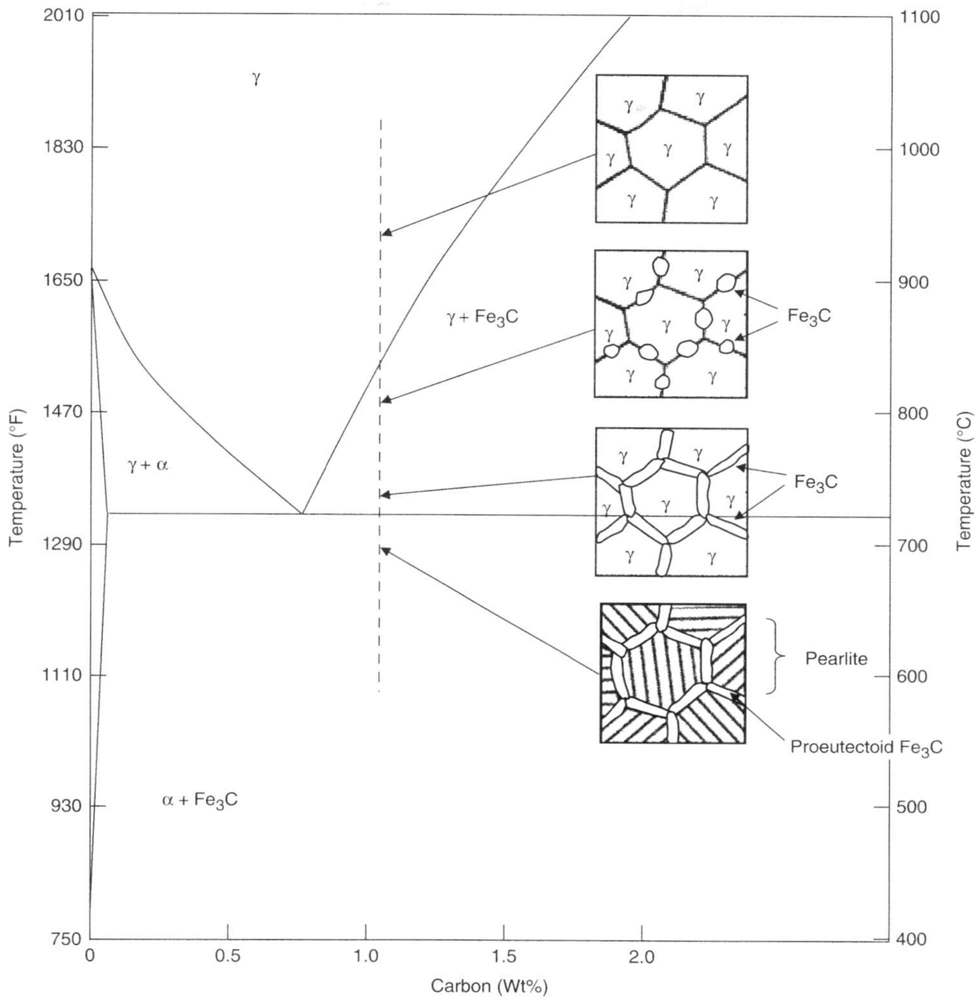

Fig. 8.21 Equilibrium cooling of a hypereutectoid steel. Source: Ref 8.3

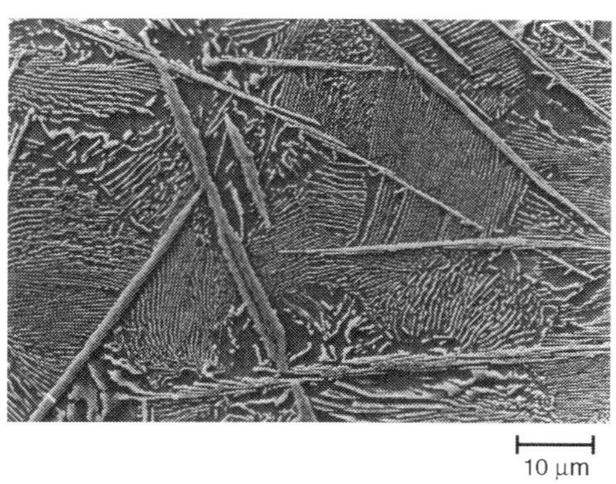

Fig. 8.22 Pearlitic microstructure with Widmanstätten cementite plates acting as nucleation sites. Source: Ref 8.11 as published in Ref 8.1

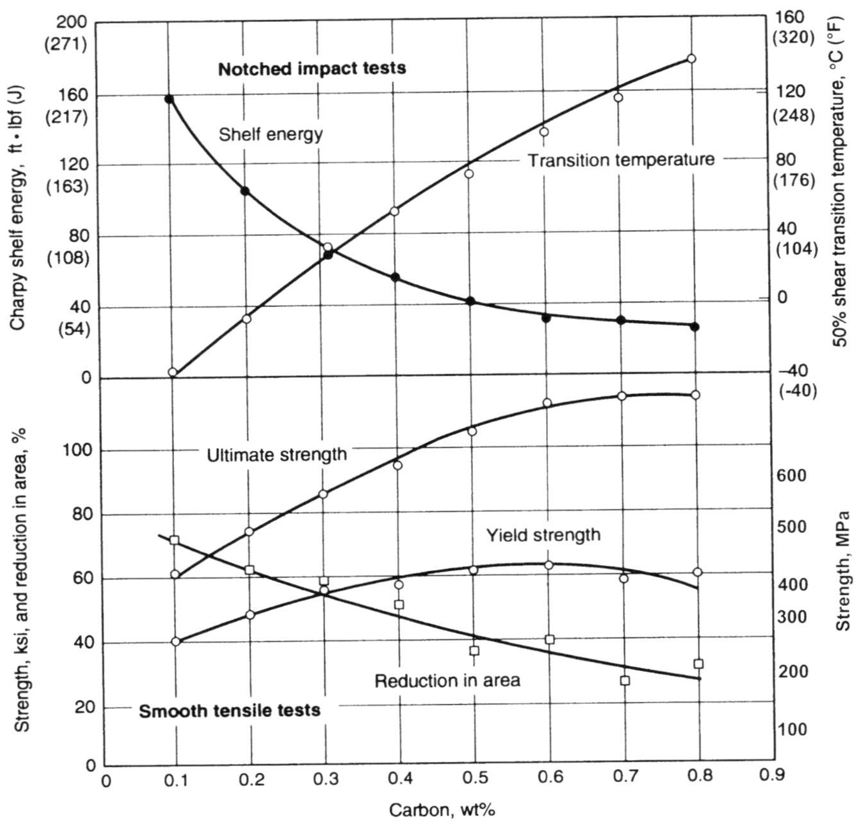

Fig. 8.23 Effects of carbon content on mechanical properties. Source: Ref 8.4 as published in Ref 8.3

is generally the continuous matrix phase in the microstructure. Therefore, pearlite plays only a minor role in yielding behavior. Ductility, as represented by reduction in area, steadily decreases with increasing carbon content. For example, a steel with 0.10 wt% C has a reduction in area of approximately 75%, whereas a steel with 0.70 wt% C has a reduction in area of only 25%.

During plastic deformation in pearlite lamellar microstructures, stress concentrations are created at the ends of pile-ups where the dislocations in the ferrite meet the ferrite-carbide interfaces, leading to carbide cracking. Cracking occurs more rapidly when the ferrite plates are thicker, because more dislocations can participate in an individual pileup for a given applied stress. The generation of cracks in cementite plates can initiate general fracture, which implies that coarse pearlite should fail before fine pearlite, because the cementite plates are expected to crack sooner when the ferrite lamellae are thicker. Also, the cracks will be longer and therefore more able to propagate. In other words, fine pearlite is both stronger and more ductile than coarse pearlite.

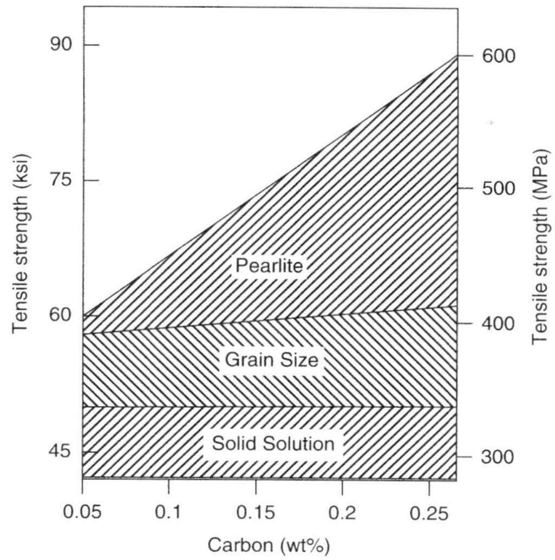

Fig. 8.24 Effect of pearlite content on mechanical properties. Source: Ref 8.15 as published in Ref 8.3

The toughness of ferrite-pearlite steels is another important consideration for applications. This is true for other bcc metals, almost all of which, unlike fcc metals, have a ductile-to-brittle transition temperature under impact loading. This is often measured by the Charpy V-notch impact test in which a notched square-section bar is broken by a swinging pendulum. Measuring the height of the pendulum at the beginning of the test and the height to which it rises after breaking the specimen enables the energy consumed in breaking the bar to be calculated. When the bar breaks in a brittle mode, the energy consumed is small. In this case, the bar breaks into two pieces and the pendulum swings onward and upward. When fracture occurs in a ductile fashion, a large amount of energy is expended by the deformation processes that precede fracture, as well as those accompanying fracture. Then, the distance traveled by the pendulum after the fracture is smaller. Extremely ductile samples may not even break, stopping the pendulum. The nature of the fracture and the amount of energy expended in an individual test can vary with the testing temperature; therefore, it is standard practice to perform Charpy tests over a range of temperatures.

It has long been known that the absorbed energy in a Charpy V-notch test decreases as the carbon content increases (Fig. 8.25). In this graph showing impact energy versus test temperature, the upper shelf energy decreases from approximately 200 J (150 ft · lbf) for a 0.11 wt% C steel to approximately 35 J (25 ft · lbf) for a 0.80 wt% C steel. Also, the transition temperature increases from approximately –50 to 150 °C (–60 to 300 °F) over this range of carbon content. The effect of carbon is due mainly to its effect on the percentage of pearlite in the microstructure. The eutectoid

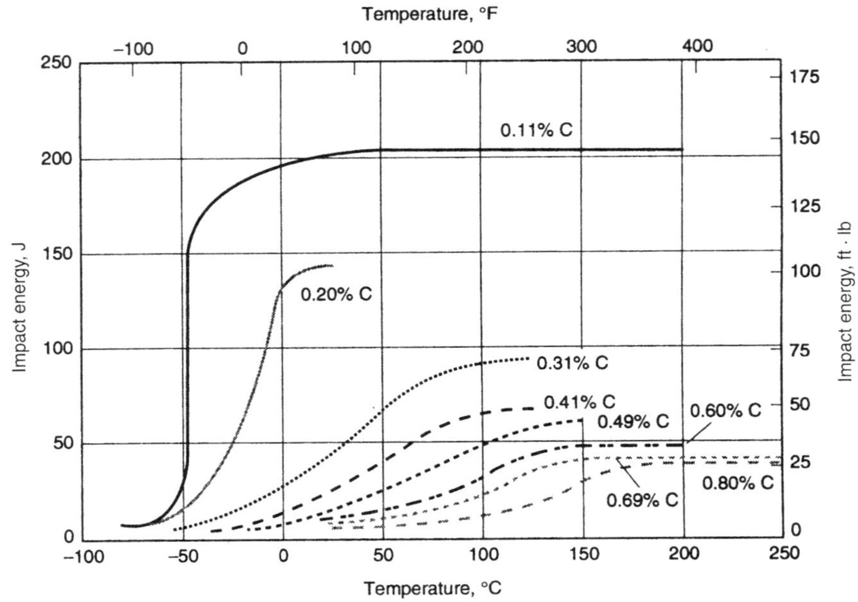

Fig. 8.25 Effect of carbon content on impact properties. Source: Ref 8.4 as published in Ref 8.3

steel, which is entirely pearlitic, shows a gradual transition of the energy consumed as a function of temperature. Presumably, the gradual transition at the high pearlite levels occurs because, even though the cementite plates crack, the total fracture path is tortuous in comparison to cleavage in the ferrite. Cleavage fracture in ferrite is a brittle fracture process in which the fracture path follows particular crystallographic planes in each grain, forming large, relatively flat facets.

Until approximately 20 years ago, hypoeutectoid steels with mixed ferrite-pearlite microstructures were commonly selected for structural applications because of their useful strength levels. Applications included beams for bridges and high-rise buildings, plates for ships, and reinforcing bars for roadways. These steels are relatively inexpensive and are produced in large tonnages. They also have the advantage of being able to be produced with a wide range of properties. Unfortunately, medium-to-high-carbon steels are not very amendable to welding and are therefore usually bolted or riveted together. In addition, their low toughness and high ductile-to-brittle transition temperatures can cause catastrophic failures. Today, high-strength low-alloy (HSLA) steels are replacing conventional hypoeutectoid steels in many structural applications. The HSLA steels, which have very low carbon contents, have an attractive combination of properties, including high strength, good toughness, a low ductile-to-brittle transition temperature, and are more readily weldable.

8.1.4 Alloying Elements

The addition of substitutional alloying elements causes the eutectoid composition and temperature to shift in the iron-carbon system. The effect of various substitutional alloying elements on the eutectoid transformation temperature and effective carbon content are shown in Fig. 8.26 and 8.27, respectively. Because the formation of pearlite is heavily dependent on long-range diffusion, the addition of these substitutional alloying elements will have a significant effect on the reaction kinetics as well. The addition of substitutional elements will decrease the diffusion rate and slow down the reaction kinetics, an effect known as partitioning and solute drag. Much like carbon, these substitutional alloying elements prefer to partition into either the ferrite or cementite lamellae of the pearlite. Because the rate of pearlite formation depends heavily on diffusion, the substitutional

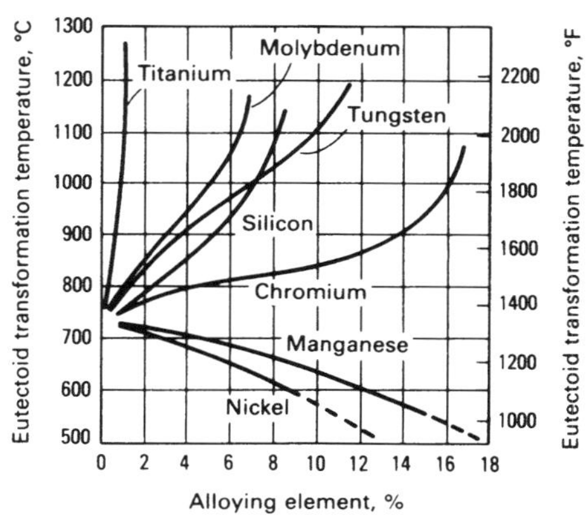

Fig. 8.26 Effect of alloying element on eutectoid temperature. Source: Ref 8.16 as published in Ref 8.1

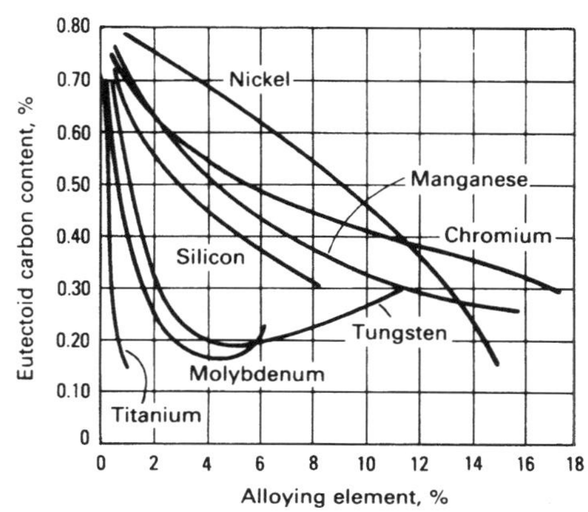

Fig. 8.27 Effect of alloying elements on effective carbon content. Source: Ref 8.16 as published in Ref 8.1

alloying elements control the rate of transformation due to substitutional diffusion being much slower than interstitial carbon diffusion. This effect is known as solute drag and can be seen by the shifting of the TTT diagram to the right, indicating increased time of reaction. The effects of alloying elements and their distribution in the pearlitic microstructure are shown in Fig. 8.28, while their effect on reaction time is evidenced by increasing reaction times in the TTT diagram (Fig. 8.29).

8.2 Peritectoid Structures

The term *peritectoid* denotes the special case of an equilibrium phase in which two or more solid phases (which are stable above the temperature T_p) react at T_p to form a new solid phase. This reaction can be written as:

$$\alpha + \beta \rightarrow \gamma$$

Peritectoid phase equilibria are very common in binary phase diagrams. Peritectoid transformations are similar to peritectic transformations, except that the initial phases are both solid. An example of a peritectoid transformation is provided by the formation of the intermetallic compound U_3Si in uranium-silicon alloys. The relevant phase diagram is shown in Fig. 8.30.

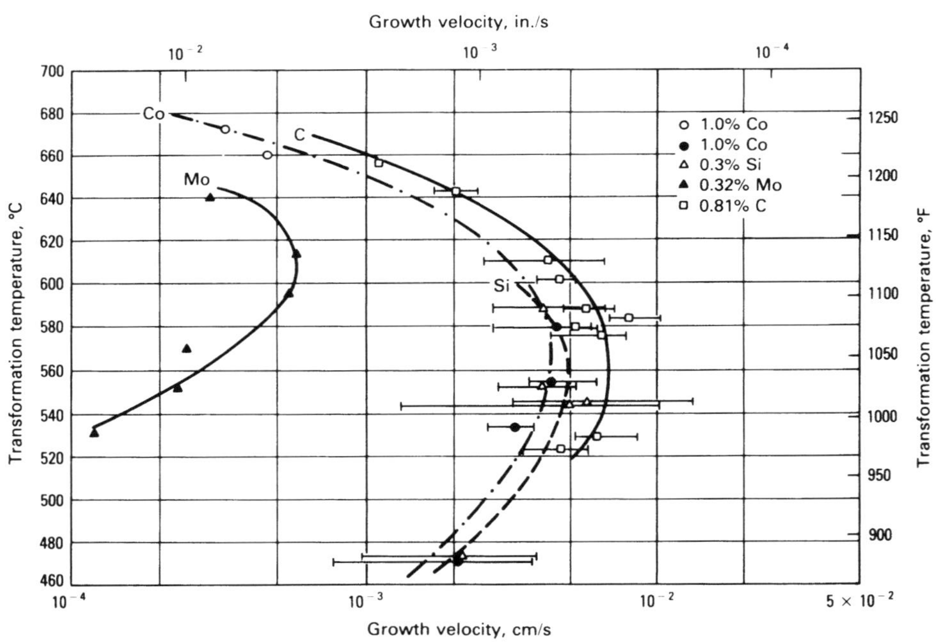

Fig. 8.28 The partitioning effect of substitutional alloying elements chromium, manganese, and silicon in pearlitic steel. Source: Ref 8.17 as published in Ref 8.1

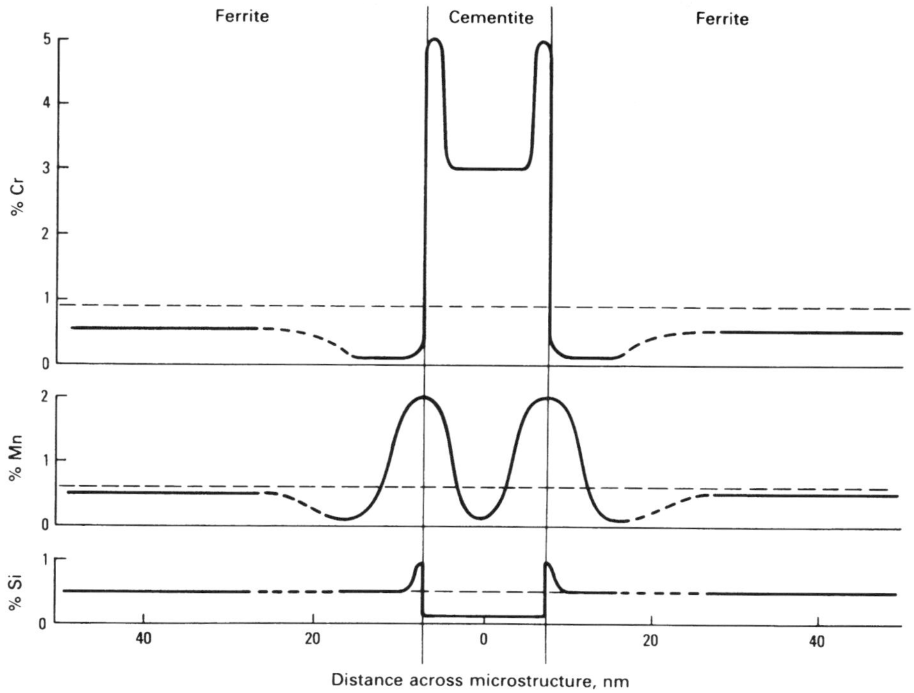

Fig. 8.29 Pearlite growth rate in Fe-C-X alloys as a function of temperature. Source: Ref 8.18 as published in Ref 8.1

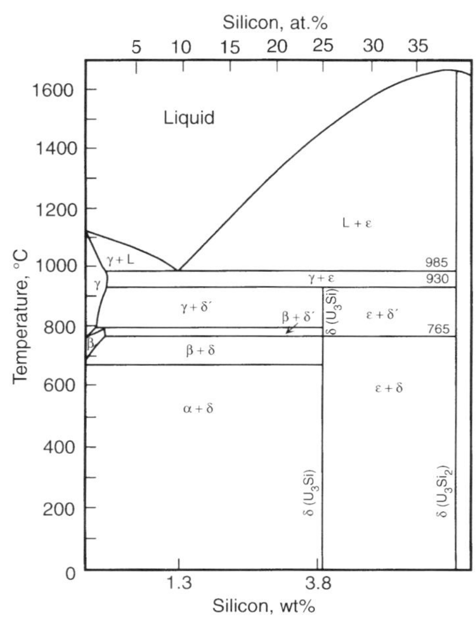

Fig. 8.30 Portion of the uranium-silicon phase diagram. Source: Ref 8.19 as published in Ref 8.1

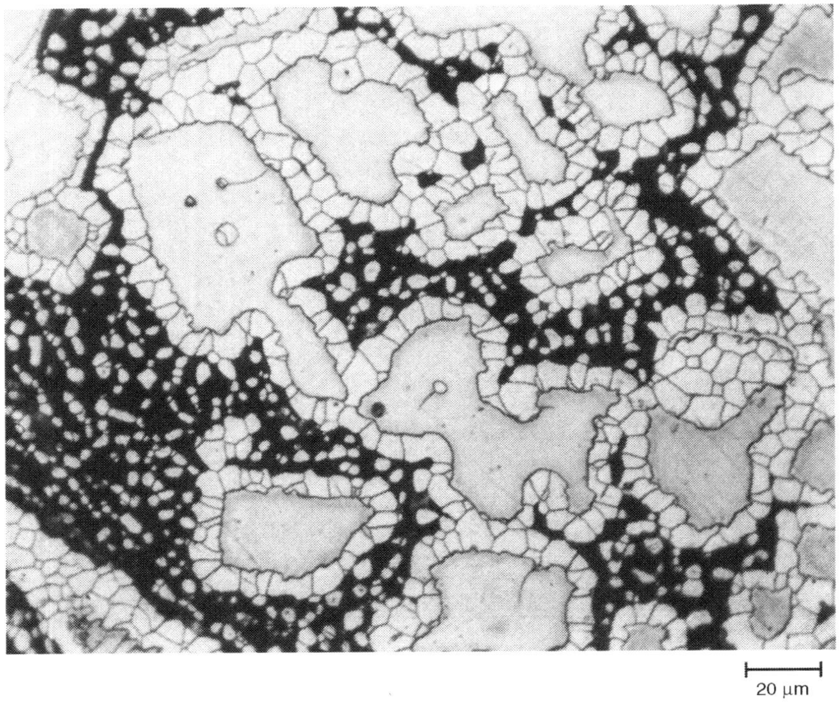

Fig. 8.31 Casting of a uranium-silicon alloy that contains 3.8% Si. Grains of U_3Si_2 are surrounded by grains of U_3Si on a background of a eutectic matrix that is a mixture of uranium and U_3Si. Original magnification: 500×. Source: Ref 8.19 as published in Ref 8.1

Casting uranium at the proper concentration of silicon (3.78%) results in a mixture of uranium and U_3Si_2. Morphologically, U_3Si_2 grains appear in a eutectic matrix, which is itself a mixture of uranium and U_3Si (Fig. 8.31). By means of thermal treatment below 930 °C (1700 °F), U_3Si grains grow at the boundary between the U_3Si_2 and the uranium phases. The reaction can be written:

$$U_3Si_2 + 3U \rightarrow 2U_3Si$$

This reaction is very slow, but when allowed to go to completion, all the material transforms to U_3Si (Fig. 8.32). The microstructure of the U_3Si phase consists mainly of transformation twins (Fig. 8.33).

ACKNOWLEDGMENT

The material for this chapter came from "Invariant Transformation Structures," in *Metallography and Microstructures,* Vol 9, *ASM Handbook*, ASM International, 2004, and *Elements of Metallurgy and Engineering Alloys* by F.C. Campbell, ASM International, 2008.

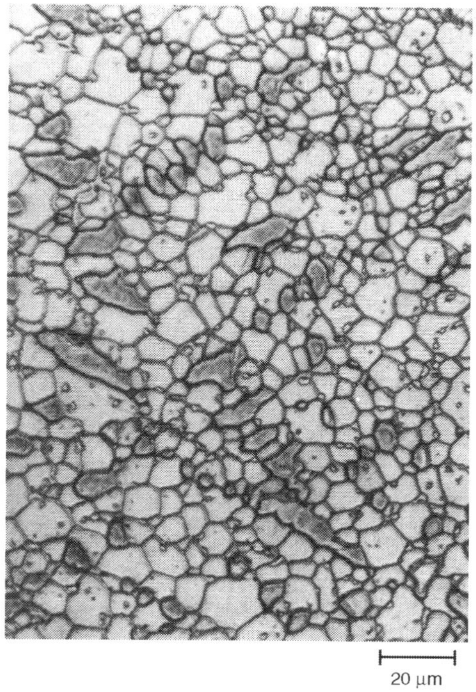

Fig. 8.32 Same uranium-silicon alloy as Fig. 8.31, but the casting has been thermally treated at 900 °C (1650 °F) for several hours. Structure is U_3Si, within which are contained the remnants of U_3Si_2. Original magnification: 500×. Source: Ref 8.1

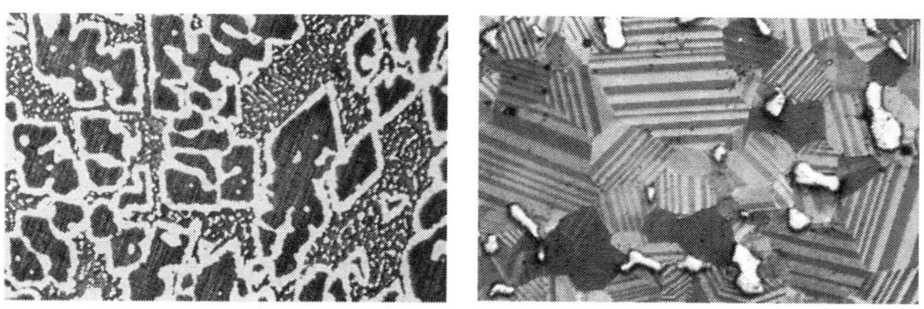

Fig. 8.33 Structure in U-3.8%Si alloy. (a) As-cast structure with U_3Si_2 (brown) surrounded by a rim of U_3Si (white) in a matrix of U-U3Si eutectic. (b) Same casting as in (a) but after heating for three days at 870 °C (1600 °F). U_3Si twinned martensite is colored; untransformed U_3Si remains uncolored. Source: Ref 8.19 as published in Ref 8.1

REFERENCES

8.1 Invariant Transformation Structures, *Metallography and Microstructures,* Vol 9, *ASM Handbook*, ASM International, 2004, p 152–164

8.2 T.B. Massalski, H. Okamoto, P.R. Subramanian, and L. Kacprzak, *Binary Alloy Phase Diagrams*, 2nd ed., ASM International, 1990

8.3 F.C. Campbell, *Elements of Metallurgy and Engineering Alloys*, ASM International, 2008

8.4 B.L. Bramfitt, Effects of Composition, Processing, and Structure on Properties of Irons and Steels, *Materials Selection and Design*, Vol 20, *ASM Handbook,* ASM International, 1997

8.5 A.G. Guy and J.R. Hren, *Elements of Physical Metallurgy*, 3rd ed., Addison-Wesley Publishing Company, 1974

8.6 D. Aliya and S. Lampman, Physical Metallurgy Concepts in Interpretation of Microstructures, *Metallography and Microstructures*, Vol 9, *ASM Handbook,* ASM International, 2004

8.7 A.R. Marder, *Phase Transformations in Ferrous Alloys,* A.R. Marder and J.I. Goldstein, Ed., TMS/AIME, 1984, p 201–236

8.8 M.A. Mangan and G.J. Shiflet, *Metall. Mater. Trans. A,* Vol 30A (No. 11), 1999, p 2767–2781

8.9 Z. Guo, T. Furuhara, and T. Maki, *Scr. Mater.,* Vol 45, 2001, p 525–532

8.10 D.V. Shtansky, K. Nakai, and Y. Ohmori, *Acta Mater.,* Vol 47 (No. 4), 1999, p 1105–1115

8.11 D.S. Zhou and G.J. Shiflet, *Metall. Trans. A,* Vol 22A (No. 6), 1991, p 1349–1365

8.12 D.A. Porter and K.E. Easterling, *Phase Transformations in Metals and Alloys,* 2nd ed., Chapman and Hall, London, 1992

8.13 A.R. Marder and B.L. Bramfitt, *Metall. Trans. A,* Vol 6, 1975, p 2009–2014

8.14 B.L. Bramfitt and S.J. Lawrence, Metallography and Microstructures of Carbon and Low-Alloy Steels, *Metallography and Microstructures*, Vol 9, *ASM Handbook,* ASM International, 2004

8.15 R.A. Higgins, *Engineering Metallurgy*, 6th ed., Arnold, 1993

8.16 E.C. Bain and H.W. Paxton, *Alloying Elements in Steel,* American Society for Metals, 1962, p 112

8.17 P.R. Williams, M.K. Miller, P.A. Beavan, and G.D.W. Smith, *Phase Transformations,* Vol 2, The Institution of Metallurgists, London, 1979, p 11.98–11.100

8.18 A.R. Marder and B.L. Bramfitt, *Metall. Trans. A,* Vol 7, 1976, p 902–906

8.19 A. Tomer, Peritectoid Transformations, *Structure of Metals Through Optical Microscopy,* ASM International, 1991

CHAPTER 9

Intermediate Phases

PHASE DIAGRAMS are often quite complex, with a number of different reactions occurring at different compositions and temperatures. In most cases, the appearance of several reactions in a binary phase diagram is the result of the presence of intermediate phases. These are phases whose chemical compositions are intermediate between the two pure metals, and whose crystalline structures are different from those of the pure metals. The difference in crystalline structure distinguishes intermediate phases from primary solid solutions, which are based on pure metals. Some intermediate phases can accurately be called intermetallic compounds, when, like Mg_2Pb, they have a fixed simple ratio of the two kinds of atoms. However, many intermediate phases that exist over a range of compositions are considered to be intermediate or secondary solid solutions.

Intermediate phases can be classified based on their melting behavior, either congruent or incongruent melting phases. On heating, an incongruent melting phase decomposes into two different phases, usually one solid and one liquid, such as a peritectic transformation. A congruent melting phase melts in the same manner as a pure metal. In this case, the phase diagram is divided into essentially independent sections. In Fig. 9.1, the congruent melting β phase divides the lead-magnesium diagram into two separate eutectic reactions that can be analyzed separately. The β phase is the intermetallic compound Mg_2Pb that has a simple fixed ratio of the two kinds of metal atoms. The details of different types of intermetallic compounds were covered in Chapter 2, "Solid Solutions and Phase Transformations" in this book. Many intermetallic compounds are avoided in alloying because they tend to be hard and brittle.

The copper-zinc phase diagram shown in Fig. 9.2 contains intermediate phases with appreciable ranges of solid solubility. In this diagram, the copper-rich α solid solution and the zinc-rich η solid solution are the two terminal phases, and the four intermediate phases are β, γ, δ, and ε. The copper-zinc system forms the basis for the industrially important brass alloys. As shown in the copper-rich portion of the copper-zinc diagram

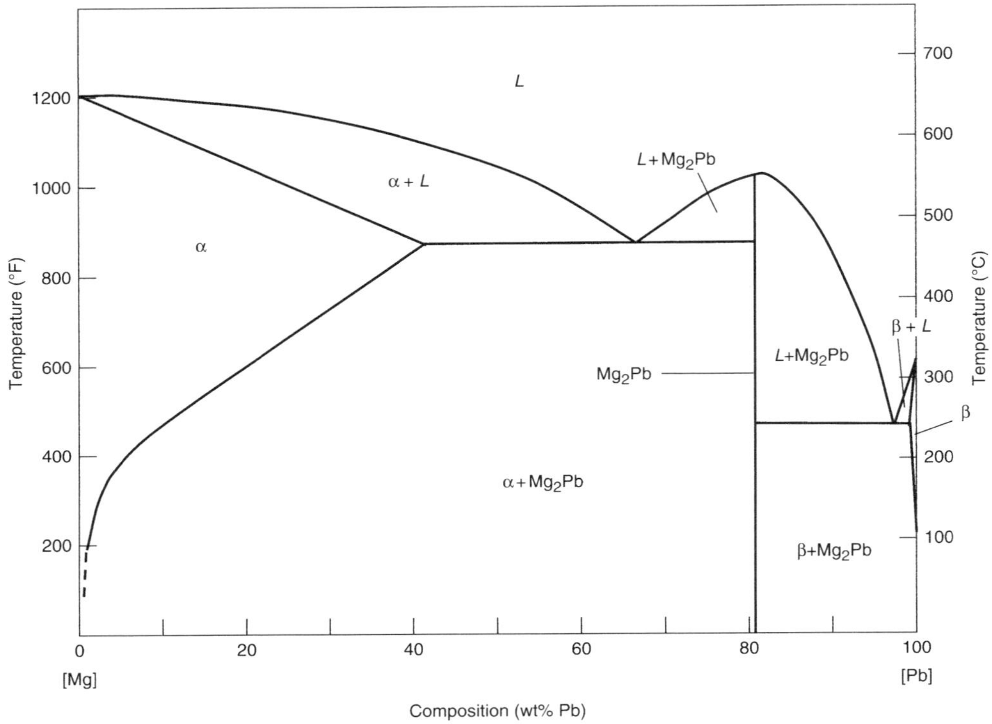

Fig. 9.1 Compound formation in magnesium-lead system. Source: Ref 9.1

of Fig. 9.3, α-copper can dissolve up to 32.5% Zn at the solidus temperature of 900 °C (1655 °F), the proportion increasing to 39.0% at 455 °C (850 °F). Although the amount of zinc decreases with further decreases in temperature, diffusion is very sluggish at temperatures below 455 °C (850 °F), and with ordinary industrial cooling rates, the amount of zinc that can remain in solid solution in copper at room temperature is approximately 39%. When the amount of zinc is increased beyond 39%, the intermediate ordered β phase, equivalent to CuZn, will form on slow cooling. This phase is hard at room temperature and has limited ductility but becomes plastic when it changes to the disordered β phase above 455 °C (850 °F). Unlike copper, which is face-centered cubic (fcc) and zinc, which is hexagonal close-packed (hcp), the β phases are body-centered cubic (bcc). The high-temperature β phase is disordered, while the lower-temperature β′ phase is ordered (Fig. 9.4). These alloys are easy to machine and hot form but are not very amendable to cold forming, because the β′ phase is brittle at room temperature. There is also a rapid reduction in the impact strength with a simultaneous increase in hardness and tensile strength. The maximum strength is attained at 44% Zn. Further increases in the zinc content beyond 50% causes the γ phase to form, which is too brittle for engineering alloys.

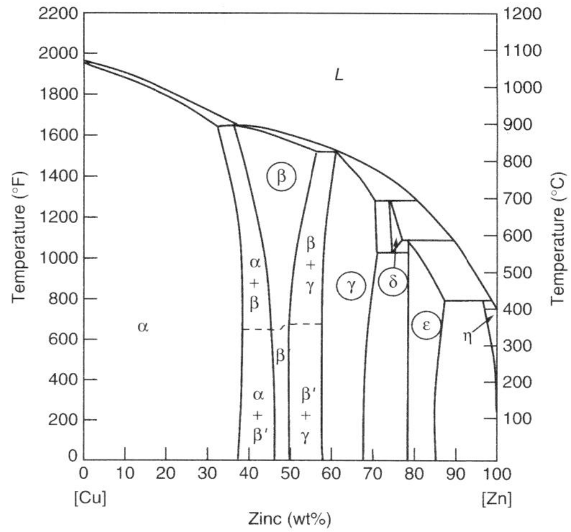

Fig. 9.2 Copper-zinc phase diagram. Source: Ref 9.1

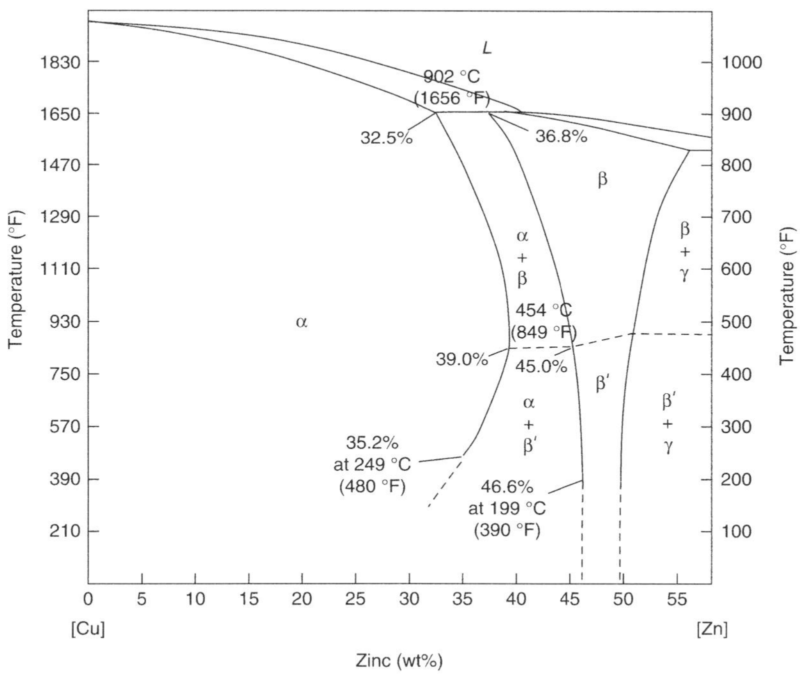

Fig. 9.3 Copper-rich section of copper-zinc phase diagram. Source: Ref 9.1

Because alloys that contain only the α phase are quite soft and ductile at room temperature, the α-brasses are very amendable to cold working. However, as a result of the hard, ordered β′ phase, brasses containing both the α + β′ phases are rather hard, with a low capacity for cold work. The α

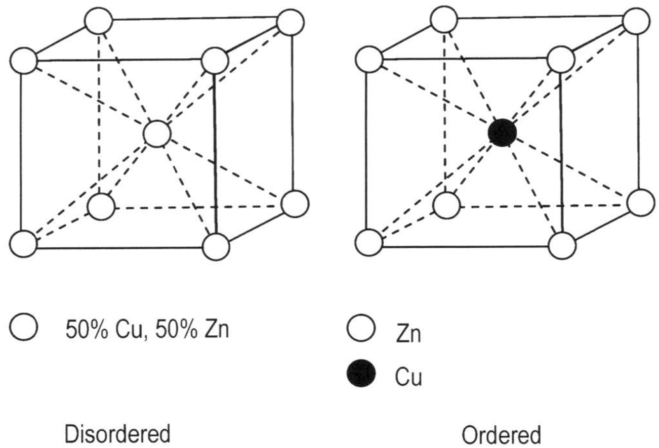

Fig. 9.4 Disordered and ordered structures in 50% Cu-50% Zn β-brasses. Source: Ref 9.1

+ β′ brasses are hot worked in the temperature range where β′ transforms β. Thus, the α-brasses are often referred to as cold work brasses and the α + β′ brasses are known as hot work brasses.

9.1 Order-Disorder Transformations

An order-disorder transformation typically occurs on cooling from a disordered solid solution to an ordered phase. During this phase transformation, there is a rearrangement of atoms from random site locations in the disordered solution to specific lattice sites in the ordered structure. When atoms periodically arrange themselves into a specific ordered array, they make up what is commonly referred to as a superlattice. Although many alloy systems may contain ordered phases, only select superlattices are discussed in this chapter. Four common superlattice structures and ordered phases that atomically arrange into the corresponding superlattice are listed in Table 9.1. As noted in the table, the superlattice types can be referred to by Strukturbericht symbols ($L1_0$, $L1_2$, B2, and $D0_3$) or by the prototype phase (CuAu I, Cu_3Au, FeAl, Fe_3Al).

Antiphase Boundaries. Most alloys that form an ordered structure are disordered at higher temperature, which means that atoms are randomly located on lattice sites. On cooling, small ordered areas will nucleate within the disordered phase and begin to grow into ordered domains. These ordered domains can also form by a continuous ordering mechanism, where local atomic rearrangements occur homogeneously throughout the disordered phase, creating ordered domains. As the temperature is decreased further, the ordered domains will grow until they impinge on or intersect each other and form antiphase boundaries (APBs). Antiphase boundaries are boundaries between two ordered domains where the periodicity of the ordered structure in one domain is out of step with the other. This is shown in Fig. 9.5, which is a schematic representing the phase transformation from a disordered structure at elevated temperature to the ordered structure, with APBs located where the domains intersect. The APBs are typically well defined within the structure and can be seen fairly easily using thin-film transmission electron microscopy (TEM).

Long-Range and Short-Range Order. A perfectly ordered structure is one that has the periodic arrangements of atoms throughout the entire crystal, without the presence of any defects. Practically, defects that disrupt the atomic sequence are typically found within the crystal lattice. For example, if an aluminum atom replaced an iron atom in the ordered FeAl phase, the structure would be less than perfect. Therefore, a parameter, *S*,

Table 9.1 Selected superlattice structures and alloy phases that order according to each superlattice

Strukturbericht symbol	Prototype phase	Base lattice type	Phases
$L1_0$	CuAu I	Face-centered cubic	AgTi, AlTi, CoPt, CrPd, CuAu, Cu_3Pd, FePd, FePt, HgPd, HgPt, HgTi, HgZr, InMg, MgTl, MnNi, Mn_2Pd_3, MnPt, NiPt, PbZn, PtZn
$L1_2$	Cu_3Au	Face-centered cubic	$AgPt_3$, Ag_3Pt, $AlCo_3$, $AlNi_3$, $AlZr_3$, $AuCu_3$ I, Au_3Pt, $CaPb_3$, $CaSn_3$, $CdPt_3$, $CePb_3$, $CeSn_3$, $CoPt_3$, Cr_2Pt, CuPd, Cu_3Au, Cu_3Pt, $FeNi_3$, $FePt_3$, Fe_3Pt, $GeNi_3$, $HgTi_3$, $InMg_3$, $LaPb_3$, $LaSn_3$, $MnNi_3$, $MnPt_3$, Mn_3Pt, $NaPb_3$, Ni_3Pt, $PbPd_3$, $PbPt_3$, Pt_3Sn, Pt_3Ti, Pt_3Zn, $TiZn_3$
B2	FeAl	Body-centered cubic	AgCd, AgCe, AgLa, AgLi, AgMg, AlCo, $AlCu_2Zn$, AuCd, AuMg, AuMn, AuZn, BeCo, BeCu, BeNi, CdCe, CeHg, CeMg, CeZn, CoFe, CoTi, CsCl, CuPd, CuZn, $CuZn_3$, FeAl, FeTi, $HgLi_2Tl$, HgMn, InNi, LaMg, LiPb, LiTl, MgPr, MgSr, MgTl, MnPt, NiAl, NiTi, RuTa, TiZn
$D0_3$	Fe_3Al	Simple cubic	$BiLi_3$, $CeMg_3$, Cu_3Sb, Fe_3Al, Fe_3Si, Mg_3Pr

Source: Ref 9.2 as published in Ref 9.3

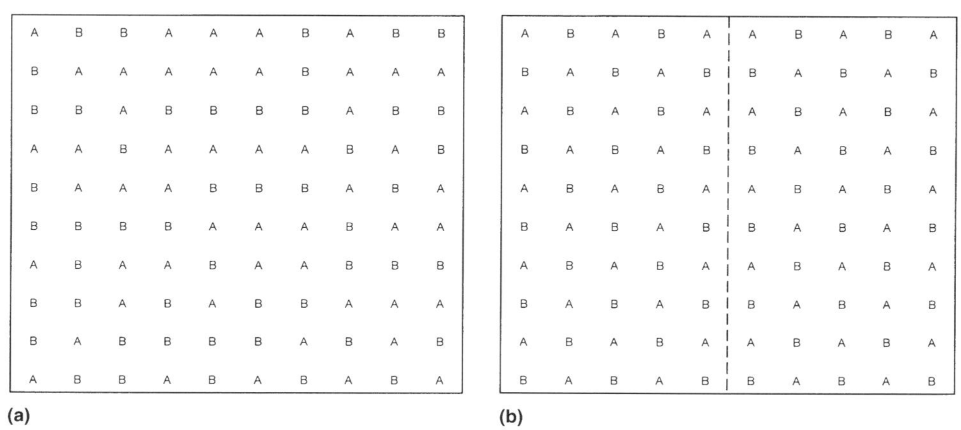

Fig. 9.5 Schematic representation of (a) a disordered solution and (b) an ordered structure with an antiphase boundary (APB) (dashed line) located where the atomic sequence is out of step. Source: Ref 9.3

was established to quantify the degree of long-range order within a crystal. For binary alloys (alloy *A-B*), if *A* atoms occupy the α sublattice, and *B* atoms occupy the β sublattice:

$$S = \frac{f_A(\alpha) - f_A}{1 - f_A}$$

where f_A is the fraction of all *A* atoms in the alloy, and $f_A(\alpha)$ is the fraction of *A* atoms that lie on the α sublattice. The term $f_A(\alpha)$ can also be described as the probability that an *A* atom occupies an α site. It can be seen from this equation that the degree of long-range order can be quantified, where a completely disordered solution is present when $S = 0$, and a perfectly ordered structure is present when $S = 1$.

Long-range order is used to describe the degree of ordering throughout an entire crystal. There is also a degree of ordering, known as short-range ordering, that is associated with a single atom and its nearest neighbors. In a crystal lattice, each atom has first- and second-order nearest neighbors. In an *A-B* alloy with a random arrangement of atoms (disordered solution), each *A* atom should have an average number of *B* nearest neighbors. In an ordered structure, the number of *B* nearest neighbors surrounding the *A* atom should increase. The amount of segregation of *B* atoms around a single *A* atom is considered the degree of short-range ordering, σ. The degree of short-range order can be described quantitatively as:

$$\sigma = \frac{q - q_r}{q_m - q_r}$$

where q is the total number of *A-B* pairs, q_r is the average number of *A-B* pairs in a disordered solution (randomly arranged), and q_m is the maximum number of *A-B* pairs that are possible. As was the case with the long-range order parameter, when the degree of short-range order is 0 ($\sigma = 0$), a disordered alloy is present, and when $\sigma = 1$, the alloy is completely ordered.

$L1_0$ Superlattice (CuAu I Structure). At elevated temperatures, the CuAu alloy is a disordered solution with a fcc lattice, where copper and gold atoms are randomly located at the face and corner sites (Fig. 9.6a). When the CuAu alloy is cooled and transforms to the ordered CuAu I structure, the gold atoms remain at the top and bottom faces and at the unit cell corners, while copper atoms are located at the side face sites, causing a slight change in the lattice parameter (Fig. 9.6b). It can be seen that the resultant superlattice is comprised of a layer of copper atoms located between two layers of gold atoms. The only other type of domain that can be present in the CuAu I phase occurs when a layer of gold atoms is located between two layers of copper atoms.

L1$_2$ Superlattice (Cu_3Au Structure). The Cu_3Au phase has the same disordered structure as the CuAu phase at high temperatures, with the exception that the gold atoms have a 25% probability of sitting on a lattice site, rather than a 50% probability for the CuAu alloy. On cooling to the ordered Cu_3Au structure, gold atoms relocate to the corner positions, while the copper atoms arrange on the faces (Fig. 9.7a). Three other atomic arrangements are possible for the ordered Cu_3Au structure and are similar in appearance (Fig. 9.7b–d). These arrangements have copper atoms at the corner sites and at two pairs of faces, while the gold atoms are located at the remaining pair of faces. Note that the unit cell still contains three copper atoms and one gold atom, regardless of the atomic locations. Because there are four possible ordered configurations for the Cu_3Au alloy, neighboring domains can be out of step in four possible ways and will lead to APBs that intersect each other to form sharp boundaries (Fig. 9.8).

B2 Superlattice (FeAl Structure). Alloys that transform to a B2 superlattice on cooling typically transform from a bcc solution, where atoms are randomly located at either the center position or at the corners. One alloy that undergoes this type of phase transformation is FeAl. At high

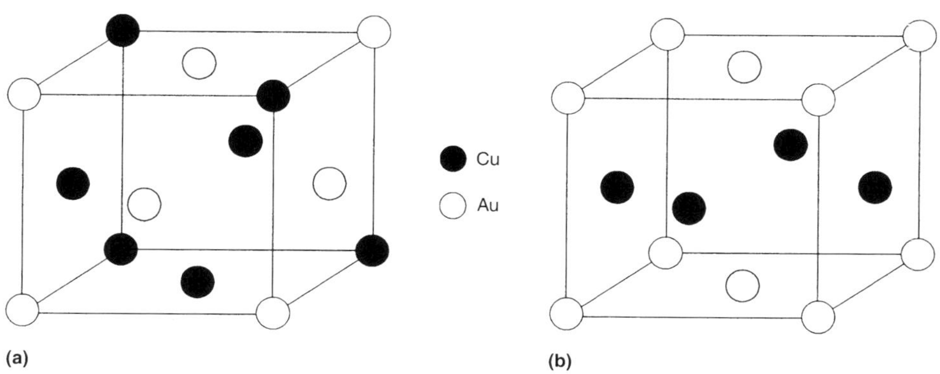

Fig. 9.6 Unit cells of (a) the disordered CuAu face-centered cubic solution at elevated temperatures and (b) the ordered CuAu I structure representing the L1$_0$ superlattice. Source: Ref 9.3

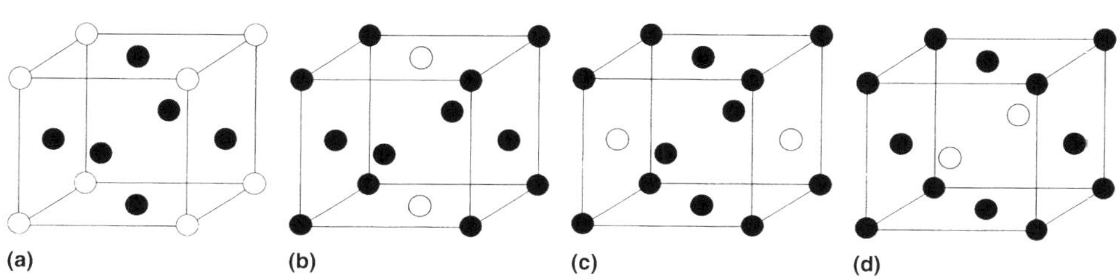

Fig. 9.7 Unit cells of Cu_3Au representing the four domain types found in the L1$_2$ superlattice: where the gold atoms (white) sit (a) at the corners and (b–d) at the face sites. Source: Ref 9.3

temperatures, the FeAl alloy starts as a disordered bcc lattice, where either atom has a 50% probability of lying on the body-centered site or at the corners. Once the transformation is complete, there are two possible types of ordered domains: where the aluminum atoms sit at the center positions and the iron atoms are at the corners; or where the aluminum atoms are at the corners and the iron atoms are at the center sites (Fig. 9.9). It can be seen from Fig. 9.10 that distinct APBs can be resolved using TEM and that these curved APBs differ from the straight APBs that were present in the ordered Cu_3Au structure.

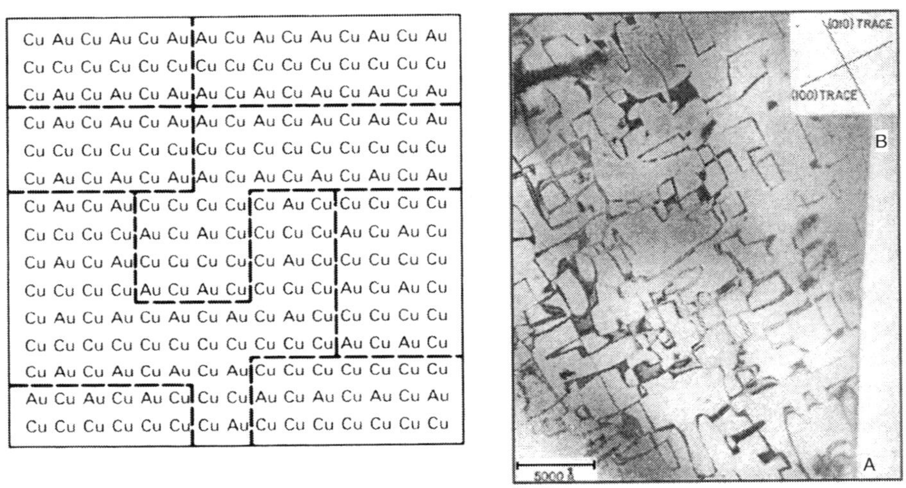

(a) (b)

Fig. 9.8 Schematic diagram (a) showing the atomic configuration in Cu_3Au that results in the formation of straight APBs, which can be seen (b) using transmission electron microscopy. Source: Ref 9.4 and 9.5, respectively, as published in Ref 9.3

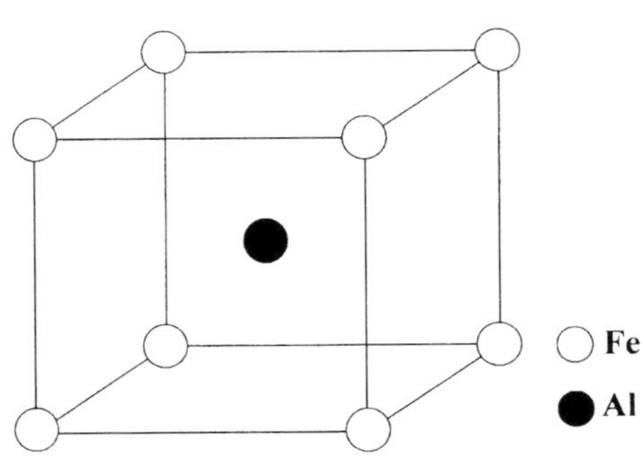

Fig. 9.9 Unit cell for one type of domain found in the ordered FeAl phase representing the B2 superlattice (switch atoms for other domain). Source: Ref 9.3

$D0_3$ Superlattice (Fe_3Al Structure). One of the most common alloys to transform into the $D0_3$ superlattice is Fe_3Al. The $D0_3$ superlattice is based on a bcc structure, which can be considered as two interpenetrating simple cubic sublattices. The corners of the bcc unit cell are taken to be one simple cubic sublattice, and the body-centered atoms make up the other. The $D0_3$ superlattice is present when half of the lattice sites on one of the simple cubic sublattices are occupied by a specific atom (such as aluminum in Fe_3Al). This can be seen in Fig. 9.11, which shows the ordered Fe_3Al structure. The transformation to Fe_3Al occurs differently from most other order-disorder transformations, because the ordered FeAl phase will typically transform from the disordered solution first, and then FeAl will undergo a transformation to Fe_3Al. This unique transformation results in two distinct types of APBs in the structure: one from FeAl (wrong first-nearest neighbors) and the other from Fe_3Al (wrong second-nearest neighbors). It can be seen from Fig. 9.12 that the small-scale APBs asso-

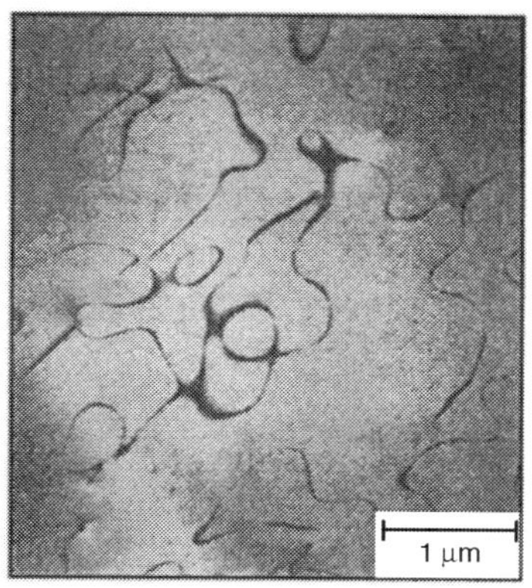

Fig. 9.10 Antiphase boundaries in the ordered FeAl phase. Source: Ref 9.6 as published in Ref 9.3

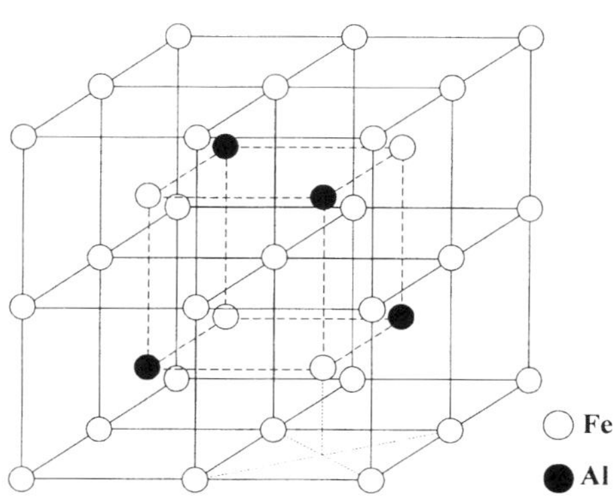

Fig. 9.11 Unit cell for the ordered Fe_3Al phase demonstrating the $D0_3$ superlattice. Source: Ref 9.3

ciated with Fe_3Al are curved and terminate at the APBs associated with FeAl.

Dislocation-Generated Antiphase Boundaries. Although APBs were previously defined as the boundary between two ordered domains where the domain atomic sequence is out of step, APBs can also be generated by dislocation motion. In an ordered structure, a defect in the atomic arrangement caused by the presence of a dislocation can cause the atomic sequence to be out of step and thus generate an APB (lower section of Fig. 9.13). The amount of dislocation-generated APBs can be minimized

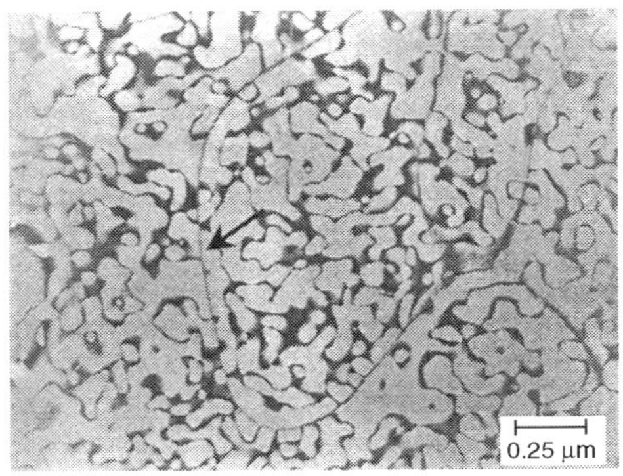

Fig. 9.12 Two types of APBs found in the ordered Fe_3Al phase: the large, curved APB from the formation of FeAl (arrow) and the small APBs within the large FeAl domains from the transformation to Fe_3Al. Source: Ref 9.7 as published in Ref 9.3

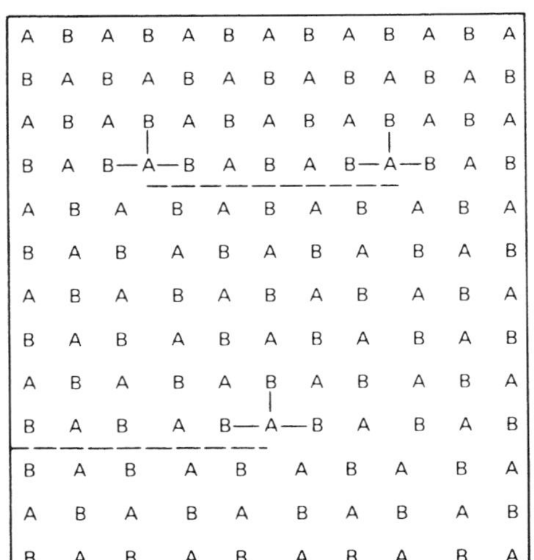

Fig. 9.13 Schematic representation of dislocation-generated APBs. The lower APB is generated by one edge dislocation, while the upper APB is terminated between a pair of edge dislocations, creating a superlattice dislocation. Source: Ref 9.8 as published in Ref 9.3

if dislocation pairs (for example, two edge dislocations) align such that the dislocation-generated APB is terminated at the other edge dislocation (upper section of Fig. 9.13). These APBs that are present between two dislocations are known as superlattice dislocations. Superlattice dislocations reduce the number of incorrect atomic bonds caused by APBs and therefore reduce the overall energy of the structure. Dislocation-generated APBs can be investigated by TEM, as shown in Fig. 9.14.

9.2 Spinodal Transformation Structures

Spinodal transformation is a phase separation reaction that occurs from kinetic behavior first described by Gibbs in his treatment of the thermodynamic stability of undercooled or supersaturated phases. It is does not involve a nucleation step, which is the mechanism of classical nucleation and growth (Fig. 9.15a) of precipitates from a metastable solid solution. Instead, spinodal reactions involve spontaneous unmixing or diffusional clustering of atoms, where a two-phase structure forms by spontaneous growth from small composition fluctuations (Fig. 9.15b). The result is homogenous decomposition of a supersaturated single phase into two phases that have essentially the same crystal structure (but different composition) as the parent phase. Spinodal structures are characterized by what has been described as a woven or "tweedy" structure. The precipitation occurs in preferential crystallographic directions, providing an obvious geometric pattern in two or three directions.

The underlying theory of spinodal decomposition involves the concept of the gradient energy. The concept of gradient energy (which is the energy

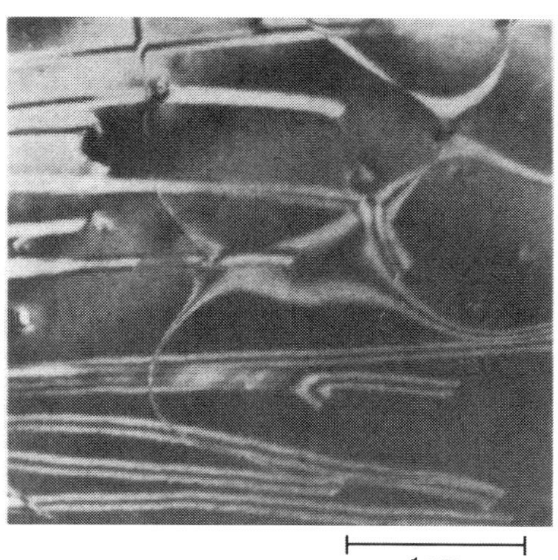

Fig. 9.14 Dislocation-generated APBs in ordered Fe_3Al. Thin-foil electron micrograph. Original magnification: 20,000×. Source: Ref 9.8 as published in Ref 9.3

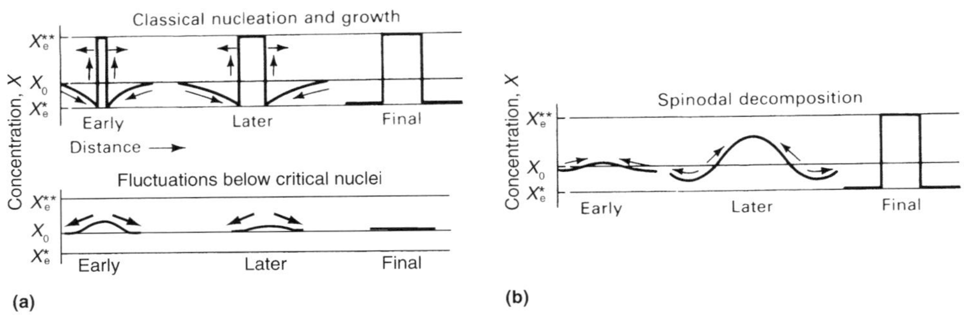

Fig. 9.15 Two sequences for the formation of a two-phase mixture by diffusion processes. (a) Classical nucleation and growth. (b) Spinodal decomposition. Source: Ref 9.9 as published in Ref 9.10

associated with a diffuse interface) in spinodal decomposition plays an equivalent role to interfacial energy in nucleation and growth reactions for setting the length scale and driving coarsening. When a composition fluctuation has a large characteristic length, growth (or amplification) of the fluctuation is sluggish because the diffusion distances are very long. When the composition fluctuation has a very short wavelength, growth of the fluctuation is suppressed by the so-called energy gradient or surface energy of the diffuse or incipient interfaces that form during phase separation. Therefore, the microstructure that develops during spinodal decomposition has a characteristic periodicity that is typically 2.5 to 10 nm (25 to 100 Å) in metallic systems.

The spinodal mechanism provides an important mode of transformation, producing uniform, fine-scale, two-phase mixtures that can enhance the physical and mechanical properties of commercial alloys. Spinodal decomposition has been particularly useful in the production of permanent magnet materials, because the morphologies favor high coercivities. The structure can be optimized by thermomechanical processing, step aging, and magnetic aging. Continuous phase separation or spinodal decomposition appears to be important in the Alnicos and Cu-Ni-Fe alloys, as well as in the newly developed Fe-Cr-Co materials. Spinodal decomposition provides a practical method of producing nanophase materials that can have enhanced mechanical and physical properties.

Theory of Spinodal Decomposition. A simple binary phase diagram with a region of spinodal decomposition is shown in Fig. 9.16(a) with a corresponding free-energy curve (Fig. 9.16b). If a composition, X_0, is heated above the critical temperature, T_c, the binary system is in the region of full solid solubility with a single-phase field, α_0, at temperature T_0. When the temperature goes below the critical temperature, a miscibility gap exists where a single-phase homogenous microstructure is no longer stable and

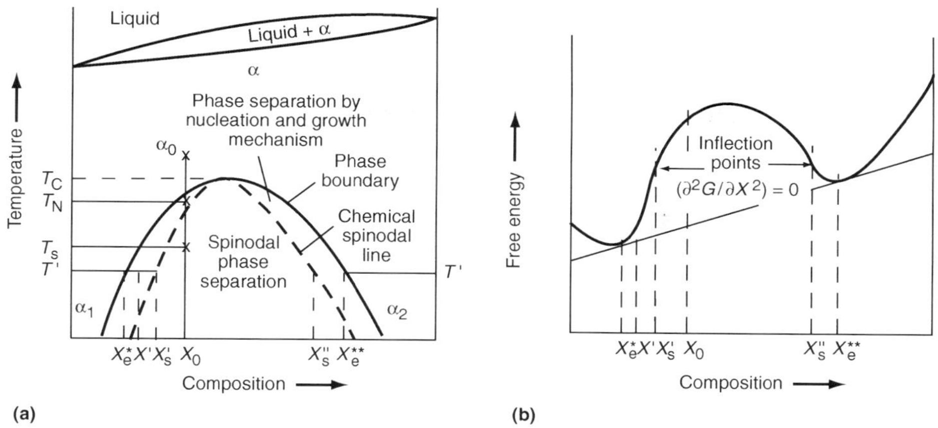

Fig. 9.16 Regions of spinodal decomposition and classical nucleation and growth of precipitates. (a) Phase diagram with a miscibility gap. (b) Variation in free energy with composition for the system shown in (a) at temperature T'. Source: Ref 9.9 as published in Ref 9.10

a two-phase, $\alpha_1 + \alpha_2$, structure forms. The phase boundary on the phase diagram at temperature T' is given by the locus of points of common tangency for the two equilibrium compositions, X_e* and X_e**, on the free-energy curve (Fig. 9.16b) for temperature T'.

When composition X_0 is at temperature T_N in a metastable (nonequilibrium) condition, the state is one of small undercooling or low supersaturation. The metastable state moves toward equilibrium by forming a second phase, but because this supersaturated condition is low (or not far from the equilibrium condition), the appearance of a second phase requires relatively large localized composition fluctuations. This is the classical nucleation process, when initiation requires a "critical nuclei" size for further growth of the new phase.

In contrast, if the α_0 solid-solution composition X_0 is at a lower temperature (e.g., T_s), then the supersaturated condition is higher, and initiation of two-phase growth may occur from smaller composition fluctuations. In particular, the area of spinodal decomposition defines a region of phase separation, where a particular kinetic process causes phase formation from very small composition fluctuations. It is not a new phase region, but rather a region with a difference in thermodynamic stability defined by the inflection points ($\partial^2G/\partial X^2 = 0$) of free energy (Fig. 9.16b). The kinetics and reaction rates of spinodal decomposition are controlled by the rate of atomic migration and diffusion distances, which depend on the scale of decomposition (undercooling).

The kinetic process of spinodal decomposition is illustrated in Fig. 9.15b, where a small fluctuation in composition becomes amplified by uphill diffusion (depicted by arrows). The reason for the uphill diffusion can be understood when considering that the direction of atomic migration is governed by the gradient of chemical potential, not by the concentration gradient as stated by Fick's first law. Atomic migration will occur from the region of high chemical potential to the region of low chemical potential. As shown in Fig. 9.17, the chemical potential inside the spinodal (inflection points of the free-energy curve) decreases as the composition increases, illustrating that the atomic species inside the chemical spinodal will migrate from low concentration to high concentration. Regions surrounding the amplified composition fluctuations will become depleted as solute diffuses up the concentration gradient. These locally depleted regions will then give rise to additional amplified regions adjacent to the locally depleted region (Fig. 9.15b).

In the region of classical nucleation and growth, decomposition into a two-phase mixture can only occur when nucleation is allowed to begin at critical nuclei size. In this two-phase region above the "chemical spinodal" line, small composition fluctuations decay by the more common process of downhill diffusion (Fig. 9.15a). The downhill diffusion can be understood by Fig. 9.17 and noting that outside of the spinodal, chemical potential decreases as composition decreases. Thus, diffusion in this region involves atomic migration from areas of high concentration to low concentration, and any small composition fluctuation decays.

Because solid-state spinodal decomposition results in two phases with the same crystal structure, the lattice must remain continuous. If the atomic radii of the species present in a spinodal structure vary appreciably, then coherency strains will be present. If the strain induced in the lattice is significant, the system can be stabilized against decomposition. This stabilization results in a displacement of the spinodal curve and the miscibility gap below the chemical spinodal, thus defining the coherent spinodal and the coherent miscibility gap (Fig. 9.18).

The wavelength, λ), of the composition fluctuations, ΔC, can be understood by considering the expression for the composition wave that was derived considering the diffusion equation and gradient energy (energy associated with diffuse interfaces):

$$\Delta C = C_0 \exp\left(R(\beta)t\right)$$

where

$$\beta = \frac{2\pi}{\lambda}$$

The amplification factor, $R(\beta)$, tends to be a maximum at intermediate wavelengths. At large wavelengths of composition fluctuation, diffusion distances are very long, and slow growth results. At very short wavelengths of composition fluctuation, a gradient energy or surface energy associated

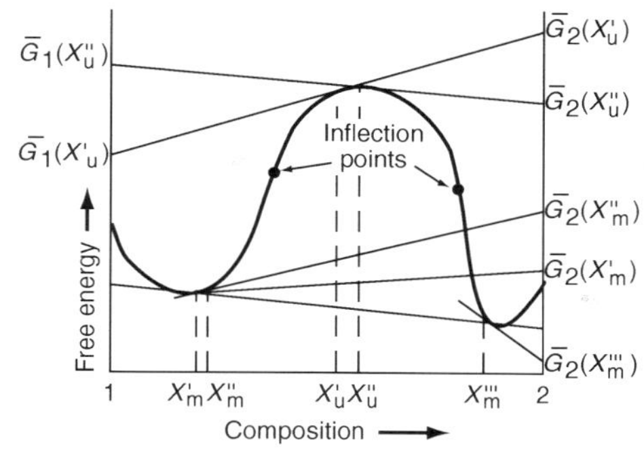

Fig. 9.17 Free-energy curve illustrating change in chemical potential with composition. Source: Ref 9.9 as published in Ref 9.10

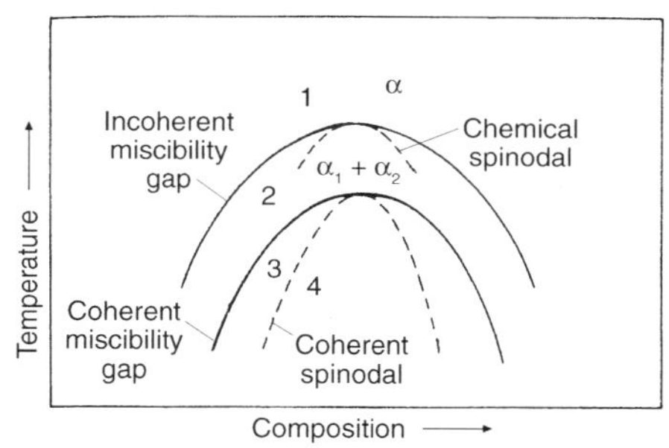

Fig. 9.18 Miscibility gap. Region 1: Homogenous α is stable. Region 2: Homogenous α is metastable; only incoherent phases can nucleate. Region 3: Homogeneous α is metastable; coherent phases can nucleate. Region 4: Homogeneous α is unstable; spinodal decomposition occurs. Source: Ref 9.11 as published in Ref 9.10

with diffuse interfaces will dominate and suppress the amplification of the composition variation.

Phase separated or spinodal structures cause diffraction effects called "satellites" or "sidebands," where the fundamental reflections are flanked by secondary intensity maxima (Fig. 9.19). The diffuse scattering that causes the satellites is a result of the periodic variation in lattice parameter and/or scattering factor. In reciprocal space, the distance between the fundamental reflection and the secondary maxima is inversely related to the wavelength of the composition waves in the solid. The kinetics of a spinodal reaction can be quantitatively studied using small-angle x-ray and neutron scattering, by monitoring the change in intensity distribution around the direct beam.

Microstructure. Spinodal structures, which are typically on the order of 2.5 to 10 nm, are too small for optical (light) microscopy. They are characterized by TEM, with typical images having a modulated homogenous appearance (Fig. 9.20). Spinodal structures can also be characterized by small-angle x-ray scattering.

When a material is elastically isotropic or if only small misfit strains exist, the spinodal structures will be isotropic, similar to the structure found in phase-separated polymers, glasses, and liquids. A spinodal-type structure from an iron-copper alloy is shown in Fig. 9.21. The two phases both have different compositions and crystal structures. Because the crys-

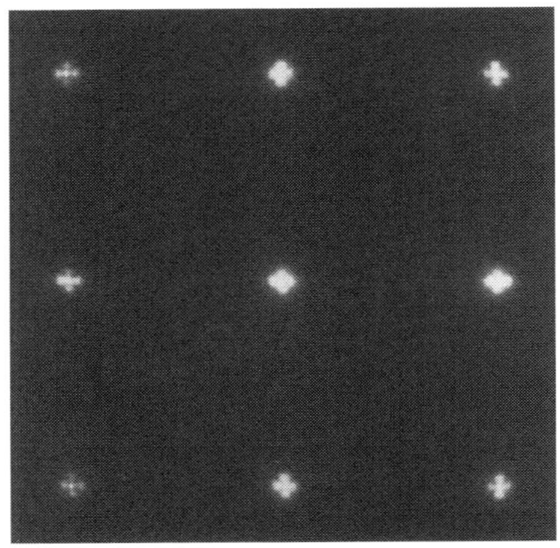

Fig. 9.19 Selected area diffraction pattern of Cu-15Ni-8Sn alloy showing satellites from structure modulation. Source: Ref 9.12 as published in Ref 9.10

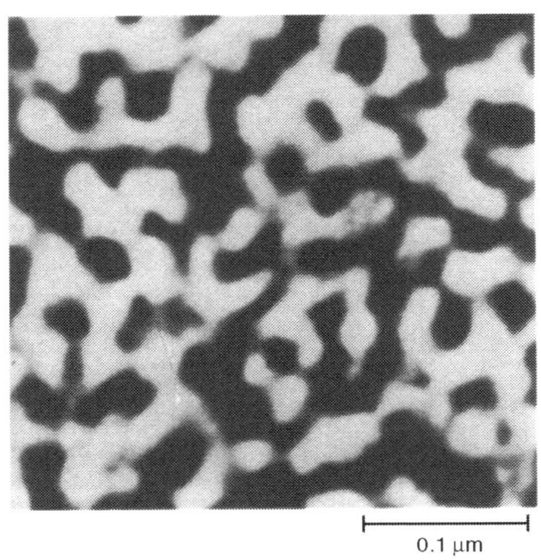

Fig. 9.20 Transmission electron micrograph (TEM) of isotropic spinodal structure developed in Fe-28.5Cr-10.6Co (wt%) alloy aged 4 h at 600 °C (1110 °F). Contrast derives mainly from structure-factor differences. Original magnification: 225,000×. Source: Ref 9.10

tal structures are different, decomposition could only have occurred in the liquid phase prior to solidification. Materials that are elastically anisotropic form spinodal structures that are developed preferentially along elastically soft directions. Figure 9.22 illustrates calculated two-dimensional and three-dimensional time developments for the structure of an iron-molybdenum alloy.

Because the iron-molybdenum system has a large lattice mismatch, the molybdenum-rich zones are aligned along the elastically soft ⟨100⟩ directions. Figure 9.23 shows spinodally decomposed structures from an Fe-25Be (at.%) system, which are typical microstructures that form in an elastically anisotropic system. It is obvious from Fig. 9.23 that the modulations in the spinodal structure coarsen as aging time increases. Because spinodal decomposition occurs homogeneously throughout the microstructure by a structure-insensitive phase separation, spinodal structures are usually uniform throughout the grains up to the grain boundaries.

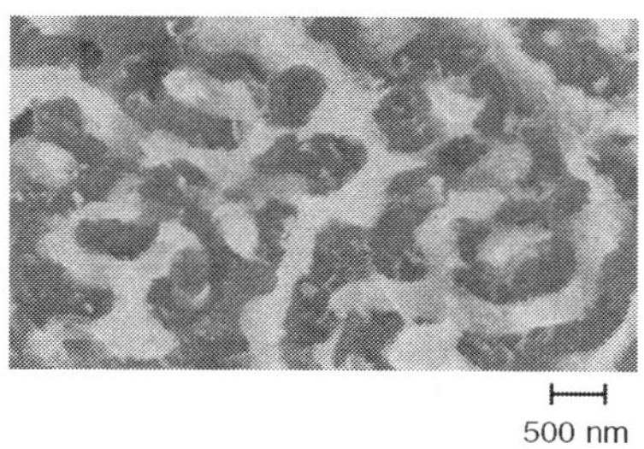

Fig. 9.21 Backscatter scanning electron micrograph of an iron-copper alloy that was rapidly solidified after undergoing liquid-phase spinodal decomposition. Source: Ref 9.12 as published in Ref 9.10

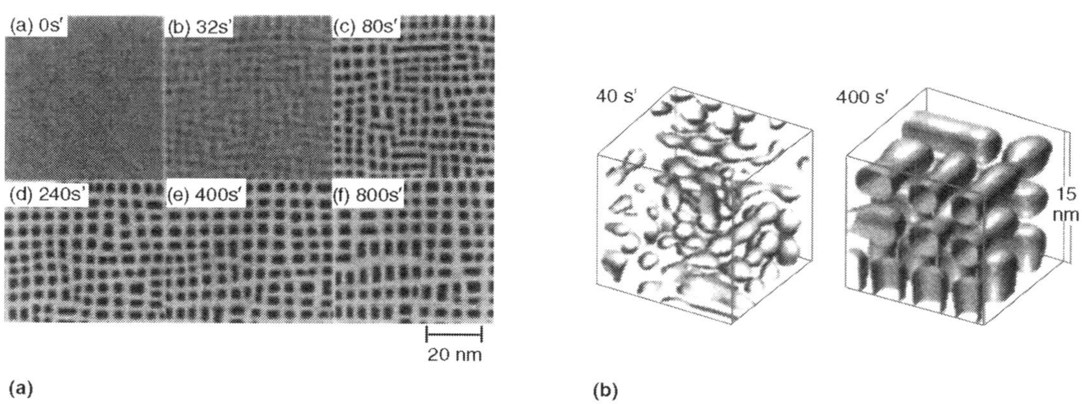

Fig. 9.22 Phase decomposition for the Fe-30Mo (at.%). (a) Two-dimensional time development. (b) Three-dimensional simulation. Source: Ref 9.13 as published in Ref 9.10

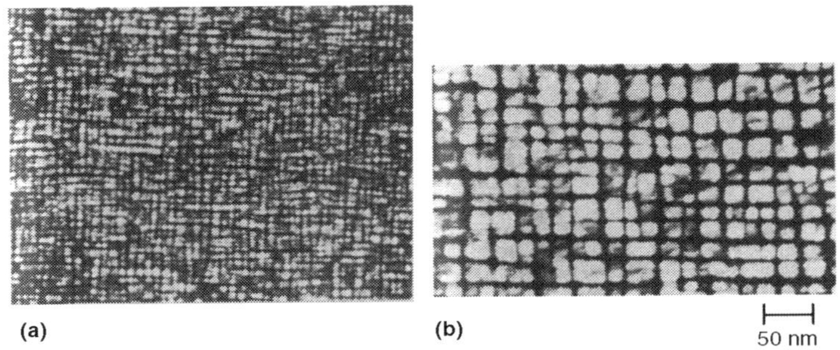

Fig. 9.23 Dark-field transmission electron micrograph of Fe-25Be (at.%) aged at 400 °C (750 °F) for (a) 0.3 h and (b) 2 h. Source: Ref 9.14 as published in Ref 9.10

ACKNOWLEDGMENTS

The material for this chapter was taken from "Ordered Structures," by J. Regina, and "Spinodal Transformation Structures," by S. Para, both in *Metallography and Microstructures,* Vol 9, *ASM Handbook*, ASM International, 2004.

REFERENCES

9.1 F.C. Campbell, *Elements of Metallurgy and Engineering Alloys*, ASM International, 2008

9.2 A. Taylor and B.J. Kagle, *Crystallographic Data on Metal and Alloy Structures,* Dover, 1963

9.3 J. Regina, Ordered Structures, *Metallography and Microstructures,* Vol 9, *ASM Handbook*, ASM International, 2004, p 144–147

9.4 M.J. Marcinkowski, *Electron Microscopy and Strength of Crystals,* G. Thomas and J. Washburn, Ed., Interscience, 1963

9.5 M.J. Marcinkowski and L. Zwell, Transmission Electron Microscopy Study of the Off-Stoichiometric Cu3Au Superlattices, *Acta Metall.,* Vol 11, 1963, p 373–390

9.6 G. Frommeyer, et al., Investigations of Phase Transformations and B2-$D0_3$ Superlattices in Ordered Iron Aluminum Alloys by FIM-Atom Probe and TEM, *Scripta Metall.,* Vol 24, 1990, p 51–56

9.7 N.S. Stoloff and V.K. Sikka, Ed., *Physical Metallurgy and Processing of Intermetallic Compounds,* Chapman & Hall, 1996

9.8 M.J. Marcinkowski, Ordered Structures, *Metallography and Microstructures,* Vol 9, *Metals Handbook,* 9th ed., American Society for Metals, 1985, p 681–683

9.9 A.K. Jena and M.C. Chaturvedi, Spinodal Decomposition, *Phase Transformations in Materials,* Prentice Hall, 1992, p 373–399

9.10 S. Para, Spinodal Transformation Structures, *Metallography and Microstructures,* Vol 9, *ASM Handbook*, ASM International, 2004, p 140–143

9.11 D.A. Porter and K.E. Easterling, *Phase Transformations in Metals and Alloys,* 2nd ed., Chapman and Hall, 1997

9.12 D.W. Zeng, C.S. Xie, and K.C. Yung, Mesostructured Composite Coatings on SAE 1045 Carbon Steel Synthesized in situ by Laser Surface Alloying, *Mater. Lett.,* Vol 6, 2002, p 680–684

9.13 T. Miyazaki, T. Koyama, and T. Kozakai, Computer Simulations of the Phase Transformation in Real Alloy Systems Based on the Phase Field Model, *Mater. Sci. Eng.,* Vol A312, 2001, p 38–49

9.14 M.G. Burke and M.K. Miller, A Combined TEM/APFIM Approach to the Study of Phase Transformations: Phase Identification in the Fe-Be System, *Ultramicroscopy,* Vol 30 (No. 1–2), 1989, p 199–209

CHAPTER 10

Ternary Phase Diagrams

TERNARY SYSTEMS are those having three components. It is not possible to describe the composition of a ternary alloy with a single number or fraction, as was done with binary alloys, but the statement of two independent values is sufficient. For example, the composition of an Fe-Cr-Ni alloy may be described fully by stating that it contains 18% Cr and 8% Ni. There is no need to say that the iron content is 74%. But the requirement that two parameters must be stated to describe ternary composition means that two dimensions must be used to represent composition on a complete phase diagram. The external variables that must be considered in ternary constitution are temperature, pressure, composition *X*, and composition *Y*. To construct a complete diagram representing all these variables would require the use of a four-dimensional space. This being out of the question, it is customary to assume pressure constant (atmospheric pressure) and to construct a three-dimensional (3-D) diagram representing, as variables, the temperature and two concentration parameters. Therefore, in any application of the phase rule, it should be recalled that one degree of freedom has been exercised in the initial construction of the 3-D diagram by electing to draw it at one atmosphere of pressure.

10.1 Space Model of Ternary Systems

To represent completely the phase equilibria at constant pressure in a ternary system, a 3-D model, commonly termed a space model, is required; the representation of composition requires two dimensions, and that of temperature, a third dimension. The model used is a triangular prism (Fig. 10.1), in which the temperature is plotted on the vertical axis, and the composition is represented on the base of the prism, which may be conveniently taken as an equilateral triangle. Thus, in Fig. 10.1, the vertical sides of the prism represent the three binary systems, *AB*, *BC*, and *AC*, that make up the ternary system, *ABC*.

A hypothetical ternary phase space diagram made up of metals *A*, *B*, and *C* is shown in Fig. 10.2. This diagram contains two binary eutectics on the two visible faces of the diagram, and a third binary eutectic between elements *B* and *C* hidden on the back of the plot. Because it is difficult to use

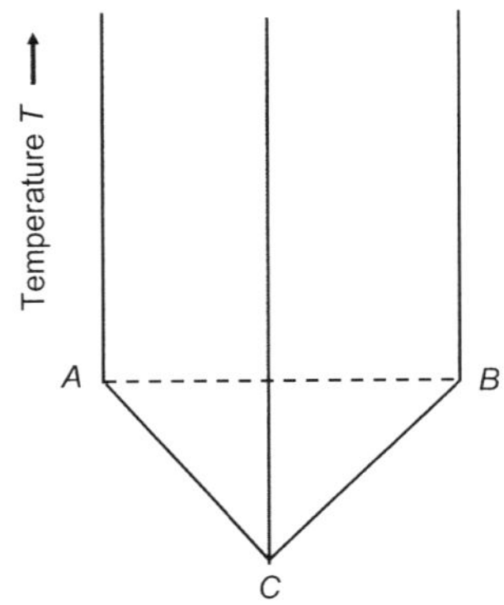

Fig. 10.1 Space model for ternary phase diagrams

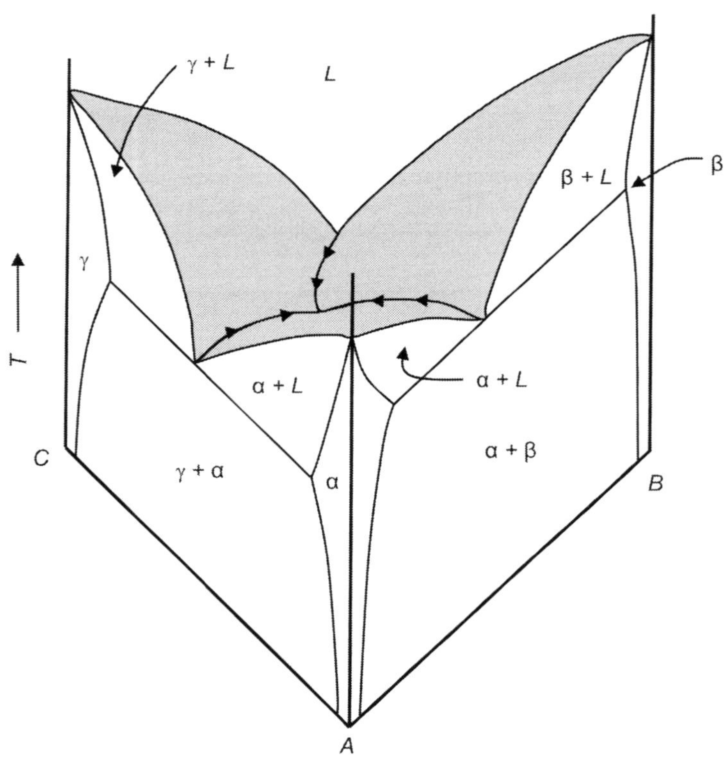

Fig. 10.2 Hypothetical ternary phase diagram. Binary phase diagrams are present along the three faces. Adapted from Ref 10.1

the 3-D ternary plot, the information from the diagrams can be plotted in two dimensions by any of several methods, including the liquidus plot, the isothermal plot, and a vertical section called an isopleth.

Liquidus Plots. The temperature at which freezing begins is shaded in Fig. 10.2. In Fig. 10.3, these temperatures for each composition are transferred onto a triangular diagram; the liquidus temperatures are plotted as isothermal contours. This presentation is helpful in predicting the freezing temperature of an alloy. Note that the liquidus lines have arrows indicating the freezing direction toward the ternary eutectic point. The liquidus plot also gives the identity of the primary phase that will form during solidification for any given alloy composition. Similar plots, known as solidus plots, showing solidus freezing are sometime presented.

Isothermal Plots. An isothermal plot shows the phases present in any alloy at a particular temperature and is useful in predicting the phases and their amounts and compositions at that temperature. An isothermal section from Fig. 10.2 at room temperature is shown in Fig. 10.4. Isothermal plots are by far the most useful because they allow compositional analysis, while liquidus and isopleth plots do not allow compositional analysis.

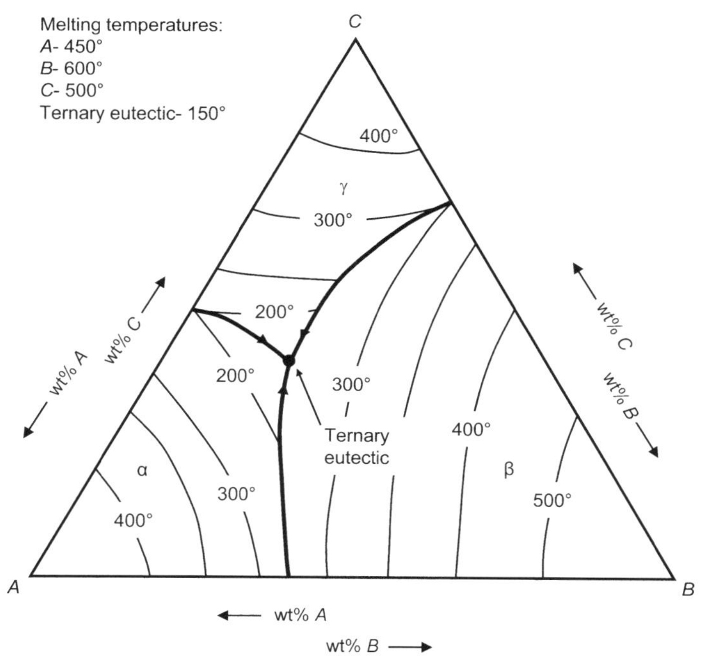

Fig. 10.3 Liquidus plot for hypothetical ternary phase diagram. Adapted from Ref 10.1

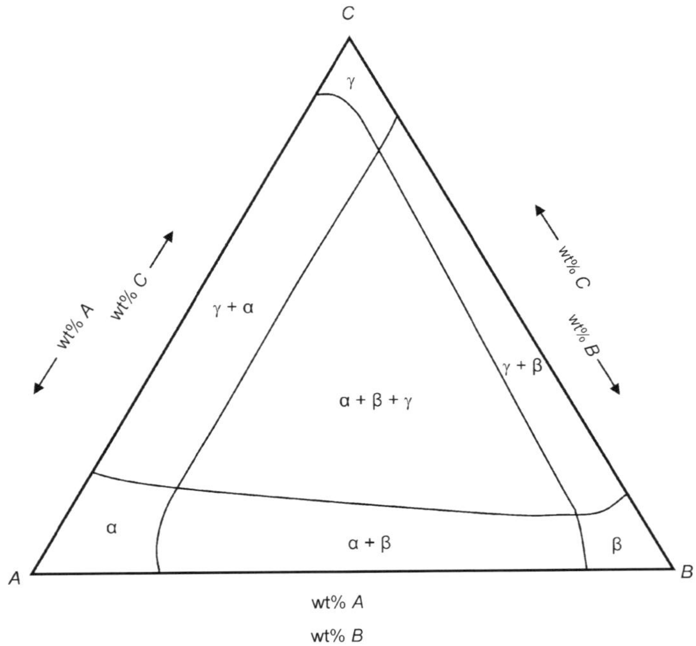

Fig. 10.4 Isothermal plot at room temperature for hypothetical ternary phase diagram. Adapted from Ref 10.1

Isopleth Plots. Certain groups of alloys can be plotted as vertical sections, also called isopleths. These sections often represent a fixed composition of one of the elements, while the amounts of the other two elements are allowed to vary. These plots show how the phases and structures change when the temperature varies and when two of the elements present change their respective amounts. Tie lines usually do not lie in the plane of a vertical section and cannot be used to obtain amounts and compositions. An isopleth through the hypothetical diagram (Fig. 10.2) at a constant 40% *C* is shown in Fig. 10.5. An alloy containing 30% *A* and 30% *B* will begin to freeze near 350 °C (660 °F), with primary β forming first. Near 275 °C (530 °F), γ will also begin to form. Finally, at approximately 160 °C (320 °F), α forms and the last liquid freezes. The final microstructure contains α, β, and γ. Isopleths are quite valuable in showing the phases that are present during equilibrium cooling and heating. They also show the temperatures at which the various phase changes occur.

Single-Phase Boundary and Zero-Phase Fraction Lines. Two-dimensional (2-D) sections of any multicomponent phase diagram, whether it is an isotherm or an isopleth, can be read by focusing on two lines that refer to one particular phase. These lines are shown in the Fig. 10.6 isopleth for Fe-17%Cr-%C alloys.

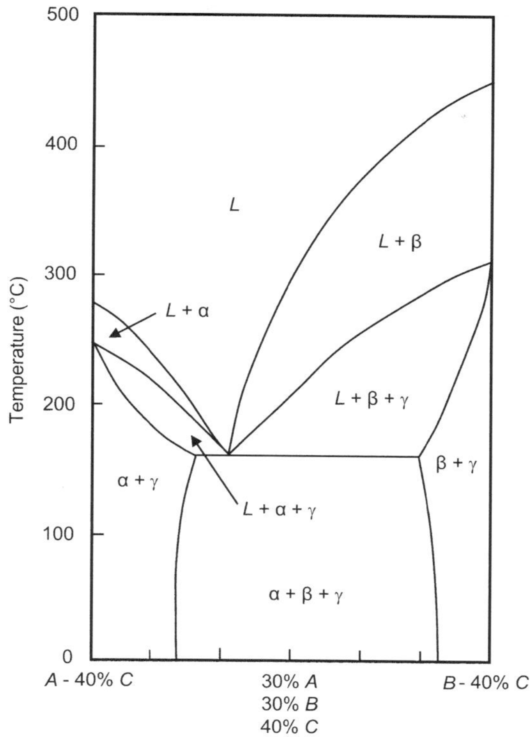

Fig. 10.5 Isopleth through hypothetical ternary phase diagram at a constant 40% C. Adapted from Ref 10.1

SPB Line. The single-phase boundary line is found on any section that contains a single-phase region. The line is what its name implies. It is the boundary line around that single-phase region. It can be used, for example, to determine compositions and temperatures where an alloy can be thoroughly solutionized.

ZPF Line. The zero phase fraction line is a line that surrounds all regions on the diagram where the phase occurs. On one side of the line there are regions with the phase, and on the other side of the line there are regions without the phase. Because the line surrounds a region of compositions and temperatures where the phase forms, it can be used to avoid a phase, for example an embrittling phase, or promote a phase, for example a precipitation-hardening phase.

By drawing these lines, the reader is able to focus on one phase at a time and ignore the lines that concern other phases. Although it is true that the lever rule cannot be used here, it can be assumed that moving closer to an SPB line will likely increase the amount of the phase, while moving closer to the ZPF line will decrease the amount of the phase. The liquidus and solidus lines are SPB and ZPF lines for the liquid, respectively. Also, it is

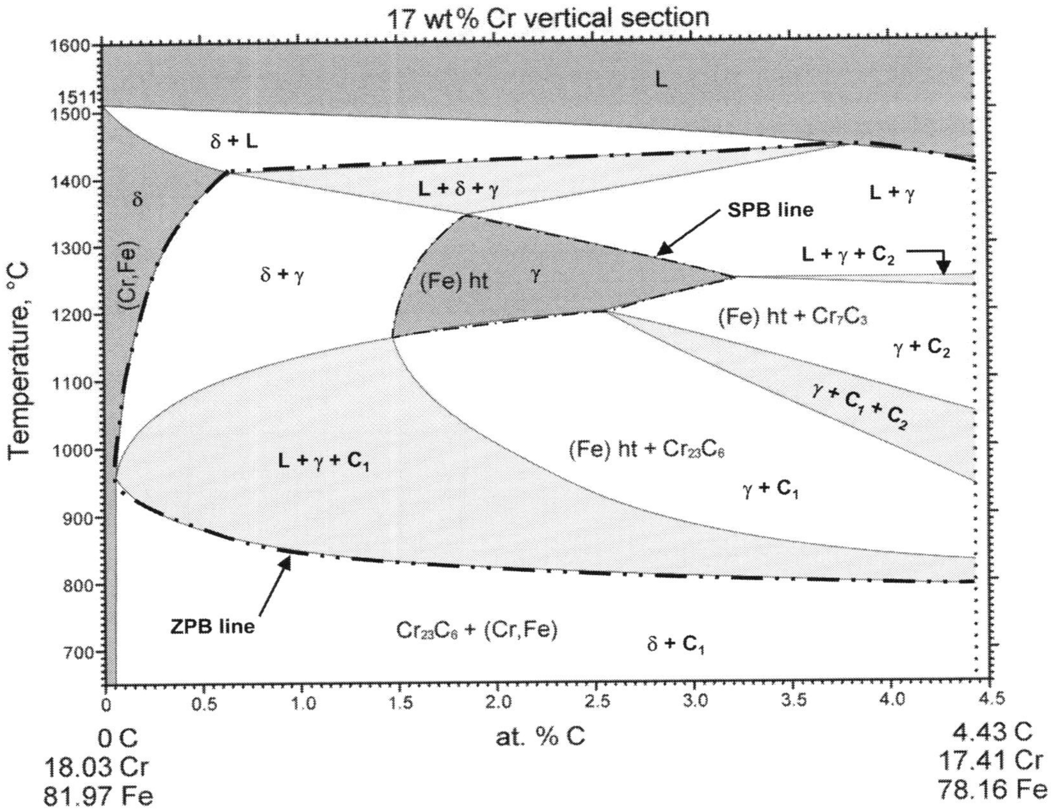

Fig. 10.6 C-Cr-Fe isopleth showing single-phase boundary (SPB) lines and zero-phase boundary (ZPB) lines. Source: Ref 10.2

worth noting that an isopleth is a collection of ZPF lines for the various phases present. Computer programs that predict phase diagrams can give a phase diagram in the form of ZPF lines alone. In this case, the lines are labeled instead of the regions.

10.2 The Gibbs Triangle

Because of its unique geometric characteristics, an equilateral triangle provides the simplest means for plotting ternary composition. On the Gibbs triangle, which is an equilateral triangle, the three pure component metals are represented at the corners, *A*, *B*, and *C*, as shown in Fig. 10.7. Binary composition is represented along the edges, that is, the binary systems *AB*, *AC*, and *BC*. And ternary alloys are represented within the area of the triangle, such as at point *P* in Fig. 10.7.

If lines are drawn through Alloy *P* parallel to each of the sides of the triangle, it will be found that these have produced three smaller equilateral

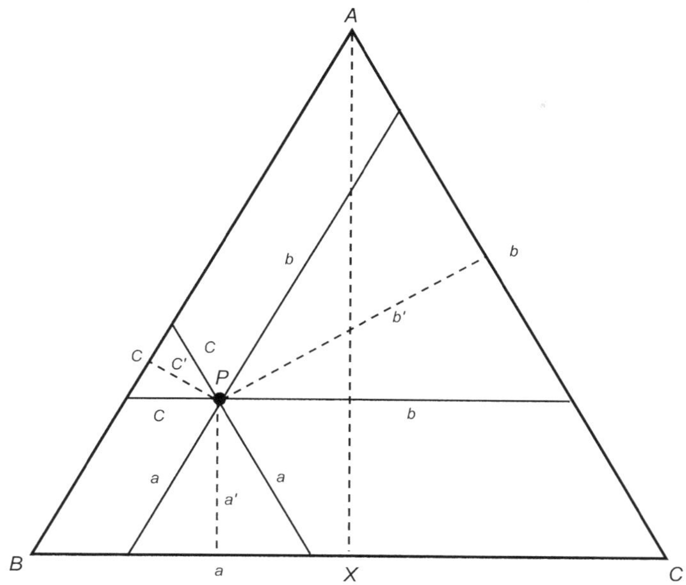

Fig. 10.7 The Gibbs triangle. Adapted from Ref 10.3

triangles: *aaa*, *bbb*, and *ccc*. The sum of the lengths of the nine sides of these three triangles is equal to the sum of the lengths of the three sides of the major triangle, *ABC*, within which they are inscribed; or the sum of the lengths of one side from each of the minor triangles is equal to the length of one side of the major triangle: $a + b + c = AB = AC = BC$. Also, the sum of the altitudes of the minor triangles is equal to the altitude of the major triangle: $a' + b' + c' = AX$.

If one side of the Gibbs triangle is divided into 100 equal parts, representing 100% on the binary composition scale, it is found that the same units can be used to measure the composition at point *P*. Let the length *a* represent the percentage of *A* in *P*, the length *b* the percentage of *B*, and the length *c* the percentage of *C*. Because these lengths total the same as one side of the Gibbs triangle, and together they must equal 100%, it is evident that 1% has the same length, whether measured along an edge of the diagram or along any inscribed line parallel to an edge. A similar result could be obtained by using altitudes, but this is less convenient. It should be noted that in either case, the percentage of *A* is measured on the side of *P* away from the *A* corner and similarly with *B* and *C*.

For convenience in reading composition, an equilateral triangle may be ruled with lines parallel to the sides (Fig. 10.8). Composition may then be read directly, for example, $P = 20\%\ A + 70\%\ B + 10\%\ C$. At point *P*, the percentage of *A* is represented by the line *Pa* (or equivalently *Pa'*), which is 20 units long; the percentage of *B* by the line *Pb* (or *Pb'*), 70 units long; and

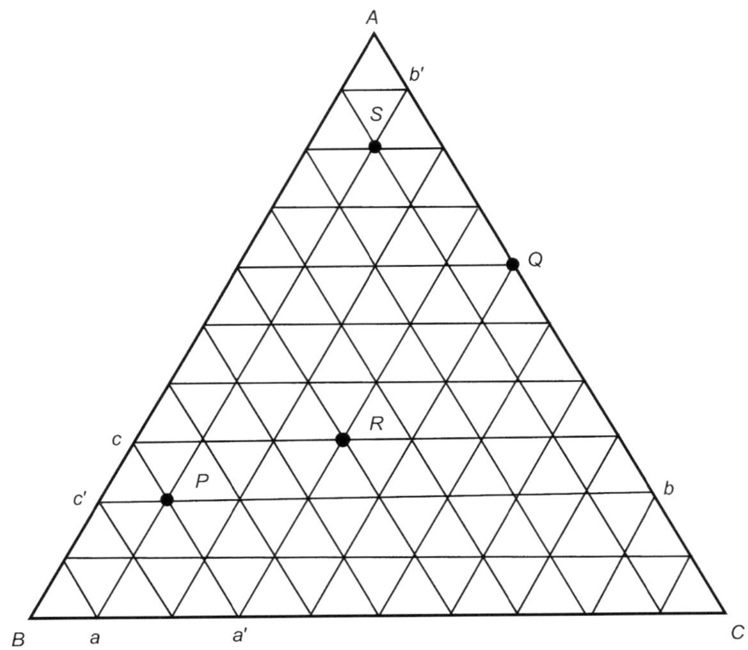

Fig. 10.8 The Gibbs triangle with composition lines. Adapted from Ref 10.3

the percentage of *C* by the line *Pc* (or *Pc′*), 10 units long. Other examples shown in Fig. 10.8 are: Alloy $R = 30\%\ A + 40\%\ B + 30\%\ C$, Alloy $S = 80\%\ A + 10\%\ B + 10\%\ C$, and Alloy $Q = 60\%\ A + 0\%\ B + 40\%\ C$.

10.3 Tie Lines

If any two ternary alloys are mixed together, tie lines can be shown. The composition of the mixture will lie on a straight line joining the original two compositions. This is true regardless of the proportions of the two alloys in the mixture. Conversely, if an alloy decomposes into two fractions of differing composition, the compositions of the two portions will lie on opposite ends of a straight line passing through the original composition point. Consider Fig. 10.9. Points *S* and *L* represent two ternary alloys of respective composition: $20\%\ A + 70\%\ B + 10\%\ C$ and $40\%\ A + 30\%\ B + 30\ \%\ C$. Suppose that one part of *S* is mixed with three parts of *L* and the mixture is analyzed. The analytical result will be:

$$0.25 \times 20\%\ A + 0.75 \times 40\%\ A = 35\%\ A$$

$$0.25 \times 70\%\ B + 0.75 \times 30\%\ B = 40\%\ B$$

$$0.25 \times 10\%\ C + 0.75 \times 30\%\ C = 25\%\ C$$

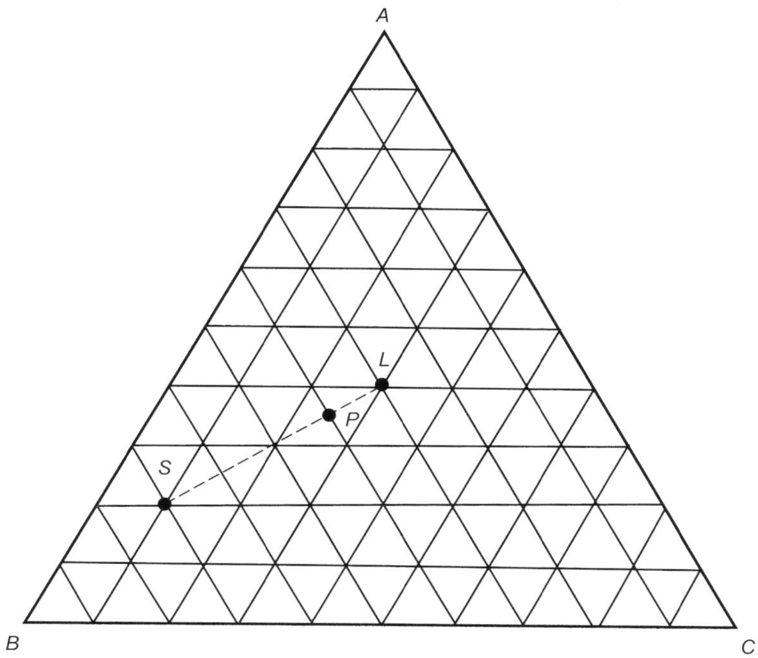

Fig. 10.9 The Gibbs triangle with tie line. Adapted from Ref 10.3

As can be seen by inspection of Fig. 10.9, this composition lies at *P*, which is a point on the straight line connecting *S* and *L*. Regardless of the compositions chosen or in what proportions they had been mixed, the total composition would have occurred on the line joining the two original compositions.

It is evident that the line *SL* has the characteristics of a tie line: It is both isobaric and isothermal, because it lies in the composition plane, which is drawn perpendicular to the temperature axis and corresponds to the case of constant atmospheric pressure (i.e., it would be drawn perpendicular to the pressure axis if a fourth dimension were available). The lever principle is applicable to this line. Therefore, the line *SL* might represent the condition of an alloy of composition *P* that is partially frozen, at the temperature under consideration, and consists of 25% solid of composition *S* and 75% liquid of composition *L*:

$$\% \; S = \frac{PL}{SL} \times 100$$

$$\% \; L = \frac{SP}{SL} \times 100$$

10.4 Ternary Isomorphous Systems

A temperature-composition (*T-X-Y*) diagram of an isomorphous system is shown in Fig. 10.10. The composition plane forms the base of the figure, and temperature is measured vertically. Here, the liquidus and solidus become surfaces bounding the $L + \alpha$ space. Above the liquidus, all alloys are fully molten; below the solidus, all are completely solid. As in binary systems, the two-phase region, $L + \alpha$, is composed of tie lines joining conjugate liquid and solid phases. In the ternary system, however, the tie lines are not confined to a 2-D area but occur as a bundle of lines of varying direction, but all horizontal (isothermal), filling the 3-D two-phase space.

Isothermal Sections. The location of the tie lines can be visualized more easily by reference to isothermal (horizontal) sections cut through the temperature-composition diagram at a series of temperature levels. The three isotherms presented in Fig. 10.11 are taken at the temperatures designated T_1, T_2, and T_3. It is seen that the first tie line on each edge of the $L + \alpha$ region is the bounding line of the figure; that is, it is the binary tie line at the temperature designated. The directions of tie lines lying within

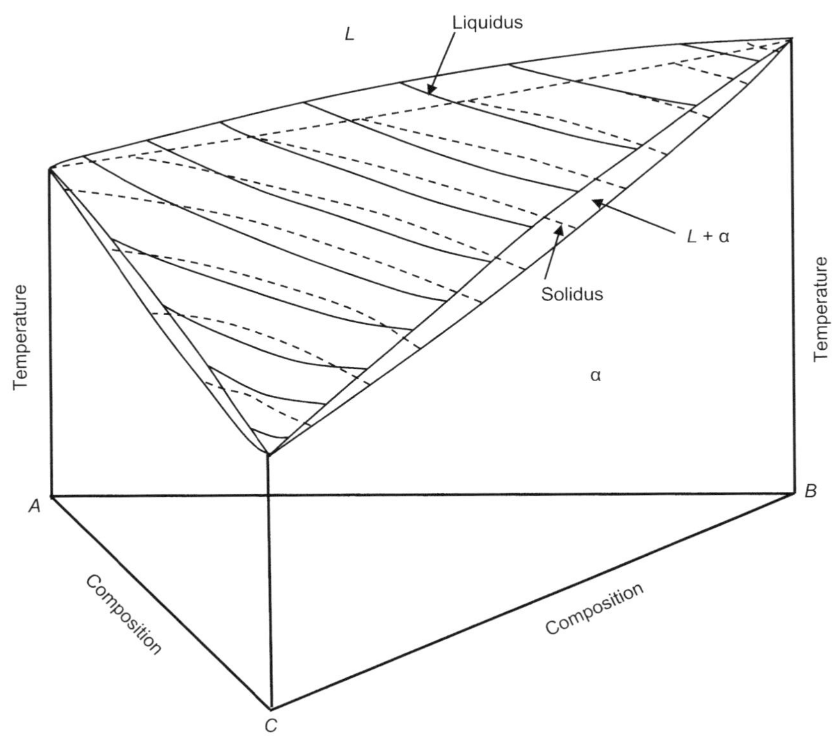

Fig. 10.10 Temperature-composition space diagram of a ternary isomorphous system. Adapted from Ref 10.3

the figure vary "fanwise" so that there is a gradual transition from the direction of one bounding tie line to that of the other. No two tie lines at the same temperature may ever cross. Beyond this nothing can be said of their direction, except, of course, that they must run from liquidus to solidus. Other than those on the edges of the diagram, none points toward a corner of the diagram unless by mere coincidence. Therefore, it is necessary to determine the position and direction of the tie lines by experiment and to indicate them on the ternary phase diagram. Isothermal sections such as those shown in Fig. 10.11 generally provide the most satisfactory means for recording ternary equilibria in two dimensions. In them the various structural configurations assume their simplest forms.

Liquidus and Solidus Plots. It is possible in certain cases to telescope a group of isotherms into a single diagram. For example, the isothermal lines that were used to delineate the liquidus surface in Fig. 10.10 may be

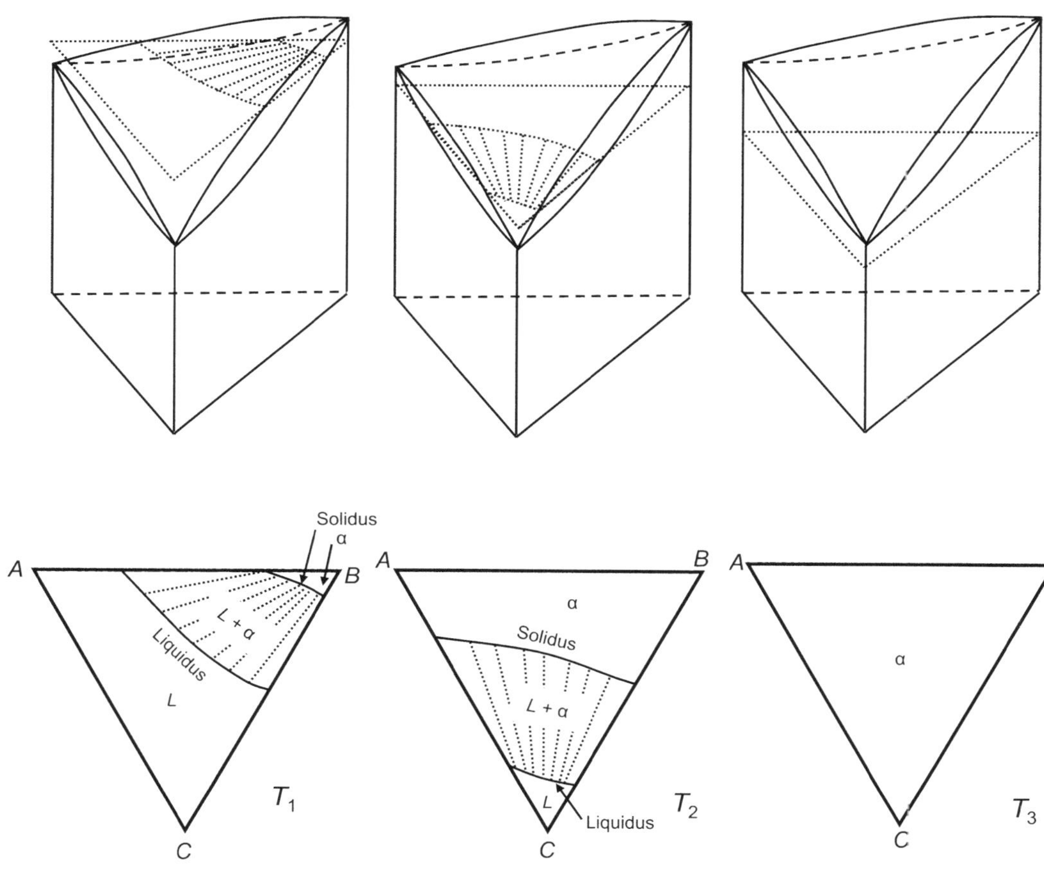

Fig. 10.11 Isotherms through a ternary isomorphous phase diagram. Adapted from Ref 10.3

projected onto a plane, such as the base of the diagram, giving the liquidus projection presented in Fig. 10.12. Each line is derived from a separate isotherm and its temperature should therefore be indicated on the line. In like manner, the solidus may be represented as a projection. Both suffer from the handicap that tie lines cannot be used.

Vertical Sections (Isopleths). Because of their resemblance to binary diagrams, vertical sections, also known as isopleths, have been widely used. The sections selected are usually (1) those radiating from one corner of the space diagram and, therefore, representing a fixed ratio of two of the components (Fig. 10.13a), or (2) those parallel to one side of the space diagram, representing a constant fraction of one of the components (Fig. 10.13b). It is observed that the $L + \alpha$ region is open at its ends except where it terminates on the B component. From vertical sections, the liquidus and solidus temperatures for any of the alloys represented can be read. However, it is not possible to represent tie lines on these sections, because no tie line can lie in either section except by coincidence. In general, the tie lines pass through the $L + \alpha$ region at an angle to the plane of the section. Consequently, it is not possible to record equilibria within the $L + \alpha$ region by the use of vertical sections. Sometimes this is a matter of little consequence, when, for example, the section lies so close to one side of the space diagram that the tie lines may be presumed to lie approximately in the section. For this reason, vertical sections continue to be of some use, though they have been largely displaced by isothermal sections in the lit-

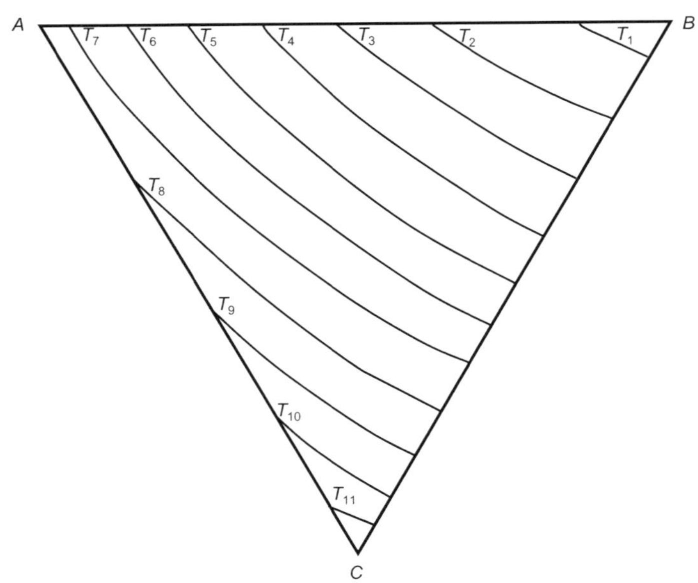

Fig. 10.12 Liquidus projection of the diagram shown in Fig. 10.10. Adapted from Ref 10.3

erature of alloy constitution. Certain vertical sections in complex diagrams (the quasi-binary sections) do contain tie lines.

Application of the Phase Rule. Within the one-phase spaces of the ternary diagram, the L and α regions of Fig. 10.10, the equilibria are quadrivariant; that is, there are four degrees of freedom:

$$F = C + 2 - P$$
$$F = 3 + 2 - 1 = 4$$

These are pressure, temperature, concentration X, and concentration Y. The pressure was chosen when the diagram was taken at 1 atmosphere of pressure. It remains to verify the correctness of representing the one-phase regions by volumes in the space diagram by ascertaining if an independent

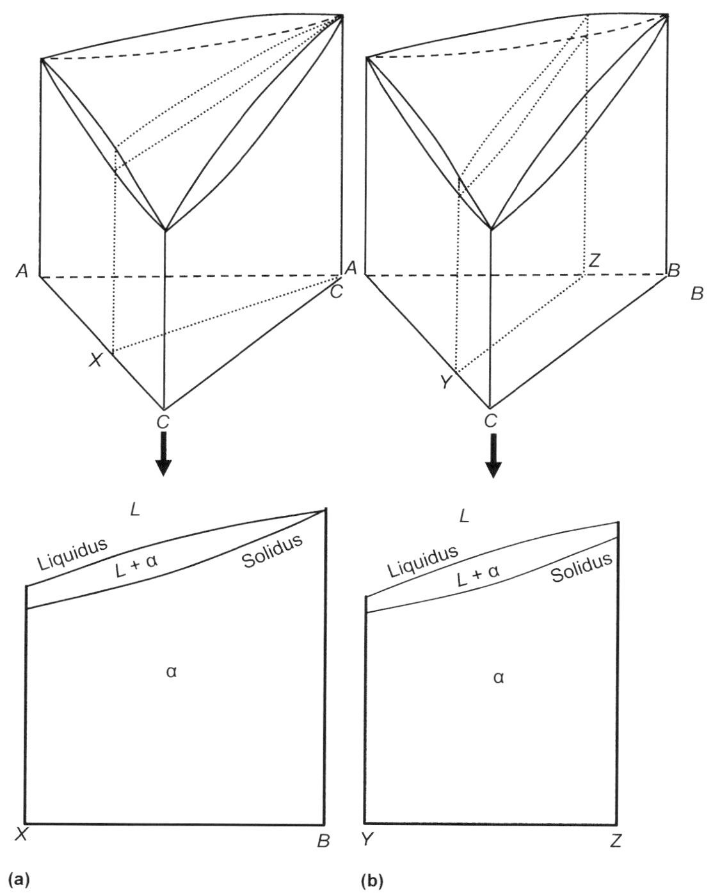

Fig. 10.13 Isopleths through an isomorphous system. Adapted from Ref 10.3

choice of temperature and two concentration values will describe a fixed equilibrium. Consider the diagram in Fig. 10.14; this figure represents a portion of the α field of Fig. 10.10, the liquidus and solidus having been removed. Let it be required that in addition to adjusting the pressure at 1 atmosphere, the temperature is established at *T*, and the composition at 20% *A* and 40% *B*. A horizontal plane, *abc*, representing a choice of temperature, is passed through the diagram at T_1. Next, one condition of composition is selected by erecting a vertical plane, *defg*, parallel to the *BC* side and one-fifth of the distance from this side to the *A* corner. This plane intersects the T_1 isotherm along the line *fg*. At this juncture, it can be seen that the equilibrium is not yet fully described, because any one of many alloys having 20% *A* can exist as the α phase at 1 atmosphere of pressure and at temperature T_1. If a second vertical plane, *hijk*, parallel to the *AC* side of the diagram and two-fifths of the distance from this side to the *B* corner, is erected, it will intersect the T_1 isotherm along the line *jk* and will cross the line *fg* at *P*. Thus, a unique point, point *P*, has been designated by the exercise of four degrees of freedom. It will be evident from the argument that the phase rule in no way limits the shape of the one-phase regions.

When two phases coexist at equilibrium in a ternary system, there can be only three degrees of freedom:

$$F = C + 2 - P$$
$$F = 3 + 2 - 2 = 3$$

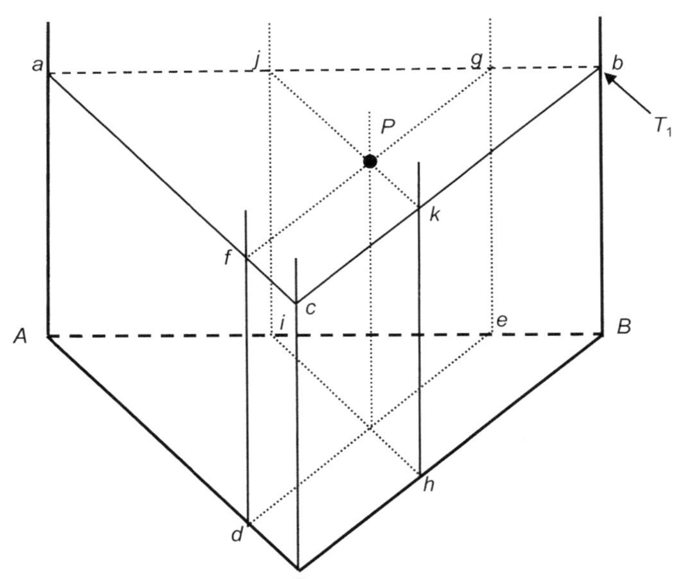

Fig. 10.14 Portion of the α field of Fig. 10.10. The liquidus and solidus have been removed. Adapted from Ref 10.3

The system is tervariant. Pressure being fixed, there remain but two degrees of freedom, which can be temperature and one concentration value or two concentration values. Both of these possibilities are explored. First, the choice of temperature and one concentration value *A* is examined. Let a temperature be selected such that two-phase equilibrium may occur, for example, T_1 in Fig. 10.11. This isotherm is reproduced in Fig. 10.15. Two degrees of freedom have now been exercised, namely, pressure and temperature; a single statement concerning the composition of one of the phases should complete the description of a specific equilibrium. Therefore, let it be specified that the alloys containing 15% *A* are represented on the isotherm by the dashed line *de*. This line includes, however, only one composition of liquid that can be in equilibrium with the solid phase, namely, the composition *g*, where the dashed line crosses the liquidus. Thus, the composition of the liquid phase is described completely, and the composition of the conjugate solid is identified by the corresponding tie line at point *i* on the solidus. Had it been specified, instead, that the α phase should contain 15% *A*, then the composition of the solid would have been given at point *f* and that of the conjugate liquid at *h*.

Two statements concerning the composition should serve also to identify the temperature and composition of both phases. This case is illustrated in Fig. 10.16. It has been specified that the liquid phase shall contain 5% *B* and 10% *C*. Two vertical planes have been erected, one at a constant *B* content of 5% and the other at a constant *C* content of 10%. These intersect in the vertical line L_1P, which intersects the liquidus at L_1. This point on the liquidus lies in the isothermal section *abcd* and is associated with a specific composition of the α phase at α_1, by means of the tie line $L_1\alpha_1$. Thus, the composition of the liquid phase, the composition of the solid phase, and

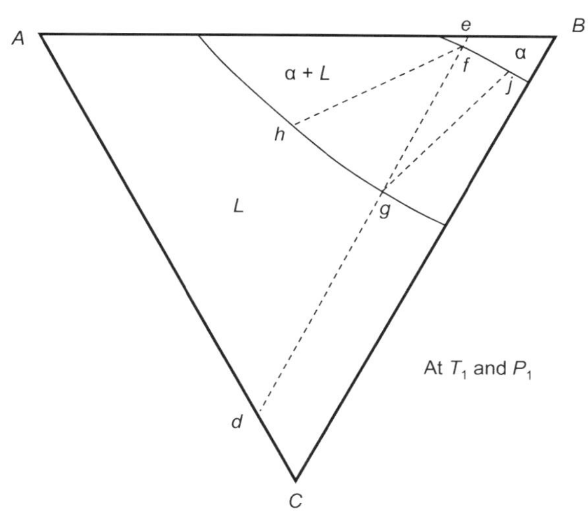

Fig. 10.15 Isotherm at T_1 from Fig. 10.11. Adapted from Ref 10.3

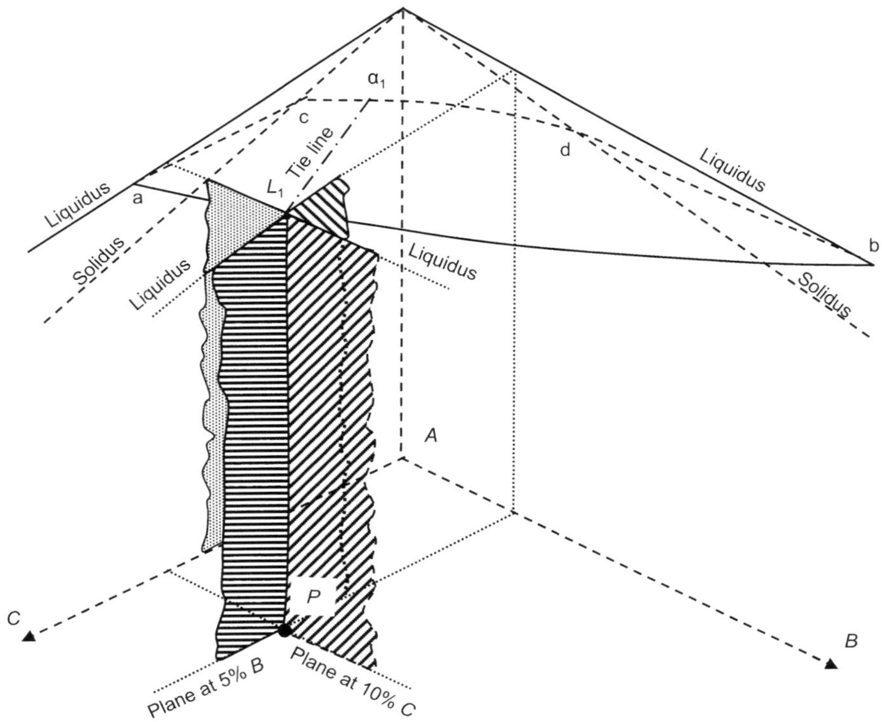

Fig. 10.16 Vertical planes at T_1 isotherm from Fig. 10.11: one at a constant B content of 5% and the other at a constant C content of 10%. Adapted from Ref 10.3

the temperature of equilibrium have all been established. The same result could have been obtained by designating two concentration parameters for the solid phase or one each for the liquid and solid phases.

From these demonstrations it is apparent that the only construction of a two-phase region consistent with the phase rule is one in which there are two conjugate bounding surfaces existing within the same temperature range and on which every point of one is connected by a horizontal tie line with a unique point on the other.

Equilibrium Freezing of Solid-Solution Alloys. The course of equilibrium freezing of a typical alloy of a ternary isomorphous system may be followed by reference to Fig. 10.17. Beginning at a temperature within the liquid field and cooling at a rate that is infinitely slow, to permit the maintenance of equilibrium, the alloy of composition X will begin to freeze when the liquidus surface is reached at temperature T_1. Here the liquid L_1 is in equilibrium with crystals of composition α_1. As the solid grows, the liquid composition will change in a direction away from the α_1 composition, following a curved path down the liquidus surface $L_1L_2L_3L_4$. At the

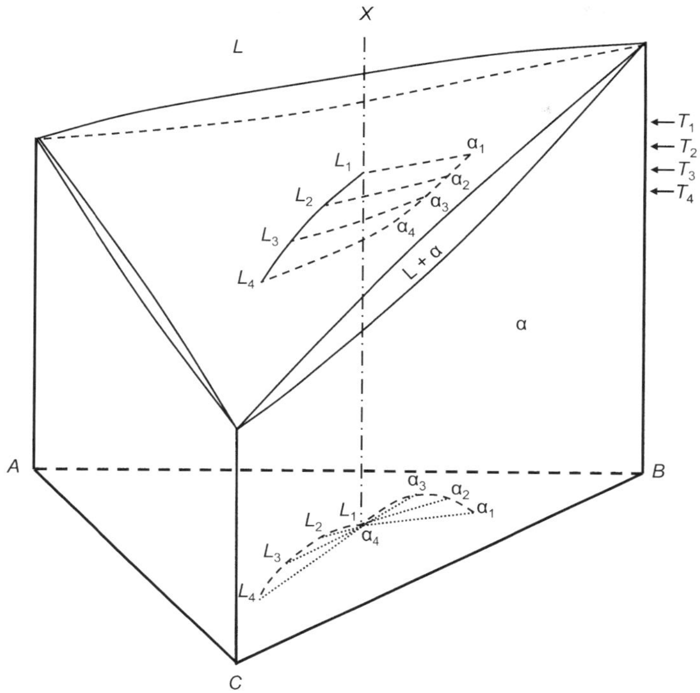

Fig. 10.17 Path of composition change of the liquid L and solid α phase during the freezing of a solid-solution alloy. Adapted from Ref 10.3

same time, the α composition approaches the gross composition X along curved path $\alpha_1\alpha_2\alpha_3\alpha_4$. The tie line, at each temperature, passes through the X composition, joining the conjugate compositions of L and α. Hence, the tie line lies in a different direction at each successive temperature.

Inscribed on the base of the diagram of Fig. 10.17 is a projection of the tie lines and the paths of composition variation of the liquid and solid phases. The liquidus path starts at the X composition and ends at L_4, while the solidus path starts at α_1 and ends at X. All tie lines, α_1L_1, α_2L_2, α_3L_3, and α_4L_4 pass through X. The same information is presented in somewhat more realistic fashion in Fig. 10.18, where the complete isotherm for each of the temperatures from T_1 to T_4 is given. In these drawings it becomes apparent that the curved paths of the solid and liquid compositions result from the turning of the tie lines to conform with the liquidus and solidus isotherms at successively lower temperatures.

10.5 Ternary Three-Phase Phase Diagrams

Three-phase equilibrium in ternary systems occurs over a temperature range and not, as in binary systems, at a single temperature. It is bivariant:

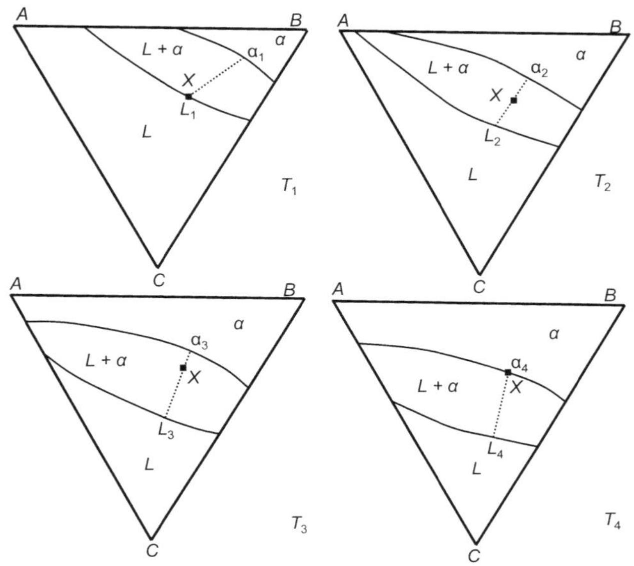

Fig. 10.18 Progress of equilibrium freezing of a ternary isomorphous alloy. Adapted from Ref 10.3

$$F = C + 2 - P$$
$$F = 3 + 2 - 3 = 2$$

After the pressure has been established, only the temperature, or one concentration parameter, may be selected in order to fix the conditions of equilibrium. The representation of three-phase equilibrium on a phase diagram requires the use of a structural unit that will designate, at any given temperature, the fixed compositions of three conjugate phases. Such a structural unit is found in the tie triangle.

Tie Triangles. If any three alloys of a ternary system are mixed, the composition of the mixture will lie within the triangle produced by connecting the three original composition points with straight lines (Fig. 10.19). For example, take three compositions *R*, *S*, and *L*:

$$R = 20\%\ A + 70\%\ B + 10\%\ C$$
$$S = 40\%\ A + 40\%\ B + 20\%\ C$$
$$L = 10\%\ A + 30\%\ B + 60\%\ C$$

and mix two parts of *R* with three parts of *S* and five parts of *L*:

$$0.2\ (20) + 0.3\ (40) + 0.5\ (10) = 21\%\ \text{A}$$
$$0.2\ (70) + 0.3\ (40) + 0.5\ (30) = 41\%\ \text{B}$$
$$0.2\ (10) + 0.3\ (20) + 0.5\ (60) = 38\%\ \text{C}$$

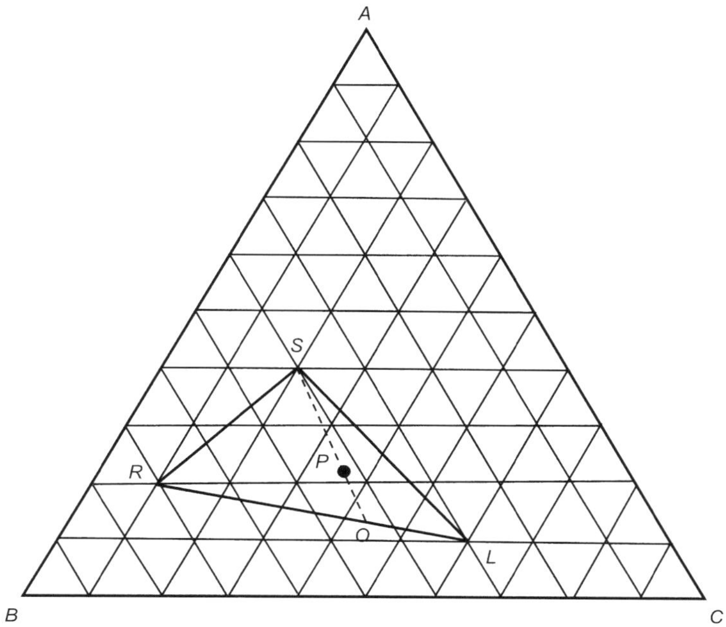

Fig. 10.19 Analysis of a tie triangle. Adapted from Ref 10.3

The total composition (21% *A* + 41% *B* + 38% *C*) lies within the triangle *RSL* at point *P* in Fig. 10.19. The same would have been true regardless of the proportions taken from the three alloys.

Once more the lever principle can be applied. This time the lever is the weightless plane triangle *RSL* supported on a point fulcrum at *P*. If 20% of the alloy *R* is placed on point *R*, 30% of alloy *S* on point *S*, and 50% of alloy *L* on point *L*, then the lever plane will balance exactly. For purposes of calculation, it is convenient to resolve the planar lever into two linear levers, such as *SPO* and *ROL* in Fig. 10.19, by drawing a straight line from any corner of the triangle through point *P* to its intersection with the opposite side. The quantity of the *S* alloy in the mixture *P* is then:

$$\%\ S = \left(\frac{PQ}{SO}\right)100 \text{ and } \%\ O = \left(\frac{SP}{SO}\right)100$$

Composition *O* represents the mixture of alloys *R* and *L*, so that:

$$\%\ R = \left(\frac{OL}{RL}\right)\left(\frac{PQ}{SO}\right)100 \text{ and } \%\ L = \left(\frac{RO}{RL}\right)\left(\frac{SP}{SO}\right)100$$

The triangle *RSL* may be employed as a tie triangle connecting three phases that associate to form an intermediate gross composition or phase or, conversely, three phases into which *P* decomposes. There are then two kinds of "tie elements" that appear in ternary diagrams, namely, the tie line and the tie triangle.

10.6 Eutectic System with Three-Phase Equilibrium

An example of three-phase equilibrium in a ternary system with a eutectic reaction is shown in Fig. 10.20. One of the binary systems involved in this ternary system is isomorphous; the other two are of the eutectic type. There are three one-phase regions: L, α, and β; three two-phase regions: $L + \alpha$, $L + \beta$, and $\alpha + \beta$; and one three-phase region: $L + \alpha + \beta$.

Five isotherms are taken at temperatures designated T_1 to T_5 in Fig. 10.21, and isothermal sections are shown in Fig. 10.22. Returning to the requirements of the phase rule, it is now a simple matter to demonstrate that the one degree of freedom remaining after the establishment of a fixed pressure is sufficient to complete the definition of all variables in three-phase equilibrium. If a specific temperature is chosen, such as T_2 in Fig. 10.22, then the compositions of all of the three conjugate phases are fixed, respectively, at α_2, β_2, and L_2. Or similarly, if a concentration parameter is defined by requiring that the α phase in equilibrium with β and L shall

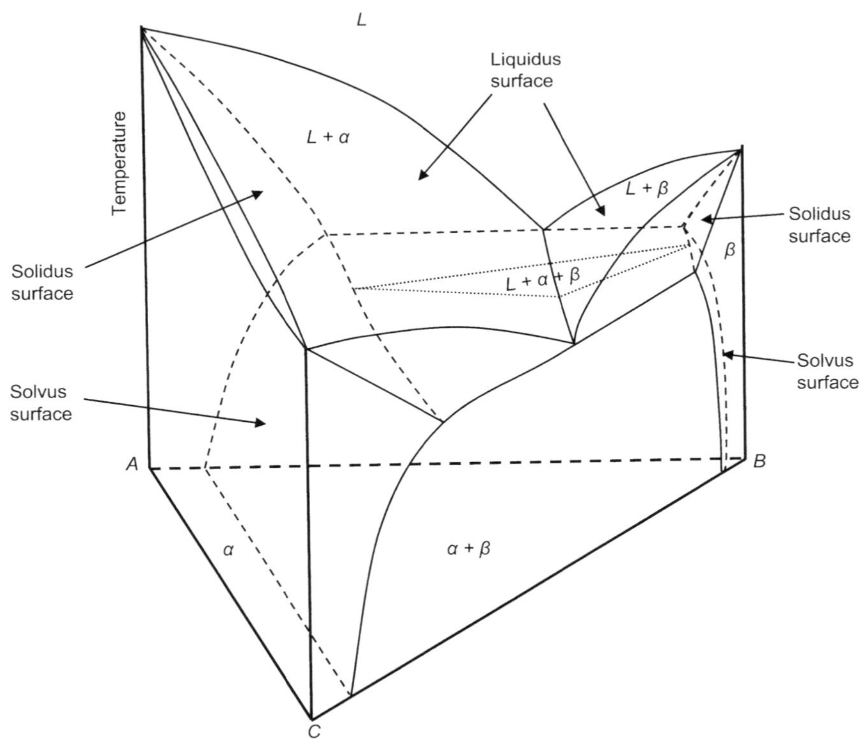

Fig. 10.20 Three-phase equilibria in a ternary system with a eutectic reaction. Adapted from Ref 10.3

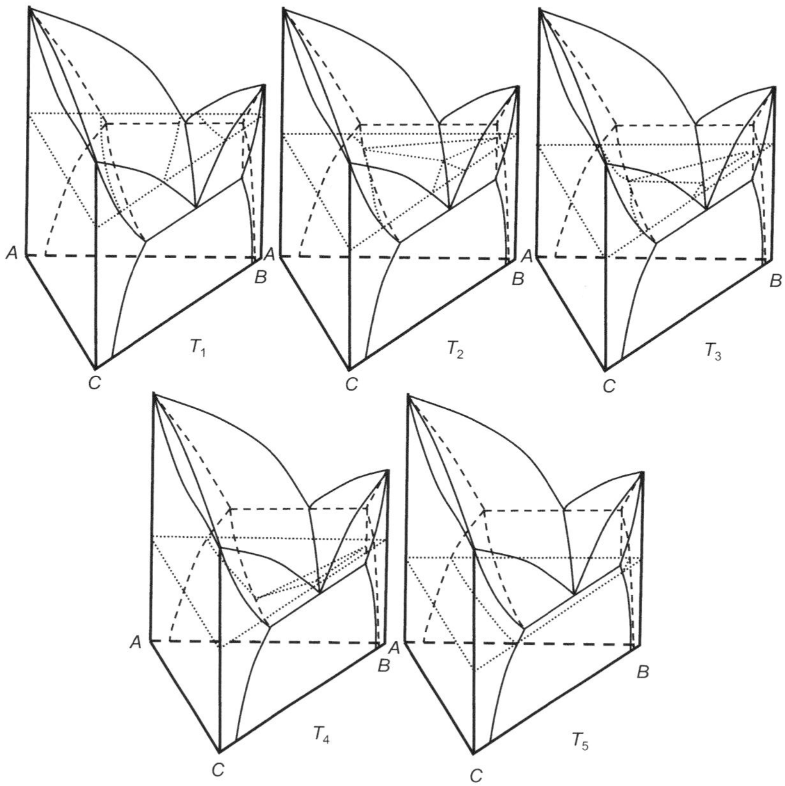

Fig. 10.21 Development of isotherms shown in Fig. 10.22. Adapted from Ref 10.3

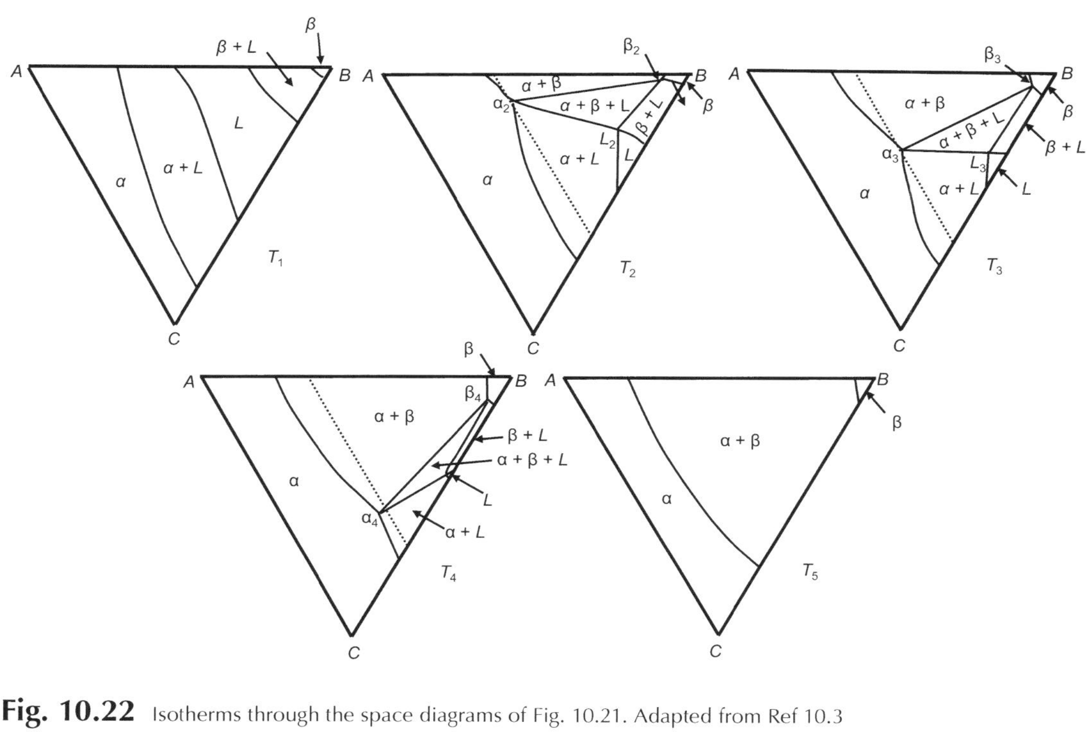

Fig. 10.22 Isotherms through the space diagrams of Fig. 10.21. Adapted from Ref 10.3

contain 35% B (indicated by the dotted line in isotherms T_2, T_3, and T_4), there will be only one isotherm (T_3) in which this condition is realized. The temperature and the compositions of the conjugate β_3 and L_3 phases are fixed by establishing one concentration parameter of α.

Freezing of an Alloy. The equilibrium freezing of any alloy can be considered by reference to the series of isotherms shown in Fig. 10.23. The black dot designates in each isotherm the gross composition X of the alloy under observation. At temperature T_1 (the first isotherm), the X composition lies in the liquid field and the alloy is fully molten.

At T_2, the next lower temperature represented, freezing is just beginning. Composition X lies on the liquidus and is joined with the first solid to appear by the tie line $L_2\alpha_2$.

At T_3, a substantial quantity of α is present:

$$\%\alpha_3 = \left(\frac{XL_3}{\alpha_3 L_3}\right)(100) \approx 20\%$$

and

$$\% L_3 = \left(\frac{\alpha_3 X}{\alpha_3 L_3}\right)(100) \approx 80\%$$

At T_4, as the gross composition passes into the $L + \alpha + \beta$ field, the first particles of the β phase begin to appear; the α and β now crystallize simultaneously from the liquid.

At T_5, all three phases are present in substantial quantity:

$$\%\alpha_5 = \left(\frac{XS_5}{\alpha_5 S_5}\right)(100) \approx 60\%$$

$$\%\alpha_5 = \left(\frac{L_5 S_5}{L_5 \beta_5}\right)\left(\frac{\alpha_5 X}{\alpha_5 S_5}\right)(100) \approx 20\%$$

$$\% L_5 = \left(\frac{S_5 \beta_5}{L_5 \beta_5}\right)\left(\frac{\alpha_5 X}{\alpha_5 S_5}\right)(100) \approx 20\%$$

At T_6 the last of the liquid disappears and there remains only:

$$\%\alpha_6 = \left(\frac{X\beta_6}{\alpha_6 \beta_6}\right)(100) \approx 75\%$$

$$\%\beta_6 = \left(\frac{\alpha_6 X}{\alpha_6 \beta_6}\right)(100) \approx 25\%$$

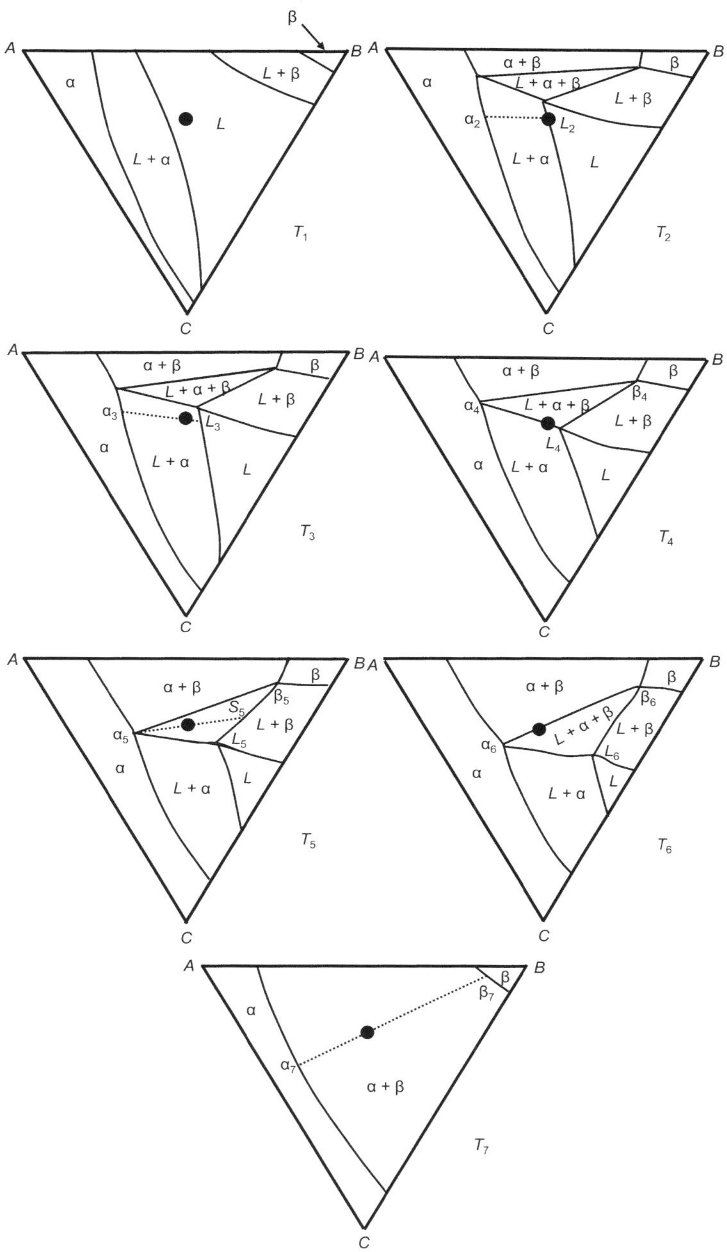

Fig. 10.23 Sequence of equilibria involved in freezing of an alloy, showing gross composition in each isotherm. Adapted from Ref 10.3

At T_7 the compositions of the two solid phases have changed slightly in response to the curvature of the α and β solvus surfaces, and their relative proportions will again be given by the tie line $\alpha_7\beta_7$.

Freezing has proceeded in two steps, neither of which is isothermal; primary freezing of the α phase has been followed by a secondary separation of α + β over the temperature range T_4 to T_6.

Wherever transformation occurs over a temperature range at a finite rate (i.e., natural freezing), it is anticipated that the products of that transformation will exhibit a coring effect. Thus, both the primary and secondary constituents should be cored. When α and β are crystallizing together, both phases should be cored.

10.7 Peritectic System with Three-Phase Equilibrium

If two binary peritectic systems are substituted for the two binary eutectic systems of the foregoing example, the resulting ternary space model will appear, as in Fig. 10.24. This diagram has the same number of fields as in the previous example, and they are similarly designated. The most evident difference is in the $L + \alpha + \beta$ field, which has been inverted to produce the peritectic reaction.

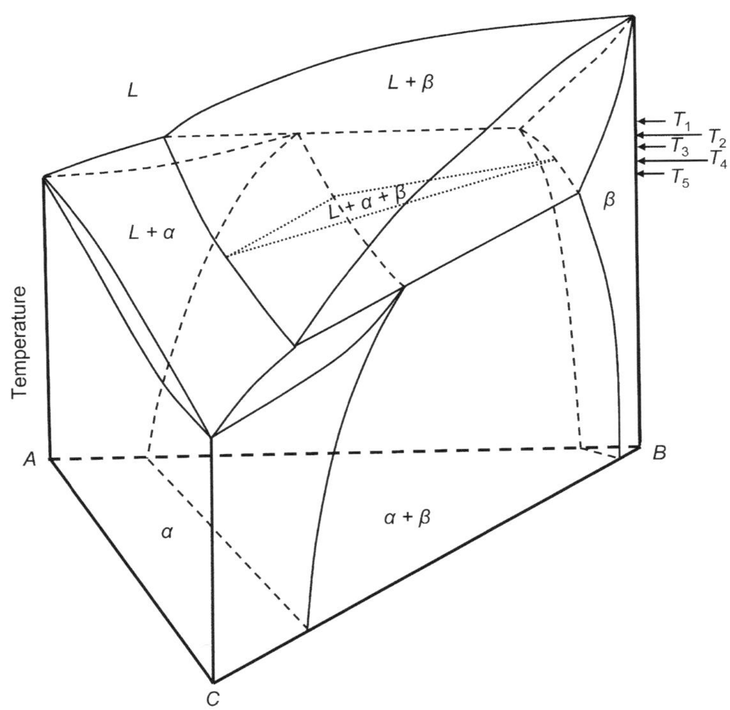

Fig. 10.24 Three-phase equilibria in a ternary system with a peritectic reaction. Adapted from Ref 10.3

These relationships are clearly evident in the isotherms of Fig. 10.25, which should be compared with those of the previous example (Fig. 10.22); it is seen that the tie triangles are reversed. The course of freezing can again be followed by the use of the isothermal sections. That the peritectic reaction results from the reversal of the tie triangles can be seen by analyzing the process in detail. Consider first an alloy of composition X shown on tie triangles at three temperatures in Fig. 10.26. At T_2, the primary separation of β crystals is complete and the precipitation of α is about to begin. The quantity of primary β is:

$$\%\beta(\text{primary}) = \left(\frac{L_2X}{L_2\beta_2}\right)(100) \approx 30\%$$

When the temperature falls to T_3, the quantity of β is reduced to:

$$\%\beta(\text{total}) = \left(\frac{mX}{m\beta_3}\right)(100) \approx 20\%$$

This change is accompanied by a corresponding decrease in the quantity of liquid:

$$\%L_2 = \left(\frac{X\beta_2}{L_2\beta_2}\right)(100) \approx 70\%$$

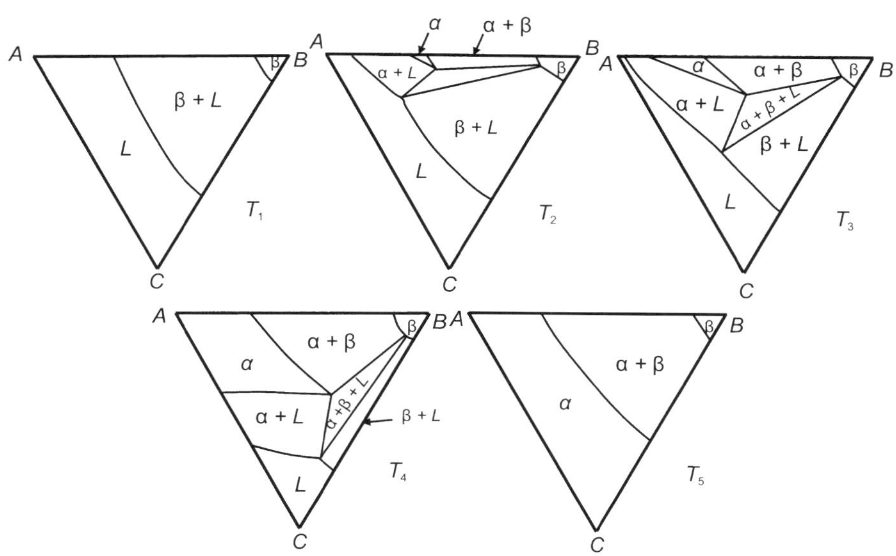

Fig. 10.25 Isotherms through the space diagram of Fig. 10.24. Adapted from Ref 10.3

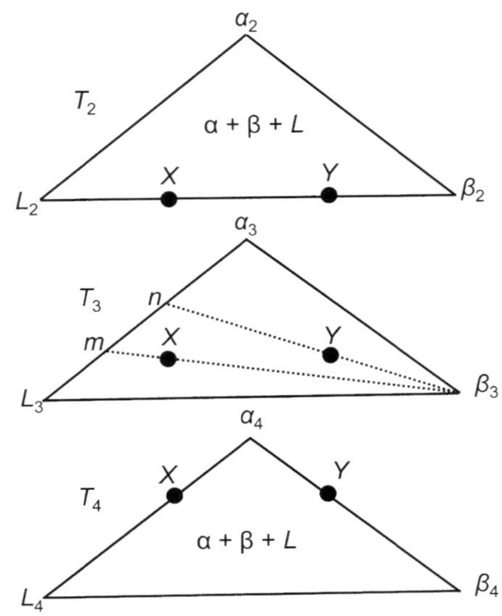

Fig. 10.26 Tie triangles at three different temperatures. Adapted from Ref 10.3

$$\%\, L_3 = \left(\frac{m\alpha_3}{L_3\alpha_3}\right)\left(\frac{X\beta_3}{m\beta_3}\right)(100) \approx 60\%$$

and an increase in the amount of α phase:

$$\%\ \alpha_2 = 0$$

$$\%\ \alpha_3 = \left(\frac{\mathrm{mL}_3}{\alpha_3 L_3}\right)\left(\frac{X\beta_3}{m\beta_3}\right)(100) \approx 20\%$$

Just as in binary peritectic reaction, the primary constituent is consumed by reaction with the liquid to form the secondary constituent. At T_4, the β phase has been consumed altogether and only liquid and α remain.

Had the alloy been of composition *Y* in Fig. 10.26, the same course of events would have been observed, except that all the liquid would have been consumed in forming α, leaving a residue of unreacted β. In natural freezing these processes are retarded by failure to maintain equilibrium. All constituents are cored, envelopment occurs, and the relative quantities of the crystalline phases deviate from the ideal, just as in binary peritectic alloys. The microstructures of cast ternary alloys of this kind are indistinguishable from their binary counterparts, the binary peritectic alloys.

10.8 Ternary Four-Phase Equilibrium ($L \rightarrow \alpha + \beta + \gamma$)

Ternary eutectic reaction occurs by the isothermal decomposition of liquid into three different solid phases:

$$L \rightarrow \alpha + \beta + \gamma$$

In the phase diagram shown in Fig. 10.27, the ternary eutectic plane has been shaded to make its location more apparent. The three corners of the triangle touch the three one-phase regions α, β, and γ and are so labeled. The liquid composition occurs at point L within the triangle, where the liquidus surfaces from the three corners of the space diagram meet at the lowest melting point of the ternary system.

According to the phase rule, four-phase equilibrium in ternary systems should be univariant:

$$F = C + 2 - P$$
$$F = 3 + 2 - 4 = 1$$

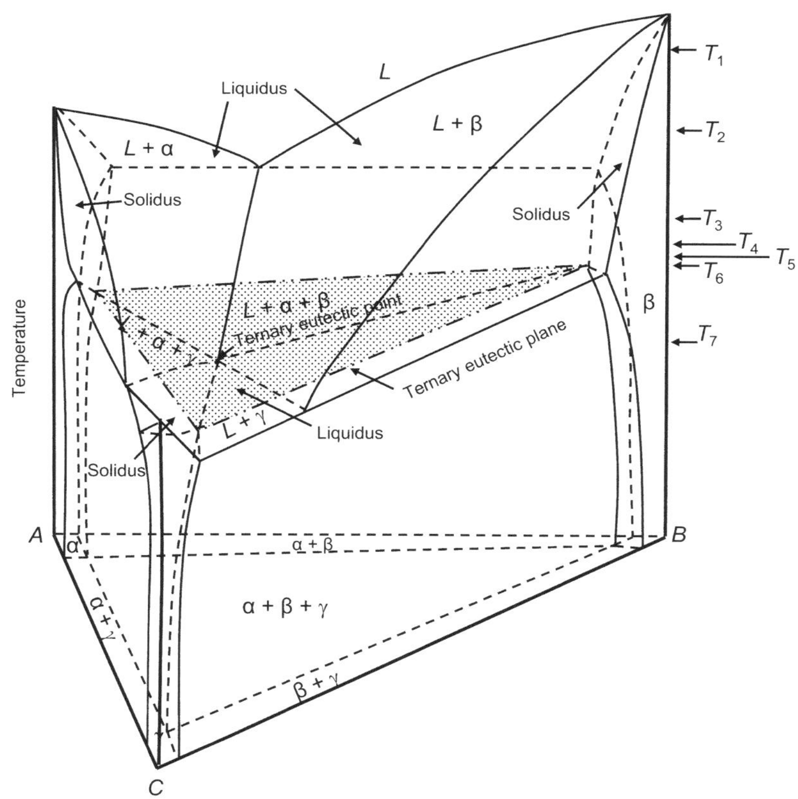

Fig. 10.27 Temperature-composition space model of a ternary eutectic system with the reaction $L \rightarrow \alpha + \beta + \gamma$. Adapted from Ref 10.3

Having established the pressure, the temperature of four-phase equilibrium and the compositions of each of the four phases should be fixed. The construction employed in Fig. 10.27 meets these requirements; the ternary eutectic plane is isothermal, and the compositions of the four phases are designated at four fixed points on the eutectic plane.

The seven isotherms of Fig. 10.28 are taken at the temperatures designated T_1 to T_7 in Fig. 10.27. In the first section, T_1, the liquidus and solidus adjacent to B, the highest melting component, are intersected, revealing fields of L, β, and $L + \beta$. Successively at T_2 and T_3 the liquidus and solidus pairs adjacent to the A and C components are intersected. The third section, T_3, is taken just below the eutectic temperature of the binary system AB and intersects the three-phase region $L + \alpha + \beta$ that originates on the binary eutectic line. Similar constructions were found in each of the phase diagrams of the preceding section; for example, compare with section T_2 in Fig. 10.22. From each of the binary eutectic lines there issues a three-phase field so that at T_4, just below the binary eutectic of the system AC, the field $L + \alpha + \gamma$ appears, and at T_5, just below the BC eutectic, the $L + \beta + \gamma$ field appears. The liquid region is now confined to a small three-cornered area in the middle of the diagram, and the liquid plus solid regions have grown narrow. As the temperature falls, these fields continue to shrink, permitting the three three-phase tie-triangles to meet and to form the ternary eutectic plane at T_6, the ternary eutectic temperature. At this temperature the liquid phase disappears and with it the six other fields involving liquid, so that below the ternary eutectic temperature, only solid phases remain. As the extent of solid solubility decreases with falling temperature T_7, the $\alpha + \beta + \gamma$ field grows larger, while the one-phase fields contract.

10.9 Ternary Four-Phase Equilibrium ($L + \alpha \rightarrow \beta + \gamma$)

There is another type of ternary four-phase equilibrium that has no direct equivalent in binary systems. It may be thought of as being intermediate between the eutectic and peritectic reaction:

$$L + \alpha \rightarrow \beta + \gamma$$

During heating or cooling, two phases interact to form two new phases. This four-phase equilibrium is shown in the Fig. 10.29 space diagram, and the isothermal sections through the diagram are shown in Fig. 10.30. It can be seen that two three-phase regions, $L + \alpha + \beta$ and $L + \alpha + \gamma$, descend from higher temperature toward the four-phase reaction plane, where they meet to form a horizontal trapezium (kite-shaped region) at the four corners of which the four phases in equilibrium are represented: $\alpha + \beta + \gamma + L$. Beneath the four-phase reaction plane, two new three-phase regions, $\alpha \rightarrow \beta + \gamma$ and $L + \beta + \gamma$, originate and descend to lower temperature. Thus, the two tie triangles, $L + \alpha + \beta$ and $L + \alpha + \gamma$ in Fig. 10.31, join along a common tie line, αL, to form the four-phase reaction trapezium, and this figure

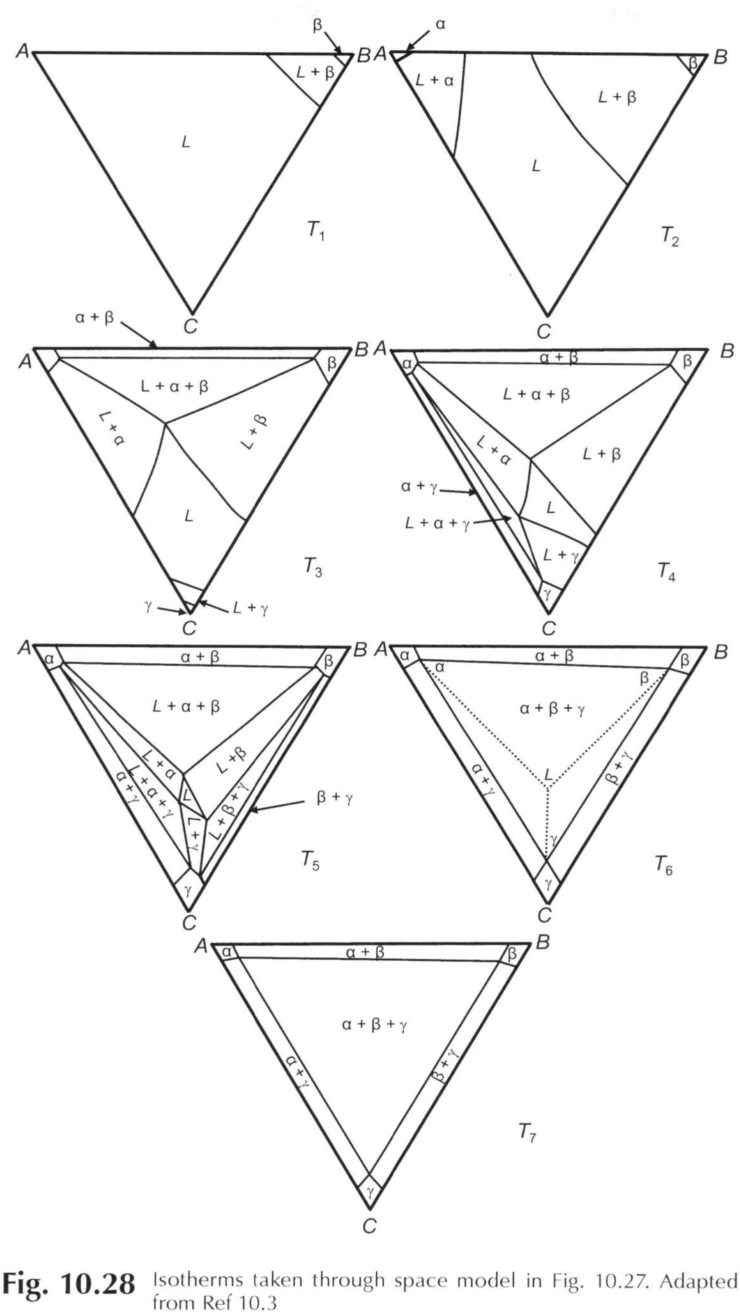

Fig. 10.28 Isotherms taken through space model in Fig. 10.27. Adapted from Ref 10.3

divides along the tie line βγ to form two new tie triangles, α + β + γ and *L* + β + γ. An alloy occurring at the composition at which these tie lines cross (intersection of the dashed lines in the central sketch of Fig. 10.31) would be composed entirely of *L* and α just above the reaction temperature and entirely of β and γ just below the reaction temperature.

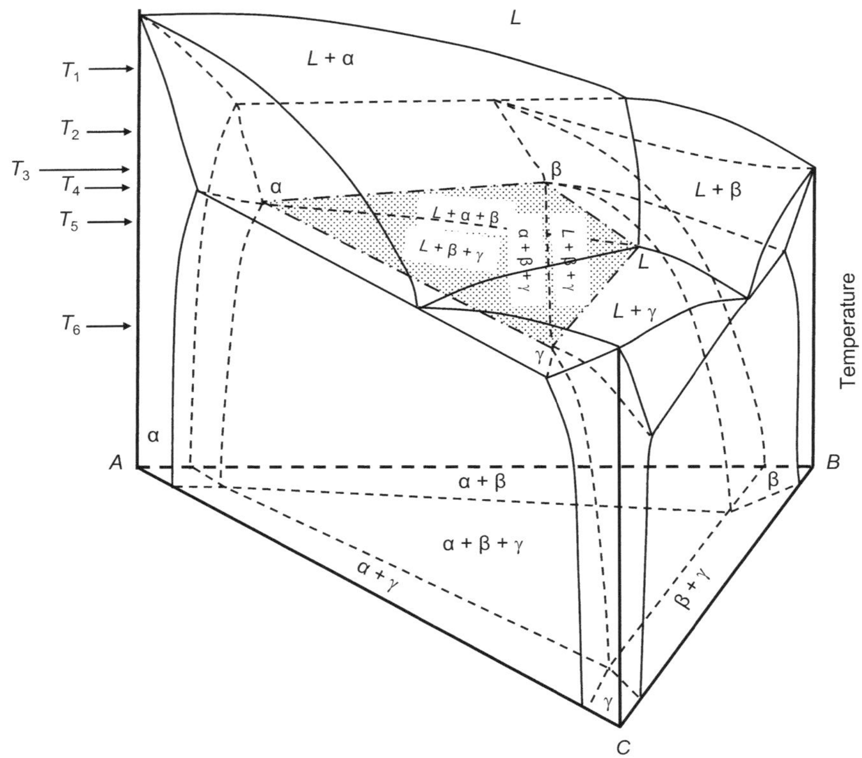

Fig. 10.29 Temperature-composition space model of a ternary eutectic system with the reaction $L + \alpha \rightarrow \beta + \gamma$. Adapted from Ref 10.3

This construction conforms to the phase rule. A four-phase reaction should be univariant; the temperature and compositions of the participating phases should be constant at fixed pressure. These conditions are met, because the four-phase trapezium is horizontal (isothermal) and the one-phase regions touch it only at its four corners where a unique composition is thereby designated for each of the four phases.

Other than the four-phase reaction trapezium, this diagram involves no regions that have not been discussed in previous sections. The one-, two-, and three-phase regions in this diagram are identical in number and designation with those found in the ternary eutectic diagram.

Course of Freezing of Alloys. An examination of the course of freezing of several typical alloys should prove helpful in gaining an understanding of the nature of this four-phase reaction. Four alloys designated 1, 2, 3, and 4 in the central drawing of Fig. 10.31 are considered. It can be seen at once that these change on descending through the four-phase reaction temperature:

1. $L + \alpha + \beta \rightarrow \alpha + \beta + \gamma$
2. $L + \alpha + \gamma \rightarrow \alpha + \beta + \gamma$
3. $L + \alpha + \beta \rightarrow L + \beta + \gamma$
4. $L + \alpha + \gamma \rightarrow L + \beta + \gamma$

By reference to Fig. 10.30, it can be seen that Alloy 1 begins freezing with a primary separation of the α phase followed by a secondary deposition of α + β. Just above the four-phase reaction temperature, the relative

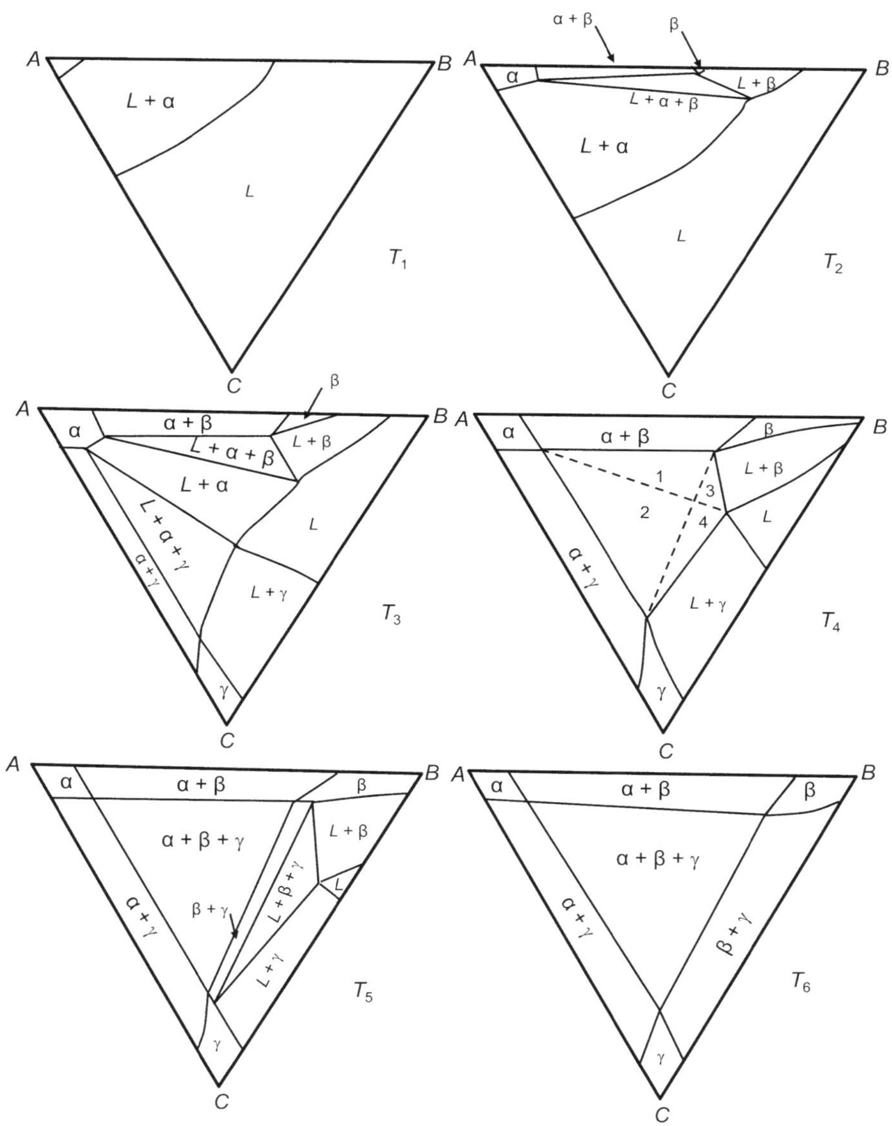

Fig. 10.30 Isotherms through the space diagram of Fig. 10.29. Adapted from Ref 10.3

proportions of the three phases may be computed from the tie triangle in solid lines in Fig. 10.32:

$$\%\beta = \left(\frac{XI}{X\beta}\right)(100) \approx 50\%$$

$$\%\alpha = \left(\frac{XL}{\alpha L}\right)\left(\frac{I\beta}{X\beta}\right)(100) \approx 28\%$$

$$\%L = \left(\frac{\alpha X}{\alpha L}\right)\left(\frac{I\beta}{X\beta}\right)(100) \approx 22\%$$

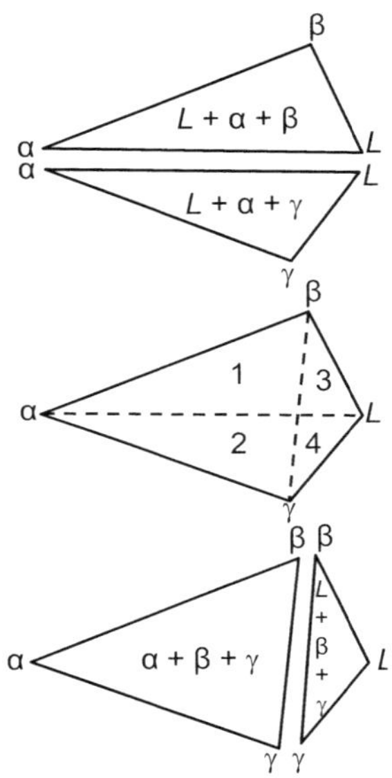

Fig. 10.31 Tie triangles for four-phase ternary equilibrium example. Adapted from Ref 10.3

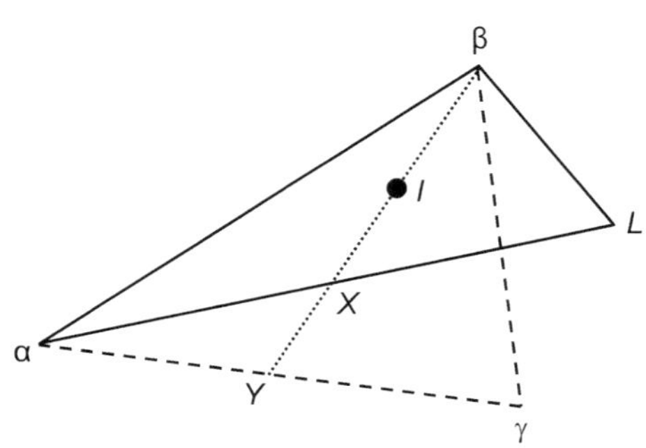

Fig. 10.32 Tie triangle for Alloy 1. Adapted from Ref 10.3

Just below the four-phase reaction temperature:

$$\%\beta = \left(\frac{YI}{Y\beta}\right)(100) \approx 66\%$$

$$\%\alpha = \left(\frac{Y\gamma}{\alpha\gamma}\right)\left(\frac{I\beta}{X\beta}\right)(100) \approx 18\%$$

$$\%\gamma = \left(\frac{\alpha Y}{\alpha\gamma}\right)\left(\frac{I\beta}{Y\beta}\right)(100) \approx 16\%$$

The percentage of α has decreased sharply, and the liquid has disappeared altogether, while the quantity of β has increased, and a substantial quantity of γ has appeared as a new phase. Some of the previously existing α has been consumed by reaction with liquid, as in peritectic reaction. The increase in the quantity of β, coincident with the appearance of γ, more nearly resembles a eutectic reaction. This is what was meant by the opening statement to the effect that this four-phase equilibrium occupies a position midway between the eutectic and peritectic types. Alloy 2 (Fig. 10.31) also begins freezing with a primary separation of the α phase, but the secondary crystallization will be of α + γ in this case, instead of α + β. If the tie triangles were analyzed for this example, it would be found that at the four-phase reaction temperature, the liquid phase is wholly consumed and the α phase is partly redissolved to form a tertiary precipitation of β + γ.

The two alloys that lie closer to the liquid corner of the four-phase trapezium, Alloys 3 and 4, differ from the foregoing pair of alloys in that the primary and secondary α crystals should be totally consumed if four-phase reaction goes to completion, while some of the liquid phase should remain to freeze as β + γ at a lower temperature. In both alloys the primary constituent is α, although its quantity may be small because these compositions are close to the lower edge of the $L + \alpha$ field. The secondary constituent in Alloy 3 will be α + β, while that in Alloy 4 will be α + γ. Both will suffer a loss of α and of some liquid to form β + γ in four-phase reaction, and the simultaneous crystallization of β and γ will continue at lower temperature until the supply of liquid is exhausted.

As with peritectic alloys, the formation of reaction layers (envelopment) on the α phase may be expected to interfere with the completion of reaction by hindering the diffusion that must take place to establish equilibrium. Consequently, an excess of the α phase is likely to persist, and some liquid is likely to survive through four-phase reaction in all the alloys of this system. This will have the effect of minimizing the differences in structure among the several alloys.

10.10 Ternary Four-Phase Equilibrium ($L + \alpha + \beta \rightarrow \gamma$)

In ternary four-phase equilibrium, the ternary equivalent of peritectic equilibrium is:

$$L + \alpha + \beta \rightarrow \gamma$$

Three phases interact isothermally on cooling to form one new phase, or conversely, on heating, one phase decomposes isothermally into three new phases.

The four-phase equilibrium diagram with the peritectic reaction is illustrated in Fig. 10.33. Three phases, α, β, and L, located at the corners of a

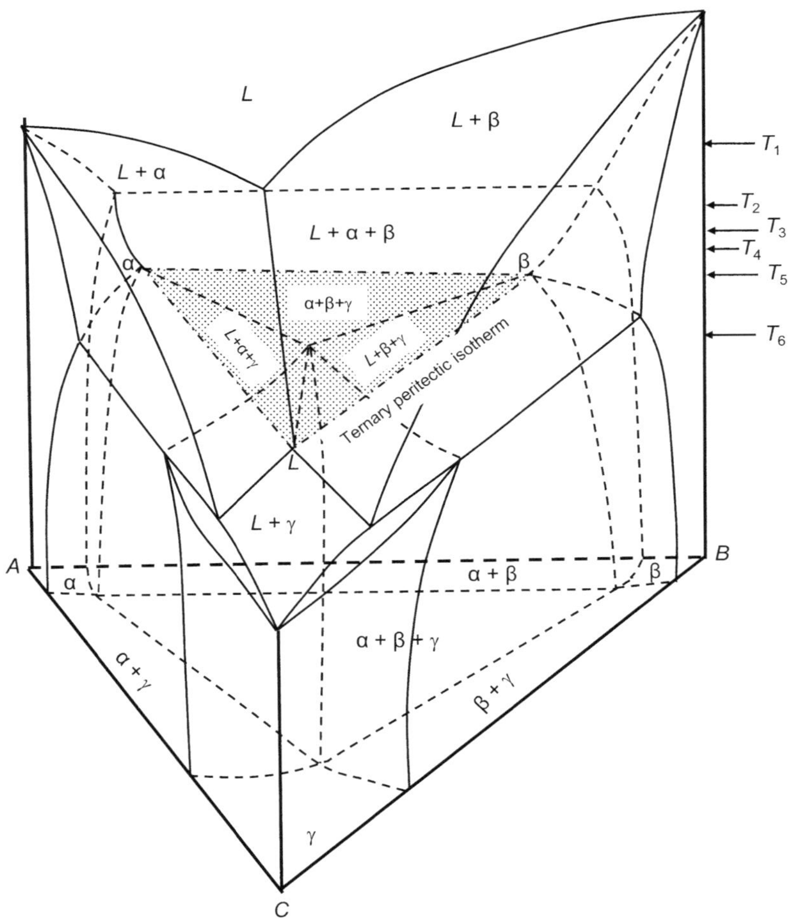

Fig. 10.33 Temperature-composition space model of a ternary peritectic system with the reaction $L + \alpha + \beta \rightarrow \gamma$. Adapted from Ref 10.3

horizontal triangular reaction plane, combine to form the γ phase, whose composition lies within the triangle. One three-phase field, $L + \alpha + \beta$, descends from higher temperature to the ternary peritectic temperature, and three three-phase regions, $\alpha + \beta + \gamma$, $L + \alpha + \gamma$, and $L + \beta + \gamma$, issue beneath and proceed to lower temperature (Fig. 10.34). The obvious equivalence of this construction to that of the ternary eutectic disposes of any necessity of further demonstrating its conformity with the phase rule.

Freezing of Alloys. Three essentially distinct types of four-phase reactions should occur during cooling:

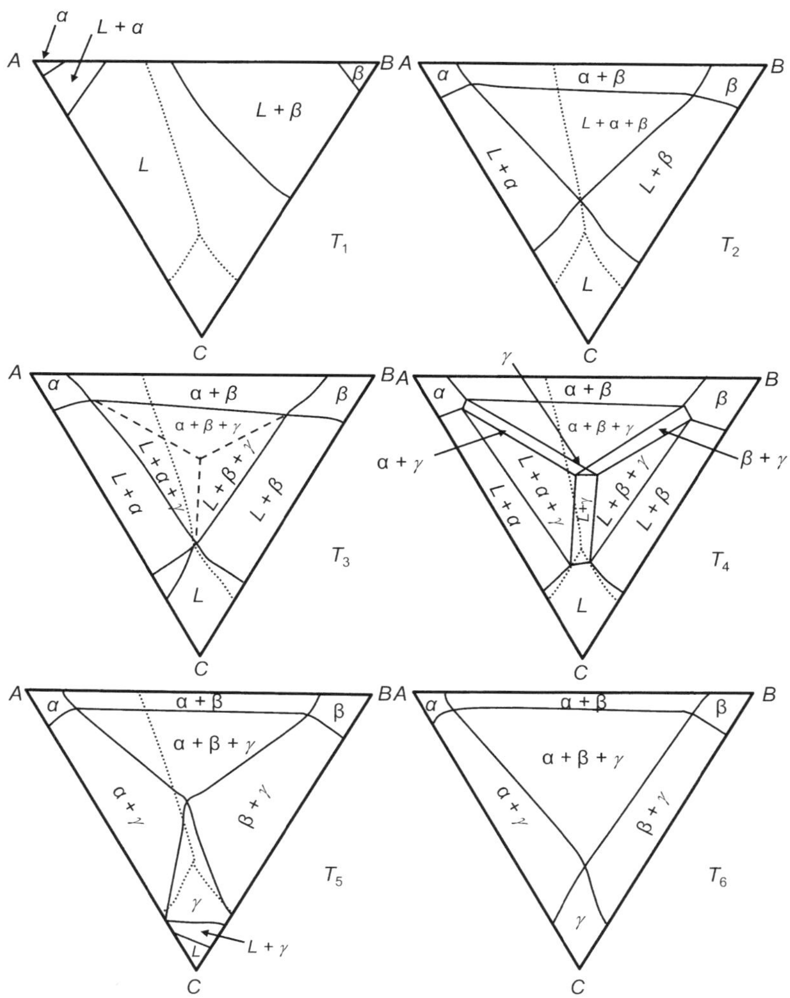

Fig. 10.34 Isotherms through the space model of Fig. 10.33. Adapted from Ref 10.3

1. $L + \alpha + \beta \rightarrow \alpha + \beta + \gamma$
2. $L + \alpha + \beta \rightarrow L + \alpha + \gamma$
3. $L + \alpha + \beta \rightarrow L + \beta + \gamma$

These reactions correspond to alloys falling within the zones numbered 1, 2, and 3 in Fig. 10.35. In each case, the phases present above the ternary peritectic temperature all either diminish in quantity or disappear altogether. This can be demonstrated by analyzing the tie triangles by the use of the lever principle. Consider Alloy *X* (Fig. 10.36) an alloy of the first type. Above the peritectic temperature, the relative proportions of the three phases should be:

$$\%\beta = \left(\frac{WX}{W\beta}\right)(100) \approx 42\%$$

$$\%\alpha = \left(\frac{WL}{\alpha L}\right)\left(\frac{X\beta}{W\beta}\right)(100) \approx 42\%$$

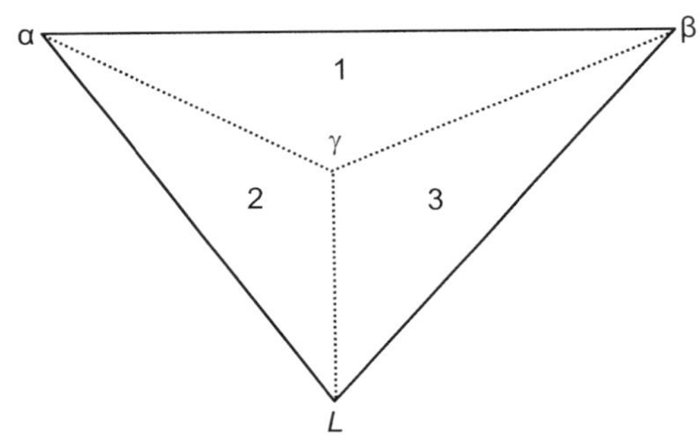

Fig. 10.35 Alloys 1, 2, and 3 illustrating different freezing behaviors. Adapted from Ref 10.3

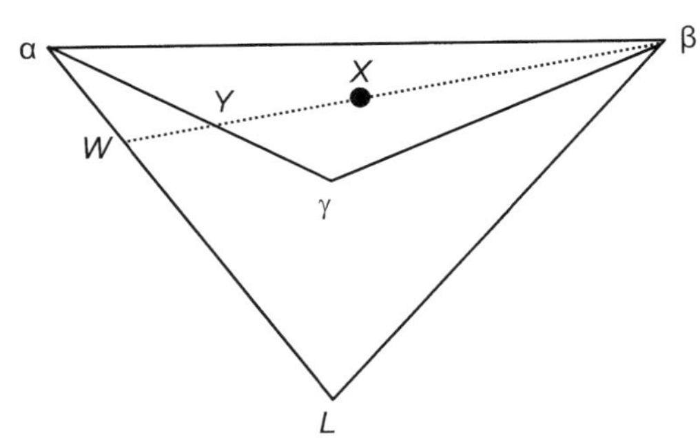

Fig. 10.36 Tie triangle for Alloy *X*. Adapted from Ref 10.3

$$\% L = \left(\frac{\alpha W}{\alpha L}\right)\left(\frac{X\beta}{W\beta}\right)(100) \approx 16\%$$

and just below this temperature:

$$\% \beta = \left(\frac{YX}{Y\beta}\right)(100) \approx 34\%$$

$$\% \alpha = \left(\frac{Y\gamma}{\alpha\gamma}\right)\left(\frac{X\beta}{Y\beta}\right)(100) \approx 33\%$$

$$\% \gamma = \left(\frac{\alpha Y}{\alpha\gamma}\right)\left(\frac{X\beta}{Y\beta}\right)(100) \approx 33\%$$

The liquid phase has vanished, and the quantities of α and β are sharply reduced. A similar condition is obtained with alloys of the second and third types.

Such reaction requires that the γ phase be formed preferentially at three-way junctions of L, α, and β in the structure of the alloy, because the composition of the phase requires the borrowing of A component from the α, B component from the β, and α component from the liquid. As with binary peritectic reaction, it is to be expected that the formation of the γ phase should retard its own further growth by lengthening the path over which solid-phase diffusion must act to supply the necessary components. Thus, incomplete reaction should be common; any remaining liquid would, of course, freeze to cored γ.

10.11 Example: The Fe-Cr-Ni System

Many commercial cast irons and steels contain ferrite-stabilizing elements (such as silicon, chromium, molybdenum, and vanadium) and/or austenite stabilizers (such as manganese and nickel). The diagram for the binary iron-chromium system is representative of the effect of a ferrite stabilizer (Fig. 10.37). At temperatures just below the solidus, body-centered cubic (bcc) chromium forms a continuous solid solution with bcc (δ) ferrite. At lower temperatures, the γ-iron phase appears on the iron side of the diagram and forms a "loop" extending to approximately 11.2% Cr. Alloys containing up to 11.2% Cr, and sufficient carbon, are hardenable by quenching from temperatures within the loop.

At still lower temperatures, the bcc solid solution is again continuous bcc ferrite, but this time with α-iron. This continuous bcc phase field confirms that δ-ferrite is the same as α-ferrite. The nonexistence of γ-iron in iron-

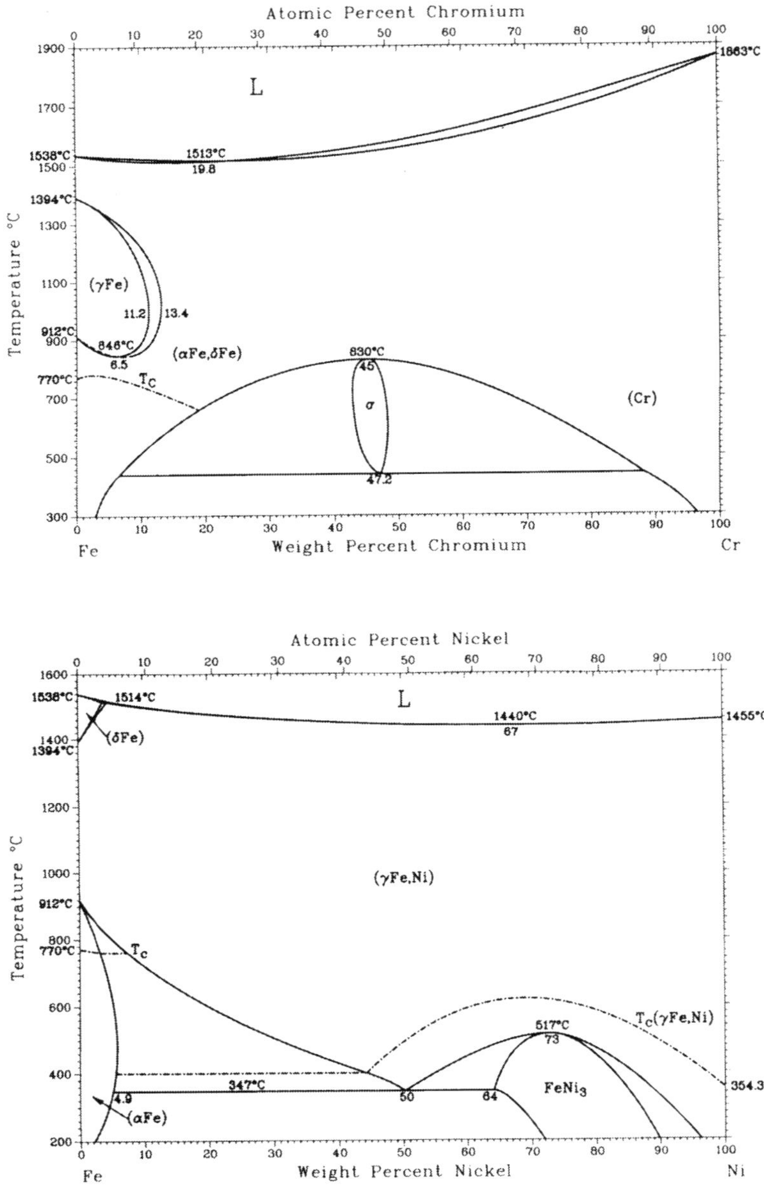

Fig. 10.37 Two representative binary iron phase diagrams, showing ferrite stabilization (iron-chromium) and austenite stabilization (iron-nickel). Source: Ref 10.4 as published in Ref 10.5

chromium alloys having more than approximately 13% Cr, in the absence of carbon, is an important factor in both the hardenable and nonhardenable grades of iron-chromium stainless steels. At these lower temperatures, a material known as σ phase also appears in different amounts from approximately 14 to 90% Cr. This is a hard, brittle phase and usually should be avoided in commercial stainless steels. Formation of σ, however, is time dependent; long periods at elevated temperatures are usually required.

The diagram for the binary iron-nickel system is representative of the effect of an austenite stabilizer (Fig. 10.37). The face-centered cubic (fcc) nickel forms a continuous solid solution with fcc (γ) austenite that dominates the diagram, although the α-ferrite phase field extends to approximately 6% Ni. The diagram for the ternary Fe-Cr-Ni system shows how the addition of ferrite-stabilizing chromium affects the iron-nickel system (Fig. 10.38). As can be seen, the popular 18-8 stainless steel, which contains approximately 8% Ni, is an all-austenite alloy at 900 °C (1652 °F), even though it also contains approximately 18% Cr.

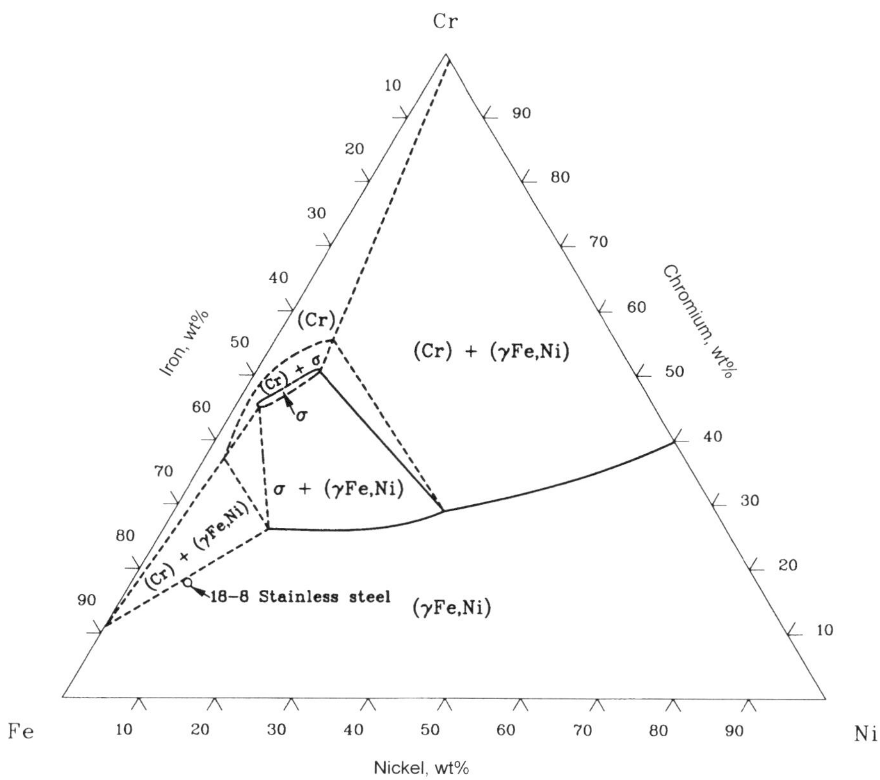

Fig. 10.38 The isothermal section at 900 °C (1652 °F) of the Fe-Cr-Ni ternary phase diagram, showing the nominal composition of 18-8 stainless steel. Source: Ref 10.6 as published in Ref 10.5

ACKNOWLEDGMENT

The majority of the material in this chapter was adapted from *Phase Diagrams in Metallurgy* by F.N. Rhines, McGraw-Hill, 1956.

REFERENCES

10.1 D.R. Askeland, *The Science and Engineering of Materials*, 2nd ed., PWS-KENT, 1989

10.2 C-Cr-Fe Phase Diagram (1966 Tricot R.), ASM Alloy Phase Diagrams Center, P. Villars, editor-in-chief; H. Okamoto and K. Cenzual, section editors, ASM International

10.3 F.N. Rhines, *Phase Diagrams in Metallurgy*, McGraw-Hill, 1956

10.4 T.B. Massalski, Ed., *Binary Alloy. Phase Diagrams,* 2nd ed., ASM International, 1990

10.5 H. Baker, Introduction to Alloy Phase Diagrams, *Alloy Phase Diagrams*, Vol 3, *ASM Handbook*, ASM International, 1992

10.6 G.V. Raynor and V.G. Rivlin, *Phase Equilibria in Iron Ternary Alloys,* Vol 4, The Institute of Metals, London, 1988

SELECTED REFERENCES

- H. Baker, Introduction to Alloy Phase Diagrams, *Alloy Phase Diagrams*, Vol 3, *ASM Handbook*, ASM International, 1992; reprinted in *Desk Handbook: Phase Diagrams for Binary Alloys*, 2nd ed., H. Okamoto, Ed., 2010
- D.R.E. West, *Ternary Equilibrium Diagrams*, 2nd ed., Chapman & Hall, 1982

CHAPTER **11**

Gas-Metal Systems

GAS-METAL SYSTEMS are important both during metallic processing and during exposure when placed in service, particularly in the evaluation of high-temperature oxidation and hot-gas corrosion. Engineering alloys react chemically when exposed to air or other more aggressive gases. Whether they survive or not depends on how fast they react. For a few metals, the reaction is so slow that they are virtually unattacked, but for others, the reaction can be disastrous. High-temperature service is especially damaging to most metals because of the exponential increase in reaction rate with temperature. The driving force for reaction of a metal with a gas is the Gibbs energy change, ΔG.

The driving force, ΔG, for a reaction such as $aA + bB = cC + dD$ can be expressed in terms of the standard Gibbs energy change, ΔG^0, by:

$$\Delta G = \Delta G^0 + RT \ln \frac{a_C^{c} a_D^{d}}{a_A^{a} a_B^{b}}$$

where the chemical activity, a, of each reactant or product is raised to the power of its stoichiometric coefficient, and R is the gas constant. For example, the oxidation of a metal can be expressed by the reaction:

$$x\mathrm{M} + \frac{y}{2}\mathrm{O}_2 = \mathrm{M}_x\mathrm{O}_y$$

where M is the reacting metal, M_xO_y is its oxide, and x and y are the moles of metal and oxygen, respectively, in 1 mol of the oxide.

The Gibbs energy change for the reaction is:

$$\Delta G = \Delta G^0 + RT \ln\left[\frac{a\mathrm{M}_x\mathrm{O}_y}{(a\mathrm{M})^x (a\mathrm{O}_2)^{y/2}}\right]$$

In most cases, the activities of the solids (metal and oxide) are invariant; that is, their activities = 1 for pure solids, and for the relatively high temperatures and moderate pressures encountered in oxidation reactions, a can be approximated by its pressure. Therefore, at equilibrium where $\Delta G = 0$:

$$\Delta G^0 = -RT\left[\frac{a_{\text{prod}}}{a_{\text{react}}}\right] = +\frac{y}{2}RT\ln pO_2$$

where pO_2 is the partial pressure of oxygen.

In solid solutions, such as an alloy, the partial molar Gibbs energy of a substance is usually called its chemical potential. If 1 mol of pure A is dissolved in an amount of solution so large that the solution concentration remains virtually unchanged, the Gibbs energy change for the mole of A ($\Delta\bar{G}_A$) is:

$$\Delta\bar{G}_A = \mu_A - \mu_A^0 = RT\ln a_A$$

where μ_A^0 is the chemical potential of 1 mol of pure A, μ_A is the chemical potential of A in the solution, and a_A is the activity of A in the solution.

Thermodynamically unstable oxides are often formed in corrosion by gases. The Gibbs energy of formation of the oxide, ΔG, is less negative than for a stable oxide, but in fact an unstable oxide can often exist indefinitely with no measurable transformation. A common example is wustite (FeO), which is formed during the hot rolling of steel. Thermodynamically, it is unstable below 570 °C (1060 °F), but it remains the major component of mill scale at room temperature because the decomposition kinetics are extremely slow.

Rapid kinetics can also favor the formation of less stable oxide on an alloy. An alloy, AB, could oxidize to form oxides AO and BO, but if BO is more stable than AO, then any AO formed in contact with B should in theory convert to BO by the reaction:

$$B + A\text{O} \rightarrow B\text{O} + A$$

Nevertheless, if AO grows rapidly compared with BO and the conversion reaction is slow, then AO can be the main oxide found on the alloy.

Thermodynamically unstable crystal structures of oxides are also sometimes found. A growing oxide film tends to try to align its crystal structure in some way with that of the substrate from which it is growing. This epitaxy can cause the formation of an unstable structure that fits the substrate best. For example, cubic aluminum oxide (Al_2O_3) may form on aluminum alloys instead of the stable rhombohedral Al_2O_3.

11.1 Free Energy-Temperature Diagrams

Metal oxides become less stable as the temperature increases. The relative stabilities of oxides are usually shown on a Gibbs energy-temperature diagram, sometimes called an Ellingham diagram (Fig. 11.1), for common metals in equilibrium with their oxides. Similar diagrams are available for sulfides, nitrides, and other gas-metal reactions. In Fig. 11.1, the reaction plotted in every case is:

$$\frac{2x}{y}M + O_2 = \frac{2}{y}M_xO_y$$

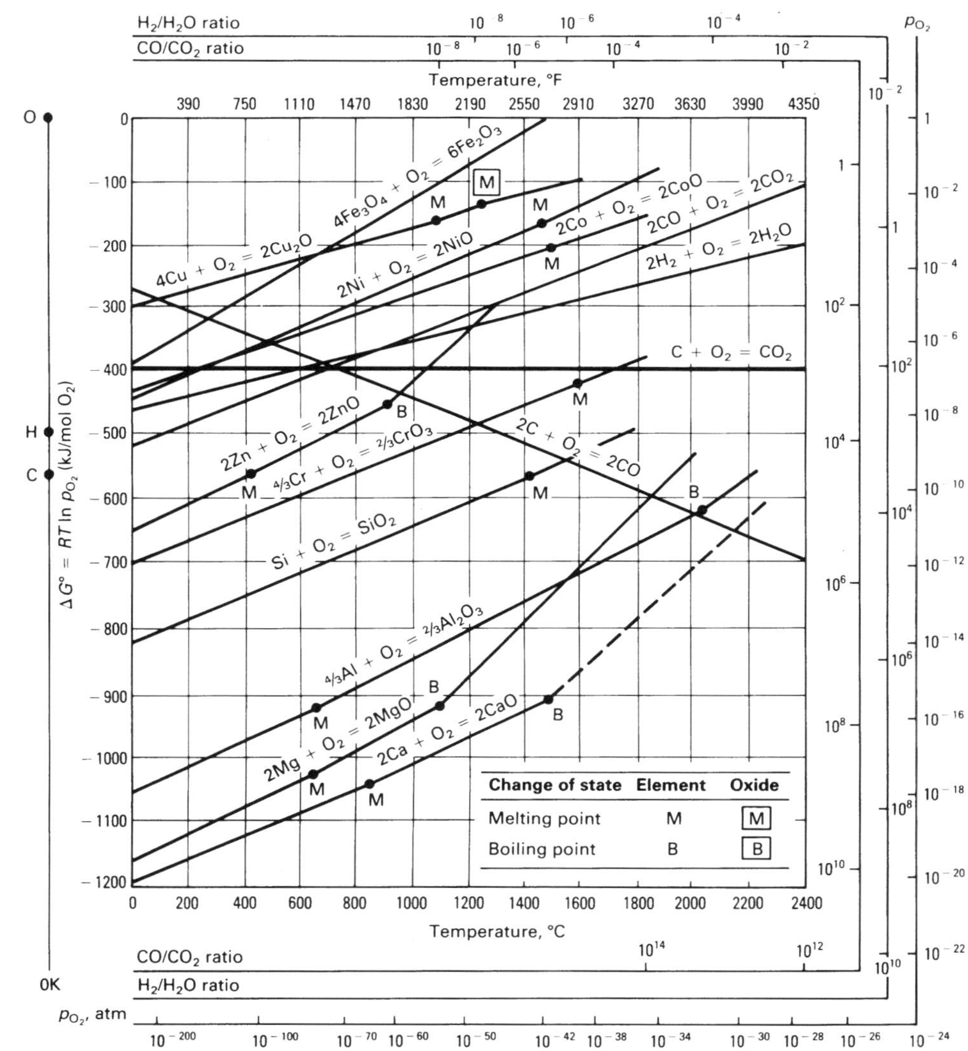

Fig. 11.1 Standard Gibbs energies of formation of selected oxides as a function of temperature. Source: Ref 11.1 as published in Ref 11.2

That is, 1 mol of O_2 gas is always the reactant, so that:

$$\Delta G^0 = RT \ln pO_2$$

For example, the Gibbs energy of formation of Al_2O_3 at 1000 °C (1830 °F), as read from Fig. 11.1, is approximately –840 kJ (–200 kcal) for 2⁄3 mol of Al_2O_3.

The equilibrium partial pressure of O_2 is:

$$pO_2 = \exp\left(\frac{\Delta G^0}{RT}\right)$$

and can also be read directly from Fig. 11.1 without calculation by use of the *p* scale along the bottom and right side of the diagram. A straight line drawn from the index point labeled "0" at the upper left of the diagram, through the 1000 °C (1830 °F) point on the Al/Al_2O_3 line intersects the *p* scale at approximately 10^{-35} atm, which is the O_2 partial pressure in equilibrium with aluminum and Al_2O_3 at 1000 °C (1830 °F). This means that any O_2 pressure greater than 10^{-35} atm tends to oxidize more aluminum, while Al_2O_3 would tend to decompose to Al + O_2 only if the pressure could be reduced to below 10^{-35} atm. Obviously, Al_2O_3 is an extremely stable oxide.

The oxidation of a metal by water vapor can be determined in the same way. The reaction is:

$$x\text{M} + y\text{H}_2\text{O} = \text{M}_x\text{O}_y + y\text{H}_2$$

The equilibrium pH_2:pH_2O ratio for any oxide at any temperature can be found by constructing a line from the H index point on the left side of Fig. 11.1. For example, for the reaction:

$$2\text{Al(l)} + 3\text{H}_2\text{O(g)} = \text{A}_2\text{O}_3\text{(s)} + 3\text{H}_2\text{(g)}$$

at 1000 °C (1830 °F), the equilibrium H_2:H_2O ratio is 10^{10}. A ratio greater than this will tend to drive the reaction to the left, reducing Al_2O_3 to the metal. A ratio less than 10^{10} produces more oxide.

Similarly, the oxidation of metals by carbon dioxide (CO_2) is also shown on Fig. 11.1. For the reaction:

$$x\text{M} + y\text{CO}_2 = \text{M}_x\text{O}_y + y\text{CO}$$

the equilibrium carbon monoxide (CO):CO_2 ratio is found from the index point marked "C" on the left side of the diagram. Oxidation of aluminum by CO_2 has an equilibrium CO:CO_2 ratio approximately 10^{10} at 1000 °C (1830 °F).

11.2 Isothermal Stability Diagrams

For situations that are more complicated than a single metal in a single oxidizing gas, it is common to fix the temperature at some practical value and plot the other variables of gas pressures or alloy composition against each other. This produces isothermal stability diagrams, or predominance area diagrams, which show the species that will be most stable in any set of circumstances.

One Metal and Two Gases. These diagrams, often called Kellogg diagrams, are constructed from the standard Gibbs energies of formation, ΔG^0, of all elements and compounds likely to be present in the system. For example, for the Ni-OS system, the ΔG^0 values of nickel monoxide (NiO) (s), nickel monosulfide (NiS) (l), nickel sulfate (NiSO4) (s), sulfur dioxide (SO_2) (g), sulfur trioxide (SO_3) (g), and S (l) are needed.

In Fig. 11.2, the boundary between the Ni (s) and NiO (s) regions represents the equilibrium Ni (s) + O_2 (g) = NiO (s); therefore, the diagram shows that at 977 °C (1790 °F), any O_2 pressure above approximately 10^{-11} atm will tend to form NiO from metallic nickel if pS_2 is low. Similarly, S_2 gas pressure greater than approximately 10^{-7} atm will form NiS from nickel at low pS_2. Also, a mixed gas of 10^{-5} atm each of S_2 and O_2 should form nearly the equilibrium ratio of NiO (s) and $NiSO_4$ (s).

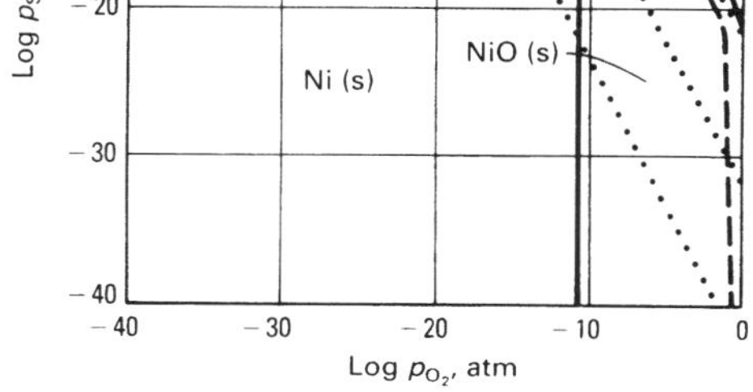

Fig. 11.2 The Ni-O-S system at 977 °C (1790 °F). Source: Ref 11.3 as published in Ref 11.2

If the principal gases of interest were SO_2 and O_2, the same ΔG^0 data could be used to construct a diagram of log p versus pSO_2, or as in Fig. 11.2, pSO_2 isobars can be added to the figure (the dotted lines). Thus, a mixed gas of 10^{-5} atm each of SO_2 and O_2 will form only NiO at 977 °C (1790 °F), with neither the sulfide nor sulfate being as stable.

When nickel metal is heated to 977 °C (1790 °F) in the open air with sulfur-containing gases, $pSO_2 + pS_2 + pO_2 \approx 0.2$ atm. The situation is shown by the dashed line in Fig. 11.2 labeled $p = 0.2$ atm.

An Alloy System and a Gas. Isothermal stability diagrams for oxidation of many important alloy systems have been worked out, such as that for the Fe-Cr-O system shown in Fig. 11.3. In this diagram, the mole fraction of chromium in the alloy is plotted against log pO_2 so that for any alloy composition the most stable oxide or mixture of oxides is shown at any gas pressure.

For an alloy system in gases containing more than one reactive component, the pressures of all but one of the gases must be fixed at reasonable values to be able to draw an isothermal stability diagram in two dimensions. Figure 11.4 shows an example of such a situation: the Fe-Zn system in equilibrium with sulfur and oxygen-containing gases, with SO_2 pressure set at 1 atm and temperature set at 890 °C (1635 °F).

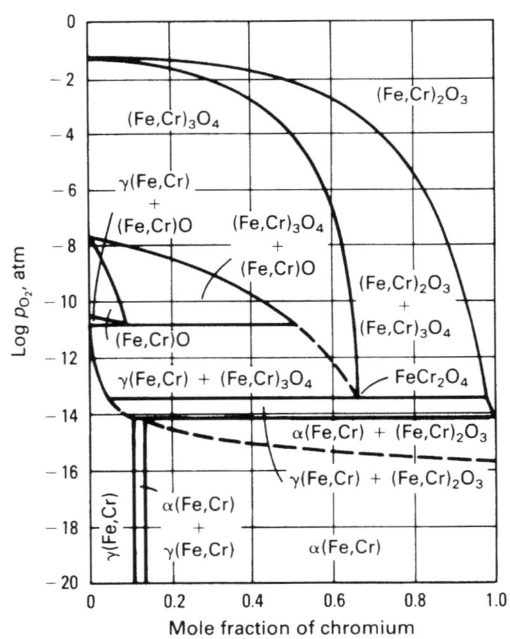

Fig. 11.3 Stability diagram for the Fe-Cr-O system at 1300 °C (2370 °F). Source: Ref 11.2

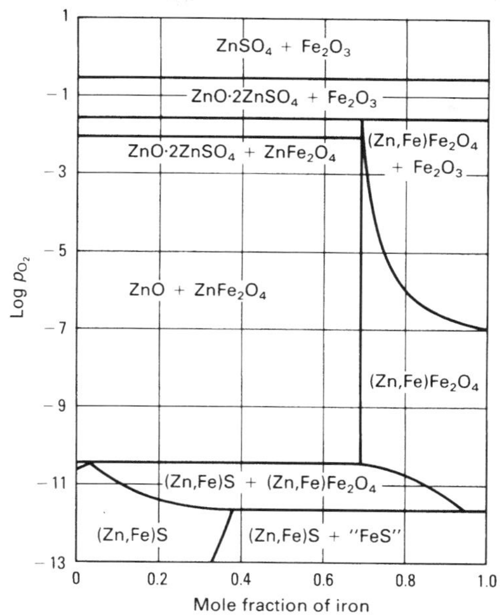

Fig. 11.4 The Fe-Zn-S-O system for $p = 1$ atm at 890 °C (1635 °F). Source: Ref 11.4 as published in Ref 11.2

11.3 Limitations of Predominance Area Diagrams

Isothermal stability diagrams, like all predominance area diagrams, must be read with an understanding of their rules:

- Each area on the diagram is labeled with the predominant phase that is stable under the specified conditions of pressure or temperature. Other phases may also be stable in that area, but in smaller amounts.
- The boundary line separating two predominance areas shows the conditions of equilibrium between the two phases.

Also, the limitations of the diagrams must be understood to be able to use them intelligently:

- The diagrams are for the equilibrium situation. Equilibrium may be reached quickly in high-temperature oxidation, but if the metal is then cooled, equilibrium is often not reestablished.
- Microenvironments, such as gases in voids or cracks, can create situations that differ from the situations expected for the bulk reactant phases.
- The diagrams often show only the major components, omitting impurities that are usually present in industrial situations and may be important.

- The diagrams are based on thermodynamic data and do not show rates of reaction.

ACKNOWLEDGMENT

The material in this chapter is from "Fundamentals of Corrosion in Gases," by S.A. Bradford in *Corrosion*, Vol 13, *ASM Metals Handbook*, ASM International, 1987.

REFERENCES

11.1 N. Birks and G.H. Meier, *Introduction to High Temperature Oxidation of Metals,* Edward Arnold, 1983

11.2 S.A. Bradford, Fundamentals of Corrosion in Gases, *Corrosion*, Vol 13, *ASM Metals Handbook*, ASM International, 1987

11.3 C.S. Giggins and F.S. Pettit, Corrosion of Metals and Alloys in Mixed Gas Environments at Elevated Temperatures, *Oxid. Met.,* Vol 14 (No. 5), 1980, p 363–413

11.4 T. Rosenqvist, Phase Equilibria in the Pyrometallurgy of Sulfide Ores, *Metall. Trans. B,* Vol 9B, 1978, p 337–351

CHAPTER 12

Phase Diagram Determination

THE VAST COLLECTION of phase diagrams is the product of the painstaking labor of a very large number of skillful investigators working in all parts of the world. It is a constantly growing body of information, improving both in the extent of its coverage and in the precision of its content. Few, if any, diagrams may be considered complete and final. Repeated investigations, with refinements in apparatus and techniques, lead to their frequent revision. The development of new analytical techniques leads to ever more precise measurements that allow continual refinement of phase diagrams.

Accurate chemical analysis is of extreme importance in the determination of phase diagrams. Wherever possible, it is best to put this part of the work in the hands of a skilled analytical chemist. Regardless of the analytical method or the skill with which it is employed, the result can be no more representative of the composition of the alloy investigated than is the sample taken for analysis. Segregation can easily cause a variation of several percent from point to point in castings of some materials. The analytical sample should be taken in such a way as to obtain an average (or representative) analysis. Drastic working of the metal, alternating with homogenizing heat treatments, will sometimes eliminate composition differences and simplify the problem of sampling. It is sometimes found that the composition of the alloy changes in the course of remelting or during heat treatment, owing to the selective oxidation or vaporization of one or more of its components. Measures should be taken to avoid composition variations by the adjustment of the conditions of the experiment, but when this cannot be done, repeated sampling and analysis throughout the course of the constitutional study become necessary. Of equal importance is the maintenance of purity and initial composition of the material that is being investigated. Only the purest obtainable metals should be used in constitutional studies. Just what limits of impurity are tolerable can be ascertained only by experience with the specific alloy system; sometimes, as little as 0.001% of a certain impurity becomes significant, while much larger quantities of other impurities produce no detectable effects.

The use of high-purity metals is of little avail unless the purity is maintained in the production and manipulation of the experimental alloys. Any apparatus that comes in contact with the research sample must be inert at the temperature concerned, so that no substantial contamination will occur to vitiate the observations. This requirement becomes difficult to meet when the temperatures are very high or the metals are highly reactive and accounts for much of the uncertainty concerning the constitution of high-melting-point alloy systems. In addition, it is important that the temperature-measuring element be in good thermal contact with the specimen and that it be not so large in comparison with the sample that it carries away a significant quantity of heat.

Anything that comes in contact with the samples, particularly when the metals are molten, is a possible source of contamination. Crucibles should be selected with great care, using only materials that are insoluble and nonreactive with the alloy. Evidence of attack of the crucible by the alloy is usually grounds for rejecting the experimental results, but the absence of such evidence does not guarantee the absence of contamination. When suitable crucibles are unobtainable, it is often possible to coat ordinary crucibles with an inert material. Failing this, it is sometimes possible, as a last resort, to use the experimental alloy as its own "crucible" by melting a pool in its top surface with a fine torch or other concentrated source of radiated heat. For constitutional studies at temperatures below the melting point, alloys that cannot be maintained in a satisfactory state of purity, if made by melting, or that cannot be melted to a homogeneous liquid can sometimes be produced by powder metallurgy techniques.

Contamination from the atmosphere is often very damaging. In extreme cases, the liquidus temperature has been observed to be depressed several hundred degrees simply as a result of the absorption of the constituents of air. Usually, the effects are much smaller and are confined to the selective oxidation of one or more of the components of the alloy, thereby altering its composition. Protection from gaseous contamination may sometimes be achieved by the use of flux covers or by the use of inert gas atmospheres, argon and helium usually being satisfactory. Vacuum melting and heat treatment are hazardous from the standpoint of maintenance of composition unless the vapor pressures of all components concerned are very low. Vapor losses can be minimized by the use of positive pressures of inert gases or suitable flux covers.

No phase diagram can be considered fully reliable until corroborating observations have been made by at least two independent methods. Even so, the diagram cannot be accepted if its construction violates the phase rule or any of the other rules of construction that have been derived by thermodynamic reasoning. If violations of these rules are encountered, it may be concluded with assurance that the experimental observations or their interpretation are at fault.

Cooling curves using simple thermocouple arrangements or other pyrometric methods were originally used to determine phase diagrams. These methods have largely been replaced by the more modern methods of equilibrated alloys and diffusion couples. All of these methods are supported by both conventional methods and newer advanced analytical methods.

12.1 Cooling Curves

One of the most widely used early methods for the determination of phase boundaries is cooling curves. The temperature of a sample is monitored while allowed to cool naturally from an elevated temperature in the liquid field. The shape of the resulting curves of temperature versus time are then analyzed for deviations from the smooth curve found for materials undergoing no phase changes (Fig. 12.1).

When a pure element is cooled through its freezing temperature, its temperature is maintained near that temperature until freezing is complete (Fig. 12.2). However, the true freezing/melting temperature is difficult to determine from a cooling curve because of the nonequilibrium conditions inherent in such a dynamic test. This is illustrated in the cooling and heat-

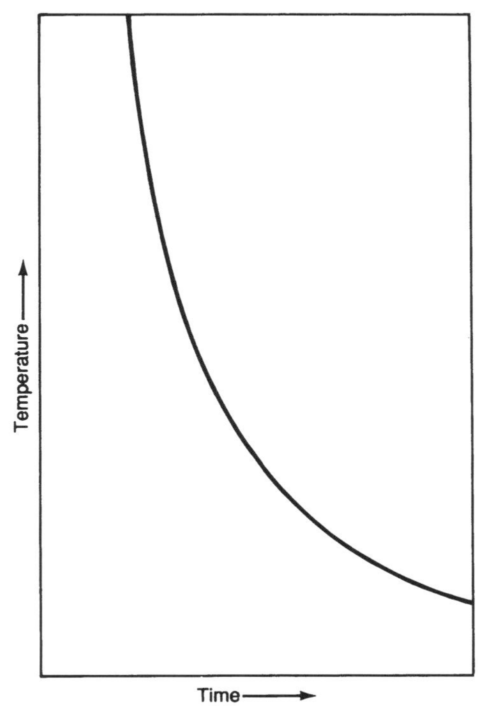

Fig. 12.1 Ideal cooling curve with no phase change. Source: Ref 12.1

ing curves shown in Fig. 12.3, where the effects of both supercooling and superheating can be seen. The dip in the cooling curve often found at the start of freezing is caused by a delay in the start of crystallization.

The continual freezing that occurs during cooling through a two-phase liquid-plus-solid field results in a reduced slope to the curve between the liquidus and solidus temperatures, as shown in Fig. 12.4. By preparing samples having compositions across the diagram, the shape of the liquidus

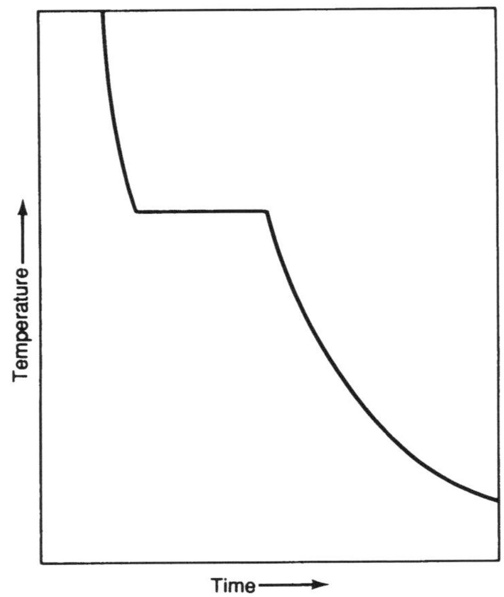

Fig. 12.2 Ideal freezing curve of a pure metal. Source: Ref 12.1

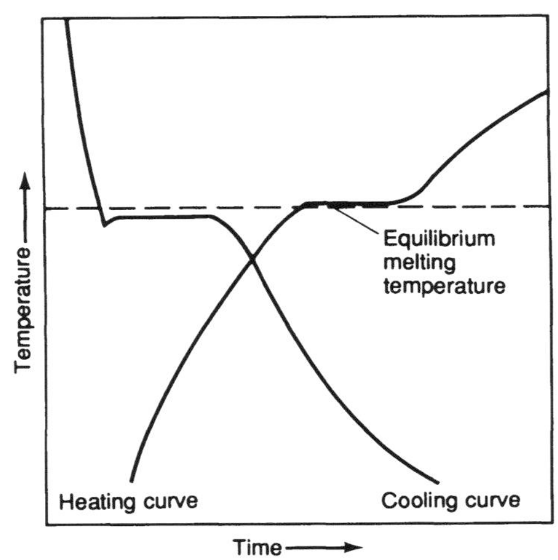

Fig. 12.3 Natural freezing and melting curves of a pure metal. Source: Ref 12.2 as published in Ref 12.1

curves and the eutectic temperature of the system can be determined (Fig. 12.5). Cooling curves can be similarly used to investigate all other types of phase boundaries.

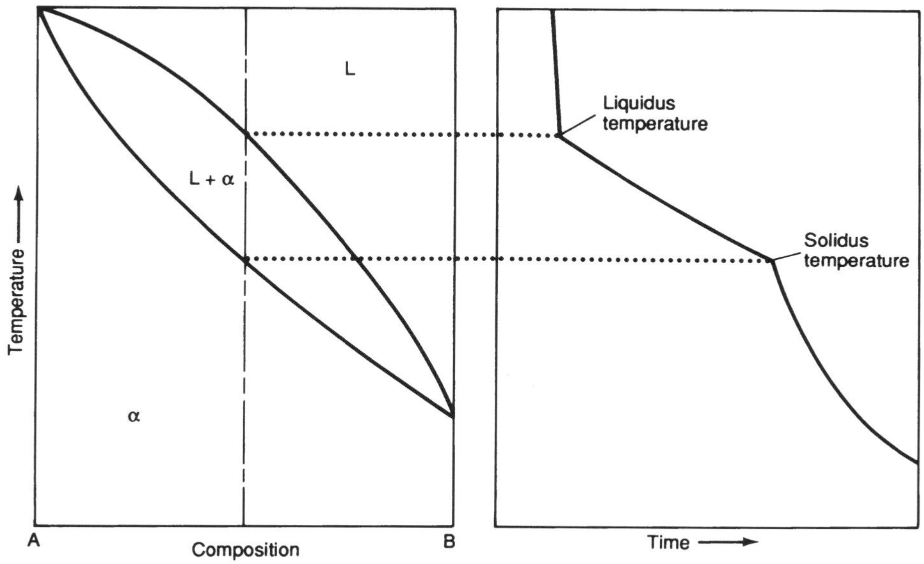

Fig. 12.4 Ideal freezing curve of a solid-solution alloy. Source: Ref 12.3 as published in Ref 12.1

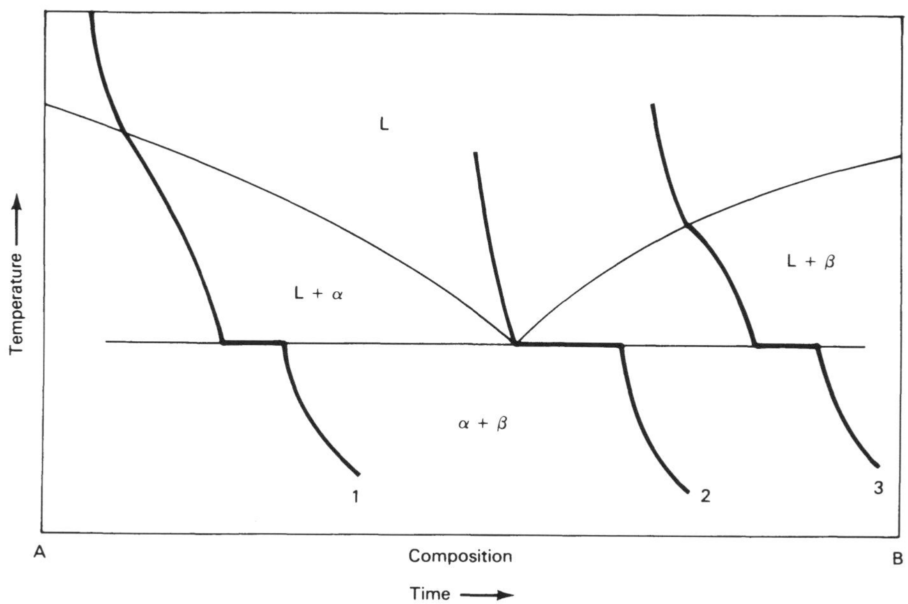

Fig. 12.5 Ideal freezing curves of (1) a hypoeutectic alloy, (2) a eutectic alloy, and (3) a hypereutectic alloy superimposed on a portion of a eutectic phase diagram. Source: Ref 12.3 as published in Ref 12.1

12.2 Equilibrated Alloys

A phase diagram can be constructed by preparing alloys of the required constituents, heat treating them at high temperatures to reach equilibria, and then identifying the phases to determine liquidus temperatures, solidus temperatures, solubility lines, and other phase transition lines. Along with equilibrated alloys, several techniques are used to determine phase diagrams. These include thermal analysis (TA), metallography, X-ray diffraction (XRD), dilatometry, electrical resistance measurement, and magnetic analysis methods, among others. All of these methods are based on the principle that when a phase transition in an alloy occurs, its physical and chemical properties, phase composition, and/or structure will vary. By analyzing the temperature, composition, and property changes associated with phase transitions, phase boundaries can be constructed according to the phase rule.

Equilibrated alloys are made from extremely pure constituent metals that are prepared using techniques such as arc melting, induction melting, or by powder metallurgy. Strict atmospheric control is maintained by using inert gases or vacuum. Homogenization heat treatments are conducted by putting the samples into a furnace at high temperatures, below the solidus temperature of the alloys, for an extended period of time. The samples are then quenched from elevated temperatures to room temperature. The quenching process is used to freeze the phases to room temperature for analysis. Alternatively, the samples may be slowly cooled to room temperature. This process generates less thermal stress in the samples, making lattice parameter measurements more accurate.

Homogenization is achieved through diffusion, thus the annealing time and temperature are two important factors. Because the diffusion coefficient increases greatly with temperature, the heat treatment temperature should be as high as possible to accelerate the homogenization process. The homogenization temperature can be as high as approximately 50 °C (90 °F) below the solidus temperature. The homogenization process is carried out in vacuum or an inert gas atmosphere in order to avoid/reduce oxidization and unwanted environmental interaction.

Equilibrium can be approached from two directions: (1) slow cooling and heating, or (2) long-term isothermal heat treatment. In general, the longer the time to approach equilibrium, the closer to equilibrium one can get. For the cooling and heating process, the liquidus temperature measured from cooling using differential thermal analysis (DTA) is usually lower than the equilibrium liquidus line. When the heating and cooling rates are reduced, the degree of overheating and overcooling could be reduced, thus approaching equilibrium more closely; however, the DTA peak-to-background ratio is degraded.

To investigate phase equilibria at elevated temperatures, two approaches can be used. The first one is to study the phase equilibria directly at the temperature of interest using a hot-stage microscope or high-temperature in situ XRD. This method has the advantages of (1) having no complications from quenching, (2) allowing more straightforward interpretation of the results, and (3) allowing a study in a continuous temperature range, which is impractical for quenching experiments. Its disadvantages include less availability of equipment, more difficult experiments, and potentially greater preferential evaporation, oxidation, or other environmental interactions. Sometimes in situ high-temperature experiments are essential to study complex phase equilibria in narrow temperature and composition ranges.

The other approach is to retain the high-temperature phase equilibria to room temperature by quenching. This requires that the quenching/cooling rate is high enough to preserve the high-temperature phases and their structures to room temperature. Liquid nitrogen, water, iced water, saltwater solutions (brine), and oil are common quenching media. Occasionally, some phase transitions take place very fast, and it is not possible to retain the high-temperature phase to room temperature even by a fast quench. One example is the γ-iron (face-centered cubic, fcc) to α-iron (body-centered cubic, bcc) allotropic transition in pure iron and dilute iron alloys. In such cases, care needs to be exercised in interpreting the result.

To determine a phase diagram with equilibrated alloys, two methods are used: (1) the static method or analysis of quenched samples to construct isothermal sections, and (2) the dynamic method or analysis of samples by heating and cooling experiments to construct vertical sections and liquid projections.

Analysis of Quenched Samples to Construct Isothermal Sections (Static Method). Metallographic analysis is one of the key tools for determining phase diagrams. Microstructure examinations are routinely used to investigate the number of phases (single phase or multiphase) and invariant reaction types (such as the eutectic or peritectic). In particular, the characteristics of each phase, such as composition, size, shape, distribution, color, orientation, and hardness can be examined. Metallographic analysis is based on the assumption that the observed microstructure represents the true structure of the samples.

Metallography has an advantage over measurements that involve changes in physical properties in that the microstructures can offer clues concerning invariant reactions. For example, a lamellar structure is indicative of the presence in the system of a eutectic reaction, and rimming is indicative of a peritectic reaction. Quantitative metallography can be used with the lever rule to determine phase boundaries. Optical metallography is still widely used but has now been expanded to include results from electron microscopy. Electron microscopy offers the capability of greater magnification and, in addition, allows quantitative determination of the compositions of individual grains, provided that the microscope is equipped with suitable attachments for composition analysis.

Optical microscopy (OM) is a basic technique for phase identification. There is tremendous value in examining a sample optically because many phases can be easily differentiated by light microscopy. Scanning electron microscopy (SEM) is another widely used tool for phase diagram determination. In SEM, a focused and collimated electron beam impinges on the surface of a sample, creating backscattered electrons, secondary electrons, characteristic X-rays, and Auger electrons, among other signals. Scanning electron microscopes are often coupled with energy-dispersive X-ray microanalysis (EDX), a microanalytical technique that uses the characteristic spectrum of X-rays emitted by the different elements in a specimen after excitation by high-energy electrons to obtain quantitative or qualitative compositional information about the samples. When employed to measure a phase diagram, EDX can provide information on compositions of individual phases and the distribution of alloying elements and microinhomogeneity.

The major shortcoming of the metallographic method in phase diagram determination is its difficulty in detecting fine precipitates. Many terminal solid solutions have decreasing solubility with decreasing temperature. After an alloy is homogenized, annealed at a lower temperature, and subsequently quenched, precipitates may be so small that even SEM cannot detect them. Only careful transmission electron microscopy (TEM) examination can clearly identify them. Ample annealing time should be given to make sure the precipitation takes place.

Electron probe microanalysis (EPMA) is another important technique for phase diagram determination. It is essentially a dedicated SEM with wavelength dispersive spectrometers (WDS) attached. As an elemental analysis technique, it uses a focused beam of high-energy electrons (5 to 30 KeV) in the SEM to impinge on a sample to induce emission of characteristic X-rays from each element. Its spatial resolution for X-ray microanalysis depends mainly on the accelerating voltage of the electron beam and the average atomic weight of a phase in a specimen, and usually ranges from one to several microns.

Another powerful tool for the determination of phase diagrams is diffraction. Initially this was XRD, but during the past century electron and neutron diffraction have been developed. Each of the three has its own advantages and limitations. X-ray diffraction is the least costly and can be used in a variety of ways in the establishment of phase diagrams. A simple XRD pattern can be used for phase identification. X-ray diffraction can also be used for the establishment of the loci of phase boundaries. For instance, in a two-component system, lattice parameters can vary with composition within a single-phase region but are invariant in a two-phase region. Additionally, with measurement of precision lattice parameters, XRD can be used to determine atomic volumes.

Electron diffraction requires operation in a vacuum atmosphere and can do the same things that XRD does. Its advantage is that it can focus on a small area so that it can be used to examine individual grains within a microstructure and, as noted previously, with the proper attachments can quantitatively determine grain compositions. While X-rays and electrons are scattered by the electron distribution in a solid, neutrons are scattered by nuclei and by magnetic dipoles. Thus, neutrons can be used to determine magnetic structures and can distinguish other structures that are difficult or impossible to see with either X-ray or electron diffraction. For instance, with X-ray or electron diffraction patterns from systems involving very light elements with heavy elements (e.g., hydrogen in niobium), the scattering power of hydrogen atoms with 1 electron is so much less than that of niobium with 41 electrons that the niobium masks the hydrogen contribution in the diffraction pattern. This is not true for neutron diffraction patterns. A similar difficulty arises when the components of a system are closely comparable in atomic number, so that it is difficult to distinguish one species from the other with X-ray or electron diffraction but not with neutron diffraction.

X-ray diffraction is widely used to determine the presence of different phases in a sample and thus place the alloy in the right phase region, with known composition of the alloy either from the nominal compositions, chemical analysis, or EPMA. The other essential application of XRD is the determination of the crystal structure of a new phase. Based on the XRD results, two methods—peak intensity method and lattice parameter method—are typically used for the determination of phase boundaries. The peak intensity method is widely used in phase diagram determination.

Different phases have different crystal structures that are characterized by their distinctive XRD peaks. The phases in an alloy can be determined by its XRD patterns to have either a single phase or multiple phases. In a single-phase region, there is only one type of XRD spectrum observed. If the crystal structure is known, the diffraction intensity of each peak can be calibrated by its position. In this case, the position of diffraction peaks may vary with alloying composition, but additional peaks should not appear.

The lattice parameter method for the determination of the phase boundary of a solid solution is straightforward. The lattice parameter is plotted against the compositions of the alloys, and the composition at which the lattice parameter first becomes constant is located. When solute element B is added to solvent A, the lattice parameter of A phase may decrease or increase due to the difference in their atomic sizes. In the single-phase region, lattice parameter changes continuously with the composition of B. However, in a two-phase field, because the composition of each phase remains constant at a given temperature, the lattice parameter does not vary with the alloying composition. Thus, the lattice parameter can be plotted against composition. The composition at which the lattice parameter first becomes constant corresponds to the solid solubility at this temperature.

The use of XRD to determine the point on the solvus curve using the lattice parameter method is illustrated in Fig. 12.6. If the solid solubility at different temperatures is measured by this method, then the phase boundary can be established.

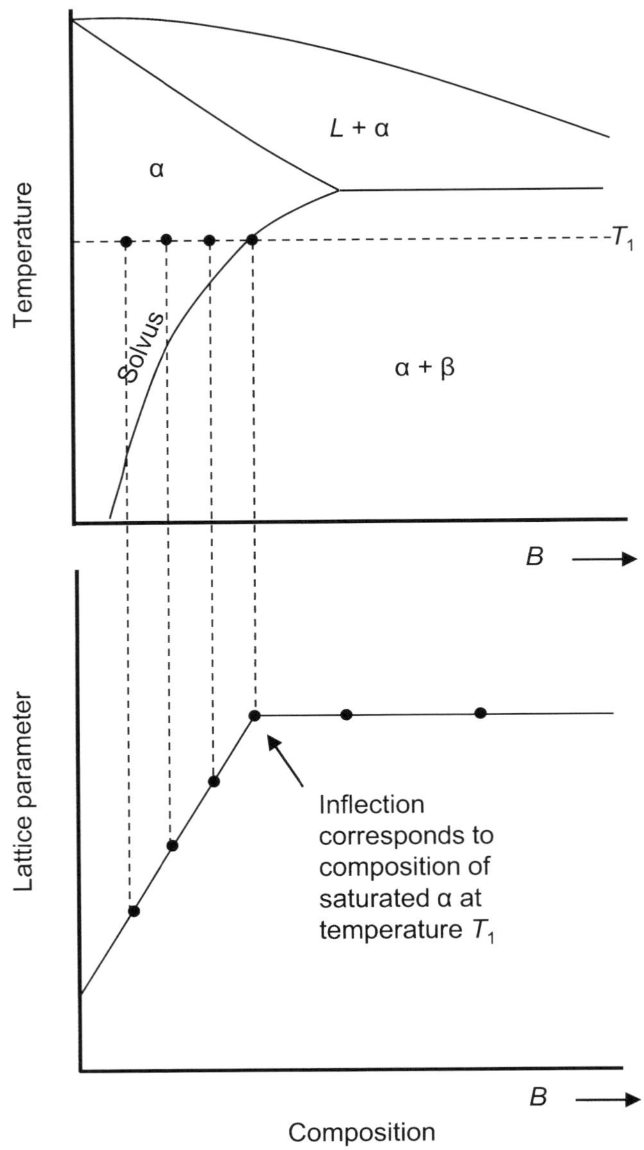

Fig. 12.6 Use of X-ray diffraction (XRD) measurements of the lattice parameter to determine a point on a solvus curve. Adapted from Ref 12.2

Analysis of Samples by Heating and Cooling Experiments to Construct Vertical Sections and Liquid Projections (Dynamic Method). In the dynamic method, phase equilibrium is determined when an alloy is heated and/or cooled while a certain property, such as heat flow or electrical resistance, is monitored continuously to detect the temperature at which a phase transition takes place. There are two advantages over the static methods: (1) it is possible to quickly explore the limits of phase stability because quenching is not involved, and (2) the possibility of a phase change occurring during quenching is avoided. However, the dynamic method is not suitable for investigation of systems with slow phase transitions. Any physical or chemical property that changes with composition and temperature may be used as a parameter for the dynamic method to determine a phase diagram. Analysis methods include high-temperature (hot-stage) metallography, high-temperature in situ XRD analysis, dilatometry, electrical resistance, and TA including both DTA and differential scanning calorimetry (DSC).

A phase transition usually involves an enthalpy change (evolution or absorption of heat); therefore, thermal properties are commonly monitored to detect phase changes. When a specimen is heated or cooled under uniform conditions, a structural change will be identified with a temperature anomaly by plotting time versus temperature. Two major techniques to detect phase transitions are developed based on enthalpy change. The first one is TA, where temperature versus time curve shows a thermal arrest at a phase transition point. The second technique is DTA, in which a test sample and an inert reference sample are heated and cooled under identical conditions, and a temperature difference between the test sample and reference sample are recorded. Because the signal is differential, it can be amplified with a suitable diffusion couple (DC) amplifier to increase sensitivity. As a result, DTA is more sensitive than TA. The differential temperature is then plotted either against time or against temperature. The DTA signal versus temperature for a pure metal is shown in Fig. 12.7. When a phase transition takes place in a sample that involves release of heat, the test sample temperature rises temporarily above that of the reference sample, resulting in an exothermic peak. Conversely, a transition accompanied by absorption of heat reduces the temperature of the test sample compared to that of the reference sample, leading to an endothermic peak. For example, in the case of a binary system, if the overall specimen composition does not vary during a heating process, a phase transition results in a slope change on the TA curve and a thermal spike on the corresponding DTA curve. A schematic of enthalpy versus temperature and the associated DTA curves for melting and freezing of a pure metal are shown in Fig. 12.8.

Differential scanning calorimetry (DSC) is a TA technique that measures the energy absorbed or emitted by a sample as a function of temperature or time. When a phase transition occurs, DSC provides a direct calorimetric measurement of the transition energy at the transition temperature by sub-

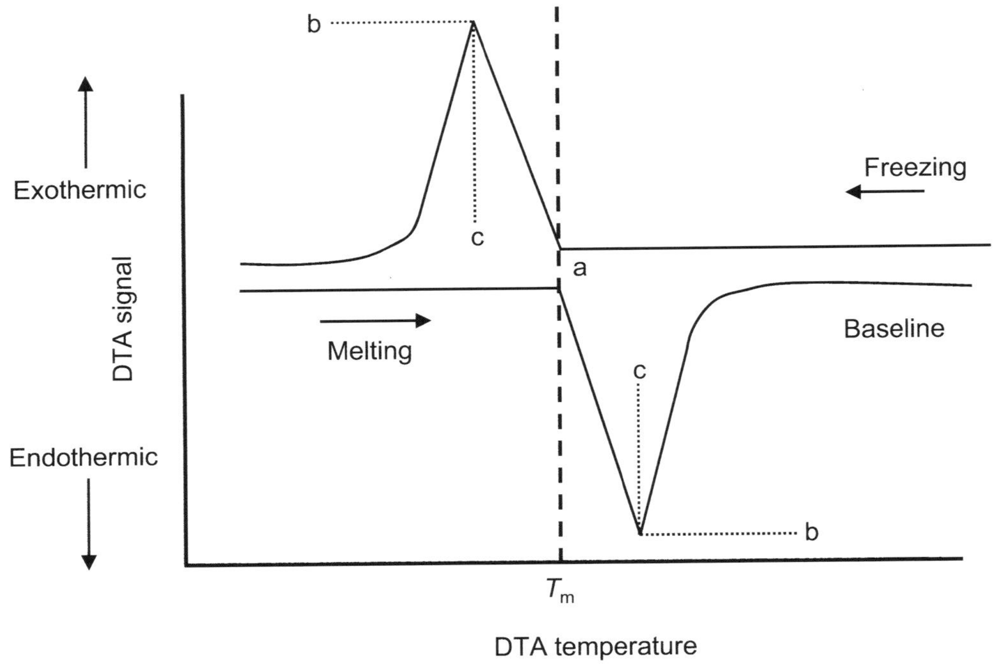

Fig. 12.7 Differential thermal analysis (DTA) responses to melting and freezing of a pure metal under ideal conditions. a, onset temperature; b, peak signal; c, peak temperature. Adapted from Ref 12.4

jecting the sample and an inert reference material to identical temperature regimes in an environment heated or cooled at a controlled rate. Differential scanning calorimetry equipment can be used not only to determine the liquidus line, solidus line, and other phase transition points on a phase diagram, but also to measure some thermodynamic parameters such as enthalpy, entropy, and specific heat, which are important for investigation of second-order phase transitions. One deficiency of DSC is that its usage temperature is usually in the range of 175 to 1100 °C (350 to 2010 °F), much lower than that of DTA. There are two types of DSC systems in common usage. In power-compensation DSC, the temperatures of the sample and reference are controlled independently using separate but identical furnaces. The temperatures of the sample and the reference are made identical by varying the power input to the two furnaces. The energy required to do so is a measure of the enthalpy or heat capacity changes in the sample compared to the reference. In heat-flux DSC, the sample and reference are connected with a low-resistance heat flow path (a metal disk). The assembly is enclosed in a single furnace. Enthalpy or heat capacity changes in the sample cause a difference in its temperature relative to the reference.

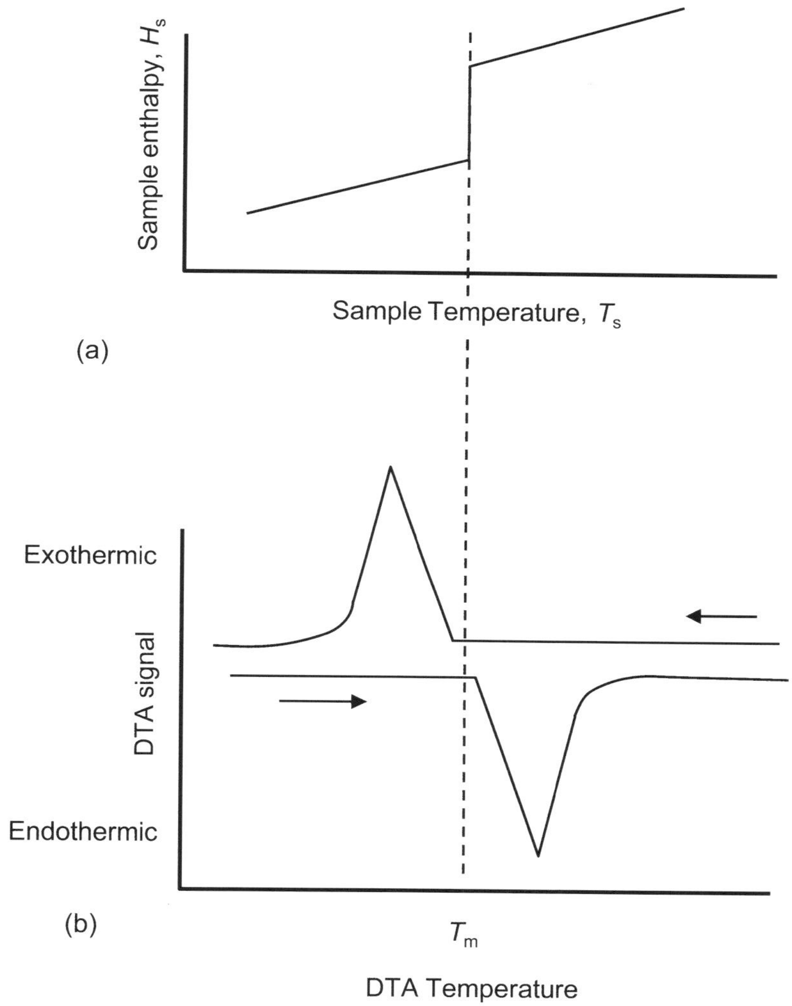

Fig. 12.8 Schematic diagram showing (a) enthalpy vs. temperature for a pure metal, and (b) DTA signal for melting (bottom) and freezing (top). Adapted from Ref 12.4

The resulting heat flow is smaller compared to that in DTA because the sample and the reference are in good thermal contact. The temperature difference is recorded and related to enthalpy change in the sample using calibration experiments.

Dilatometry is another technique used to study phase transitions in alloys. This technique uses the change in volume associated with nearly all transitions and measures the change of length of a specimen as it is heated and cooled at a fixed rate. The relationships of length versus time and temperature versus time are measured simultaneously, so as to plot the length-versus-temperature curve. The dilatometry curve is similar to the cooling curve in the TA, and the phase transition temperature can be determined. The use of dilatometry to determine solid-state phase boundaries is shown in Fig. 12.9. A dilatometer consists of a sensing element such as a transducer, variable capacitor, and dial gage that is activated by the specimen positioned in the furnace. Changes in length are transmitted to the sensor by means of a push rod. The specimen is enclosed in a thermal mantle, which is heated or cooled by passage of gas. Temperatures are measured by a thermocouple and plotted on one axis of an x-y recorder, and changes in length are recorded simultaneously.

Measurement of electrical resistance is a useful auxiliary technique for the determination of alloy phase diagrams. The basic procedure to determine phase regions is to plot the electrical resistance against composition or to plot the electrical resistance against temperature for a fixed composition. To investigate subsolidus equilibria, this method appears to have advantages over DTA. The measurements do not require spontaneous heat effects and can therefore be observed during arbitrarily slow rates of temperature change or on allowance of sufficient time for equilibrium. At subsolidus temperatures, formation of a new phase will, in most cases, be evidenced by a change of slope in the conductivity-versus-temperature curve. The use of electrical resistance measurements to determine points on the solvus, solidus, and lines of three-phase equilibrium is shown in Fig. 12.10.

Magnetic analysis is another technique for the determination of an alloy phase diagram. Curie temperature, T_c, and saturation magnetization, σ, are

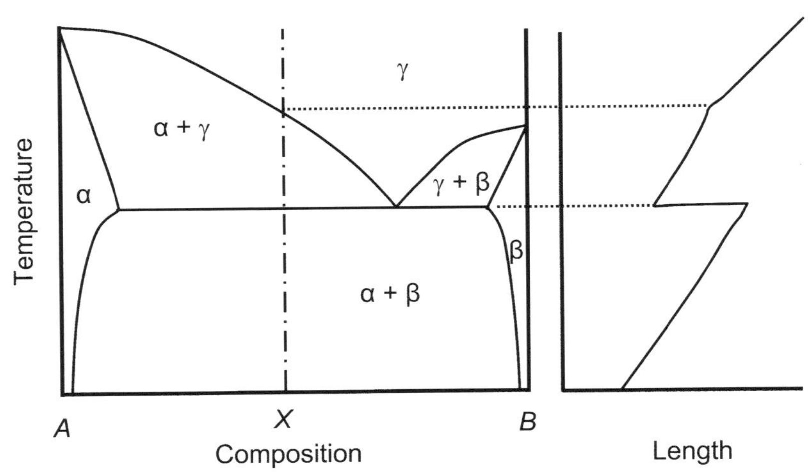

Fig. 12.9 Use of dilatometric measurements to determine points on phase boundaries in the solid state. Adapted from Ref 12.2

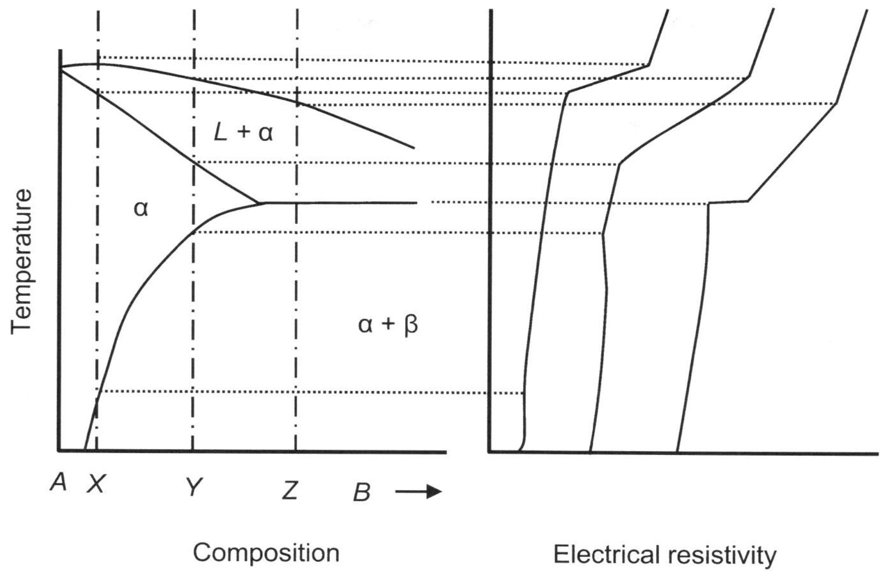

Fig. 12.10 Use of electrical resistivity measurements, as a function of temperature, to determine points on the solvus, solidus, and lines of three-phase equilibrium. Adapted from Ref 12.2

two important parameters typically used for this method. Because both T_c and σ are only affected by the atomic arrangement but not by the imperfection of crystal lattice, their values depend essentially only on an alloy composition and the corresponding phase compositions. By measuring these magnetic properties, phase relationships of alloy systems that include ferromagnetic components or compounds can be determined.

12.3 Diffusion Couples

The use of diffusion couples in phase diagram studies is based on the assumption of local equilibria at the phase interfaces in the diffusion zone. The latter implies that an infinitesimally thick layer adjacent to the interface in such a diffusion zone is effectively in thermodynamic equilibrium with its neighboring layer on the other side of the interface. In other words, the chemical potential (activity) of species varies continuously through the product layers of the reaction zone and has the same value at both sides of an interphase interface.

A diffusion-controlled interaction in a multiphase binary system will invariably result in a diffusion zone with single-phase product layers separated by parallel interfaces in a sequence dictated by the corresponding phase diagram. The reason for the development of only straight interfaces with fixed composition gaps follows directly from the phase rule. Three degrees of freedom are required to fix temperature and pressure and to vary the composition. Reaction morphologies consisting of two-phase structures (i.e., precipitates or wavy interfaces) are, therefore, thermodynamically forbidden for binary systems, assuming that only volume diffusion takes place. In a ternary system, on the other hand, it is possible to develop two-phase areas in the diffusion zone because of the extra degree of freedom. The diffusion zone morphology, which develops during solid-state interaction in a ternary couple, is defined by type, structure, number, shape, and topological arrangement of the newly formed phases.

Several techniques are available to make solid-state diffusion couples, that is, to bring two (or more) materials in such intimate contact that one diffuses into the other. Most commonly, the bonding faces of the couple components are ground and polished flat, clamped together, and annealed at the temperature of interest. Depending on the initial materials, various protective atmospheres can be used (e.g., vacuum and inert gas). After the heat treatment, quenching of the sample is desirable in order to freeze the high-temperature equilibrium.

Sometimes phases present in the reaction zone of diffusion couples can be detected simply using an optical microscope. Different measurement techniques can be used to determine the chemical compositions on both sides of the interfaces. These are Auger electron spectroscopy (AES), secondary ion mass spectrometry (SIMS), Rutherford backscattering spectrometry (RBS), EPMA, and analytical electron microscopy (AEM).

In EPMA, high-energy electrons are focused to a fine probe and directed at the point of interest in the diffusion couple. The incident electrons interact with the atoms in the sample and generate, among other signals, characteristic X-rays. These X-rays are detected and identified for qualitative analysis, and with the use of suitable standards, they can be corrected for matrix effects in order to perform quantitative analysis. Electron probe microanalysis is used to investigate bulk solid samples, while AEM is dealing with electron transparent thin films. The procedures used for X-ray quantification are quite different for both techniques. Presently, two different techniques, wavelength dispersive spectrometry (WDS) and energy-dispersive spectrometry (EDS) can be used to collect X-ray spectra from samples being analyzed.

As an example, magnesium-aluminum diffusion couples were prepared using pure magnesium (99.9%) and pure aluminum (99.99%). Small pieces of each were cut and shaped into discs approximately 1.3 cm (0.5 in.) in diameter and 0.32 to 0.64 cm (0.125 to 0.25 in.) thick. One side of each disc was polished down to 0.05 μm. For each temperature, one aluminum disc and one magnesium disc were placed inside a stainless steel tubular clamp, with the polished faces of the discs facing each other. Each clamp was individually sealed in a Pyrex tube filled with helium at one-third atmosphere pressure to prevent oxidation during the heat treatment. Each tube was placed in a furnace and isothermally heat treated for a period of 6 to 24 days at temperatures ranging from 360 to 420 °C (680 to 790 °F). Several different methods were used to characterize the intermediate phases, including light microscopy, microhardness measurements, and EPMA. A light micrograph of a diffusion couple made at 367 °C (693 °F) is shown in Fig. 12.11. Electron probe microanalysis was used to identify the β, ε, and γ phases.

A diffusion multiple is an assembly of three or more different metal blocks in intimate interfacial contact and subjected to a high temperature to promote thermal interdiffusion to form solid solutions and intermetallic compounds. For the purpose of phase diagram determination, a diffusion multiple is nothing more than a sample with multiple diffusion couples and diffusion triples in it. An example is schematically shown in Fig. 12.12, which contains a five-metal diffusion multiple. The local equilibrium at the phase interfaces allows the extraction of phase equilibrium information from diffusion multiples in the same way as that from diffusion couples.

The biggest advantage of the diffusion multiple approach in phase diagram determination is its high efficiency in both time and raw materials usage. An entire ternary phase diagram can be obtained from a tri-junction region of a diffusion multiple. By creating several tri-junctions in one sample, isothermal sections of multiple ternary systems can be determined without making dozens or even hundreds of individual alloys, thus saving the usage of raw materials. The diffusion multiple approach can also save EPMA examination time, because there is no need to exchange many alloy samples in and out of the EPMA system, which is very time consuming; that is, one needs to wait for a good vacuum to start the analysis each time.

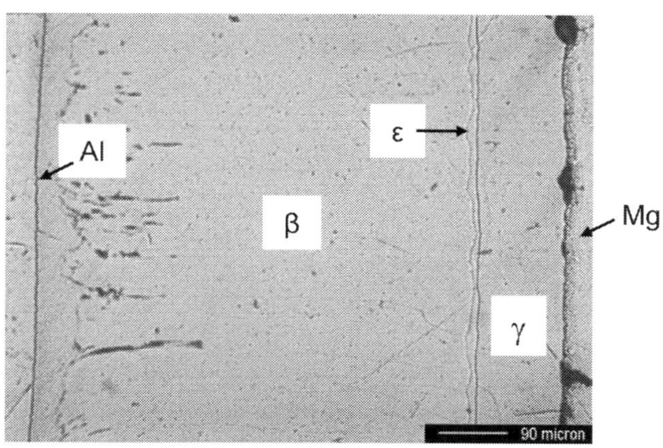

Fig. 12.11 Light micrograph of 367 °C (693 °F) diffusion couple between pure aluminum and pure magnesium. Electron probe microanalysis (EPMA) was used to identify the β, ε, and γ phases. Source: Ref 12.5

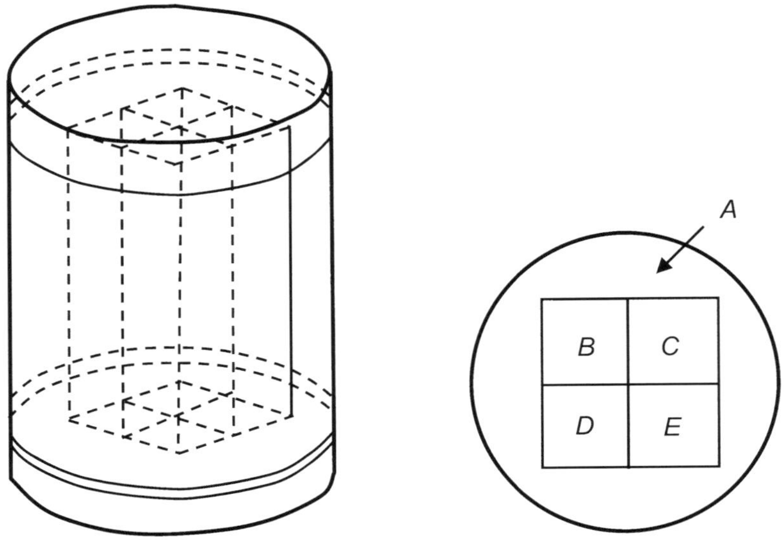

Fig. 12.12 Fabrication procedure for five-metal diffusion multiple. Adapted from Ref 12.4

Simple optical microscopy is usually used first to examine the phase formation and the integrity of a diffusion multiple. Most phases can usually be seen with an optical microscope, although sometimes it is difficult to identify each when many phases are present. Optical examination from low magnifications to high magnifications, along with information on the existing binary phase diagrams, can give clues to the phases present. After optical examination, SEM is often performed to obtain backscattered electron (BSE) images. The atomic number contrast, along with some EDS analysis, which is often available with SEM, can help to further define some or most of the phases. Sometimes the contrast from different grain orientations can confound the atomic number contrast. A high-quality and high-contrast BSE image provides good information about the phases and related equilibria.

Electron backscatter diffraction (EBSD) is a very useful tool to perform crystal structure analysis to aid phase identification. It can be used to identify crystal structures of micron-size phases in a regularly polished sample without going through the trouble of making TEM thin-foil specimens. Commercial EBSD systems are available as an attachment to regular SEMs. As a focused electron backscatter impinges on a phase, it generates BSEs in addition to secondary electrons. Sophisticated algorithms have been developed to automatically capture and index the EBSD patterns.

The best way to check the reliability of the diffusion multiple approach is to compare the phase diagrams obtained from diffusion multiples to those obtained from equilibrated alloys and diffusion couples.

12.4 Phase Diagram Construction Errors

Hiroaki Okamoto and Thaddeus Massalski have prepared the hypothetical binary shown in Fig. 12.13, which exhibits many typical errors of construction (marked as points 1 to 23). The explanation for each error is given in the accompanying text; one possible error-free version of the same diagram is shown in Fig. 12.14. Typical phase-rule violations in Fig. 12.13 include:

1. A two-phase field cannot be extended to become part of a pure-element side of a phase diagram at zero solute. In example 1, the liquidus and the solidus must meet at the melting point of the pure element.
2. Two liquidus curves must meet at one composition at a eutectic temperature.
3. A tie line must terminate at a phase boundary.
4. Two solvus boundaries (or two liquidus, or two solidus, or a solidus and a solvus) of the same phase must meet (i.e., intersect) at one composition at an invariant temperature. (There should not be two solubility values for a phase boundary at one temperature.)
5. A phase boundary must extrapolate into a two-phase field after crossing an invariant point. The validity of this feature, and similar features related to invariant temperatures, is easily demonstrated by constructing hypothetical free-energy diagrams slightly below and slightly above the invariant temperature and by observing the relative positions of the relevant tangent points to the free-energy curves. After intersection, such boundaries can also be extrapolated into metastable regions of the phase diagram. Such extrapolations are sometimes indicated by dashed or dotted lines.

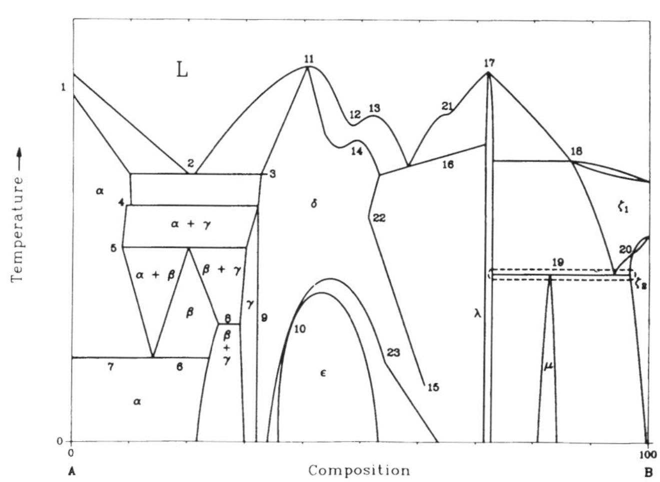

Fig. 12.13 Hypothetical binary phase diagram showing many typical errors of construction. See accompanying text for discussion of the errors at points 1 to 23. Source: Ref 12.6 as published in Ref 12.1

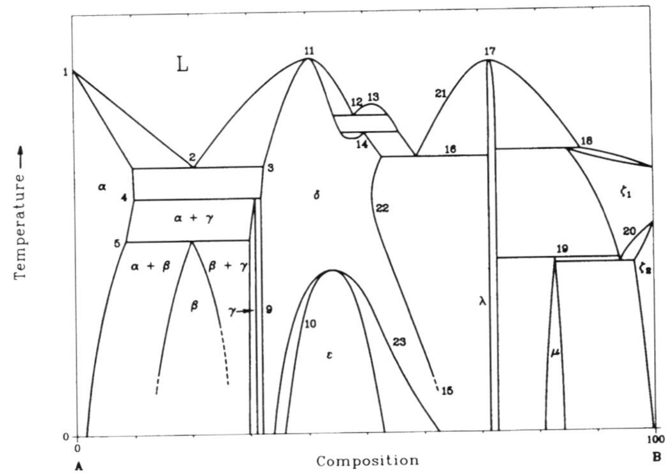

Fig. 12.14 Error-free version of the phase diagram shown in Fig. 12.13. Source: Ref 12.6 as published in Ref 12.1

6. Two single-phase fields (α and β) should not be in contact along a horizontal line. (An invariant temperature line separates two-phase fields in contacts.)
7. A single-phase field (α, in this instance) should not be apportioned into subdivisions by a single line. Having created a horizontal (invariant) line at point 6 (which is an error), there may be a temptation to extend this line into a single-phase field, α, creating an additional error.
8. In a binary system, an invariant-temperature line should involve equilibrium among three phases.
9. There should be a two-phase field between two single-phase fields (Two single phases cannot touch except at a point. However, second-order and higher-order transformations may be exceptions to this rule.)
10. When two phase boundaries touch at a point, they should touch at an extremity of temperature.
11. A touching liquidus and solidus (or any two touching boundaries) must have a horizontal common tangent at the congruent point. In this instance, the solidus at the melting point is too "sharp" and appears to be discontinuous.
12. A local minimum point in the lower part of a single-phase field (in this instance, the liquid) cannot be drawn without additional boundary in contact with it. (In this instance, a horizontal monotectic line is most likely missing.)
13. A local maximum point in the lower part of a single-phase field cannot be drawn without a monotectic, monotectoid, systectic, and sintectoid reaction occurring below it at a lower temperature. Alternatively, a solidus curve must be drawn to touch the liquidus at point 13.
14. A local maximum point in the upper part of a single-phase field cannot be drawn without the phase boundary touching a reversed monotectic,

or a monotectoid, horizontal reaction line coinciding with the temperature of the maximum. When a point 14 type of error is introduced, a minimum may be created on either side (or on one side) of point 14. This introduces an additional error, which is the opposite of point 13, but equivalent to point 13 in kind.

15. A phase boundary cannot terminate within a phase field. (Termination due to lack of data is, of course, often shown in phase diagrams, but this is recognized to be artificial.
16. The temperature of an invariant reaction in a binary system must be constant. (The reaction line must be horizontal.)
17. The liquidus should not have a discontinuous sharp peak at the melting point of a compound. (This rule is not applicable if the liquid retains the molecular state of the compound, i.e., in the situation of an ideal association.)
18. The compositions of all three phases at an invariant reaction must be different.
19. A four-phase equilibrium is not allowed in a binary system.
20. Two separate phase boundaries that create a two-phase field between two phases in equilibrium should not cross each other.
21. Two inflection points are located too closely to each other.
22. The boundary direction reverses abruptly (more abrupt than a typical smooth “retro-grade”). This particular change can occur only if there is an accompanying abrupt change in the temperature dependence of the thermodynamic properties of either of the two phases involved (in this instance, δ or λ in relation to the boundary). The boundary turn at point 22 is very unlikely to be explained by a realistic change in the composition dependence of the Gibbs energy functions.
23. An abrupt change in the slope of a single-phase boundary. This particular change can occur only by an abrupt change in the composition dependence of the thermodynamic properties of the single phase involved (in this instance, the δ phase). It cannot be explained by any possible abrupt change in the temperature dependence of the Gibbs energy function of the phase. (If the temperature dependence was involved, there would also be a change in the boundary of the ε phase.)

Problems Connected with Phase-Boundary Curvatures. Although phase rules are not violated, their additional unusual situations (points 21, 22, and 23 in Fig. 12.13) have also been included in Fig. 12.15. In each instance, a more subtle thermodynamic problem may exist related to these situations. Examples are discussed in which several thermodynamically unlikely diagrams are considered. The problems with each of these situations involve an indicated rapid change of slope of a phase boundary. If such situations are to be associated with realistic thermodynamics, the temperature (or the composition) dependence of the thermodynamic functions of the phase (or phases) involved would be expected to show corresponding abrupt and unrealistic variations in the phase diagram regions where such abrupt phase-boundary changes are proposed, without any clear reason for them. Even the onset of ferromagnetism in a phase does not normally cause an abrupt change of slope of the related phase boundaries.

Congruent Transformations. The congruent point on a phase diagram is where different phases of the same composition are equilibrium. The Gibbs-Konovalov Rule for congruent points, which was developed by Dmitry Konovalov from a thermodynamic expression given by J. Willard Gibbs, states that the slope of phase boundaries at congruent transformations must be zero (horizontal). Examples of correct slope at the maximum and minimum points on liquidus and solidus curves can be seen in Fig. 12.16.

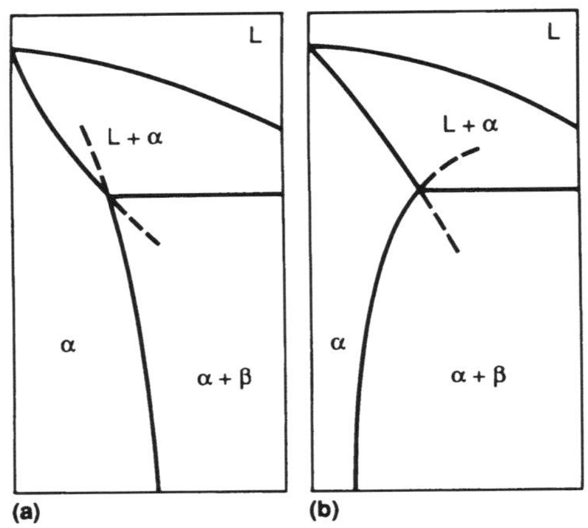

Fig. 12.15 Examples of acceptable intersection angles for boundaries of two-phase fields. Source: Ref 12.2 as published in Ref 12.1

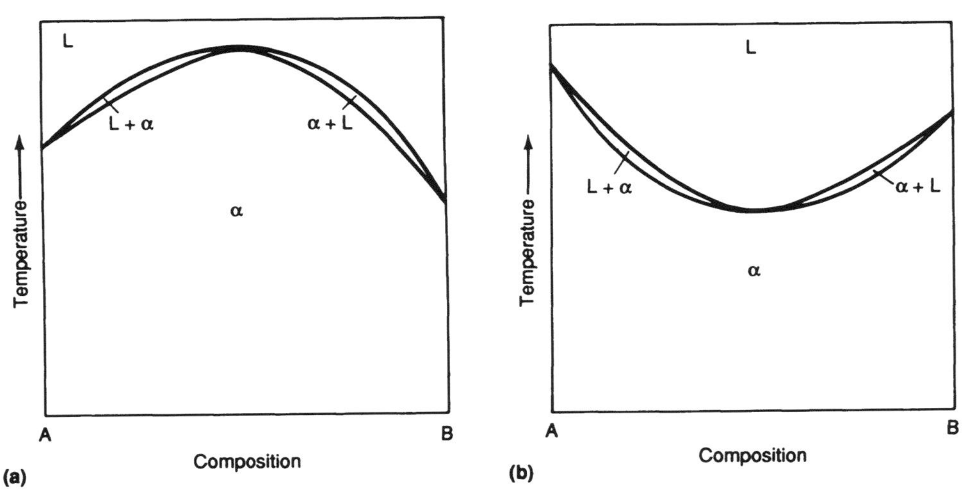

Fig. 12.16 Schematic binary phase diagrams with solid-state miscibility where the liquidus shows (a) a maximum and (b) a minimum. Source: Ref 12.1

Often, the inner curve on a diagram such as that shown in Fig. 12.16 is erroneously drawn with a sharp inflection (Fig. 12.17).

A similar common construction error is found in the diagrams of systems containing congruently melting compounds (such as the line compounds shown in Fig. 12.18) but having little or no association of the component

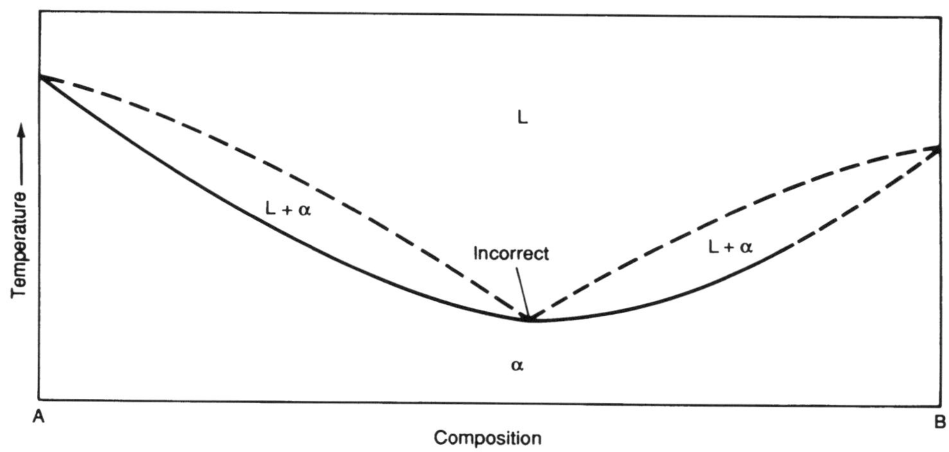

Fig. 12.17 Example of a binary phase diagram with a minimum in the liquidus that violates the Gibbs-Konovalov Rule. Source: Ref 12.7 as published in Ref 12.1

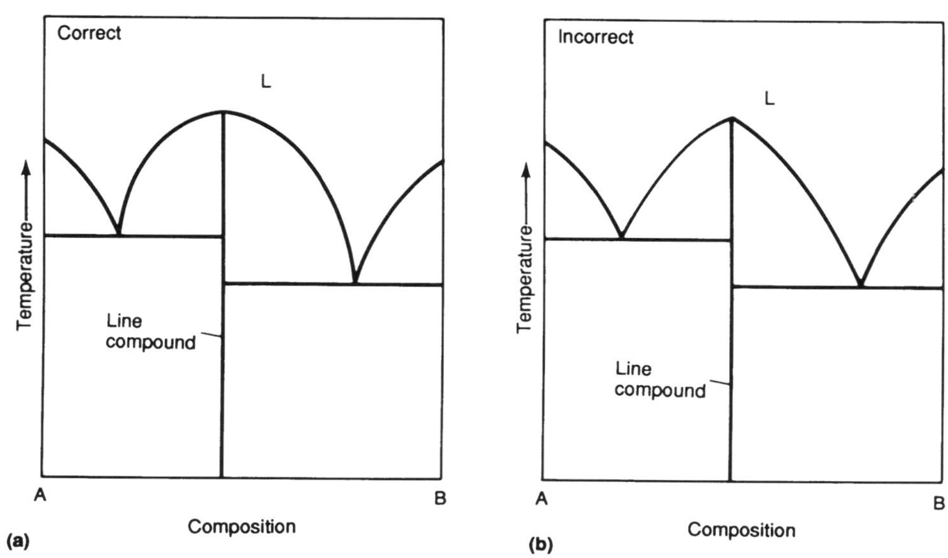

Fig. 12.18 Schematic diagrams of binary systems containing congruent melting compounds but having no association of the component atoms in the melt common. The diagram in (a) is consistent with the Gibbs-Konovalov Rule, whereas that in (b) violates the rule. Source: Ref 12.7 as published in Ref 12.1

atoms in the melt (as with most metallic systems). This type of error is especially common in partial diagrams, where one or more system components is a compound instead of an element. However, the slope of liquids and solidus curves must *not* be zero when they terminate at an element or at a compound having complete association in the melt.

ACKNOWLEDGMENT

Portions of this chapter came from "Introduction to Alloy Phase Diagrams," by H. Baker in *Alloy Phase Diagrams*, Vol 3, *ASM Handbook*, ASM International, 1992, reprinted in *Desk Handbook: Phase Diagrams for Binary Alloys*, 2nd ed., H. Okamoto, Ed., ASM International, 2010.

REFERENCES

12.1 H. Baker, Introduction to Alloy Phase Diagrams, *Alloy Phase Diagrams*, Vol 3, *ASM Handbook*, ASM International, 1992, reprinted in *Desk Handbook: Phase Diagrams for Binary Alloys*, 2nd ed., H. Okamoto, Ed., ASM International, 2010

12.2 F.N. Rhines, *Phase Diagrams in Metallurgy,* McGraw-Hill, 1956

12.3 A. Prince, *Alloy Phase Equilibria,* Elsevier, 1966

12.4 J.-C. Zhao, *Methods for Phase Diagram Determination*, Elsevier, 2007

12.5 C. Brubaker and Z.-K. Liu, Diffusion Couple Study of the Mg-Al System, *Magnesium Technology 2004*, TMS, 2004

12.6 H. Okamoto and T.B. Massalski, Thermodynamically Improbable Phase Diagrams, *J. Phase Equilibria,* Vol 12 (No. 2), 1991, p 148–168

12.7 D.A. Goodman, J.W. Cahn, and L.H. Bennett, The Centennial of the Gibbs-Konovalov Rule for Congruent Points, *Bull. Alloy Phase Diagrams,* Vol 2 (No. 1), 1981, p 29–34

CHAPTER 13

Computer Simulation of Phase Diagrams

PHASE DIAGRAMS have traditionally been determined purely by experimentation, which is costly and time consuming. While the experimental approach is feasible for the determination of binary and simple ternary phase diagrams, it is less efficient for the complicated ternaries and becomes practically impossible for higher-order systems over a wide range of compositions and temperatures. Commercial alloys are multicomponent in nature, and a more efficient approach is needed in the determination of multicomponent phase diagrams. In recent years, a phenomenological approach, the CALPHAD approach—after the acronym CALculation of PHAse Diagrams—has been widely used for the study of phase equilibria of multicomponent systems.

While thermodynamic models that represent nonstoichiometric solution phases have existed for many decades, it was not until the advent of computer technology that it became possible to calculate thermodynamic equilibria between such phases. The seminal works of Kaufman and coworkers in the development of computer-based methods marked the beginning of what has become known as the CALPHAD approach. As computing power increased, it became possible to use more complex descriptions of thermodynamic properties and move from binary and ternary alloys to actual multicomponent alloys used in industrial practice. Although the scientific methodology had been available to do this for some time, practical applications required a minimum of computing capability before they became useful. In the past few decades, groups have been established worldwide that specialize in both the development of software for the calculation of complex multicomponent phase equilibria and the development of thermodynamic databases that describe the properties of the relevant phases in the alloy of interest. It is the linking of software to thermodynamic databases, especially developed and validated for multicomponent alloys, that has held the key to the practical and everyday use of CALPHAD methods in industry. As computational speeds further increased, it has become possible to use CALPHAD methods as a general tool that could be linked to material models for phase transformations and subsequently to the modeling of more general material properties, such as physical and mechanical properties.

There are two main requirements for the calculation of phase equilibria:

- The thermodynamic properties of the phases in an alloy must be represented.
- A Gibbs energy minimization must be performed so that the phase equilibria between the requisite phases can be calculated.

13.1 Thermodynamic Models

Thermodynamic modeling of solution phases lies at the very core of the CALPHAD method. In metallic alloys, it is only rarely that calculations involve purely stoichiometric compounds. Solution phases are defined here as any phase in which there is solubility of more than one component.

Random substitutional models are used for phases such as the gas phase or a simple metallic liquid and solid solutions, where components can mix on any spatial position that is available to the phase. For example, in a simple body-centered cubic phase, any of the components could occupy any of the atomic sites that define the cubic structure, as shown in Fig. 13.1. In a gas or liquid phase, the crystallographic nature of structure is lost, but otherwise, positional occupation of the various components relies on random substitution rather than the occupation of any preferred site by a particular component.

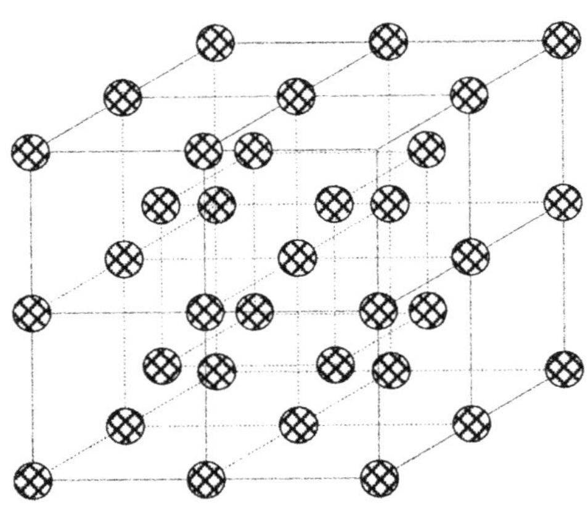

Fig. 13.1 Simple body-centered cubic structure with random occupation of atoms and all sites consisting of eight-unit cells. Source: Ref 13.1

Dilute Solution Models. There are a number of areas in materials processing where low levels of alloying are important, for example, in refining and some age-hardening processes. In such cases, it is possible to deal with solution phases by dilute solution models. These have the advantage that there is substantial experimental literature that deals with the thermodynamics of impurity additions, particularly for established materials such as ferrous- and copper-base alloys. However, there are fundamental limitations in handling concentrated solutions.

General solution models include the ideal and nonideal models.

Ideal Solution Model. The simplest solution model is an ideal substitutional solution, which is characterized by the random distribution of components on a lattice with an interchange energy equal to zero.

Nonideal Solutions—Regular and Subregular Solution Models. The regular solution model is the simplest of the nonideal models and basically considers that the magnitude and sign of interactions between the components in a phase are independent of composition. However, it has been realized for a long time that the assumption of composition-independent interactions is too simplistic. This led to the development of the subregular solution model, where interaction energies are considered to change linearly with composition.

Most methods of extrapolating the thermodynamic properties of alloys into multicomponent systems are based on the summation of the binary excess parameters. The relevant formulae are based on various geometrical weightings of the mole fractions.

One of the assumptions of this approach is that the ternary interactions are small in comparison to those that arise from the binary terms. This may not always be the case, and where the need for higher-order interactions is evident, these can be taken into account with an excess ternary interaction parameter. There is little evidence of the need for any higher-order interaction terms, and prediction of the thermodynamic properties of substitutional solution phases in multicomponent alloys is usually based on an assessment of binary and ternary values. Various other polynomial expressions for the excess term in multicomponent systems have been considered. However, these are based on predicting the properties of the higher-order system from the properties of the lower-component systems.

More complex representations of the thermodynamic properties of solution phases exist, for example, the sublattice model, where the phase can be envisioned as composed of interlocking sublattices (Fig. 13.2) on which the various components can mix. It is usually applied to crystalline phases, but the model can also be extended to consider ionic liquids, where mixing on ionic sublattices is considered. The model is phenomenological in nature and does not define any crystal structure within its general mathematical

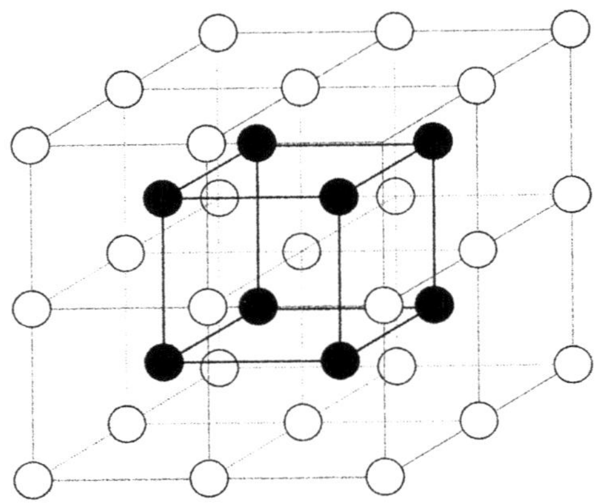

Fig. 13.2 Simple body-centered cubic structure sites consisting of eight-unit cells with preferential occupation of atoms in the body center and corner positions. Source: Ref 13.1

formulation. It is possible to define internal parameter relationships that reflect structure with respect to different crystal types, but such conditions must be externally formulated and imposed on the model. Equally, special relationships apply if the model is to be used to simulate order-disorder transformations.

Sublattice modeling is now one of the most predominant methods used to describe solution and compound phases. It is very flexible and can account for a variety of different phase types ranging from interstitial phases, such as austenite and ferrite in steels, to intermetallic phases, such as sigma and Laves, which have wide homogeneity ranges.

13.2 Computational Methods

The previous section dealt with thermodynamic models, because these are the basis of the CALPHAD method. However, it is the computational methods and software that allow these models to be applied in practice. In essence, the issues involved in computational methods are less diverse and mainly revolve around Gibbs energy minimization.

It is also worthwhile to make some distinctions between methods of calculating phase equilibrium. For many years, equilibrium constants have been used to express the abundance of certain species in terms of the amounts of other arbitrarily chosen species. Such calculations have significant disadvantages in that some prior knowledge of potential reactions is often necessary, and it is difficult to analyze the effect of very complex reactions involving many species on a particular equilibrium reaction. Furthermore, unless equilibrium constants are defined for all possible chemical reactions, a true equilibrium calculation cannot be made, and, in the case of a reaction with 50 or 60 substances present, the number of possible reactions is massive.

CALPHAD methods attempt to provide a true equilibrium calculation by considering the Gibbs energy of all phases and minimizing the total Gibbs energy of the system (G). The CALPHAD methodology has the significant advantage that, because the total Gibbs energy is calculated, it is possible to derive all of the associated functions and characteristics of phase equilibria, that is, phase diagrams, chemical potential diagrams, and so on.

13.3 Calculation of Phase Equilibria

The actual calculation of phase equilibria in a multicomponent, multiphase system is a complex process involving a high level of computer programming. Details of programming aspects are too lengthy to go into detail, but most of the principles by which Gibbs energy minimization is achieved are conceptually quite simple. This section therefore concentrates on the general principles rather than going into detail concerning the currently available software programs, which, in any case, often contain proprietary code.

Essentially, the calculation must be defined so that the number of degrees of freedom is reduced, the Gibbs energy of the system can be calculated, and some iterative technique can be used to minimize the Gibbs energy. The number of degrees of freedom is reduced by defining a series of constraints, such as the mass balance, electroneutrality in ionic systems, composition range in which each phase exists, and so on.

Most thermodynamic software uses local minimization methods. As such, preliminary estimates for equilibrium must be given so that the process can begin and subsequently proceed smoothly to completion. Such estimates are usually set automatically by the software and do not need to accurately reflect the final equilibrium. However, the possibility that phases may have multiple minima in their Gibbs energy formulations should be recognized and start points automatically set so that the most stable minima are taken into account. Local minimization tools have the advantage of being rapid in comparison to global minimization methods, which automatically search for multiple Gibbs energy minima. As such, local minimization methods were invariably favored in the early days of CALPHAD. However, with the advent of faster computers, such an advantage becomes less tangible when dealing with relatively simple calculations, and the user will notice little effective difference in speed. Codes such as Thermo-Calc and PANDAT now offer global minimization methods as part of their calculation capability. However, if the problem to be solved involves multicomponent alloys with numerous multiple sublattice phases containing a potentially large number of local minima, speed issues will still arise.

Whether to use local or more global methods is a pertinent question if reliability of the final calculation is an issue. For the case of calculation of multicomponent alloys, such as those used by industry and where the composition space is reasonably prescribed, local minimization methods have been used successfully for many years and have proved highly reliable.

One of the earliest examples of Gibbs energy minimization applied to a multicomponent system was by White et al., who considered the chemical equilibrium in an ideal gas mixture of oxygen, hydrogen, and nitrogen with the species H, H_2, H_2O, N, N_2, NH, NO, O, O_2, and OH being present. The problem here is to find the most stable mixture of species. These authors presented two methods of Gibbs energy minimization, one of which used a linear programming method, and the other is based on the method of steepest descent using Lagrange's method of undetermined multipliers. The method of steepest descent provides a rapid solution to the minimization problem and was later used in the software codes SOLGAS and SOLGASMIX. SOLGAS, the earlier code, treated a mixture of stoichiometric condensed substances and an ideal gas mixture, while SOLGASMIX was further able to include nonideal solution phases.

The minimization methods used by later programs such as ChemSage, the successor to SOLGASMIX; F*A*C*T; FactSage, the recent combining of ChemSage and F*A*C*T; Thermo-Calc; and MTDATA are, in the broadest sense, similar in principle to that described previously, although there are clear differences in their actual operation. Thermodynamic models are now more complex, which may make it necessary to consider further degrees of freedom. However, constraints are still made to the system such that the Gibbs energy may be calculated as a function of extensive variables, such as the amount of each phase present in the system. Initial estimates are made for the Gibbs energy as a function of phases present, their amounts and composition, and so on. The Gibbs energy is then calculated and some numerical method is used, whether it be through Lagrangian multipliers or a Newton-Raphson method, by which new values can be estimated and which will cause the Gibbs energy to be decreased. When the difference in calculated Gibbs energy between the iterative steps reaches some small-enough value, the calculation is taken to have converged.

As mentioned earlier, recent progress has been made in developing more global minimization codes, for example, Thermo-Calc and PANDAT. Such codes use mathematical methods to search out all possible minima and compare Gibbs energies of the local minima so that the most stable equilibrium is automatically calculated.

13.4 Application of CALPHAD Calculations to Industrial Alloys

From the beginning, one of the aims of CALPHAD methods has been to calculate phase equilibria in the complex multicomponent alloys that are used regularly by industry. Certainly the necessary mathematical formulations to handle multicomponent systems have existed for some time, and they have been programmed into the various software packages for calculation of phase equilibria since the middle of the last century. However, it is interesting to note that, until the 1990s (with the exception of steels), there had been very little actual application to the complex systems that exist in technological or industrial practice, other than through calculations using simple stoichiometric substances, ideal gas reactions, and dilute solution models. Dilute solution modeling has been used for some time because it is not very intensive in computational terms, and some industrially important materials, although containing many elements, are actually quite low in total alloy or impurity content (e.g., high-strength low-alloy steels). Although useful for certain limited applications, they could not with any accuracy begin to handle highly alloyed materials such as stainless steels or nickel-base superalloys. Substance calculations, while containing large numbers of species and condensed phases, are in many ways even more limited in their application to alloys, because they do not consider interactions in phases involving substantial mixing of the components.

The main areas of application for more generalized models were, until the 1990s, mainly restricted to binary and ternary systems or limited to "ideal industrial materials," where only major elements were included. The key to the general application of CALPHAD methods in multicomponent systems was the development of sound, validated thermodynamic databases for use in the available computing software. Until then, there had been a dearth of such databases. Steels were a notable exception to this trend and, in particular, stainless and high-speed steels, where alloy contents can rise to well above 20 wt%. For such alloys, a concentrated solution database (Fe-Base) has existed since 1978. However, although far more generalized than dilute solution databases, its range of applicability is limited in temperature to between 700 and 1200 °C (1290 and 2190 °F). The lack of similar databases for other material types presented severe problems for CALPHAD calculations with any of the other commonly used materials and led to a concentration of application to steels. However, during the 1990s, further multicomponent databases were developed for use with aluminum alloys, steels, nickel-base superalloys, and titanium- and TiAl-base alloys. These databases were created mainly for use with industrial, complex alloys, and the accuracy of computed results was validated to an extent not previously attempted. Simple, statistical analysis of average deviation of calculated result from experimental measurement in "real," highly alloyed, multicomponent alloys has demonstrated that CALPHAD methods can provide predictions for phase equilibria whose accuracy lies close to that of experimental measurements.

The importance of validation of computed results cannot be stressed too highly. Computer models, such as those used to simulate materials processing, rely on input data that can be time consuming to measure but readily predicted via CALPHAD and related methods. For example, it is possible to model the processing of a steel at all stages of manufacture, starting from the initial stages in a blast furnace, through the refinement stages to a casting shop, followed by heat treatment and thermomechanical processing to the final product form. Such a total modeling capability requires that confidence can be placed in the predictions of each of the building blocks, and, in the case of CALPHAD methods, the key to success is the availability of high-quality, validated databases.

13.5 Databases

Substance Databases. From a simple point of view, substance databases have little complexity because they are assemblages of assessed data for stoichiometric condensed phases and gaseous species. They have none of the difficulties associated with nonideal mixing of substances, which is the case for a "solution" database. However, internal self consistency still must be maintained. For example, thermodynamic data for $C_{(s)}$, $O_{2<g>}$, and $CO_{2<g>}$ are held as individual entries, which provide their requisite properties as a function of temperature and pressure. However, when put together in a calculation, they must combine to give the correct Gibbs energy change for the reaction $C_{(s)} + O_{2<g>} \leftrightarrow CO_{2<g>}$. This is a simple example, but substance databases can contain more than 10,000 different substances; therefore, it is a major task to ensure internal self-consistency so that all experimentally known Gibbs energies of reaction are well represented.

Solution databases, unlike substance databases, contain thermodynamic descriptions for phases that have potentially very wide ranges of existence, both in terms of temperature and composition. For example, the liquid phase usually extends across the whole of the compositional space encompassed by complete mixing of all of the elements. Unlike an ideal solution, the shape of the Gibbs energy space arising from nonideal interactions can become extremely complex, especially if nonregular terms are used.

Although it may seem an obvious statement, it is important to remember that thermodynamic calculations for multicomponent systems are multidimensional in nature. This means that it becomes impossible to envision the types of Gibbs energy curves that are illustrated in many teaching texts on thermodynamics and which lead to the easy conceptualization of miscibility gaps, invariant reactions, and so on. In multicomponent space, such things are often very difficult to understand, let alone conceptualize.

Miscibility gaps can appear in ternary and higher-order systems, even though no miscibility gap exists in the lower-order systems, and the Gibbs phase rule becomes vitally important in understanding reaction sequences. Partition coefficients that apply in binary systems are usually altered in a multicomponent alloy and may change sign. Also, computer predictions can be surprising at times, with phases appearing in temperature/composition regimes where an inexperienced user may well not expect. In practice, many scientists and engineers who require results from thermodynamic calculations are not experts in CALPHAD. As such, it is necessary to validate the database for multicomponent systems so that the user can have confidence in the calculated result.

13.6 Industrial Applications

Although industrial processes rarely reach an equilibrium state, knowledge of phase stabilities, phase transformation temperatures, phase amounts, and phase compositions in an equilibrium state is critical in the determination of processing parameters. In this section, four examples are presented to show how multicomponent phase diagram calculation can be readily useful for industrial applications.

In Example 1, phase diagram calculation is used to predict bulk metallic glass formability.

In Example 2, calculations are made for nickel alloys, focusing on the major concerns in the development of nickel-base superalloys, such as γ' solvus temperature, matrix (γ)/precipitate (γ') misfit, and the formation of deleterious topologically close-packed (tcp) phases.

In Example 3, phase diagram calculation is applied to commercial titanium alloys. The β transus (temperature at which a starts to precipitate from β) and β approach curve (fraction of β phase as a function of temperature) are calculated for a variety of titanium alloys and compared with experimental data. The effect of interstitial elements, such as oxygen, carbon, nitrogen, and hydrogen, on the β transus temperatures is also discussed.

In Example 4, phase diagram calculation is used to predict dilation of iron alloys during phase transformations from austenite to ferrite and from austenite to martensite.

Example 1: Prediction of Bulk Metallic Glass Formability. Bulk metallic glass (BMG) materials exhibit unique properties, such as high strength, wear and corrosion resistance, castability, and fracture toughness. These properties make them extremely attractive as materials having great potential for practical applications in the areas of sports and luxury goods, electronics, medical, defense, and aerospace. Regardless of their importance, BMGs are mainly developed by experimental trial-and-error approaches involving, in many cases, the production of hundreds to thousands of different alloy compositions. Therefore, there is a need to either formulate a theoretical model or develop a computational approach for predicting families of alloy compositions with a greater tendency for glass formation. In recent years, Chang and his colleagues (Ref 13.2) have used multicomponent phase diagram calculation to identify the alloy compositions that have high BMG formability. The theoretical basis of their approach is that the reduced glass transition temperature (T_r), which is defined as the ratio of the glass formation temperature and the liquidus temperature, should be on the order of 0.6 or higher. According to this argument, BMGs are often found to form near deep eutectic invariant reactions where high T_r can be achieved. In addition, due to the low liquidus temperature at a deep eutectic region, the liquid has an exceptionally high viscosity, further suppressing nucleation by inhibiting mass transport. It has been found that the formation of BMG usually requires three or more elements, even though they do form in some binary systems. This implies that, in principle, if phase diagrams of multicomponent systems are known as functions of temperature and composition, potential alloy compositions that favor BMG formation can readily be identified. Unfortunately, accurate liquidus projections and phase diagrams of multicomponent systems, which are required to guide the exploration of BMGs, are rarely available, and this is true even for many ternary systems. Multicomponent phase diagram calculation thus becomes a desirable approach for efficiently scanning the vast number of possible compositions that may yield BMGs. The methodology is to develop a thermodynamic database for the system to be studied, from which the lowering liquidus valley can be calculated. Key experiments will then be arranged to validate the BMG formability for alloy compositions being identified by the calculation.

To demonstrate the capability of this approach, two examples are given here. The first one is for the identification of BMG formers in the Zr-Ti-Cu-Ni system, and the second is for the effect of titanium on the glass formability of the Zr-Cu-Ni-Al alloys. In the first example, a thermodynamic database for the Zr-Ti-Cu-Ni system was developed. The liquidus projection was then calculated for this quaternary using Pandat software, and 18 five-phase invariant equilibria were identified that show low-lying liquidus temperatures. The compositions of the liquid phase at these invariant temperatures, shown as solid circles in Fig. 13.3, are found to be located in two areas in the composition diagram, which is in excellent agreement with those identified experimentally. Doing the calculations first would save a tremendous amount of experimental work.

In the second example, the effect of titanium on the glass-forming ability (GFA) of Zr-Cu-Ni-Al quaternary alloys was studied first by thermodynamic calculation and then validated by experimental approach. The quaternary $Zr_{56.28}Cu_{31.3}Ni_{4.0}Al_{8.5}$ alloy was first found to be a BMG-forming alloy based on the calculated low-lying liquidus surface of the quaternary Zr-Cu-Ni-Al system. By calculating the liquidus projection of the Zr-Cu-Ni-Al-Ti system, the liquidus temperature was found to decrease rapidly from the quaternary alloy $Zr_{56.28}Cu_{31.3}Ni_{4.0}Al_{8.5}$ (at.%) due to the addition of titanium. Further calculation of several isopleths revealed that the liquidus temperature decreased the fastest when zirconium was replaced by titanium and reached a minimum at 4.9 at.% Ti, as shown in Fig. 13.4. Alloy ingots of nominal compositions $Zr_{56.28-c}Ti_cCu_{31.3}Ni_{4.0}Al_{8.5}$ (c = 0 ~ 10.0 at.%) with different amounts of titanium were then prepared for experimental study of the GFA of these alloys.

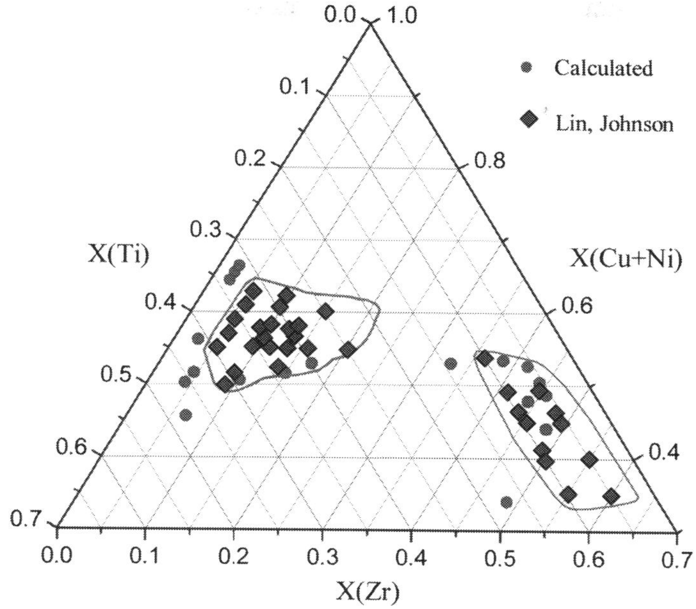

Fig. 13.3 Comparison of thermodynamically calculated compositions of liquid alloys at the five-phase invariant equilibria with those identified experimentally as bulk metallic glass formers for the Zr-Cu-Ni-Ti system. Source: Ref 13.3 as published in Ref 13.2

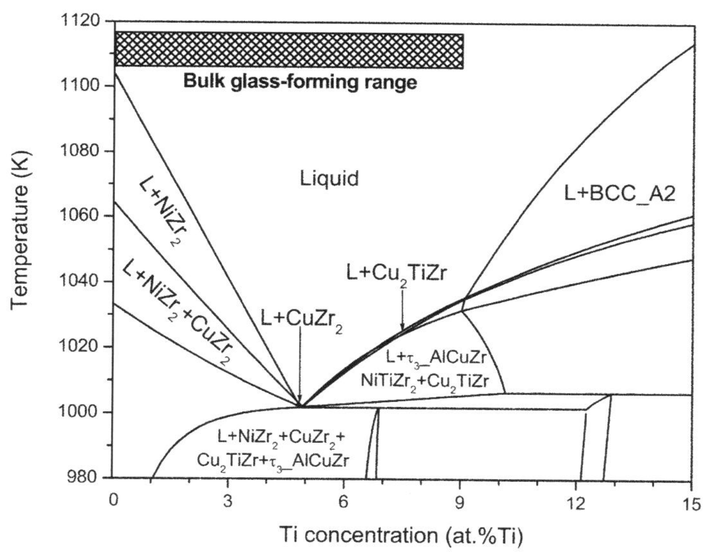

Fig. 13.4 Calculated isopleth of the quinary Al-Cu-Ni-Ti-Zr system expressed in terms of temperature as a function of titanium concentration (at.%). The composition of titanium at the origin is 0; corresponding to $Zr_{56.2}Cu_{31.3}Ni_{4.0}Al_{8.5}$, the compositions of copper, nickel, and aluminum are fixed at 31.3, 4.0, and 8.5 at.%, respectively. The shaded area denotes the experimentally observed bulk glass-forming range. Source: Ref 13.2

As shown in Fig. 13.5, the critical diameters of the amorphous rods formed increased from 6 mm (0.24 in.) for the base quaternary alloy to 14 mm (0.55 in.) for the quinary alloy with 4.9 at.% Ti and then decreased again. It is worth pointing out that 14 mm (0.55 in.) was the diameter limit of the casting mold used in the experiment. An amorphous rod with larger diameter could have been obtained if a larger casting mold had been used. It is not difficult to imagine how many alloys must be studied in a five-component system to identify the alloys with high GFA if a purely trial-and-error experimental method is used. These two examples thus demonstrate the great power of thermodynamic calculation in quickly locating the alloy chemistries with a high potential of GFA and guiding the experimental study.

Example 2: Design and Development of Nickel-Base Superalloys. Nickel-base superalloys are a unique class of metallic materials that possess an exceptional combination of high-temperature strength, toughness, and resistance to degradation under harsh operating environment. These materials are widely used in aircraft and power-generation turbines, rocket engines, and other challenging environments, including nuclear power and chemical processing plants. To improve the performance of nickel-base superalloys, alloy composition and processing conditions must be carefully optimized to promote the formation of desired phases and microstructure and to avoid the formation of deleterious phases.

Multicomponent phase diagram calculation therefore plays an important role in nickel-base superalloy design. With this approach, not only equilibrium phases but also phase amounts and phase transformation temperatures can be predicted, given alloy chemistry. As an example, the liquidus, solidus, and γ' solvus for a number of Rene′ N6 alloys in the specification range are calculated and compared with those determined by experiments. As shown in Fig. 13.6, the liquidus varies in the temperature range of 1660

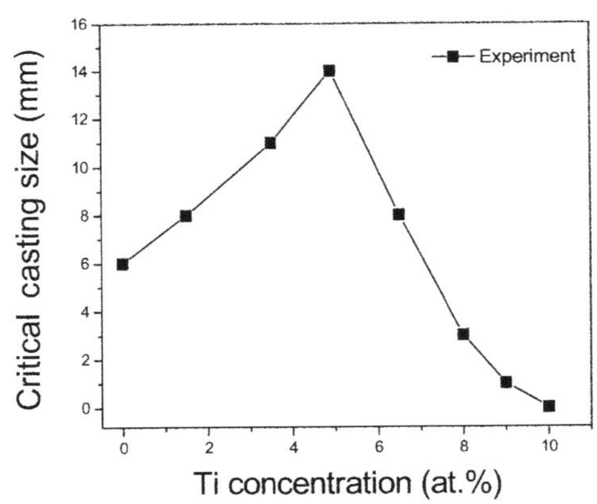

Fig. 13.5 Critical diameters of the cast glassy rods as a function of the titanium concentration in atomic percent. Source: Ref 13.2

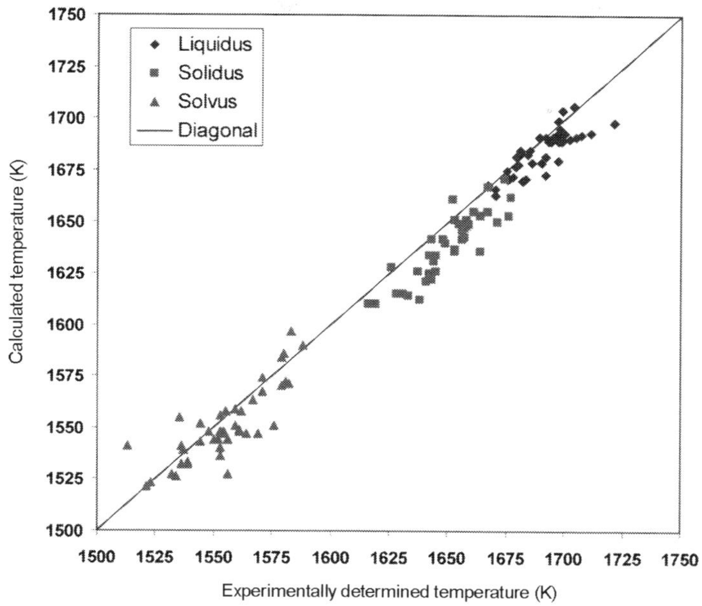

Fig. 13.6 Comparison of calculated liquidus, solidus, and γ′ solvus temperatures with those determined by experiment. Source: Ref 13.4 as published in Ref 13.2

to 1700 K (1385 to 1425 °C), the solidus of 1615 to 1675 K (1340 to 1400 °C), and the γ′ solvus of 1525 to 1585 K (1250 to 1310 °C). Obviously, even for the same group of alloys whose compositions are all within the specification range of Rene′ N6, their phase transformation temperatures can differ significantly, depending strongly on the alloy chemistry. It is important that these temperatures can be calculated, because they are critical parameters needed for the determination of processing conditions.

Another example is shown in Fig. 13.7, which compares the calculated and experimentally determined volume percent of γ′ phase for a number of nickel-base superalloys containing aluminum, chromium, molybdenum, titanium, and tungsten. The two major phases of nickel-base superalloys are γ and γ′, and the presence of a high volume fraction of γ′ is the key to strengthening. This figure clearly shows that a high volume fraction of γ′ can be obtained through the optimization of alloy composition. Calculations were carried out using the PanNi thermodynamic database (Ref 13.6) and experimental data. Good agreement was obtained.

Because the matrix phase (γ) and the precipitate phase (γ′) of nickel-base superalloys are crystallographically coherent, the precipitate-matrix misfit results in internal stresses that influence the shape of the precipitates and thus the final mechanical properties. The precipitate-matrix misfit (δ) can be calculated as:

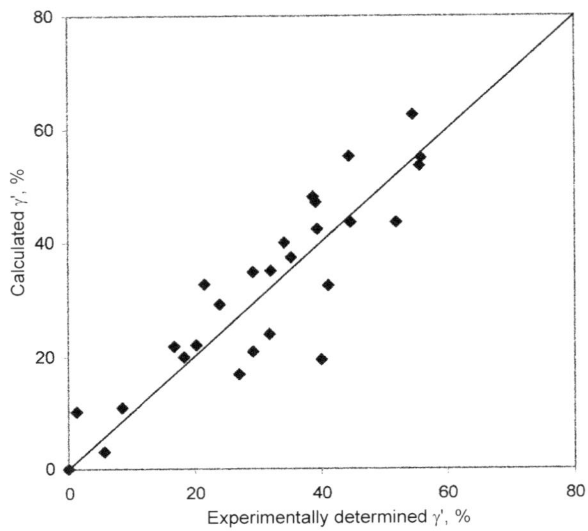

Fig. 13.7 Comparison of the calculated and experimentally determined percent of γ' in the $\gamma + \gamma'$ two-phase mixture for a number of Ni-Al-Cr-Mo-Ti-W alloys. Source: Ref 13.5 as published in Ref 13.2

$$\delta = \frac{a_{\gamma'} - a_{\gamma}}{0.5\left(a_{\gamma'} + a_{\gamma}\right)}$$

where $a_{\gamma'}$ and a_{γ} are the lattice parameters of the γ' and γ phases, respectively. Obviously, the precipitate-matrix misfit is determined by the lattice parameters of the γ' and γ phases, which depend on the compositions of these two phases. Elemental partitioning is therefore an important alloy design consideration, because the compositions of the constituent phases will directly impact both the mechanical and environmental characteristics of the alloy. With the computational tool for multicomponent phase diagram calculation, the phase composition and elemental partitioning can be readily calculated if a reliable thermodynamic database is available for the alloy system. Figure 13.8(a) compares the calculated and experimentally determined aluminum contents in the γ and γ' phases for a number of nickel-base superalloys, while Fig. 13.8(b to e) are for those of cobalt, chromium, molybdenum, and titanium. All calculations are carried out using the PanNi database, and experimental data from the literature are used for comparison. In all cases, there is good agreement between the calculated and experimental data.

In general, refractory alloying elements, such as molybdenum, niobium, rhenium, and tungsten, are added for solid-solution strengthening of the γ phase and to provide high-temperature creep resistance. A major concern with these alloying elements is the formation of tcp phases. The tcp phases are typically rich in refractory alloying elements and are detrimental because they deplete strengthening elements from the matrix phase. Examples of tcp phases include the orthorhombic *P* phase, the tetragonal σ phase, the rhombohedral *R* phase, and the rhombohedral μ phase. Being able to predict the correlation between the stabilities of the tcp phases and the alloy composition, thereby avoiding the formation of deleterious phases by wisely adjusting the alloy composition, is of critical importance in the design of new alloys and the modification of specification ranges for existing alloys. The CALPHAD approach has been used for such purposes. One example is shown in Fig. 13.9. In this example, the effect of rhenium content on the stability of σ is presented for the CMSX-4 alloy. This figure shows that the σ precipitation temperature and its amount strongly depends on the content of rhenium in the alloy, and the higher the rhenium content, the more σ will form. The predicted trend is correct, while the accuracy of the phase transformation temperature and phase amount depends on the quality of the thermodynamic descriptions for the tcp phases.

Example 3: Prediction of β Transus and β Approach Curves of Commercial Titanium Alloys. Pure titanium has two allotropic forms: the high-temperature body-centered cubic (β) structure and the low-temperature hexagonal close-packed (α) structure. The β transus, defined as the temperature at which α starts to form from β during solidification, is 883 °C (1621 °F) for pure titanium. However, this temperature can vary by several hundred degrees due to the addition of alloying elements. Elements such as aluminum, gallium, germanium, oxygen, carbon, and nitrogen are referred to as α stabilizers because they increase the β transus temperature, while vanadium, iron, molybdenum, chromium, niobium, nickel,

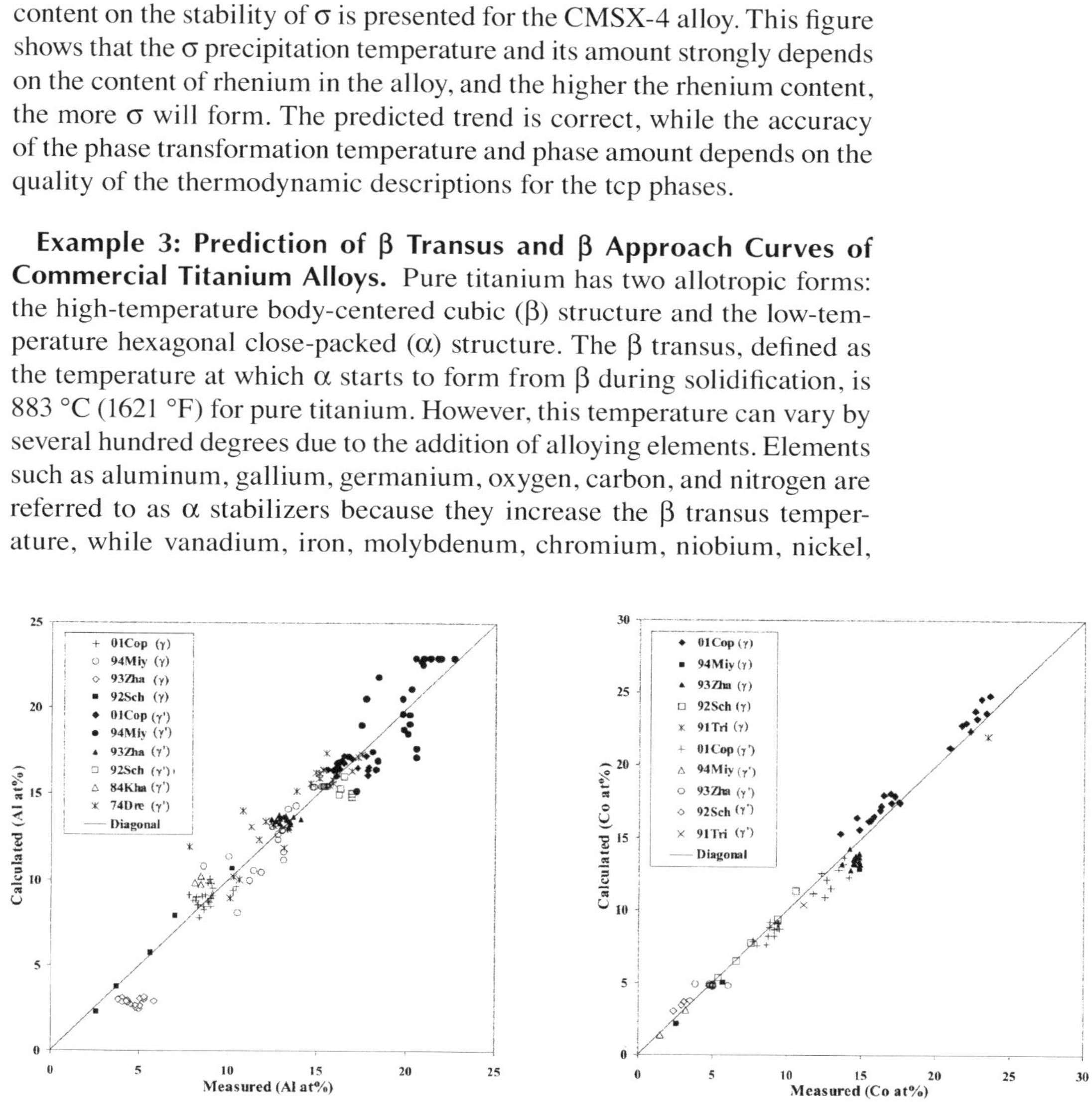

(a) Aluminum (b) Cobalt

Fig. 13.8 Comparison between the calculated and measured equilibrium compositions of a variety of elements in γ and γ'. Source: Ref 13.2

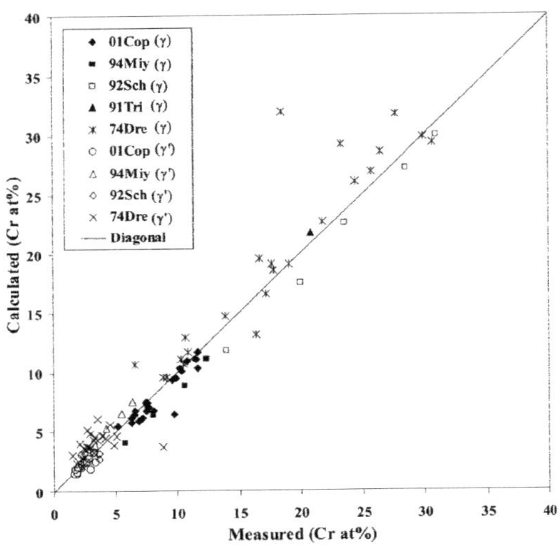

(c) Chromium

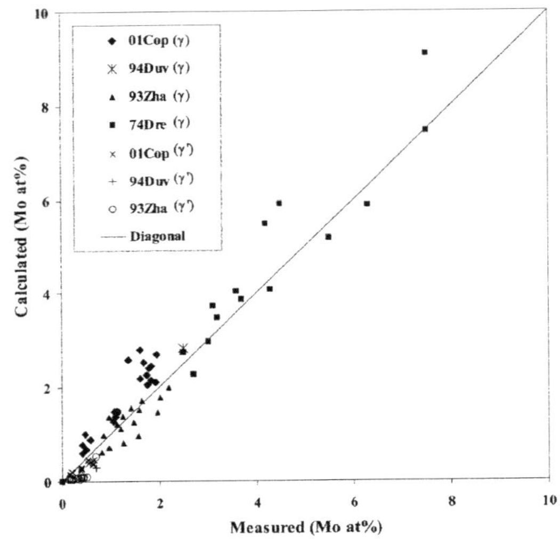

(d) Molybdenum

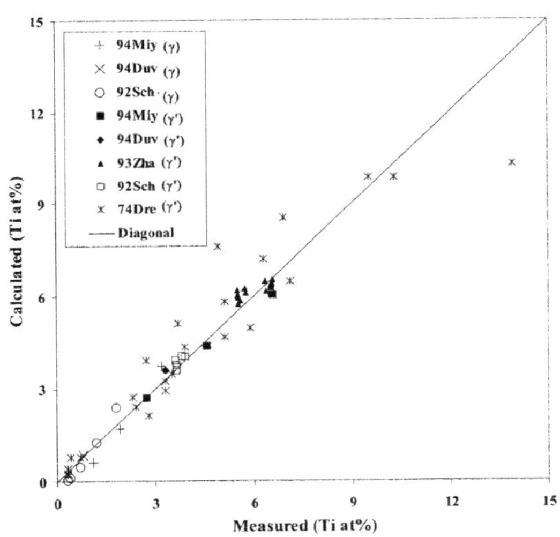

(e) Titanium

Fig. 13.8 (continued) Comparison between the calculated and measured equilibrium compositions of a variety of elements in γ and γ′. Source: Ref 13.2

copper, tantalum, and hydrogen are β stabilizers because they decrease the β transus. Technical multicomponent titanium alloys are classified as α, β, and α-β alloys, depending on the relative amounts of α and β stabilizers in the alloy. Within the last category are the subclasses near-α and near-β, referring to alloys whose compositions place them near the α/(α + β) or (α + β)/β phase boundaries, respectively.

According to different applications, alloy chemistry and heat treatment conditions must be carefully optimized to achieve the desired microstruc-

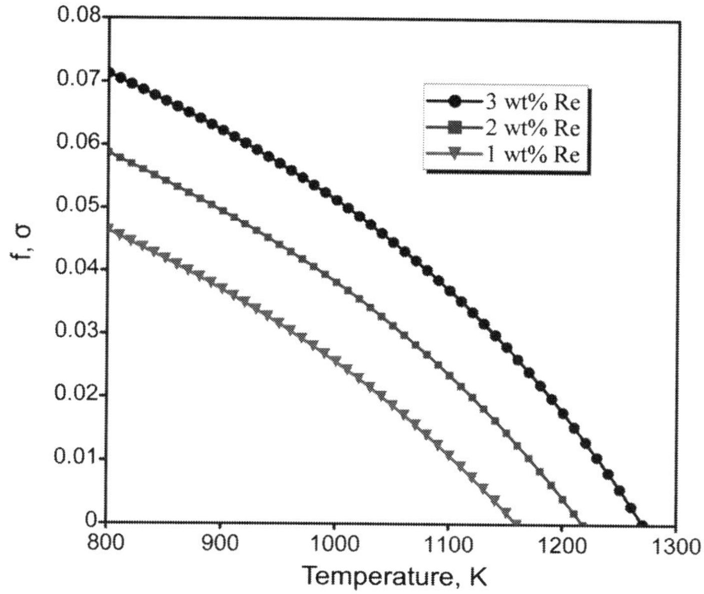

Fig. 13.9 Effect of rhenium content on the σ phase precipitation temperature and its amount at different temperatures, as calculated by the PanNi database. Source: Ref 13.6 as published in Ref 13.2

ture and mechanical properties. The β transus temperature is an important reference parameter used by metallurgical engineers in the selection of specified heat treatment conditions for titanium alloys. Current aerospace industrial practice calls for measurement of the β transus on every heat of titanium material, which is costly and time consuming. These operations and their associate costs can be significantly reduced with the help of a thermodynamic modeling tool, from which the β transus temperature can be easily calculated, given alloy chemistry.

In the following, examples are presented that apply thermodynamic calculations to two α-β alloys: Ti-6Al-4V (Ti-64) and Ti-6Al-2Sn-4Zr-6Mo (Ti-6246). It is noteworthy to point out that the preceding chemistries are the nominal compositions (wt%). For commercial alloys, the actual composition in each alloy varies. In addition, it is inevitable that commercial alloys always contain minute amounts of impurities, such as iron, silicon, carbon, oxygen, nitrogen, and hydrogen. Ti-64 is the most widely used titanium alloy in the world, with 80% of its usage going toward the aerospace industry. The β transus temperatures of Ti-64 range from 950 to 1050 °C (1740 to 1920 °F), depending on the amounts of components and interstitial elements, such as oxygen, carbon, nitrogen, and hydrogen. This temperature can be readily calculated by the CALPHAD approach if the thermodynamic database is available for multicomponent titanium alloys. In this study, PanTi (Ref 13.7) was used to calculate the β transus for a large number of Ti-64 heats with slightly different chemistries, and Fig.

13.10 compares the calculated and experimentally measured temperatures. Good agreement between the calculation and measurement was obtained, with an average difference of 5.2 °C. To understand the difference between the predicted values and the experimental data, a sensitivity analysis was carried out for Ti-64. It was found that the calculated β transus is very sensitive to the interstitial impurities, such as oxygen, carbon, nitrogen, and hydrogen (Table 13.1). In this table, the nominal chemistry of Ti-64 is given as Ti-6Al-4V-0.12O-0.01C-0.01N-0.005H (wt%). It is seen that 10 weight parts per million (wppm) of additional interstitial impurities (oxygen, carbon, nitrogen, hydrogen) will change the β transus of Ti-64

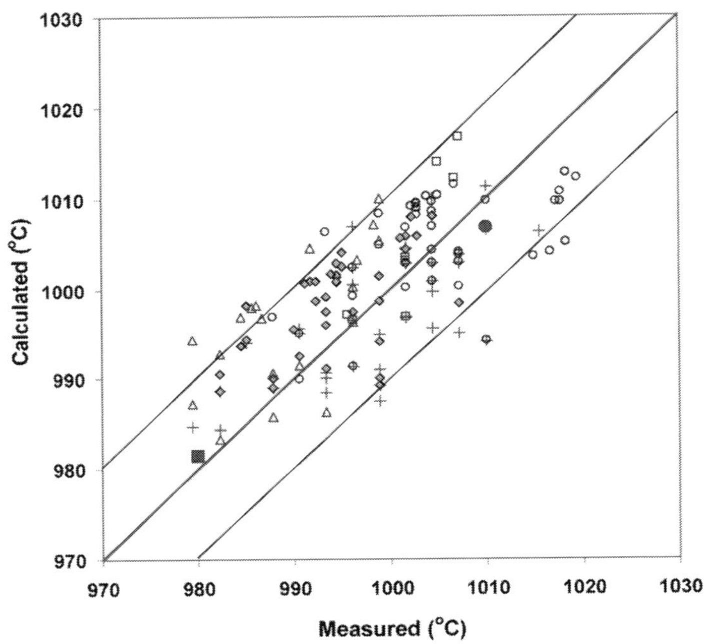

Fig. 13.10 Comparison of calculated and experimentally determined β transus for a number of Ti-64 alloys with slightly different chemistries. The calculation was carried out using the PanTi database and experimental data. Source: Ref 13.7 and 13.8 as published in Ref 13.2

Table 13.1 Effects of 10 wppm interstitial elements on the β transus of Ti-64

Alloy chemistry (wt%)	β transus (°C)	ΔT (°C) (compare to Ti-64)	ΔC (wppm) needs to change ΔT by 5 °C
Ti-64: Ti-6Al-4V-0.12O-0.01C-0.01N-0.005H	982.59		
Ti-64+10 wppm O	982.86	0.27	185.2
Ti-64+10 wppm C	983.08	0.49	102.0
Ti-64+10 wppm N	983.62	1.03	48.5
Ti-64+10 wppm H	981.51	−1.08	46.3

Source: Ref 13.2

alloy by 0.27, 0.49, 1.03, and –1.08 °C, respectively. The small variation in the composition of these interstitial impurities (measured in wppm) that changes the β transus of Ti-64 by 5 °C is listed in the last column.

As an α-β alloy, the relative amount of α and β in the microstructure plays a key role in the determination of the mechanical properties of the final product. With multicomponent phase equilibrium calculation, the volume fractions of α and β can be calculated as a function of temperature, which provides valuable information for the selection of proper heat treatment conditions to achieve the desired microstructure. Figure 13.11 shows such an example, in which the α approach curve, that is, the fraction of a phase as a function of temperature, is calculated and compared with the experimental data for one Ti-64 alloy. In addition to Ti-64, the thermodynamic modeling tool is also applied to other titanium alloys. Figure 13.12 shows the calculated β approach curve, that is, the fraction of β phase as a function of temperature of one Ti-6242 alloy, while Fig. 13.13 calculates the distribution of aluminum and molybdenum in α and β for the same alloy. Both calculations agree very well with the experimental data. Calculations shown in Fig. 13.11 to 13.13 provide valuable guidance in the optimization of processing conditions of titanium alloys, and the good agreement between calculation and experimentation suggests that the PanTi database can be used for commercial titanium alloys beyond Ti-64.

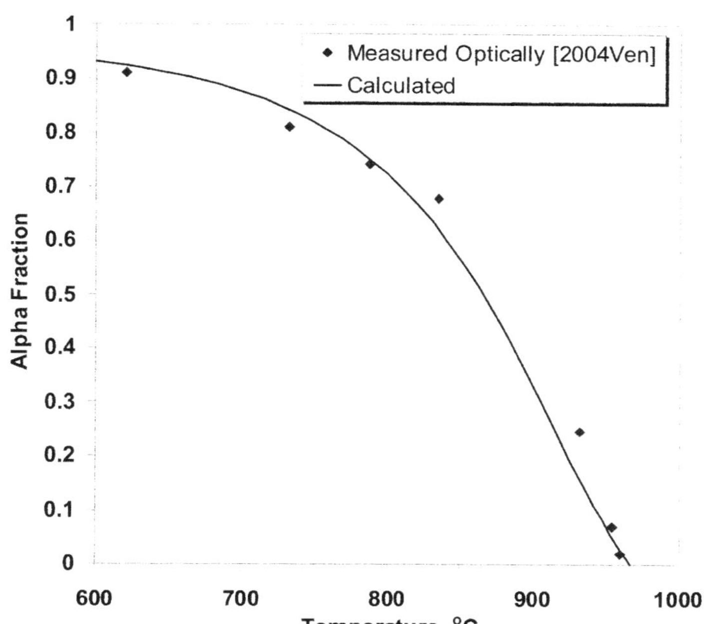

Fig. 13.11 Plot of the phase fraction of α as a function of temperature for one Ti-64 alloy. The line was calculated by PanTi and experimental data. Source: Ref 13.7 and 13.9 as published in Ref 13.2

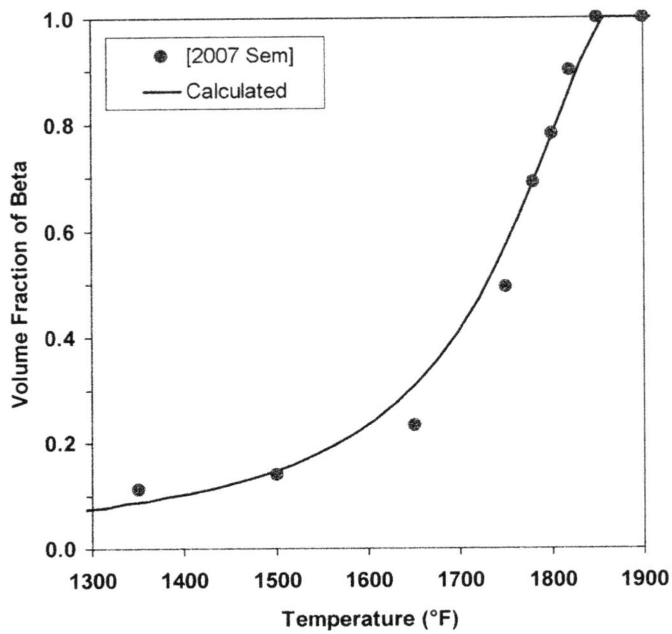

Fig. 13.12 Plot of the phase fraction of β as a function of temperature for one Ti-6242 alloy. The line was calculated by PanTi and experimental data. Source: Ref 13.7 and 13.10 as published in Ref 13.2

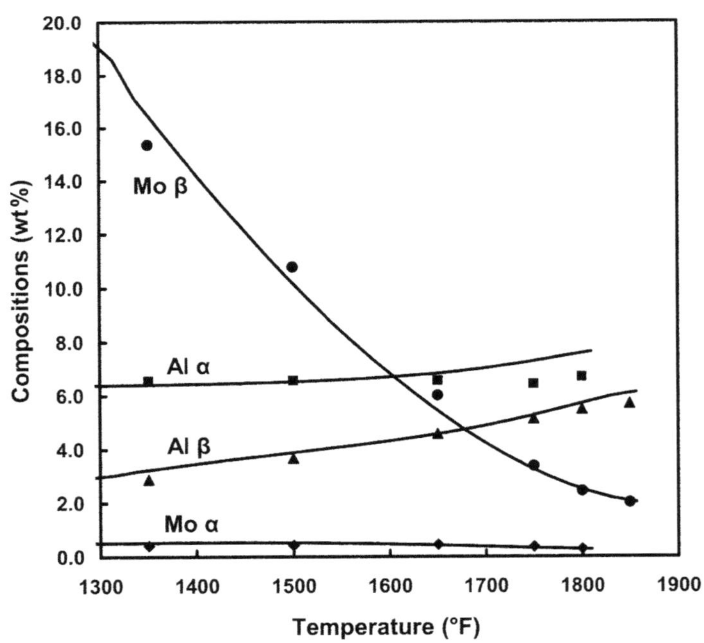

Fig. 13.13 Distribution of aluminum and molybdenum in the α and β phases for the same Ti-6242 alloy as in Fig. 13.12. The lines were calculated from PanTi and experimental data. Source: Ref 13.7 and 13.10 as published in Ref 13.2

Example 4: Calculation of Dilation during Phase Transformation of Austenite to Ferrite and Austenite to Martensite. Ferrous alloys that contain many alloy elements, such as cast iron and stainless steel, have been used worldwide as structural materials. Even though they are considered to be mature materials, there is a continuing demand to further improve their properties, with simultaneous cost reduction. Phase diagrams of this class of multicomponent materials provide the guide for processing optimization, leading ultimately to the improvement of their performance in service. For example, it is known that ferrous alloys undergo dimensional change, referred to as dilation, when subjected to heat treatment. The dilation could cause distortion and lead to cracking of a material component. The origin of this dilation is the result of phase transformation from austenite to ferrite as well as austenite to martensite during heat treatment. Experimental determination of dilation is not only tedious but also very challenging, even for an experienced experimentalist, because the accuracy of the measured data depends strongly on the cooling rate. This is particularly true when a multicomponent alloy is considered, because the phase transformation temperatures are usually unknown. On the contrary, the computational approach is more efficient and provides more consistent results. The following example demonstrates how computational thermodynamics can be used to predict the degree of dilation of a chosen ferrous alloy prior to processing this material.

The dilation can be calculated using:

$$\frac{\Delta l}{l} = \frac{V_i - V_A}{3V_A}$$

where l is the length, V_i is the instantaneous volume, and V_A is the volume of austenite at a specified temperature. The instantaneous volume (V_i) is calculated from the phase fraction and molar volume of each phase involved in the phase transformation:

$$V_i = \sum_m f_V^{\phi} \cdot V_m^{\phi}$$

where f_V^{ϕ} represents the volume fraction, and V_m^{ϕ} is the molar volume of the j phase. For the phase transformation from austenite to ferrite, the fractions of austenite and ferrite at a certain temperature and composition can be readily obtained by thermodynamic calculation if a reliable thermodynamic database is available for ferrous alloys. The PanFe database is suitable for such a purpose (Ref 13.11). As shown in Fig. 13.14, the calculated values of f_V for austenite are in good accordance with known experimental data for a variety of stainless alloys, an indication of the reliability of the PanFe database. These data, when combined with the molar volumes of these two phases, can then be used to calculate the dilation for

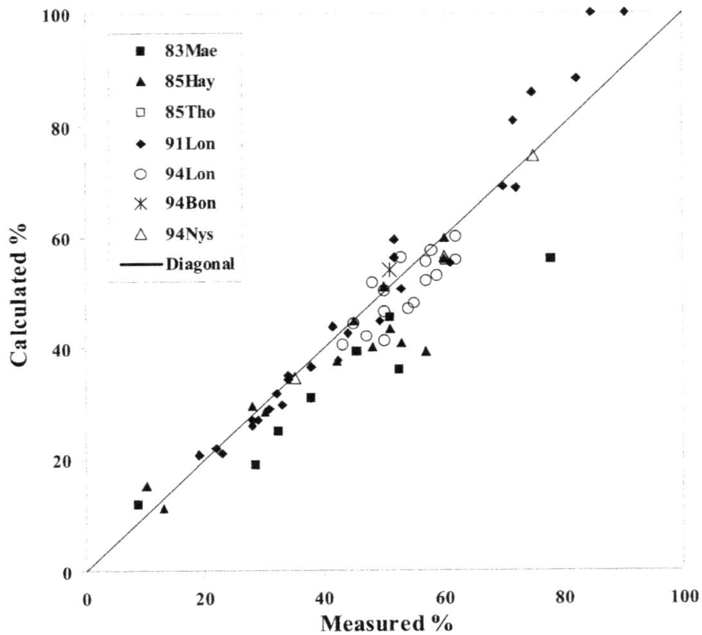

Fig. 13.14 Comparison between the calculated and measured amounts of austenite for a variety of stainless steels. Source: Ref 13.2

the phase transformation from austenite to ferrite. On the other hand, the calculation for the transformation from austenite to martensite is more challenging. Martensite, a supersaturated ferrite with the composition of the parent austenite, is a metastable phase. The total Gibbs energy change for transforming austenite to martensite is the sum of the bulk Gibbs energy difference due to the transformation from austenite to ferrite, the strain energy, the surface energy, and the interfacial work. The first part can be calculated easily because the Gibbs energies of the austenite and ferrite are well developed in the thermodynamic database, while the other three parts must be estimated by empirical equations. The martensite transformation temperature, the fraction of martensite, and the retained austenite are usually calculated from semiempirical and empirical equations. However, with the thermodynamic calculation, when the Gibbs energy for the martensite is formulated, the martensite transformation temperature for a given alloy and the fractions of martensite and austenite as a function of temperature can be calculated. These quantities, in combination with the molar volume of each phase, can be used to calculate the dilation due to the phase transformation from austenite to martensite. As an example, the dilation of the high-speed tool steel 6-6-2 is calculated and compared with experimental data (Fig. 13.15). Steel 6-6-2 is a multicomponent alloy with a composition of Fe-0.81C-0.24Mn-0.26Si-5.95W-4.10Cr-1.64V-4.69Mo (wt%). The dashed line and solid squares represent the calculated and

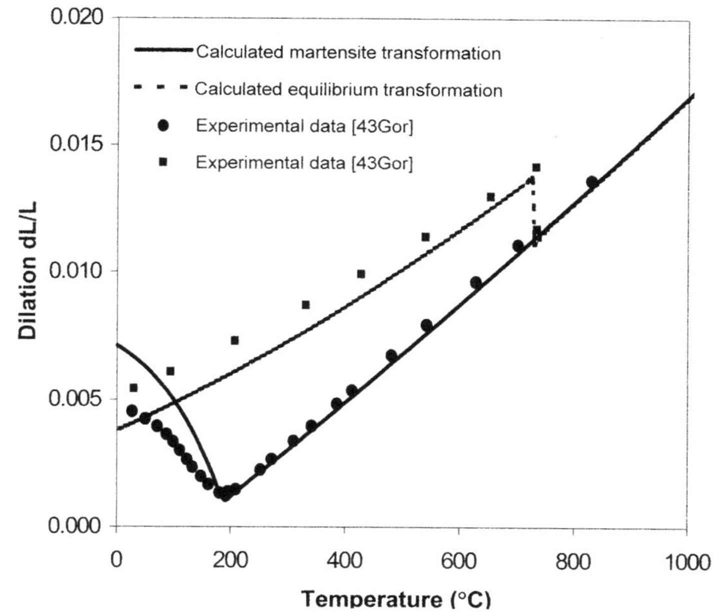

Fig. 13.15 Comparison between the calculated dilation versus temperature and the experimental data for high-speed steel 6-6-2. Source: Ref 13.12 as published in Ref 13.2

experimentally determined dilation for the transformation from austenite to ferrite, respectively, while the solid line and the solid circles are those for the transformation from austenite to martensite. By integrating the calculated dilation data with mechanical property prediction models, the strain, distortion, cracking, residual stress, and related properties can be predicted.

13.7 Limitations of the CALPHAD Approach

The essence of the CALPHAD method is that it provides an efficient approach to convert the experimental information of lower-order systems to self-consistent thermodynamic model parameters, on the basis of which a thermodynamic database of a multicomponent system can be established. Only a limited amount of experimental effort is needed to confirm the reliability of the database thus obtained before it can be used for industrial applications. There is no doubt of the potential power of the CALPHAD approach to save a tremendous amount of experimental work. However, like every other modeling approach, this approach has its limitations, especially when handling new phases that appear in the higher-order systems but not in the constituent lower-order systems.

Topologically closed-packed phases in the technologically important nickel-base superalloys are one example. These phases, including *P*, σ,

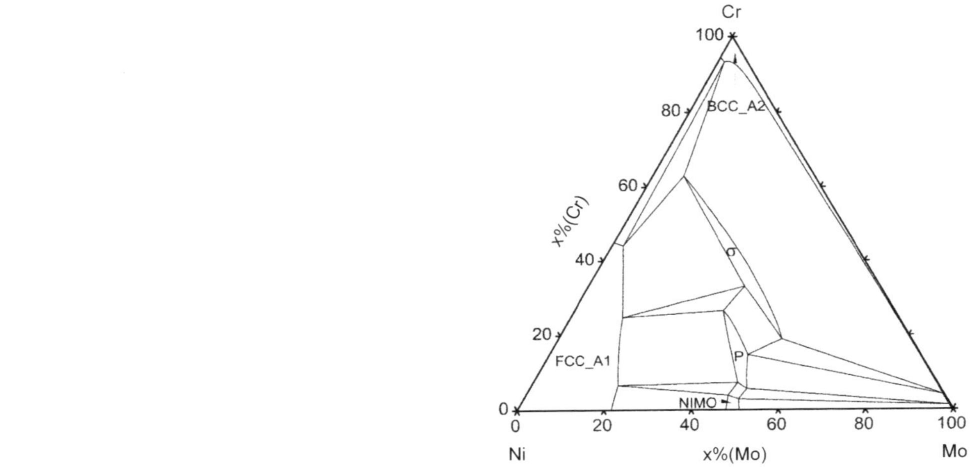

Fig. 13.16 Isothermal section of the Ni-Cr-Mo ternary system at 1000 °C (1832 °F). Source: Ref 13.2

μ, and *R*, are typically rich in refractory alloying elements and possess complex crystal structures characterized by close-packed layers of atoms. As discussed earlier, the tcp phases are detrimental because they deplete strengthening elements from the matrix phase and serve as crack-initiation sites during cyclic loading. If the stabilities of these tcp phases can be accurately predicted as a function of temperature-given alloy chemistry, formation of the deleterious tcp phases can be avoided by deliberately adjusting the alloy chemistry. Thermodynamic descriptions of the tcp phases in the current nickel database are based on the available but limited experimental information in the binaries and some ternaries where they are stable, while their Gibbs energies in other subsystems are estimated. One difficulty in the development of accurate thermodynamic descriptions for the tcp phases is that they may form in the middle of a ternary system similar to a new ternary phase while not stable in any of the three constituent binaries. One typical example is the Ni-Cr-Mo ternary system, in which σ and *P* are stable in the ternary but not any of the constituent binaries, as shown in Fig. 13.16. In such a situation, the experimentally well-established phase relationship in the ternary is used to optimize the model parameters in the three constituent binaries. In reality, however, the phase stabilities of these tcp phases are not known in many subsystems.

Gibbs energies thus developed for the multicomponent system should therefore be used with caution. First-principles-calculated enthalpy values may be used to improve the Gibbs energy descriptions of these phases, but it is not an easy task to obtain these values in every constituent subsystem due to the complicated structure of the tcp phases. It is also found that the stabilities of these tcp phases are comparable, which means a small difference in the Gibbs energy may lead to a totally different phase relationship. This makes the modeling of these tcp phases even more challenging. The status of the current nickel database is that it exhibits a lack of capability to accurately predict the stabilities of the tcp phases, but it does provide reasonable trends in most cases, such as the example shown in Fig. 13.9. Although more basic research is needed in this area for additional improvements of the thermodynamic descriptions of the topological phases, computational thermodynamics has been successfully used to obtain phase equilibria for practical applications.

ACKNOWLEDGMENTS

The material for this chapter is from "The Application of Thermodynamic and Material Property Modeling to Process Simulation of Industrial Alloys," by N. Saunders and Commercial Alloy Phase Diagrams and Their Industrial Applications by F. Zhang, Y. Yang, W.S. Cao, S.L. Chen, and K.S. Wu, and Y.A. Chang. Both articles are from *Metals Process Simulation*, Vol 22B, *ASM Handbook*, ASM International, 2010.

REFERENCES

13.1 N. Saunders, The Application of Thermodynamic and Material Property Modeling to Process Simulation of Industrial Alloys, *Metals Process Simulation*, Vol 22B, *ASM Handbook*, ASM International, 2010

13.2 F. Zhang, Y. Yang, W.S. Cao, S.L. Chen, K.S. Ku, Y.A. Chang, Commercial Alloy Phase Diagrams and Their Industrial Applications, *Metals Process Simulation*, Vol 22B, *ASM Handbook*, ASM International, 2010

13.3 X.H. Lin and W.L. Johnson, Formation of Ti-Zr-Cu-Ni Bulk Metallic Glasses, *J. Appl. Phys.*, Vol 78, 1995, p 6514

13.4 F. Ritzert, D. Keller, and V. Vasudevan, "Investigation of the Formation of Topologically Close Packed Phase Instabilities in Nickel-Base Superalloys, Rene N6," NASA/TM, 1999

13.5 R.L. Dreshfield and J.F. Wallace, The Gamma-Gamma Prime Region of the Ni-Al-Cr-Ti-W-Mo System at 850C, *Metall. Trans.*, Vol 5, 1974, p 71–78

13.6 PanNi—Thermodynamic Database for Multicomponent Nickel Alloys, CompuTherm, LLC, Madison, WI, 2000

13.7 PanTi—Thermodynamic Database for Multicomponent Titanium Alloys, CompuTherm, LLC, Madison, WI, 2000

13.8 D. Furrer, Ladish Co., Inc., Cudahy, WI, beta-transus of titanium alloys, private communication, 2003

13.9 V. Venkatesh, TIMET, Henderson, NV, titanium alloys, private communication, 2004

13.10 S.L. Semiatin, T.M. Lehner, J.D. Miller, R.D. Doherty, and D.U. Furrer, Alpha/Beta Heat Treatment of a Titanium Alloy with a Non-Uniform Microstructure, *Metall. Mater. Trans. A,* Vol 38, 2007, p 910

13.11 PanFe—Thermodynamic Database for Multicomponent Iron Alloys, CompuTherm, LLC, Madison, WI, 2000

13.12 P. Gordon, M. Cohen, and R.S. Rose, The Kinetics of Austenite Decomposition in High Speed Steel, *Trans. ASM,* Vol 3, 1943, p 161–216

Phase Diagrams—Understanding the Basics
F.C. Campbell, editor

CHAPTER 14

Phase Diagram Applications

ALLOY PHASE DIAGRAMS are useful to metallurgists, materials engineers, and materials scientists in four major areas: (1) development of new alloys for specific applications, (2) fabrication of these alloys into useful configurations, (3) design and control of heat treatment procedures for specific alloys that will produce the required mechanical, physical, and chemical properties, and (4) solving problems that arise with specific alloys in their performance in commercial applications, thus improving product predictability. In all these areas, the use of phase diagrams allows research, development, and production to be done more efficiently and cost effectively.

In the area of alloy development, phase diagrams have proved invaluable for tailoring existing alloys to avoid over design in current applications, designing improved alloys for existing and new applications, designing special alloys for special applications, and developing alternative alloys or alloys with substitute alloying elements to replace those containing scarce, expensive, hazardous, or "critical" alloying elements. Application of alloy phase diagrams in processing includes their use to select proper parameters for working ingots, blooms, and billets, finding causes and cures for microporosity and cracks in castings and welds, controlling solution heat treating to prevent damage caused by incipient melting, and developing new processing technology.

In the area of performance, phase diagrams give an indication of which phases are thermodynamically stable in an alloy and can be expected to be present over a long time when the part is subjected to a particular temperature (e.g., in an automotive exhaust system). Phase diagrams also are consulted when attacking service problems such as pitting and intergranular corrosion, hydrogen damage, and hot corrosion.

In a majority of the more widely used commercial alloys, the allowable composition range encompasses only a small portion of the relevant phase diagram. The nonequilibrium conditions that are usually encountered in practice, however, necessitate the knowledge of a much greater portion of the diagram. Therefore, a thorough understanding of alloy phase diagrams in general and their practical use will prove to be of great help to a metallurgist expected to solve problems in any of the areas mentioned.

While some of these uses have previously been discussed, phase diagrams are used to:

- *Predict the temperature at which freezing or melting begins or ends for any specific alloy composition in an alloy system.* A vertical line represents the composition of a specific alloy, for example, point X in Fig. 14.1. Its intersections with the solidus (T_2) and liquidus (T_1) indicate, respectively, the temperature below which, at equilibrium, the alloy is completely solid, and above which it is completely liquid. On heating, melting begins at T_2, and the alloy is completely liquid above temperature T_1. If an alloy is to be cast, then the temperature of the molten alloy has to be higher than T_1. In order to fill the mold completely before freezing blocks of any thin section, the alloy should be at least 50 °C (28 °F) higher than its liquidus temperature, T_1.
- *Predict the safe temperature for hot working or heat treatments.* The temperature of hot working, or heat treatment, should be at least 30 °C (17 °F) lower than its solidus temperature to allow for any impurities present and for furnace temperature fluctuations. Heating the alloy above temperature T_2 causes partial melting, called burning of the alloy. The "sweat out" molten metal leaves behind voids whose interior surfaces become oxidized at elevated temperature. Because a burnt

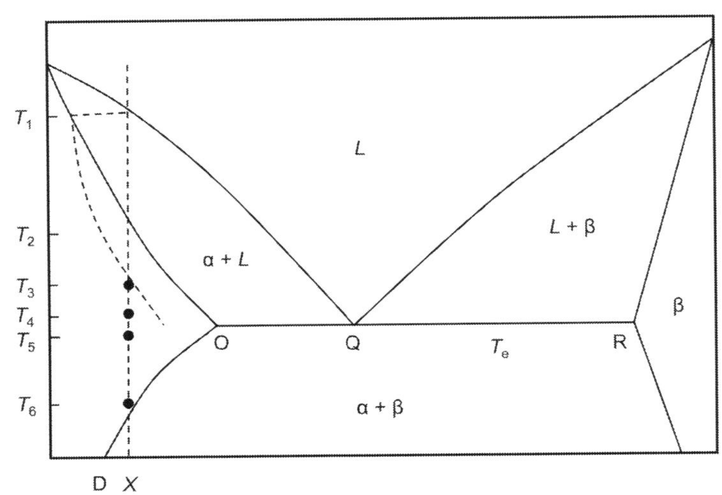

Fig. 14.1 Eutectic phase diagram with partial solid solubility

alloy cannot be repaired by welding during hot working, a burnt alloy is essentially scrap. If the alloy contains coring, then a homogenizing temperature above T_3 will also cause "burning" of the alloy. A safer temperature is T_4. During precipitation hardening, if the solutionizing temperature T_5 is chosen instead of T_6, grain growth of α phase will occur at the higher temperature.

- *Determine the number of phases, type of phases, and composition of phases present in any given alloy at a specific temperature.* One of the primary functions of an equilibrium phase diagram is to graphically show the extent and boundaries of composition-temperature regions within which an alloy exists as a single phase or as two phases. Thus, the fields of the diagram are labeled so that the number and the general nature of the phases present are indicated at a specific temperature.

  If the point with coordinates at a specific composition and a particular temperature lies in a single-phase field, then the alloy is either an unsaturated homogeneous liquid or a solid solution with the composition of the alloy. A point in a two-phase field indicates that both the phases are saturated solutions, which could be liquids, solids, or a liquid and a solid. If the composition of the alloy is changed at the same temperature but still lying within the two-phase field, the number, type, and composition of the phases do not change, but their relative amounts change.

  A horizontal line in a binary phase diagram indicates a particular temperature and a range of alloy compositions at which three phases can coexist at equilibrium. The horizontal line separates either a two-phase field from some other two-phase field that has only one phase in common with it, or a two-phase field from a one-phase field that is different from both of these two phases.
- *Calculate the relative amounts of the phases present in a two-phase alloy.* The lever rule can be used to calculate the amounts of the two phases present in a two-phase field.
- *Describe the freezing or melting of an alloy.* Cooling of an alloy from the molten state to room temperature can be observed with the help of an equilibrium diagram. During slow heating of the alloy, the changes would be exactly reversed.
- *Predict the microstructure of an alloy at a given temperature.* A major advantage of phase diagrams is to make fairly accurate predictions of the microstructure developed in an alloy at a specific temperature or after an actual or proposed heat treatment. This is important because the microstructure controls the properties of an alloy. For example, the structure of the pure element that is freezing is quite important. Zinc has hexagonal close-packed crystal structure and is strongly anisotropic. When zinc crystals are growing in liquid zinc, growth tends to be more along one or another of its close-packed directions. Acicular (needle-shaped) crystals develop. However, when zinc is part of a

eutectic reaction, it freezes as a finely distributed phase with the other phase in the eutectic mixture.

- *Predict possible heat treatments.* The presence of a solvus line in a phase diagram, which shows decrease of solid solubility (line OD in Fig. 14.1) with the decrease of temperature, indicates the chance of using a precipitation-hardening heat treatment. The presence of a eutectoid reaction in a phase diagram helps to predict possible heat treatments such as annealing, normalizing, or hardening. It is also possible to predict what heat treatments are likely to be harmful and must be avoided. For example, if an alloy is not going to be cold worked, and if no phase transformation occurs during heating (or cooling), then heating such an alloy to high temperatures can result in grain growth and inferior properties.
- *Choose the alloy composition to develop the best properties.* The composition of the alloy that gives the best properties can be chosen. An alloy having maximum solute content indicated by the solubility limit by the solvus line may develop maximum strength by precipitation hardening if it develops a coherent precipitate. If the solute content is more (or less) than optimal, then the maximum strength will be less. The castability of an alloy system is usually best at the eutectic composition.

14.1 Industrial Applications of Phase Diagrams

The following are but a few of the many instances where phase diagrams and phase relationships have proved invaluable in the efficient solving of practical metallurgical problems. The areas covered include alloy design, processing, and performance.

14.1.1 Alloy Design

Four examples of the application of phase diagrams to alloy design are given: the development of a basis for age-hardening aluminum alloys, material substitution in two types of wrought stainless steel alloys to reduce costs, and an improvement in the manufacturing process for Fe-Nd-B-base magnets.

Age-Hardening Alloys. One of the earliest uses of phase diagrams in alloy development was at the suggestion in 1919 by the U.S. Bureau of Standards that precipitation of a second phase from solid solution would harden an alloy. The age hardening of certain aluminum-copper alloys (then called "Duralumin" alloys) had been accidentally discovered in 1904, but this process was thought to be a unique and curious phenomenon. The work at the Bureau, however, showed the scientific basis of this process. This work led to the development of several families of commercial "age-hardening" alloys covering different base metals.

Austenitic Stainless Steel. In connection with a research project aimed at the conservation of always expensive, sometimes scarce, materials, the question arose: Can manganese and aluminum be substituted for nickel and chromium in stainless steels? In other words, can standard chromium-nickel stainless steels be replaced with an austenitic alloy system? The answer came in two stages—in both instances with the help of phase diagrams. It was first determined that manganese should be capable of replacing nickel because it stabilizes the γ-iron phase (austenite), and aluminum may substitute for chromium because it stabilizes the α-iron phase (ferrite), leaving only a small γ loop (see Fig. 14.2 and 14.3).

Aluminum is known to impart good high-temperature oxidation resistance to iron. Next, the literature on phase diagrams of the Al-Fe-Mn system was reviewed, which suggested that a range of compositions exists where the alloy would be austenitic at room temperature. A nonmagnetic alloy with austenitic structure containing 44% Fe, 45% Mn, and 11% Al was prepared. However, it proved to be very brittle, presumably because of the precipitation of a phase based on β-manganese. By examining the phase diagram for C-Fe-Mn (Fig. 14.4), as well as the diagram for Al-C-Fe, the researcher determined that the problem could be solved through the addition of carbon to the Al-Fe-Mn system, which would move the composition away from the β-manganese phase field. The carbon addition also would further stabilize the austenite phase, permitting reduced manganese content. With this information, the composition of the alloy was modified to 7 to 10% Al, 30 to 35% Mn, and 0.75 to 1% C, with the balance iron. It had good mechanical properties, oxidation resistance, and moderate stainlessness.

Permanent Magnets. A problem with permanent magnets based on Fe-Nd-B is that they show high magnetization and coercivity at room temperature but unfavorable properties at higher temperatures. Because hard magnetic properties are limited by nucleation of severed magnetic domains, the surface and interfaces of grains in the sintered and heat treated material are the controlling factor. Therefore, the effects of alloying additives on the phase diagrams and microstructural development of the Fe-Nd-B alloy system plus additives were studied. These studies showed that the phase relationships and domain-nucleation difficulties were very unfavorable for the production of a magnet with good magnetic properties at elevated temperatures by the sintering method. However, such a magnet might be produced from Fe-Nd-C material by some other process, such as melt spinning or bonding.

14.1.2 Processing

Two examples of the application of phase diagrams to alloy design are discussed: alloy additions to a hacksaw blade steel to allow the production

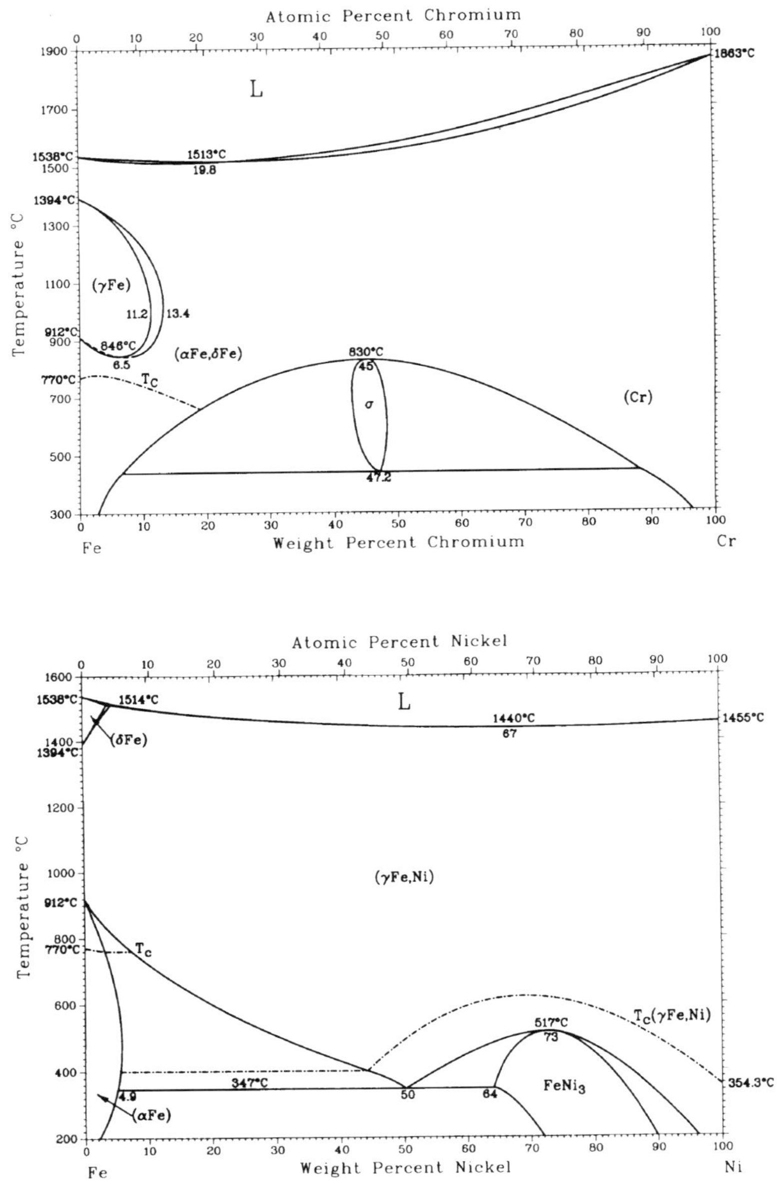

Fig. 14.2 Two binary iron phase diagrams, showing ferrite stabilization (iron-chromium) and austenite stabilization (iron-nickel). Source: Ref 14.1 as published in Ref 14.2

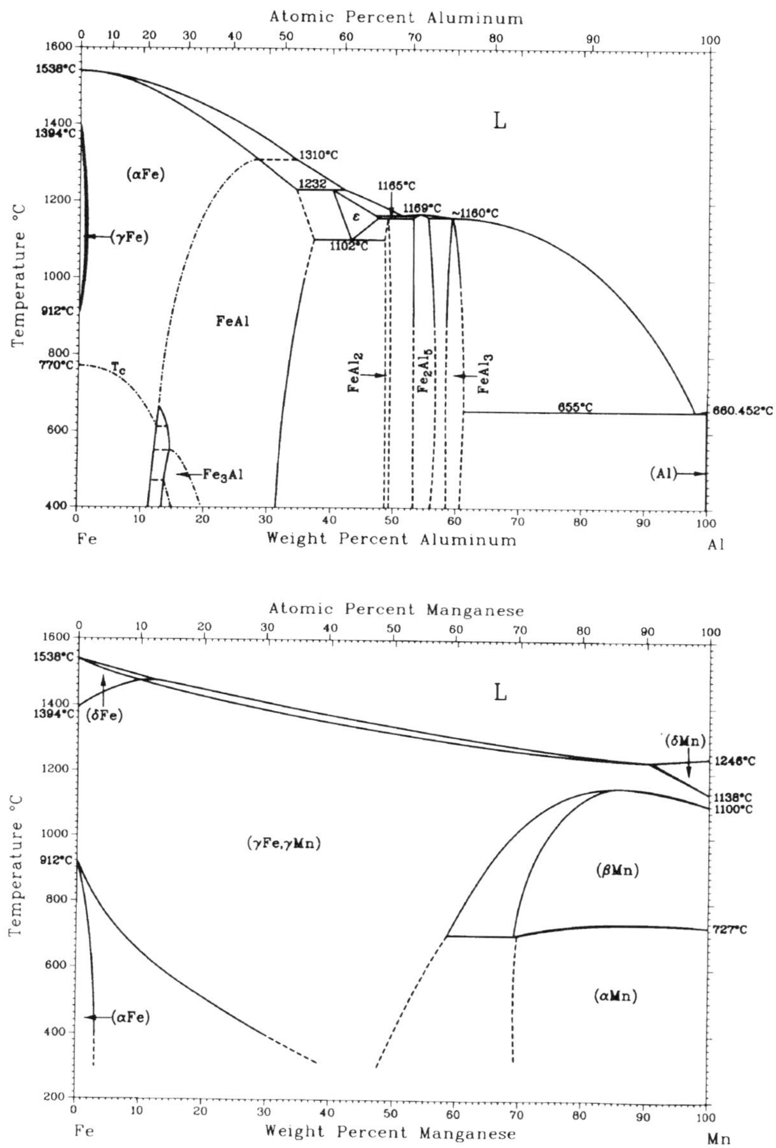

Fig. 14.3 The aluminum-iron and iron-manganese phase diagrams. Source: Ref 14.3 as published in Ref 14.2

of more cost-effective blades, and alloy additions to a hardfacing alloy that produced superior properties.

Hacksaw Blades. In the production of hacksaw blades, a strip of high-speed steel for the cutting edges is joined to a backing strip of low-alloy

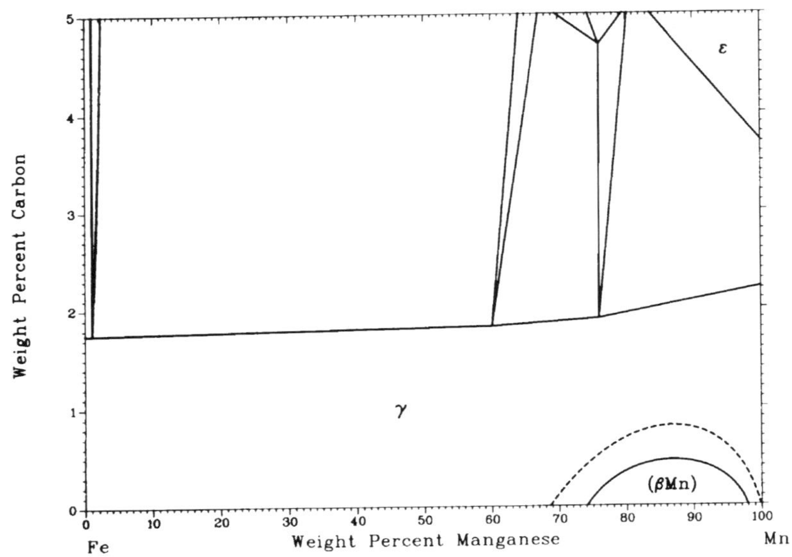

Fig. 14.4 The isothermal section at 1100 °C (2012 °F) of the Fe-Mn-C phase diagram. Source: Ref 14.4 as published in Ref 14.2

steel by laser or electron beam welding. As a result, a very hard martensitic structure forms in the weld area that must be softened by heat treatment before the composite strip can be further rolled or set. To avoid the cost of the heat treatment, an alternative technique was investigated. This technique involved alloy additions during welding to create a microstructure that would not require subsequent heat treatment. Instead of expensive experiments, several mathematical simulations were made based on additions of various steels or pure metals. In these simulations, the hardness of the weld was determined by combining calculations of the equilibrium phase diagrams and available information to calculate (assuming the average composition of the weld) the martensite transformation temperatures and amounts of retained austenite, untransformed ferrite, and carbides formed in the postweld microstructure. Of those alloy additions considered, chromium was found to be the most efficient.

Hardfacing. A phase diagram was used to design a nickel-base hardfacing alloy for corrosion and wear resistance. For corrosion resistance, a matrix of at least 15% Cr was desired; for abrasion resistance, a minimum amount of primary chromium-boride particles was desired. After consulting the B-Cr-Ni phase diagram, a series of samples having acceptable amounts of total chromium borides and chromium matrix were made and tested. Subsequent fine tuning of the composition to ensure fabricability of welding rods, weldability, and the desired combination of corrosion, abrasion, and impact resistance led to a patented alloy.

14.1.3 Performance

Four examples of the application of phase diagrams to performance are listed: the elimination of sulfur contamination from Nichrome heating elements, the elimination of lead and bismuth contaminants from extruded aluminum electric motor housings, a deficiency in the amount of carbon in sintered tungsten-carbide cutting tools, and a problem in which components were failing where the gold lead wires were fused to aluminized transistor and integrated circuits.

Heating elements made of Nichrome (a Ni-Cr-Fe alloy registered by Driver-Harris Company, Inc., Harrison, NJ) in a heat-treating furnace were failing prematurely. Reference to nickel-base phase diagrams suggested that low-melting eutectics can be produced by very small quantities of the chalcogens (sulfur, selenium, or tellurium), and it was thought that one of these eutectics could be causing the problem. Investigation of the furnace system resulted in the discovery that the tubes conveying protective atmosphere to the furnace were made of sulfur-cured rubber, which could result in liquid metal being formed at temperatures as low as 637 °C (1179 °F), as shown in Fig. 14.5. With this information, a metallurgist solved the problem by substituting neoprene for the rubber.

Electric Motor Housings. At moderately high service temperatures, cracks developed in electric motor housings that had been extruded from aluminum produced from a combination of recycled and virgin metal. Extensive studies revealed that the cracking was caused by small amounts

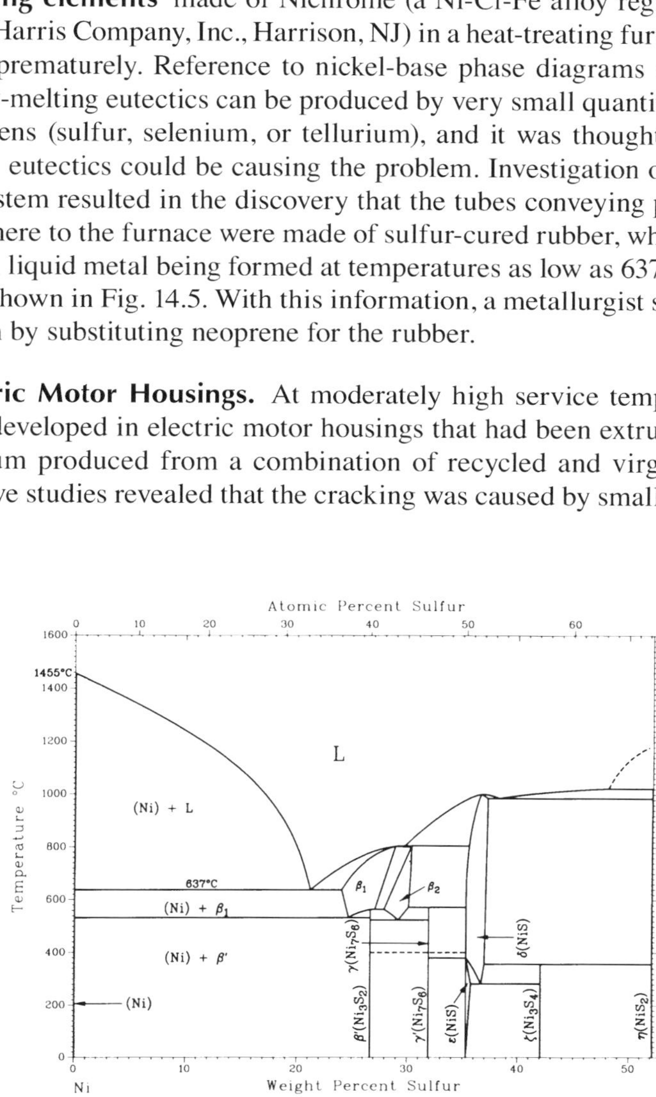

Fig. 14.5 The nickel-sulfur phase diagram. Source: Ref 14.1 as published in Ref 14.2

of lead and bismuth in the recycled metal reacting to form bismuth-lead eutectic at the grain boundaries at 327 and ~270 °C (621 and ~518 °F), respectively, much below the melting point of pure aluminum (660.45 °C, or 1220.81 °F) (Fig. 14.6). The question became: How much lead and bismuth can be tolerated in this instance? The phase diagrams showed that

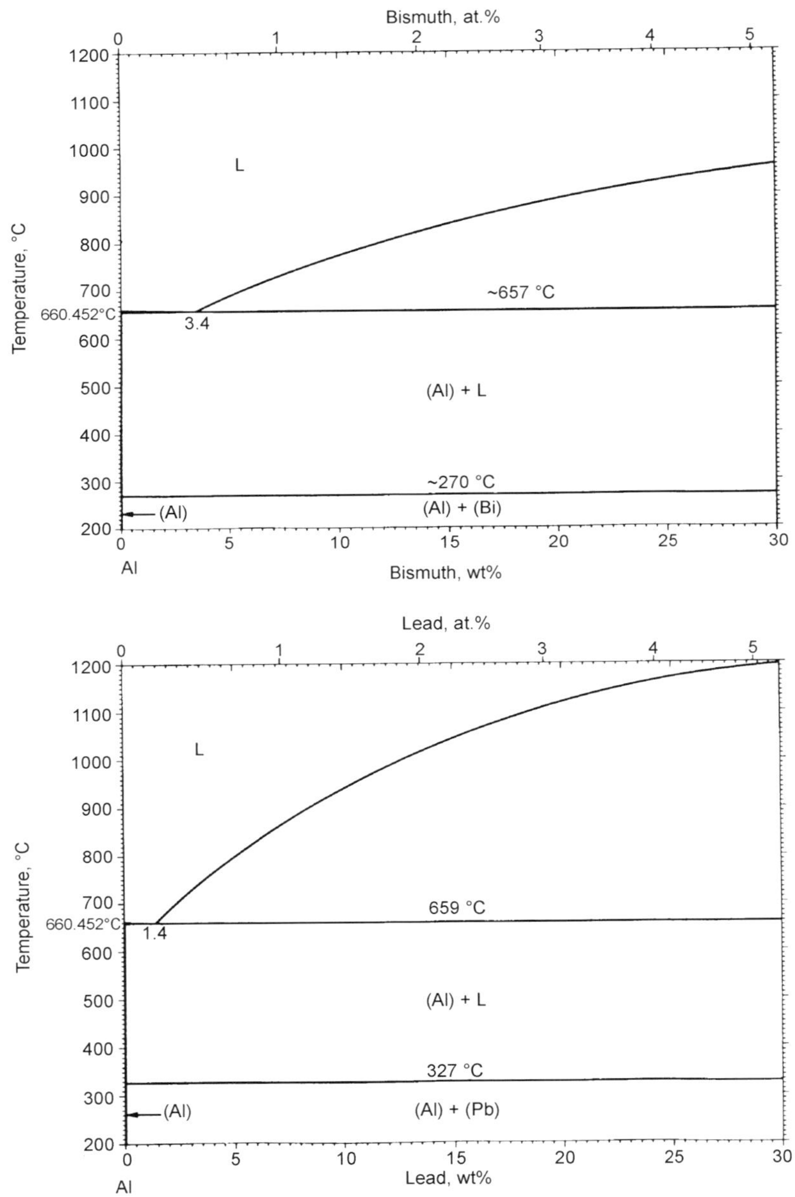

Fig. 14.6 The aluminum-bismuth and aluminum-lead phase diagrams. Source: Ref 14.1 as published in Ref 14.2

aluminum alloys containing either lead or bismuth in amounts exceeding their respective solubility limits (<0.05% and ~0.2%) can lead to hot cracking of the aluminum.

Carbide Cutting Tools. A manufacturer of carbide cutting tools once experienced serious trouble with brittleness of the sintered carbide. No impurities were found. The range of compositions for cobalt-bonded sintered carbides is shown in the shaded area of the ternary phase diagram in Fig. 14.7, along the dashed line connecting pure tungsten carbide (WC) on the right and pure cobalt at the lower left. At 1400 °C (2552 °F), materials with these compositions consist of particles of tungsten carbide suspended in liquid metal. However, when there is a deficiency of carbon, compositions drop into the region labeled WC + η + liquid, or the region labeled WC + η where tungsten carbide particles are surrounded by a matrix of η phase. The η phase is known to be brittle. The upward adjustment of the carbon content by only a few hundredths of a weight percent eliminated this problem.

Solid-State Electronics. In the early stages of the solid-state industry, a phenomenon known as the "purple plague" nearly destroyed the fledgling industry. Components were failing where the gold lead wires were fused to aluminized transistor and integrated circuits. A purple residue was formed, which was thought to be a product of corrosion. Actually, what was happening was the formation of an intermetallic compound, an aluminum-gold precipitate (Al_2Au) that is purple in color and very brittle. Millions of actual and opportunity dollars were lost in identifying the

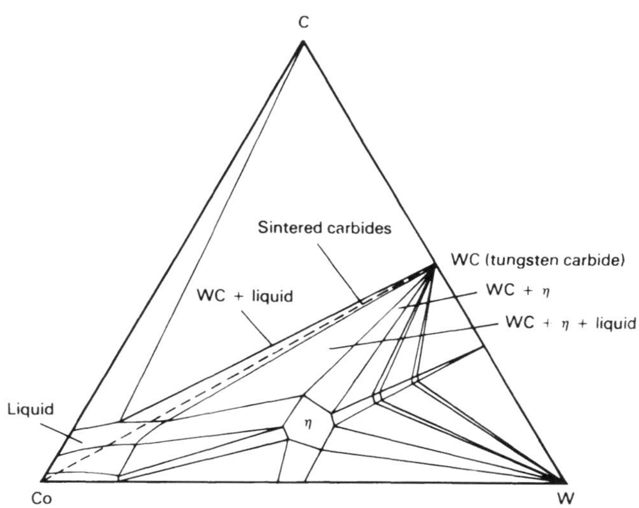

Fig. 14.7 The isothermal section at 1400 °C (2552 °F) of the Co-W-C phase diagram. Source: Ref 14.5 as published in Ref 14.2

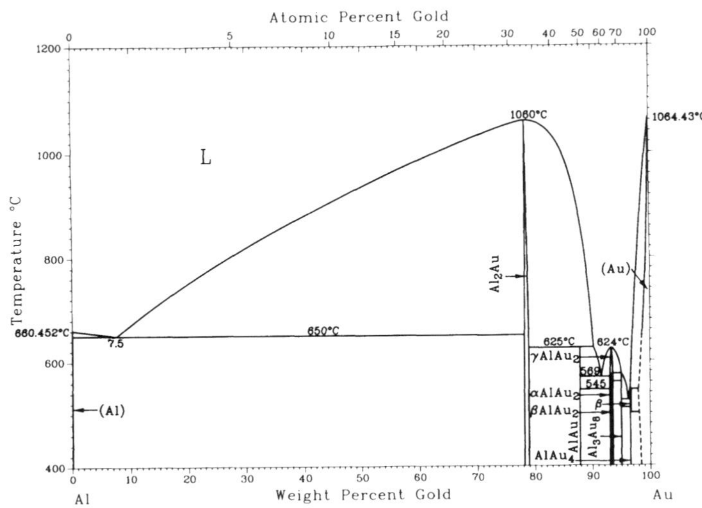

Fig. 14.8 The aluminum-gold phase diagram. Source: Ref 14.6 as published in Ref 14.2

problem and its solution, which could have been avoided had the proper phase diagram been examined (Fig. 14.8).

A question concerning purple plague problems, however, has remained unresolved: whether or not the presence of silicon near the gold-aluminum interface has an influence on the stability and rate of formation of the damaging intermetallic phase. An examination of the phase relationships in the Al-Al_2Au-Si subternary system showed no stable ternary Al-Au-Si phases. It was suggested instead that the reported effect of silicon may be due to a reaction between silicon and alumina (Al_2O_3) at the aluminum-gold interface that becomes thermodynamically feasible in the presence of gold.

14.2 Limitations of Phase Diagrams

Phase diagrams play an extremely useful role in the interpretation of the microstructures developed in alloys, but they have several limitations:

- Phase diagrams only show the equilibrium state of alloys (i.e., under very slow cooling rates); however, in normal industrial processes, alloys are rarely cooled slowly enough to approach equilibrium.
- Phase diagrams do not indicate whether a high-temperature phase can be retained at room temperature by rapid cooling.
- Phase diagrams do not indicate whether a particular transformation (e.g., a eutectoid transformation) can be suppressed, and what should be the rate of cooling of the alloy to avoid the transformation.
- Phase diagrams do not indicate the phases produced by fast cooling rates. For example, the formation of martensite is not shown in the

Fe-Fe_3C phase diagram. Thus, they do not indicate the temperature of the start of such transformations (e.g., the M_s temperature) and their kinetics of formation.

- Even under equilibrium conditions, phase diagrams do not indicate the character of the transformations. They do not indicate the rate at which the equilibrium will be attained.
- Phase diagrams only give information on the constitution of alloys, such as the number of phases present at a point, but do not give information about the structural distribution of the phases; that is, they do not indicate the size, shape, or distribution of the phases, which affects final mechanical properties. The structural distribution of phases are affected by the surface energy between phases and the strain energy produced by the transformation. For example, if the β phase, in a mixture of α and β, is present in small amounts and is totally distributed with the α grains, the mechanical properties will largely be governed by the α phase. However, if β is present around the grain boundaries of α, then the strength and ductility of the alloy is largely dictated by properties of the β phase.

Additional examples of applications of phase diagrams developed by thermodynamics modeling and computer simulation are given in Chapter 13, "Computer Simulation of Phase Diagrams," in this book.

ACKNOWLEDGMENT

The material in this chapter came from Introduction to "Alloy Phase Diagrams," by H. Baker in *Alloy Phase Diagrams*, Vol 3, *ASM Handbook*, 1992, reprinted in *Desk Handbook: Phase Diagrams for Binary Alloys*, 2nd ed., H. Okamoto, Ed., 2010.

REFERENCES

14.1 T.B. Massalski, Ed., *Binary Alloy. Phase Diagrams,* 2nd ed., ASM International, 1990

14.2 H. Baker, Introduction to Alloy Phase Diagrams, *Alloy Phase Diagrams*, Vol 3, *ASM Handbook*, ASM International, 1992, reprinted in *Desk Handbook: Phase Diagrams for Binary Alloys*, 2nd ed., H. Okamoto, Ed., ASM International, 2010

14.3 H. Okamoto, *Phase Diagrams of Binary Iron Alloys,* ASM International, 1992

14.4 R. Benz, J.F. Elliott, and J. Chipman, *Metall. Trans.*, Vol 4, 1973, p 1449

14.5 P. Rautala and J.T. Norton, *Trans. AIME,* Vol 194, 1952, p 1047

14.6 H. Okamoto, Ed., Binary Alloy Phase Diagrams Updating Service, ASM International, 1992

CHAPTER 15

Nonequilibrium Reactions—Martensitic and Bainitic Structures

WHILE MANY OF THE transformations discussed so far occur under equilibrium or near-equilibrium conditions, martensitic and bainitic structures are produced by nonequilibrium cooling. In both cases, the alloy is heated to a sufficiently high temperature and then suddenly cooled (quenched) to a much lower temperature. Both martensitic and bainitic transformations are extremely important strengthening mechanisms for iron-carbon base-alloys.

In this chapter, the topics covered include martensitic structures in steels, martensitic structures in nonferrous alloys, shape-memory alloys, and bainitic structures in steels.

15.1 Nonequilibrium Cooling—TTT Diagrams

Equilibrium cooling rates might be approximated by a very large piece of steel undergoing very slow furnace cooling. Under normal industrial processing conditions, the cooling rates are almost always much faster, and new phases that are not on the equilibrium phase diagram can appear. When the cooling rate is extremely rapid, such as when quenching steel from temperatures in the austenite phase field (>740 °C, or 1364 °F) into cold water, the iron-carbon phase diagram can no longer be used because quenching is such a radical departure from equilibrium. At rapid cooling rates, it is necessary to use a time-temperature-transformation (TTT) diagram. These diagrams are so named because they depict Transformations as a function of Time and Temperature.

To construct a TTT diagram, very small samples, approximately the size of a dime, are first heated into the single-phase austenite field; this is also referred to as austenitization. Samples are then withdrawn from the furnace and immediately quenched into salt baths maintained at a given intermediate temperature. The specimen is held in the salt bath for a precise time and then quenched in a water bath maintained at room temperature. This procedure is shown in Fig. 15.1. At each intermediate temperature, the sample is held for a precise period of time, as shown in the example in Fig. 15.2. After quenching, the samples are examined by metallography to determine the microstructure. In the example, note that the amount of pearlite in the microstructure increases as the length of time in the intermediate salt bath increases. After a series of samples has been evaluated at one temperature, other temperatures are evaluated, for example, 700, 600, 500, 400, 300 °C (1200, 1100, 1000, 900, 800 °F) and so on down toward room temperature. Once the data is collected, it is used to plot the extent of transformation

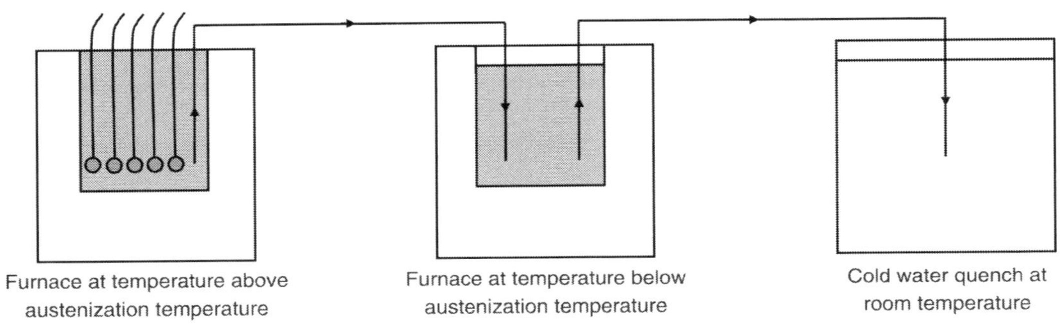

Fig. 15.1 Method for determining isothermal transformations. Source: Ref 15.1 as published in Ref 15.2

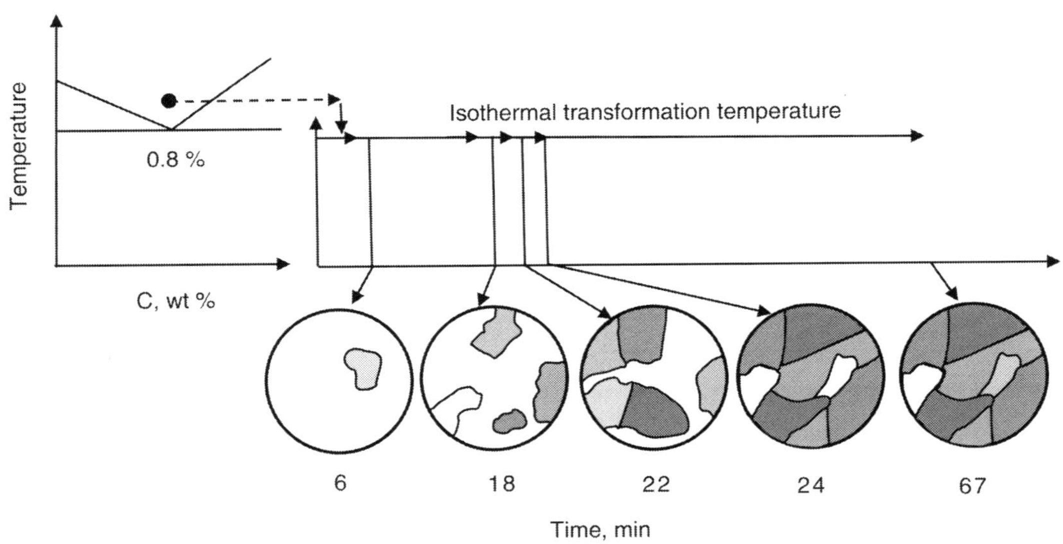

Fig. 15.2 Isothermal transformation of plain carbon steel. Source: Ref 15.1 as published in Ref 15.2

product in the microstructure on a temperature-time plot in the manner shown in Fig. 15.3. A fully developed TTT curve with the transformation products marked on the diagram is shown for the eutectoid plain-carbon steel 1080 in Fig. 15.4, which nominally contains 0.8 wt% C.

A TTT diagram is essentially a visual map that allows one to determine which constituents form as a function of temperature and time. For example, a sample rapidly cooled from the austenitic phase field to a temperature of 600 °C (1110 °F) will begin to transform to pearlite if held at the intermediate temperature for approximately 2 to 3 seconds. On further holding at 600 °C (1110 °F), it will completely transform to pearlite in approximately 10 s. In this particular steel, pearlite forms at all temperatures along the start-of-transformation curve from the A_1 temperature (725 °C, or 1340 °F) to the nose of the diagram (540 °C, or 1000 °F). At the higher transformation temperatures, the pearlite interlamellar spacing is very coarse. The interlamellar spacing is increasingly fine as the temperature decreases toward the nose of the diagram. This is significant because

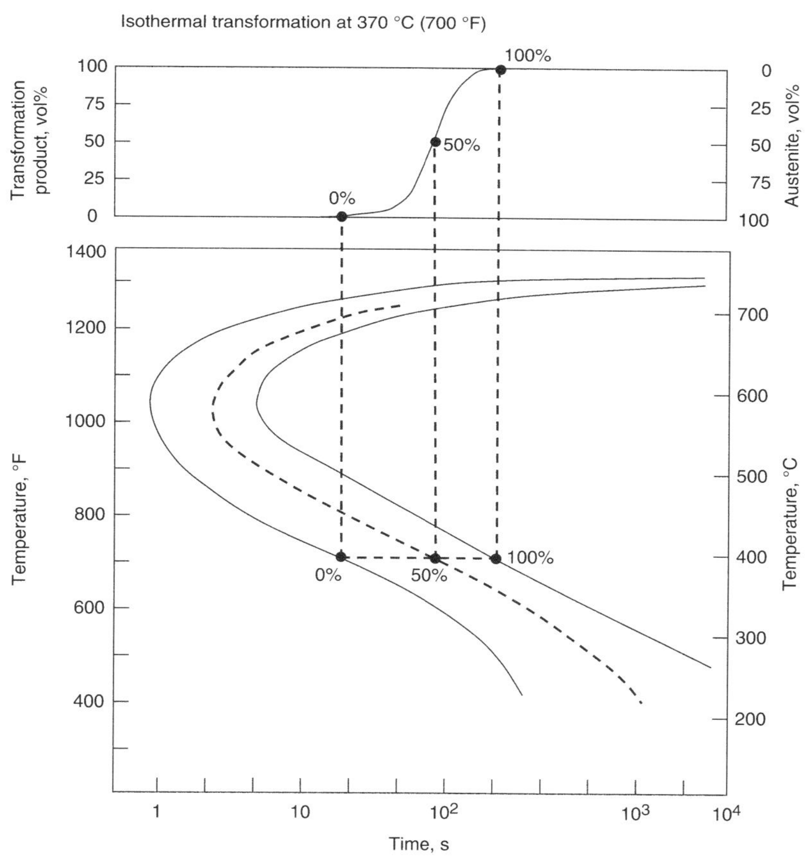

Fig. 15.3 Isothermal transformation curve development. Source: Ref 15.2

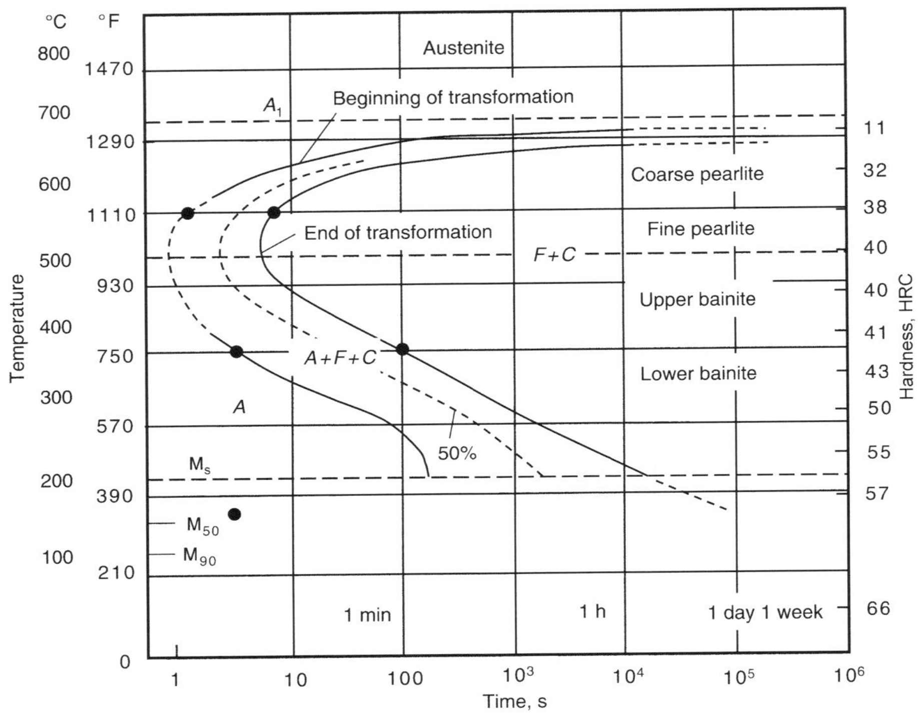

Fig. 15.4 Isothermal transformation diagram for 1080 steel. Source: Ref 15.2

a steel with a coarse pearlite interlamellar spacing is softer and of lower strength than a steel with a fine pearlite interlamellar spacing.

At small amounts of undercooling below the A_1 isotherm, the transformation to pearlite takes a relatively long time to nucleate. As the degree of undercooling increases, the incubation time decreases rapidly, even though the transformation is occurring at lower temperatures where diffusion is less rapid. Eventually, the incubation time reaches a minimum and then begins to increase again. This minimum incubation time occurs at the nose of the TTT diagram. The shape of the TTT diagram is a result of changing reaction kinetics. At high temperatures, the driving force for reaction is lower due to less undercooling, but the diffusion rates are faster, while at lower temperatures, the driving force for reaction is increasing due to greater degrees of undercooling, while the diffusion rates are decreasing due to lower temperatures.

If the steel sample is quenched to 400 °C (750 °F) and held for various times, pearlite does not form; instead, a totally new constituent called bainite forms. Like pearlite, bainite consists of ferrite and cementite, but the morphology is different. Also, note that there are two types of bainite shown on the diagram: upper bainite, which forms at higher temperatures, and lower bainite, which forms at lower temperatures.

Finally, if the specimens are quenched in a salt bath to 150 °C (300 °F), another new constituent, called martensite, forms. Martensite is quite different from both pearlite and bainite. It is not a mixture of ferrite and cementite. Instead, martensite is a form of ferrite that is supersaturated with carbon. Because of the very fast cooling rate, the carbon atoms do not have time to diffuse from their interstitial positions in the body-centered cubic (bcc) lattice to form cementite particles. Because martensite forms suddenly, there are no start-of-transformation or finish-of-transformation curves, as for pearlite and bainite. Instead, there are M_s (martensite start) and M_f (martensite finish) temperatures.

The hardness of the various transformation products is indicated on the right axis of the diagram and shows that the hardness increases with decreasing transformation temperature. Steel products produced with an as-quenched martensitic microstructure are very hard and brittle. Therefore, most martensitic products are tempered by heating to temperatures between approximately 345 and 650 °C (650 and 1200 °F). The tempering process allows some of the carbon to diffuse from the supersaturated iron lattice and form a carbide phase, softening the steel and restoring some ductility. The degree of softening is determined by the tempering temperature and the time at the tempering temperature. The higher the temperature and the longer the time of tempering, the softer the steel becomes. Thus, most steels with martensite are used in the quenched and tempered condition.

Martensitic Structures. As shown previously in the TTT diagram, martensite is a metastable structure that forms during athermal (nonisothermal) conditions. Unlike isothermal decomposition of phase constituents that approach equilibrium conditions by diffusion-controlled mechanisms, martensite does not appear on equilibrium phase diagrams. The mechanism of martensitic transformation is a diffusionless process, where rapid changes in temperature cause shear displacement of atoms and individual atomic movements of less than one interatomic spacing. The transformation also depends on the temperature: martensite begins to form at a M_s temperature, and additional transformation ceases when the material reaches a M_f temperature. The M_s and M_f temperatures depend on the alloying elements in the metal.

In general, martensitic transformation can occur in many types of metallic and nonmetallic crystals, minerals, and compounds, if the cooling or heating rate is sufficiently rapid. The most important example is martensite in steel, when the more densely packed austenite (face-centered cubic, or fcc) phase transforms to the less densely packed crystal structures of either bcc ferrite or body-centered tetragonal (bct) martensite (Fig. 15.5). When steel is slowly cooled from the austenite phase, the crystal structure (size) transforms to the less densely packed ferrite phase. At faster cooling rates, the formation of ferrite is suppressed, while formation of martensite

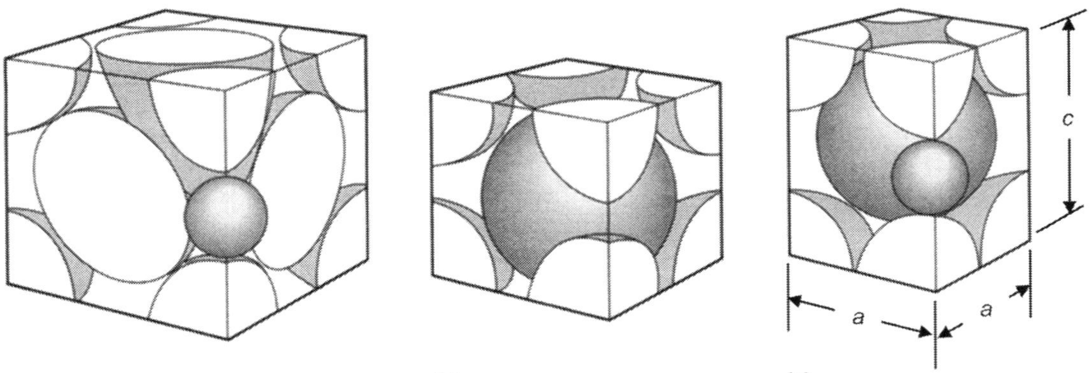

Fig. 15.5 Crystal structures. (a) Austenite (face-centered cubic, fcc). (b) Ferrite (body-centered cubic, bcc). (c) Martensite (body-centered tetragonal, bct). Source: Ref 15.3

is enhanced by the shear displacement of iron atoms into an interstitial, supersaturated solid solution of iron and carbon. This metastable state has bct structure, which is even less densely packed than austenite. This results in lattice distortion that provides strength/hardness by impeding dislocation movements and a volumetric expansion at the M_s temperature. During cooling, when the steel reaches the M_f temperature, the martensitic transformation ceases and any remaining austenite (γ) is referred to as retained austenite.

15.2 Martensite in Steels

Martensite is formed in steels when they are heated into the austenitic field and then rapidly quenched to room temperature. The martensite transformation differs from other phase transformations in several ways: (1) the reaction is a diffusionless phase transformation, (2) no chemical composition change occurs between martensite and the parent phase, (3) there is a coordinated crystal lattice structural change ("military mode") during transformation, (4) the movement of atoms in the crystal is no more than the distance between two atoms in the crystal lattice, and (5) there are certain crystallographic relationships between martensite and the parent phase. A shear mechanism controls this transformation, wherein a large number of atoms move cooperatively and almost at the same time, as opposed to the atom-by-atom movement that occurs in diffusional transformations. This shear action produces two important characteristics of the martensite transformation: orientation relationships between parent and product phases and surface tilting around the martensite crystal. Due to the absence of diffusion in the transformation, the composition of the parent austenite and product martensite are the same. Additionally, the martensite transformation is athermal—no thermal activation energy is associated with the transformation. While still a nucleation and growth mechanism, martensite forms through the diffusionless transformation of austenite. In general, a martensitic reaction can be defined as a mechanism for forming a new crystallographic structure that does not require atomic diffusion.

To form martensite, the cooling rate must be great enough to avoid the nose of the TTT diagram. Martensite starts forming when it is cooled below the M_s temperature and is completely transformed when it reaches the M_f temperature. Above the M_s temperature, martensite does not form, even at high cooling rates. The transformation only occurs at temperatures between the M_s and M_f. Because of matrix constraints, the width of the martensitic units is limited, and the transformation proceeds primarily by the successive nucleation of new crystals. This process occurs only on cooling to lower temperatures and is therefore independent of time. The transformation ceases when the steel falls below the M_f temperature. Some remaining austenite may still be present when the transformation stops, and it is referred to as retained austenite.

The martensitic transformation is effectively athermal because it occurs during cooling and begins when the sample is cooled below a particular temperature; the amount of the new phase that is formed forms depends on the temperature to which the sample is cooled rather than the time it spends at that temperature. If cooling is done quickly enough, structural rearrangement of atoms occurs by shear displacement over a small distance, on the order of approximately an interatomic spacing. This shear movement of atoms and phase growth occurs rapidly. It is not an instantaneous transformation, but it is quicker and distinct from the slower diffusion-controlled process in isothermal transformations.

The M_s temperature can be raised by the application of stress during transformation. This occurs because a crystal that forms during a martensitic transformation has a different shape and volume than that of the austenite from which it is forming. Therefore, if there is an applied stress available to achieve this shape change, the transformation will be accomplished more easily. If it can occur more easily, then it will require less thermodynamic driving force for it to occur, which means that the necessary degree of undercooling will be smaller. Thus, if an applied stress is locally oriented so that it helps the transformation to produce the shape change, martensite will be able to form at a higher temperature than for the unstressed condition. In general, tensile stresses are found to be more effective than compressive stresses, but rolling can have appreciable effects. The highest temperature to which the M_s temperature can be raised by applied stresses is defined as the M_d temperature. When the M_d temperature is above room temperature and the M_s is below room temperature, it is possible to retain the austenite at room temperature and then to form some martensite by working the metastable austenite at room temperature.

As-quenched martensite has a very high strength but a very low fracture resistance or toughness. Therefore, almost all steels that are quenched to martensite are also tempered, or reheated to some temperature below A_1 in order to increase toughness. In tempered martensite, the carbide consists of small, chunky particles rather than platelets. These finely divided particles of cementite in tempered martensite provide strengthening and restore ductility, depending on the specific tempering temperature. In general, higher tempering temperatures result in lower strength but greater ductility.

The hardness of heat treated martensite is determined by its carbon content, as shown in Fig. 15.6. Martensite attains a maximum hardness of HRC 66 at carbon contents of 0.8 to 1.0 wt%. The reason that the hardness does not monotonically increase with carbon is that retained austenite starts forming when the carbon content is above approximately 0.4 wt%, and retained austenite is much softer than martensite.

Formation of Martensite in Steels. In a martensitic transformation process, large numbers of atoms experience cooperative movements with only a slight displacement of each atom relative to its neighbors. The fcc austenite transforms to a bct structure. This structure is essentially the same as the ferrite bcc structure, except that it has been distorted into a tetragonal structure due to the entrapment of carbon that did not have time to diffuse out and form cementite. Because the martensitic transformation

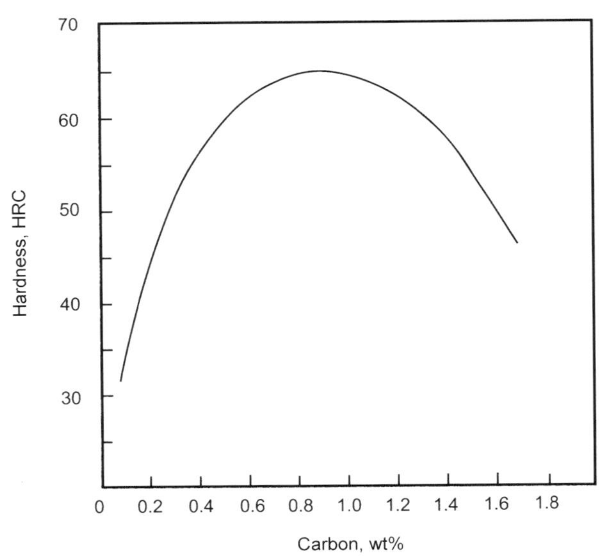

Fig. 15.6 Hardness versus carbon content for quenched steels. Source: Ref 15.4 as published in Ref 15.2

does not involve diffusion, it occurs almost instantaneously, approaching the speed of sound within the austenite matrix. All the carbon atoms remain as interstitial impurities in martensite, and as such, they constitute a supersaturated solid solution that is capable of rapidly transforming to other structures if heated to temperatures at which diffusion rates become appreciable. Because the bct crystal structure is less densely packed than the fcc structure of austenite, the transformation results in a volumetric expansion and a hardening of steel. As shown in Fig. 15.7, the *c*/*a* ratio of the tetragonal unit cell increases as the carbon content increases. To a good approximation, the variations in *c* and *a* are linear, with the *c* value increasing at a greater rate with increasing carbon contents than the rate at which the *a* parameter decreases.

Two adjacent fcc unit cells of austenite are shown in Fig. 15.8(a), in which a bct unit cell has been identified. The atoms identified in Fig. 15.8(a) have been isolated in the left-hand portion of Fig. 15.8(b). At this stage, the dimensions of the bct cell are still those derived from the austenite lattice parameter. The unit cell on the right-hand portion of Fig. 15.8(b) is that of martensite with lattice parameters *a* and *c*, corresponding to a given carbon content shown in Fig. 15.7. Note that a lattice deformation is required to produce martensite from austenite. This deformation was first identified by Bain and is referred to as the "Bain strain." As previously shown in Fig. 15.7, the Bain strain produces an expansion along the *c* axis and a contraction along the *a* axis.

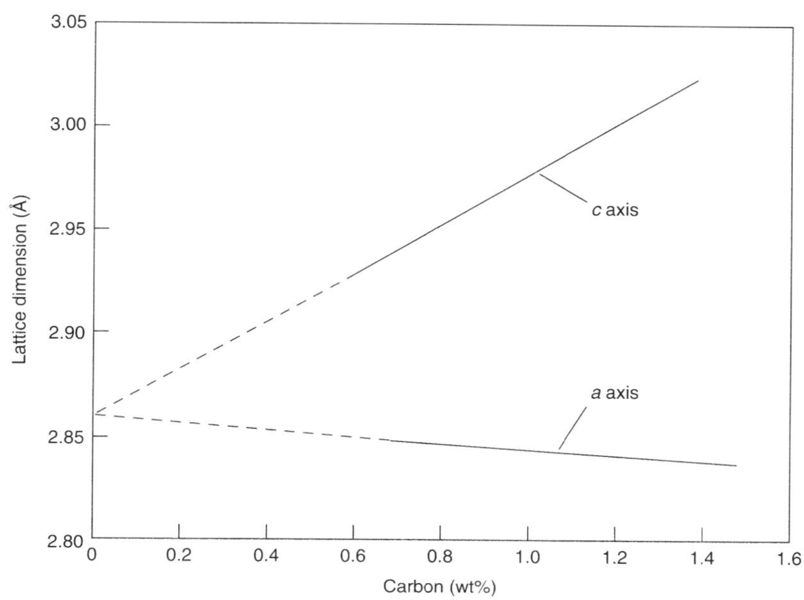

Fig. 15.7 Variation of *c*/*a* parameters with carbon content. Source: Ref 15.2

The formation of a martensitic crystal is shown schematically in Fig. 15.9. Shearing occurs parallel to a fixed crystallographic plane, termed the habit plane, and produces a uniformly tilted surface relief on a free surface. This mechanism involves shear displacement of iron atoms on specific planes, namely the {111} planes. In addition to the lattice distortion caused by the formation of the bct structure, the resulting martensite is simultaneously deformed because of the constraints created by maintaining an unrotated and undistorted habit plane within the bulk austenite. The deformation of martensite is referred to as a lattice invariant deformation,

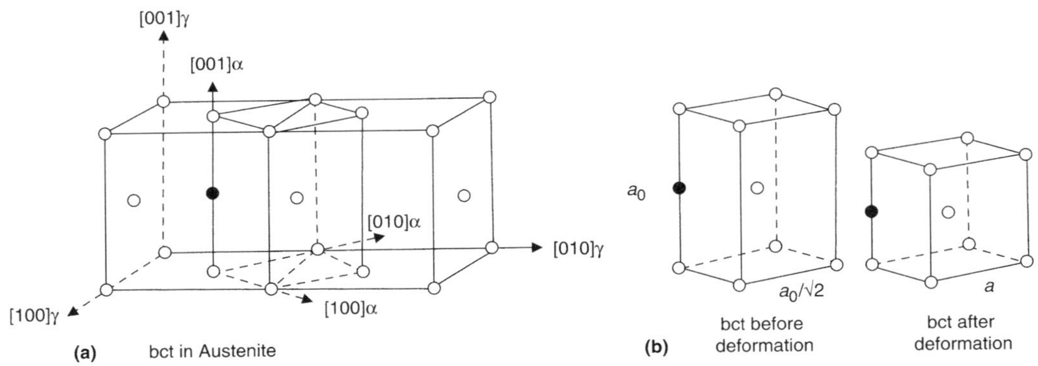

Fig. 15.8 Transformation from austenite to martensite. bct, body-centered tetragonal. Source: Ref 15.2

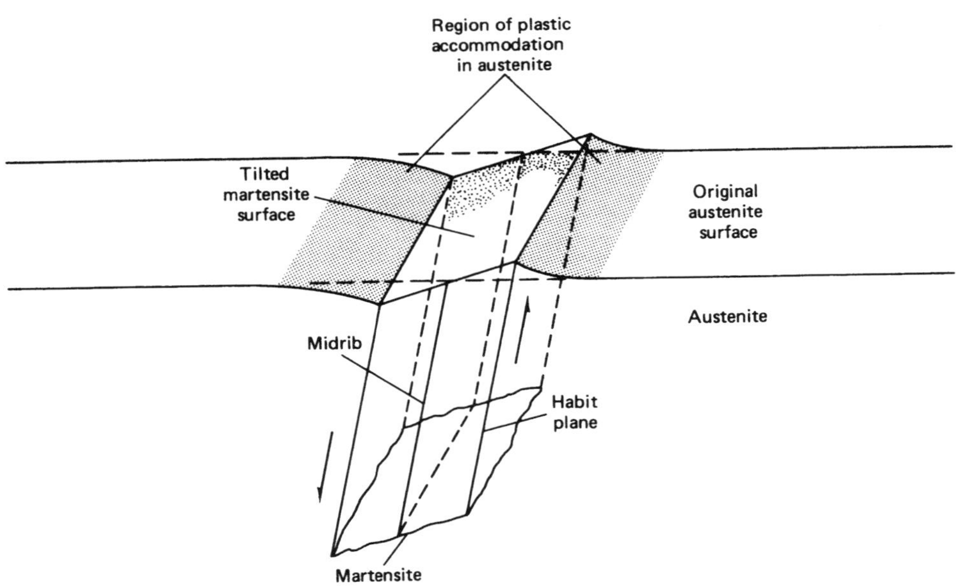

Fig. 15.9 Shear and surface tilting during martensite formation. Source: Ref 15.5 as published in Ref 15.2

and it produces a high density of dislocations or twins in martensite. The result is a bct crystal structure that is less densely packed than the fcc structure of austenite. Hence, the martensitic transformation results in a volumetric expansion and a hardening of steel. This fine structure, together with the carbon atoms trapped within the octahedral interstitial sites of the bct structure, produces the very high strength of as-quenched martensite.

Morphology of Martensite. In steels, there are two distinctly different microstructures that are found, depending on the carbon content of the steel (Fig. 15.10). For alloys containing less than approximately 0.6 wt% C, the martensite grains form as laths (i.e., long and thin plates) that form side by side and are aligned parallel to one another. Furthermore, these laths are grouped into larger structural entities, called blocks. The morphology of this lath or massive martensite is shown in Fig. 15.11. Lenticular, or plate, martensite is typically found in alloys containing greater than approximately 0.6 wt% C. With this structure, the martensite grains take on a needlelike (i.e., lenticular) or platelike appearance, which grows across the complete austenite grain shown in Fig. 15.12. Generally, plate martensite can be distinguished from lath martensite by its plate morphology with a central midrib. As the carbon content increases, twins begin to replace dislocations within the plates so that high-carbon martensite is composed mainly of twinned plates. The transformation is also associated with an appreciable volume increase, because it replaces the close-packed fcc structure with the more loosely packed bct structure. This transformation creates residual stresses that are related to the specific volume change, in addition to the strains due to the misfit of the interstitial solute atoms.

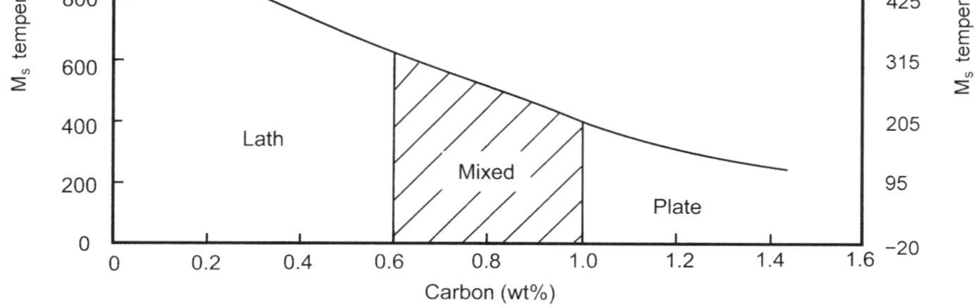

Fig. 15.10 Martensite morphology versus carbon content. Source: Ref 15.5 as published in Ref 15.2

At high carbon levels, these stresses can become so severe that the material cracks as the martensite forms. These cracks can range from small microcracks that require microscopy to be detected, to large cracks easily visible to the unaided eye. The cracks form during transformation when a growing plate impinges on an existing plate. Because of these microcracks, plate martensite is generally avoided in most applications. With carbon contents between 0.6 and 1.0 wt%, the martensite is a mixture of lath and plate morphologies.

Retained Austenite. Austenite starts transforming to martensite when it reaches the M_s temperature and continues to transform until the M_f temperature is reached. When the transformation ceases, some remaining austenite may still be present. When retained austenite is present, it is usually as small island grains surrounded by martensite (Fig. 15.13). Because martensite expands when it is formed, the remaining austenite is

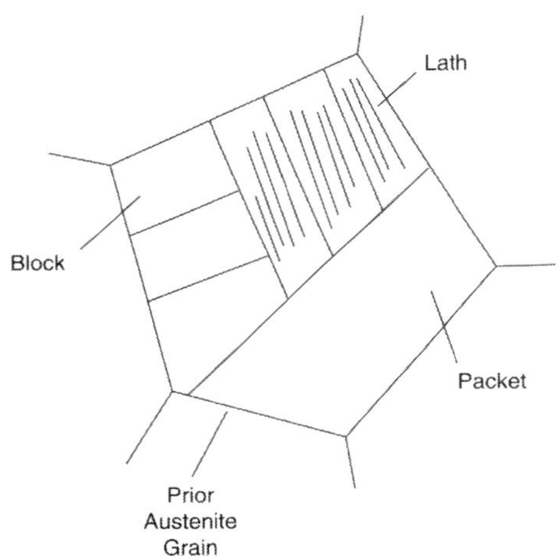

Fig. 15.11 Lath martensite. Source: Ref 15.4 as published in Ref 15.2

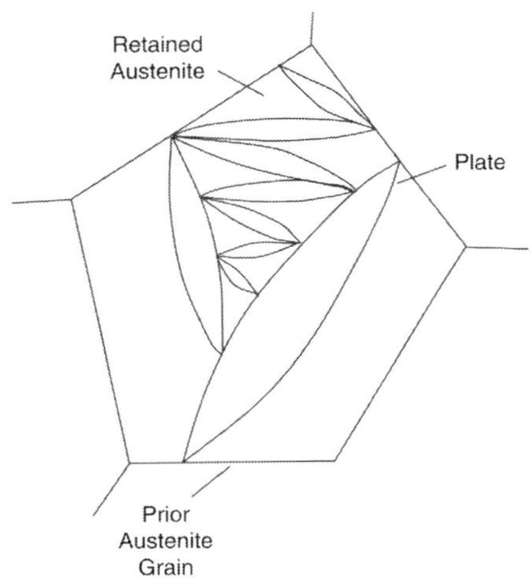

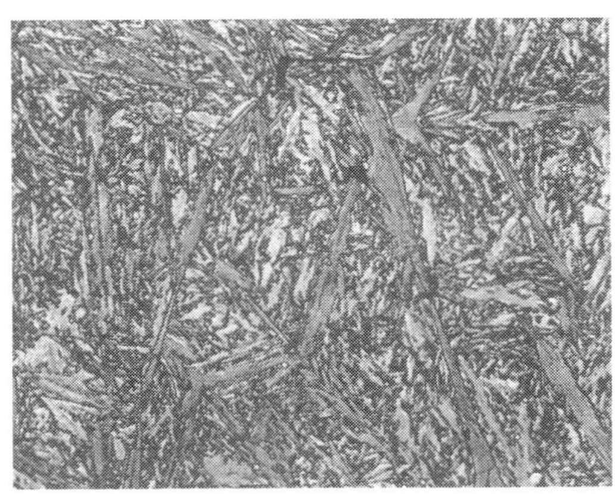

Fig. 15.12 Plate martensite. Source: Ref 15.4 as published in Ref 15.2

effectively pressurized from all sides. This makes it more difficult for it to expand, which it needs to do in order to transform into martensite. Because elastic deformation results in a volume expansion, impact-related stresses or other service loading conditions, as well as low-temperature exposure, can allow retained austenite to transform to martensite in service. Unless the service temperature is somewhat elevated, the new martensite will remain in an essentially untempered condition.

Carbon Content. The addition of carbon and other alloying elements causes significant changes in the isothermal transformation curves. As shown in Fig. 15.14, increasing amounts of carbon push the nose of the

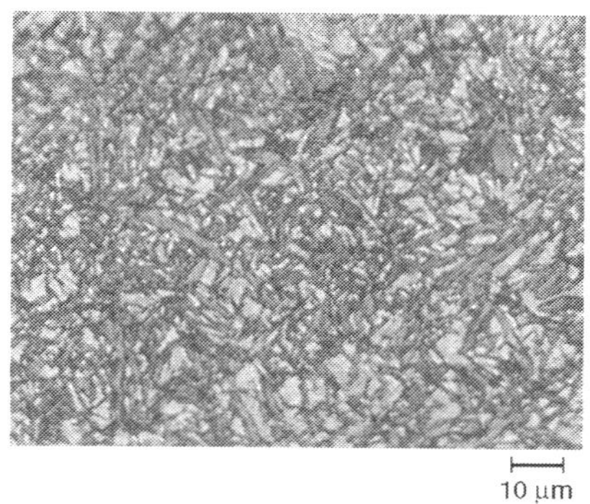

Fig. 15.13 Martensite microstructure with retained austenite (light areas). Source: Ref 15.6 as published in Ref 15.2

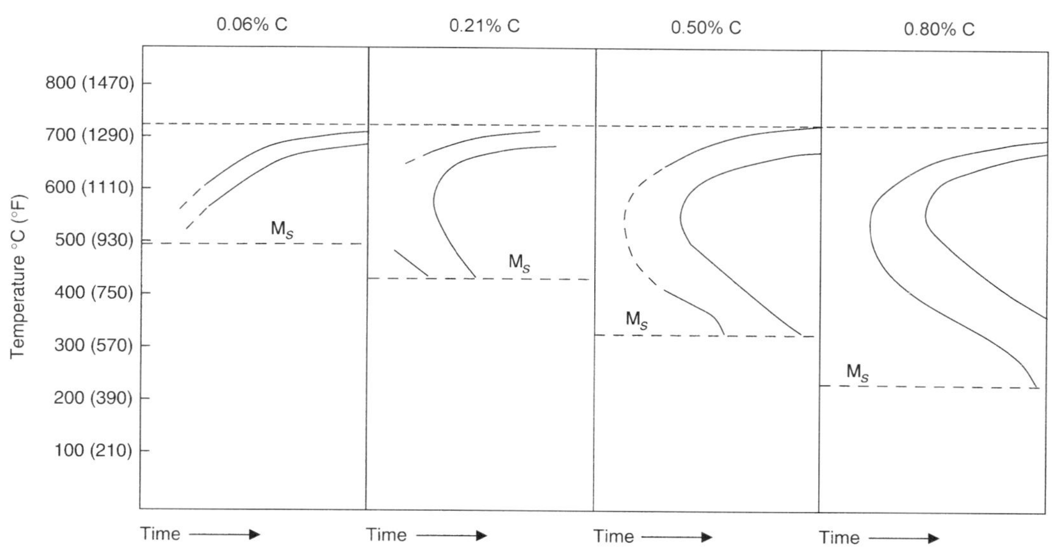

Fig. 15.14 Effect of carbon content on time-temperature-transformation (TTT) diagram. Source: Ref 15.2

TTT diagram to the right. The practical consequence is that steels with higher carbon contents can be hardened to form fully martensitic structures. For example, the steels containing 0.06 and 0.21 wt% C cannot be hardened to fully martensitic structures because there is not a quench fast enough to miss the nose of the TTT diagram. Thus, even with fast cooling, some pearlite will form. On the other hand, the steels containing 0.50 and 0.80 wt% C could, at least in thin sections, be hardened to fully martensitic structures. The main advantage of using substitutional alloying elements, rather than increasing carbon contents, is that the steel can be hardened to a greater depth without causing the brittleness associated with high carbon contents.

The M_s and M_s temperatures also depend on the carbon content of steel. When the carbon content is increased, carbide formation (Fe_3C) becomes more dominant under both isothermal and athermal conditions, and thus the M_s and M_f temperatures are lowered. Almost all other alloying elements also lower M_s. The reduction of the M_s and M_f temperatures, as function of carbon content and the alloying element manganese, are shown in Fig. 15.15. It is important to recognize that the martensitic transformation rarely goes to 100% completion. A small amount of the microstructure in a eutectoid plain carbon steel can be retained austenite. The amount of retained austenite varies with composition, being higher in steels containing high carbon contents and other alloying elements.

15.3 Tempering of Martensite in Steels

The martensitic structure that forms on rapid quenching of austenite in iron-carbon alloys is inherently brittle due to the large amount of lattice strain from supersaturation of carbon atoms, segregation of impurity atoms to grain boundaries, and residual stresses from the quench. This brittleness imparts high hardness to the steel but also results in low ductility and toughness. To regain ductility, the martensite may be tempered, which involves heating the steel to a temperature below the A_1 temperature

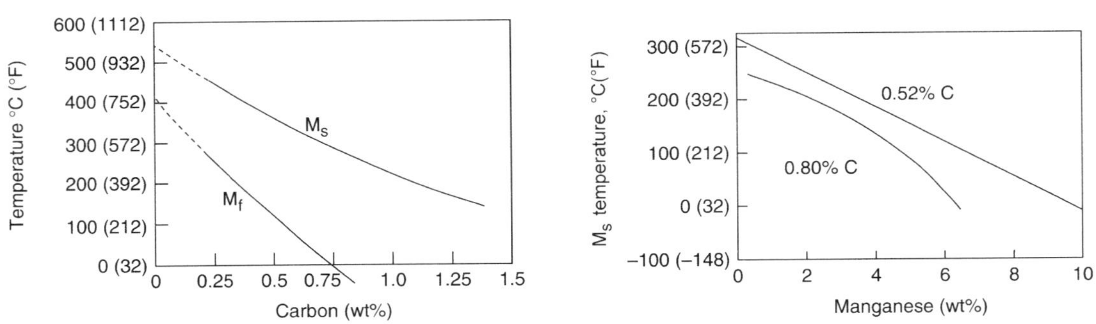

Fig. 15.15 Increasing carbon and alloy content lowers the M_s temperature. Source: Ref 15.2

(eutectoid temperature) and holding for varying amounts of time. Tempering will make the steel more ductile, but will also decrease the strength or hardness (Fig. 15.16).

The as-quenched structure of iron-carbon martensite is very unstable. Factors that contribute to this instability include supersaturation of the interstitial lattice sites with carbon atoms, strain energy from the fine structure (twins or dislocations), large amounts of interfacial energy from the laths or plates, and the presence of retained austenite. On reheating the as-quenched steel, the martensite will transform from the bct structure to a mixture of bcc iron (ferrite) and carbide (Fe_3C) precipitates. A typical tempered microstructure is shown in Fig. 15.17 for an Fe-0.2C alloy. Both the ferrite and the carbide will coarsen with increasing time and temperature, the driving force being the reduction of interfacial energy between the precipitates and the ferrite matrix.

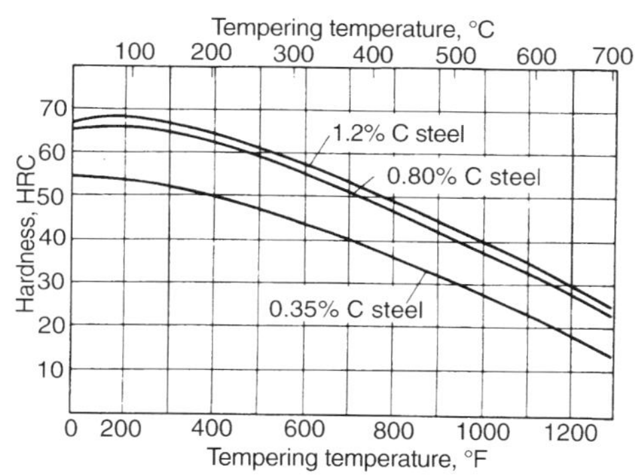

Fig. 15.16 Influence of increasing temperature (1 h) on decreasing the hardness of quenched carbon steels. Source: Ref 15.7 as published in Ref 15.3

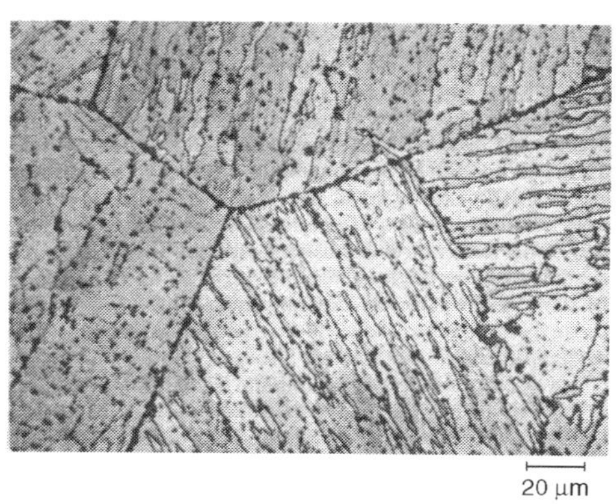

Fig. 15.17 Microstructure of lath martensite in an Fe-0.2C alloy after tempering at 700 °C (1290 °F) for 2 h. Nital etch. Original magnification: 500×. Source: Ref 15.8 as published in Ref 15.3

Early work on tempering of ferrous martensite outlined three distinct stages in the tempering process (Table 15.1). More recent research has been published that identifies the structural changes that occur during tempering (Table 15.2). The temperature ranges are approximate and are usually based on 1 h treatment times.

The fundamental mechanism responsible for tempering is a thermally activated process; both time and temperature are important variables in the tempering process. A tempering parameter is often used to describe the interaction between time and temperature: $T(20 + \log t) \times 10^{-3}$, where T is temperature in Kelvin, and t is time in hours. The tempering parameter makes it possible to create different time and temperature combinations to achieve a certain tempered structure.

The tempering time and temperature for a martensitic steel must be chosen carefully in order to obtain the required combination of strength and ductility. Overtempering may result in a loss of strength to such a degree

Table 15.1 Stages of the tempering process

Stage	Description	Temperature °C	Temperature °F
I	Formation of a transition carbide (ε or η) and lowering of the carbon content of the matrix martensite to approximately 0.25% C	100–250	210–480
II	Transformation of retained austenite to ferrite and cementite	200–300	390–570
III	Replacement of the transition carbide and low-carbon martensite by cementite and ferrite	250–350	480–660

Source: Ref 15.8 as published in Ref 15.9

Table 15.2 Tempering reactions in steel

Temperature °C	Temperature °F	Reaction and symbol (if designated)	Comments
–40–100	–40–210	Clustering of 2–4 carbon atoms on octahedral sites of martensite; segregation of carbon atoms to dislocation boundaries	Clustering is associated with diffuse spikes around fundamental diffraction spots of martensite.
20–100	70–210	Modulated clusters of carbon atoms on (102) martensite planes (A2)	Identified by satellite spots around electron diffraction spots of martensite
60–80	140–180	Long period ordered phase with ordered carbon atoms arranged (A3)	Identified by superstructure spots in electron diffraction patterns
100–200	210–390	Precipitation of transition carbide as aligned 2 nm diam particles (T1)	Recent work identifies carbides as η (orthorhombic, Fe_2C); earlier studies identified the carbides as ε (hexagonal, $Fe_{24}C$).
200–350	390–660	Transformation of retained austenite to ferrite and cementite (T2)	Associated with tempered martensite embrittlement in low- and medium-carbon steels
250–700	480–1290	Formation of ferrite and cementite; eventual development of well-spheroidized carbides in a matrix of equiaxed ferrite grains (T3)	This stage now appears to be initiated by χ-carbide formation in high-carbon Fe-C alloys.
500–700	930–1290	Formation of alloy carbides in Cr-, Mo-, V-, and W-containing steels. The mix and composition of the carbides may change significantly with time (T4).	The alloy carbides produce secondary hardening and pronounced retardation of softening during tempering or long-time service exposure near 500 °C (930 °F).
350–550	660–1020	Segregation and cosegregation of impurity and substitutional alloying elements	Responsible for temper embrittlement

Source: Ref 15.10 as published in Ref 15.9

that the component is no longer useful for the intended application. The amount of softening that occurs with tempering can be altered with the addition of alloy elements. Softening occurs by the diffusion-controlled coarsening of cementite. Strong carbide formers, such as chromium, molybdenum, and vanadium, will reduce the rate of coarsening and thus minimize the amount of softening. Additionally, at higher tempering temperatures, these elements may themselves form carbides, leading to an increase in overall hardness; this is termed secondary hardening.

Different morphologies of tempered martensite will form depending on the heat treatment and the original martensite microstructure. It has been observed that packets of aligned laths in low-carbon martensites will transform into large acicular grains, as shown in Fig. 15.18(a–c). In higher-carbon plate martensites, large martensite plates transform to equiaxed grains on tempering (Fig. 15.19a–c). Additionally, these figures show how the carbides form on the grain boundaries and how both the ferrite grains and the carbides coarsen. When tempering procedures are not carefully

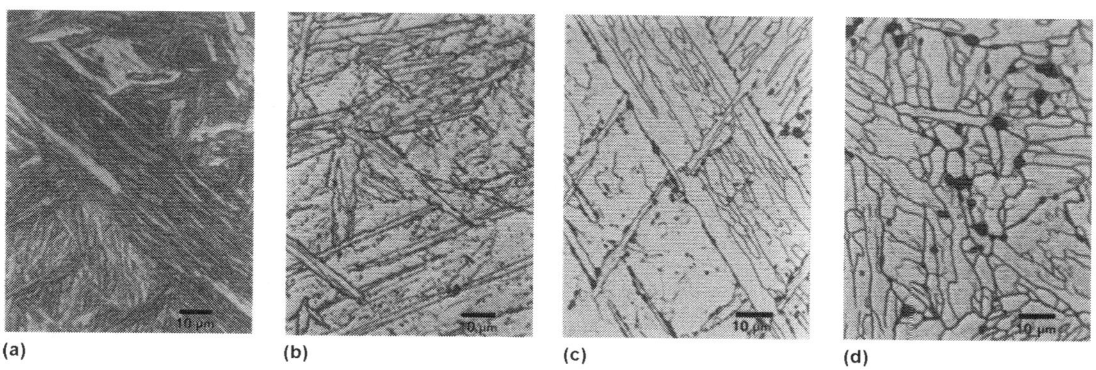

Fig. 15.18 Fe-0.2C alloy (a) in the water-quenched condition, followed by tempering at 690 °C (1275 °F) for (b) 1.5 × 10^3 s, (c) 1.03 × 10^4 s, and (d) 6.05 × 10^5 s. Source: Ref 15.11 as published in Ref 15.9

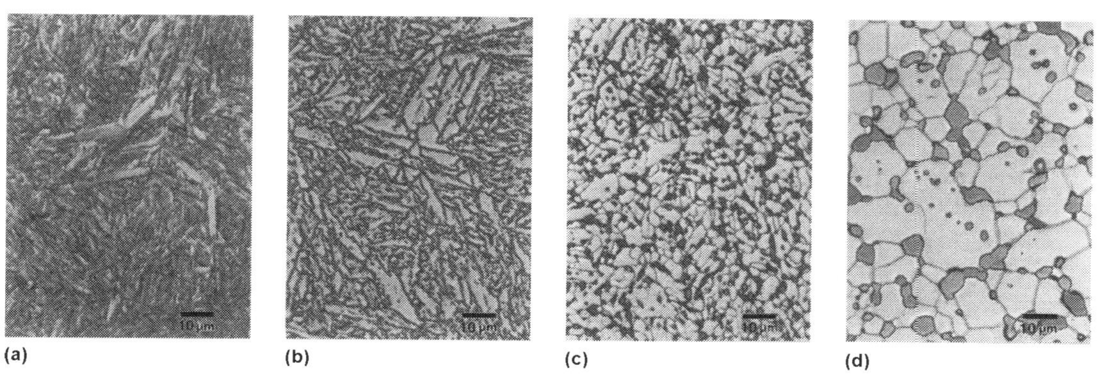

Fig. 15.19 Fe-1.2C alloy (a) in the water-quenched condition, followed by tempering at 690° C (1275 °F) for (b) 1.5 × 10^3 s, (c) 1.03 × 10^4 s, and (d) 6.05 × 10^5 s. Source: Ref 15.11 as published in Ref 15.9

chosen, spheroidization can occur, wherein the Fe_3C coalesces to form spheroid particles. The microstructures in Fig. 15.18(d) and 15.19(d) are both spheroidized.

The change in morphology of tempered martensite, as shown in Fig. 15.18 and 15.19, provides an explanation for the change in mechanical properties of martensite from the as-quenched to the as-tempered form. As-quenched martensite has a very high density of dislocations, leading to high hardness and high work hardenability due to dislocation tangles. Tempering martensite causes the carbides to coarsen, increasing their average size while simultaneously decreasing their total population. Dislocation interactions with carbides is thus reduced significantly on tempering, and work hardenability is reduced. This effect is depicted in Fig. 15.20, which shows true-stress/true-strain curves for an as-quenched and a tempered lath martensite. The work-hardening rate, indicated by the slope of the stress-strain curve, is much higher for the as-quenched steel than for the tempered steel.

15.4 Nonferrous Martensite

Formation of the martensite structure in nonferrous systems occurs by the same diffusionless displacive transformation mechanisms described for ferrous systems. Macroscopic deformation resulting from martensitic transformations will result in upheaval on the surface of polished specimens, as shown in Fig. 15.21. If there are no obstructions from a diffusion-controlled transformation (as there are for iron-carbon alloys), the martensitic surface relief formed on cooling can be removed by heating to temperatures above the start temperature of the parent phase. The

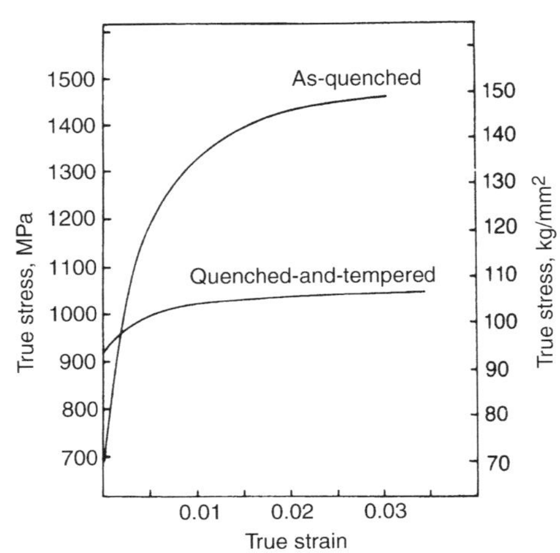

Fig. 15.20 True-stress/true-strain curves for Fe-0.2C as-quenched and quenched-and-tempered lath martensite with packet size of 8.2 μm. Source: Ref 15.8 as published in Ref 15.9

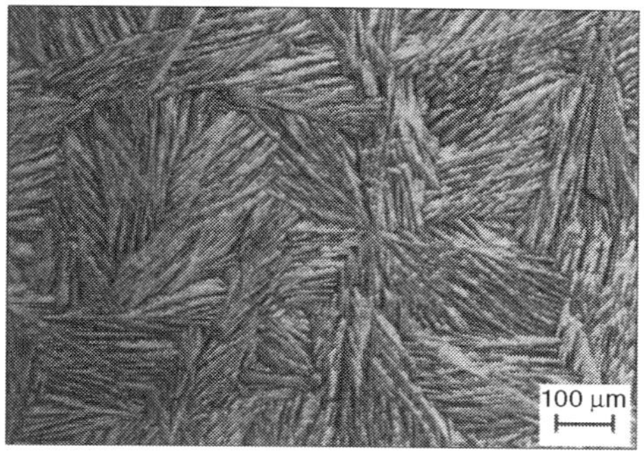

Fig. 15.21 Relief effects observable on a polished surface of Ni-50.3Ti-2W are characteristic of the martensitic transformation. Water quenched from 550 °C (1020 °F). Source: Ref 15.12 as published in Ref 15.13

transformation hysteresis is thus defined by the difference between the forward and reverse transformation temperatures. Reversibility of the transformation is typical of nonferrous systems that undergo the martensitic transformation.

Nonferrous martensitic transformations exhibit a characteristic platelike microstructure (in most cases), such that one dimension of the martensite region is much smaller than the other two. Viewed in cross section on a scale observable in light optical microscopy, these plates have a needle shape that can traverse entire grains or form various internal arrangements within grains (Fig. 15.22). Macroscopic shape changes associated with the martensitic transformation from parent to product phase, a result of the Bain strain and lattice-invariant deformation, induce stress in the surrounding matrix. Lenticular-shaped martensite plates are often observed, believed to form due to an increasing elastic stress surrounding the plate during formation. The lenticular plate shapes, and martensite plate groupings in various orientations (variants), have been found to reduce the constraining elastic stress through accommodation of the macroscopic shape changes (Fig. 15.23).

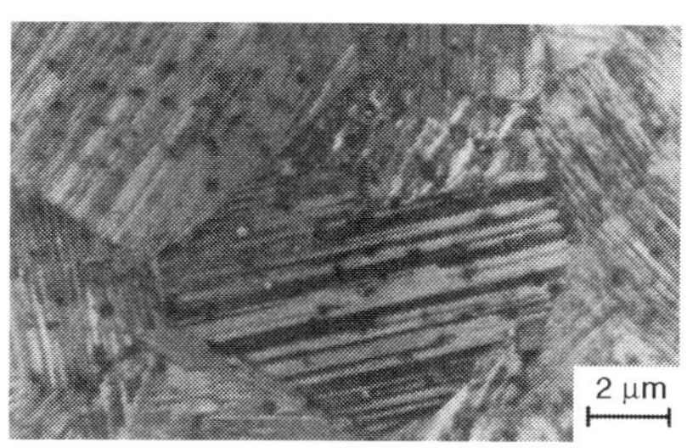

Fig. 15.22 Transmission electron microscopy image of martensite present in Cu-11.4Al-5Mn-2.5Ni-0.4Ti (wt%). Melt spun at a wheel speed of 6.5 m/s. Precipitates of Cu_2AlTi are visible, dispersed evenly across the different grains. Source: Ref 15.14 as published in Ref 15.13

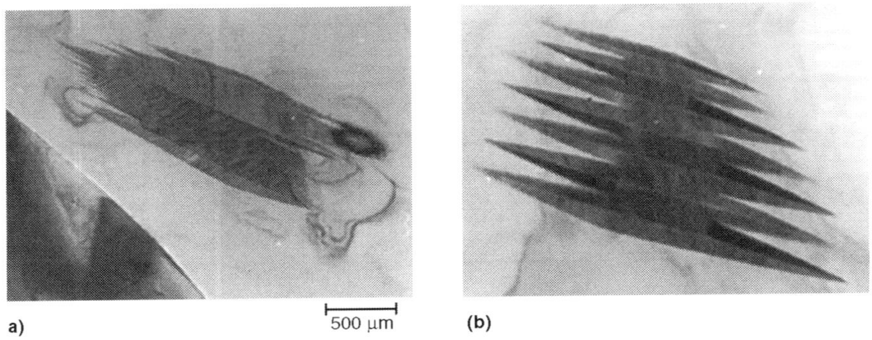

Fig. 15.23 Transmission electron microscopy images of splat-cooled Ni-37.5Al (at.%) showing accommodating martensite groupings. Source: Ref 15.15 as published in Ref 15.13

Historically, martensitic transformations in metal alloys (specifically steel) have been studied the most, but recent attention has also been given to ceramic, mineral, and organic systems that develop martensite structures. Table 15.3 lists different metal alloy systems that undergo martensitic transformations, along with specific alloys or compounds and the general changes in crystal structure for such systems.

Nonferrous martensite in metal alloys has been studied extensively. In metallic systems, nonferrous martensite generally occurs in substitutionally alloyed systems that demonstrate small transformation strains and a small transformation hysteresis. The macroscopic deformation from the resultant transformation from one crystal structure to another can be observed from the surface relief and in the plate morphology that the martensite phase usually assumes. Substructure of the martensite plates can be characterized with the use of transmission electron microscopy (TEM). Stacking faults, twins, and dislocations are commonly observed within martensite plates. Twinning deformation or slip along the shear planes can accommodate structural transformations that would otherwise result in a significant shape change. Their presence and frequency within the martensite phase is dependent on the alloy system and composition.

Table 15.3 Metal alloy martensite transformations for selected systems

Metal alloy	General parent crystal structure	General martensite crystal structure
Ag alloys (Ag-Cd, Ag-Ge, Ag-Zn)	bcc	3R, 9R, 2H
Au alloys (Au-Cd, Au-Cu-Zn, Au-Zn)	bcc	3R, 9R, 2H
Co alloys (Co, Co-Be, Co-Ni)	fcc	hcp
Cu alloys (Cu-Al, Cu-Al-Ni, Cu-Sn, Cu-Zn)	bcc	3R, 9R, 2H
In alloys (In-Tl, In-Cd)	fcc	fct
Li alloys (Li, Li-Mg)	bcc	hcp
Ni alloys (Ni-Al, Ni-Ti)	bcc	3R, 9R, 2H
Mn alloys (Mn-Cu, Mn-Ni)	fcc	fct
Ti alloys (Ti, Ti-Cu, Ti-Mn, Ti-Mo, Ti-V)	bcc	hcp
Zr alloys (Zr, Zr-Mo, Zr-Nb)	bcc	hcp

bcc, body-centered cubic; fcc, face-centered cubic; hcp, hexagonal-close packed; fct, face-centered tetragonal. Source: Ref 15.13

Martensitic transformations in metallic systems can be grouped into three categories:

- In the first group, allotropic transformations of the solvent atoms create the martensite product. Martensitic transformations in cobalt, titanium, zirconium, hafnium, and lithium alloys belong to this group. Cobalt and alloys undergo a fcc to hexagonal close-packed (hcp, ε) martensitic transformation. Cobalt and cobalt-nickel alloys can therefore have a high measure of stacking faults within the martensite plates. Titanium, zirconium, hafnium, and lithium alloys generally transform from bcc (β phase) to hcp structures, although fcc and orthorhombic martensite structures have been reported in some titanium and zirconium alloys. Lath morphologies have been observed in some of the titanium and zirconium alloys as well. Twins are commonly found in the plate martensite form of these alloys, while stacking faults are commonly observed in lath martensite.
- Copper, gold, and silver alloys belong to the β phase Hume-Rothery alloy group, whose parent phase is a bcc structure. Nickel alloys such as nickel-aluminum and nickel-titanium (~50 to 50 ratio) are also part of this alloy group, with a bcc parent phase. Because the transformations are diffusionless and lattice correspondence is maintained, order or disorder present in the parent phase is transferred to the martensite phase. Often a superlattice structure is present that is then converted to the martensite product. These alloys all belong to the second category, marking a weak first-order transformation with an intermediate stability of the martensite phase at temperatures above the M_s temperature. Alloys used in shape memory applications are found in this category.
- The third category of martensitic transformations belongs to second-order transformations (very weak first order) that have a larger mechanical instability of the martensite phase. First-order phase transformations are those phase transformations for which the first derivative of Gibbs free energy with respect to temperature and pressure are discontinuous at the equilibrium transformation temperature. Second-order phase transformations have continuous first derivatives but have discontinuous second derivatives of Gibbs free energy with respect to temperature and pressure. The fcc to face-centered tetragonal (fct) martensitic transformation of manganese and indium alloys falls into this category.

15.5 Shape Memory Alloys

Shape memory alloys (SMA) are a group of alloys that have the ability to return to some previously defined shape or size when subjected to an appropriate thermal cycle. Generally, these materials can be plastically deformed at relatively low temperature and, on exposure to some higher temperature, will return to their shape prior to the deformation. Materials that exhibit shape memory only on heating are referred to as having a one-way shape memory. Some materials also undergo a change in shape on recooling. These materials are said to have a two-way shape memory. Nitinol (50Ti-50Ni) is one of the best known and most useful of the SMAs. Nitinol has been used for blood clot filters in medical applications, hydraulic couplings for aircraft, force actuators, proportional control devices, and superelastic eyeglass frames.

A SMA can be further defined as one that yields thermoelastic martensite. In this case, the alloy undergoes a martensitic transformation of a type that allows the alloy to be deformed by a twinning mechanism below the transformation temperature. The deformation is then reversed when the twinned structure reverts on heating to the parent phase. The martensitic transformation that occurs in SMAs yields a thermoelastic martensite that develops from a high-temperature austenite phase with long-range order. The transformation does not occur at a single temperature but over a range of temperatures that varies with each alloy system. The usual way of characterizing the transformation and naming each point in the cycle is shown in Fig. 15.24. Most of the transformation occurs over a relatively narrow temperature range, although the beginning and end of the transformation during heating or cooling actually extends over a much larger temperature range. The transformation also exhibits a hysteresis in that the transformation on heating and on cooling does not overlap. This transformation hysteresis varies with the alloy system.

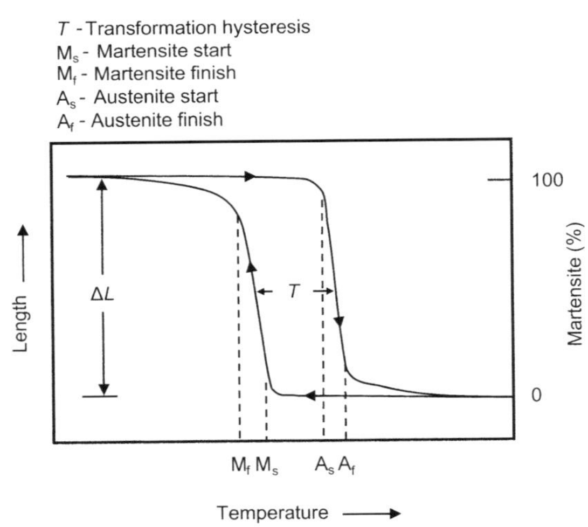

Fig. 15.24 Typical transformation-versus-temperature curve for a shape memory alloy. Source: Ref 15.16 as published in Ref 15.2

15.6 Bainitic Structures

At temperatures between those at which the eutectoid transformation of austenite to pearlite and the transformation of austenite to martensite occur, a variety of unique microstructures may form in carbon steels. Davenport and Bain showed by careful light microscopy that the microstructures formed at such intermediate temperatures were quite different from those of pearlite and martensite, and in honor of Edgar C. Bain, his colleagues termed the unique microstructures bainite. The schematic TTT diagram in Fig. 15.25 clearly shows the intermediate temperature range, between those of pearlite and martensite, for bainite formation.

Steels with carbon contents other than the eutectoid composition would of course have regions of proeutectoid phase formation at temperatures above that of pearlite formation. The schematic diagram of Fig. 15.25 shows a well-defined TTT range for bainite formation. Such a well-defined range of bainite transformation is characteristic of low-alloy steels, especially on continuous cooling. In plain carbon steels, the transformation regions for proeutectoid ferrite/pearlite and bainite are more continuous and even overlap with decreasing temperature. In alloy steels, alloying elements may even cause the arrest of bainite transformation, causing incomplete transformation at intermediate temperatures. The extreme effects of alloy-

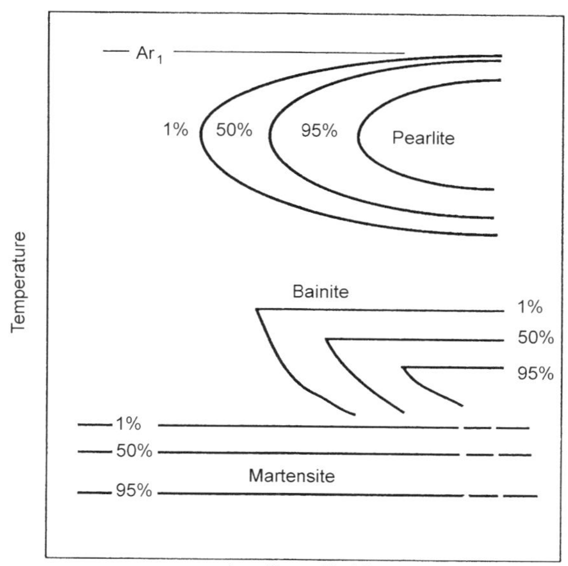

Fig. 15.25 Schematic TTT diagram for a steel with well-defined pearlite and bainite formation ranges. Source: Ref 15.17 and 15.18 as published in Ref 15.19

ing on bainitic transformation, ranging from those in plain carbon steels to those in alloyed steels, are shown schematically in the TTT diagrams in Fig. 15.26.

Austempering is the interrupted quenching process that is used to form a bainitic structure. The part is quenched into a salt bath above the M_s temperature. However, it is allowed to remain at that temperature until the transformation to bainite is complete (Fig. 15.27). Because austempering is an interrupted quenching process that produces bainite rather than martensite, parts can usually be produced with less dimensional change

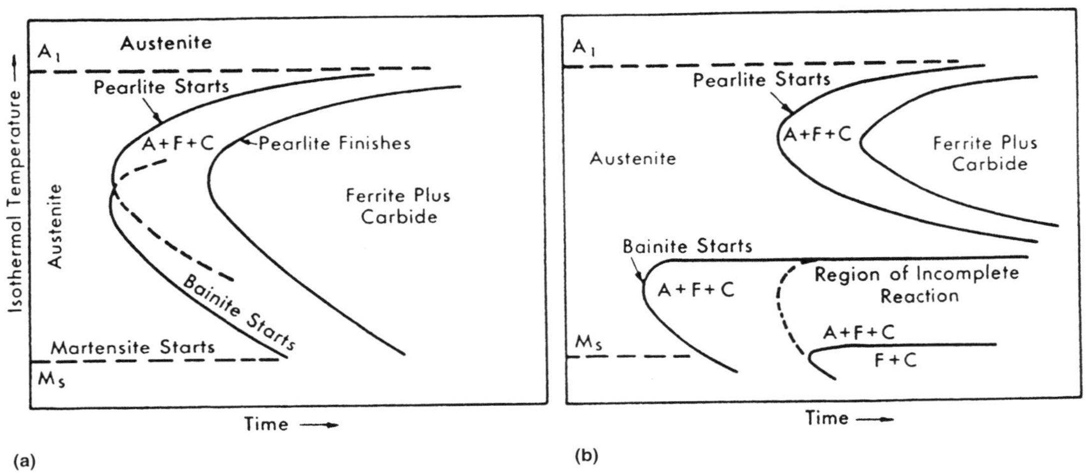

Fig. 15.26 Schematic TTT diagrams for (a) plain carbon steel with overlapping pearlite and bainite transformation and (b) alloy steel with separated bainite transformation and incomplete bainite transformation. Source: Ref 15.20 and 15.21 as published in Ref 15.19

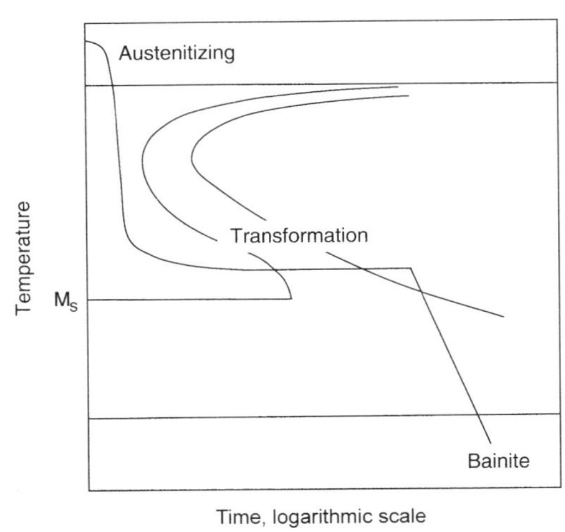

Fig. 15.27 Typical austempering heat treatment. Source: Ref 15.2

than is typical through conventional quenching and tempering. In addition, austempered parts do not normally require a tempering operation. A typical austempering cycle would be to heat to a temperature within the austenitizing range (790 to 910 °C, or 1450 to 1675 °F), quench in a salt bath maintained at a constant temperature, usually in the range of 260 to 400 °C (500 to 750 °F), allowing time for the austenite to isothermally transform to bainite, and then air cool to room temperature. For true austempering, the metal must be cooled from the austenitizing temperature to the temperature of the austempering bath fast enough so that no transformation of austenite occurs during cooling, and then held at bath temperature long enough to ensure complete transformation of austenite to bainite. Austempering offers several potential advantages, including increased ductility, toughness, and strength at a given hardness, reduced distortion, and the ability to heat treat steels to a hardness of HRC 35 to 55 without having to temper.

Bainite Transformation Start Temperatures. The temperature at which bainite transformation starts is referred to as the B_s temperature, and several empirical equations that show the effect of alloying elements on B_s have been determined. The equation for B_s as a function of composition (in wt%) for hardenable low-alloy steels containing from 0.1 to 0.55% C is:

$$B_s\ (°C) = 830 - 270(\%C) - 90(\%Mn) - 37(\%Ni) - 70(\%Cr) - 83(\%Mo) \qquad \text{(Eq 15.1)}$$

Equation 15.2 is applicable for low-carbon bainitic steels (containing between 0.15 and 0.29% C) for high-temperature applications in the electric power industry. Compositions of the alloying elements are in wt%.

$$B_s\ (°C) = 844 - 597(\%C) - 63(\%Mn) - 16(\%Ni) - 78(\%Cr) \qquad \text{(Eq 15.2)}$$

Bainite versus Ferritic Microstructures. Bainitic microstructures take many forms. In medium- and high-carbon steels, similar to pearlite, bainite is a mixture of ferrite and cementite and is therefore dependent on the diffusion-controlled partitioning of carbon between ferrite and cementite. However, unlike pearlite, the ferrite and cementite are present in nonlamellar arrays. Similar to martensite, the ferrite of bainitic microstructures may appear as acicular crystals, similar to the laths and plate-shaped crystals of martensite. Bainite is grouped into two types, upper and lower bainite, depending on the temperature range at which the transformation occurs. The effect of steel carbon content on transition temperatures between upper and lower bainite formation is shown in Fig. 15.28. In low-carbon steels, at intermediate transformation temperature ranges, austenite may transform only to ferrite, resulting in two-phase microstructures of ferrite and retained austenite. Although some features of the intermediate ferritic

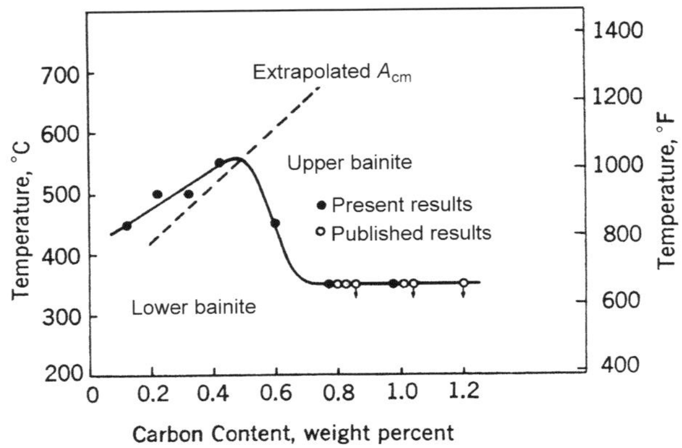

Fig. 15.28 Effect of steel carbon content on the transition temperature between upper and lower bainite. Source: Ref 15.22 as published in Ref 15.19

microstructures are similar to those of the classical bainites, the absence of cementite in ferritic microstructures makes possible a clear differentiation of intermediate-temperature-transformation products of austenite decomposition.

According to a microstructural definition of bainite in steels as a nonlamellar ferrite-cementite product of austenite transformation, six morphologies of cementite-ferrite microstructures considered to be bainite are shown schematically in Fig. 15.29. Upper and lower bainites are the most common forms found in medium-carbon steel. However, in the absence of cementite, intermediate-temperature-transformation products of austenite fall in the category of ferrites.

Upper bainite forms in the temperature range just below that at which pearlite forms, typically below 500 °C (932 °F). Figure 15.30 shows light micrographs of upper bainite formed by holding 4360 steel at 495 and 410°C (920 and 770°F). The bainite appears dark and the individual ferritic crystals have an acicular shape. The bainitic transformation was not completed during the isothermal holds at the temperatures noted, and therefore the light etching areas are martensite that formed in untransformed austenite on quenching after the isothermal holds. The bainite appears dark (i.e., has low reflectivity) because of roughness produced by etching around the cementite particles of the bainitic structure. The cementite particles, however, are too fine to be resolved in the light microscope.

The feathery appearance of the clusters of ferrite crystals is clearly shown in the light micrographs and is sometimes an important identifying feature of upper bainite. Upper bainite microstructures develop by packets or sheaves of parallel ferrite crystals growing across austenite grains, producing a blocky appearance. This latter characteristic of upper bainite in 4150 steel transformed at 460 °C (860 °F) is shown in Fig. 15.31.

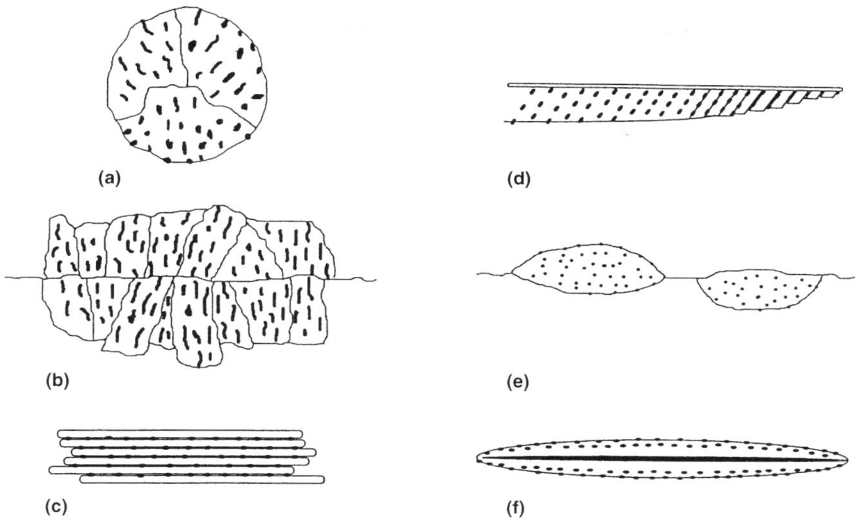

Fig. 15.29 Schematic illustrations of various ferrite (white)-cementite (black) microstructures defined as bainite according to Aaronson et al. (a) Nodular bainite. (b) Columnar bainite. (c) Upper bainite. (d) Lower bainite. (e) Grain-boundary allotromorphic bainite. (f) Inverse bainite. Source: Ref 15.20 as published in Ref 15.19

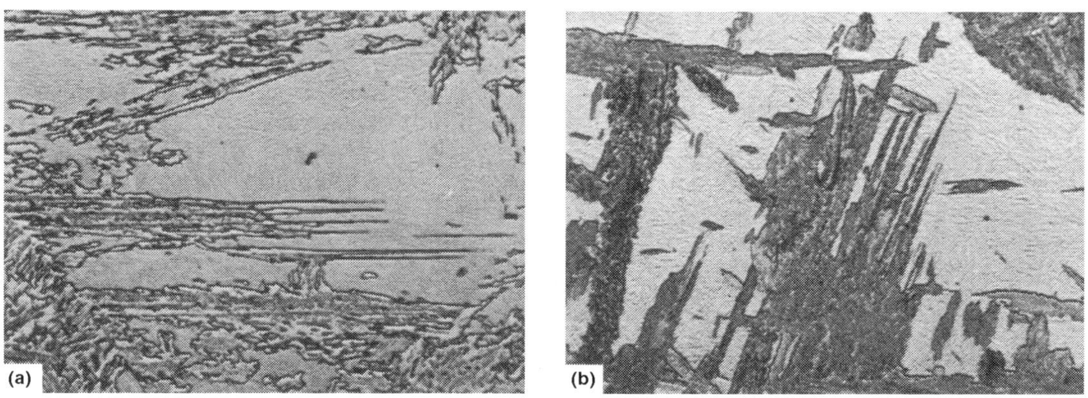

Fig. 15.30 Upper bainite in 4360 steel isothermally transformed at (a) 495 °C (920 °F) and (b) 410 °C (770 °F). Light micrographs, picral etch. Original magnification: 750×. Source: Ref 15.23 as published in Ref 15.19

The cementite particles of upper bainite form between ferrite crystals in austenite enriched by carbon rejection from the growing ferrite crystals.

Figure 15.32 is a thin-foil TEM micrograph that shows interlath cementite in a 4360 steel transformed to bainite at 495 °C (920 °F). The carbide particles, compared with those that are present in lower bainite, are relatively coarse and appear black and elongated. In some steels, especially those with high silicon content, cementite formation is retarded. As a result, the carbon-enriched austenite between the ferrite laths is quite stable and

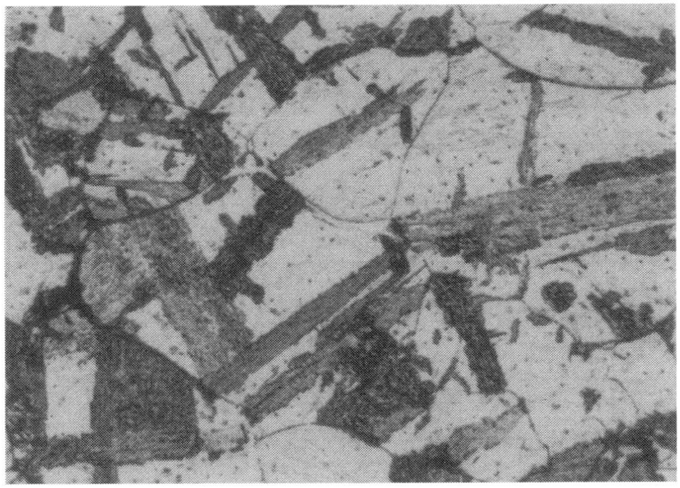

Fig. 15.31 Upper bainite (dark rectangular areas) in 4150 steel transformed at 460 °C (860 °F). Original magnification: 500×. Courtesy of Florence Jacobs, Colorado School of Mines. Source: Ref 15.19

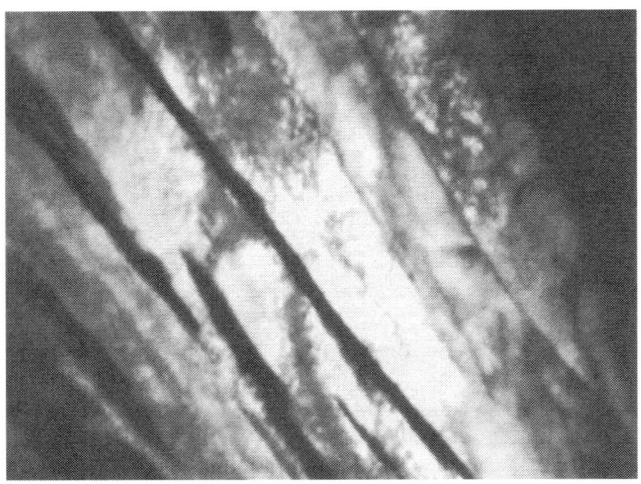

Fig. 15.32 Carbide particles (dark) formed between ferrite crystals in upper bainite in 4360 steel transformed at 495 °C (920 °F). Transmission electron micrograph. Original magnification: 25,000×. Source: Ref 15.23 as published in Ref 15.19

is retained during transformation and at room temperature. Figure 15.33 shows retained austenite in bainite formed at 400 °C (752 °F) in a steel containing 0.6% C and 2.0% Si. The austenite in this TEM image appears gray.

Lower Bainite. An example of lower bainite, obtained from a specimen of 4360 steel partially transformed at 300 °C (570 °F), is shown in Fig. 15.34. Again, the bainite etches dark, and the white-etching matrix is

Fig. 15.33 Retained austenite (gray, marked with A) between ferrite laths of upper bainite in 0.6% C steel containing 2.0% Si and transformed at 400 °C (750 °F). Transmission electron micrograph. Original magnification: 40,000×. Source: Ref 15.23 as published in Ref 15.19

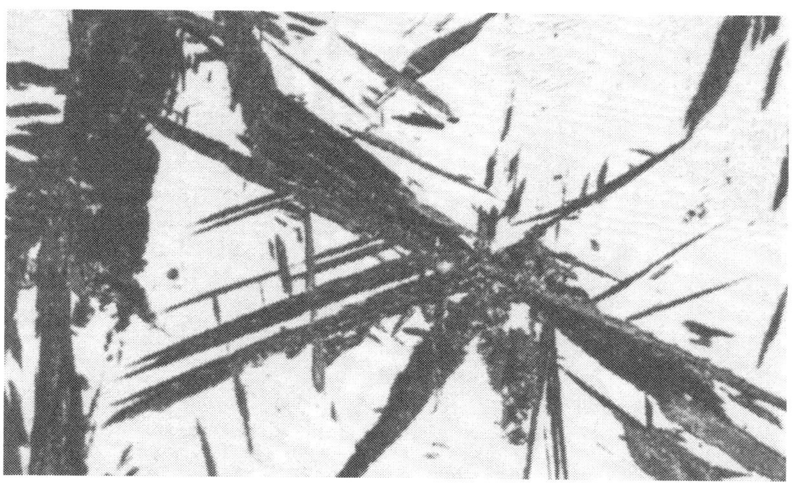

Fig. 15.34 Lower bainite in 4360 steel transformed at 300 °C (570 °F). Original magnification: 750×. Source: Ref 15.23 as published in Ref 15.19

martensite formed on cooling in the austenite not transformed to bainite at 300 °C (570 °F).

Lower bainite is composed of large ferrite plates that form nonparallel to one another and, analogous to plate martensite microstructures, is often characterized as acicular. The carbides in the ferrite plates of lower bainite are responsible for its dark etching appearance but are much too fine to be resolved in the light microscope.

Figure 15.35 shows the very fine carbides that have formed in ferrite of lower bainite in 4360 steel transformed at 300 °C (570 °F). The fine carbides typically make an angle of approximately 60° with respect to the long axis of the matrix ferrite crystal. In contrast to upper bainite, fine carbides form within ferrite crystals, rather than between plates, and are significantly finer than the interlath carbides of upper bainite. A variant of lower bainite is termed lower bainite with midrib and forms isothermally at lower temperatures, 150 to 200 °C (300 to 350 °F), than the temperatures at which conventional lower bainite forms, 200 to 350 °C (390 to 660 °F). Figure 15.36 shows light and TEM micrographs of lower bainite

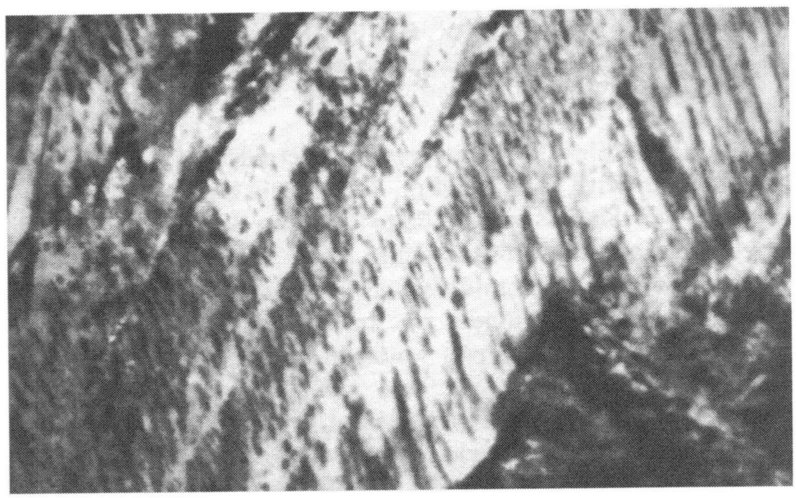

Fig. 15.35 Lower bainite with fine carbides within ferrite plates in 4360 steel transformed at 300 °C (572 °F). Transmission electron micrograph. Original magnification: 24,000×. Source: Ref 15.23 as published in Ref 15.19

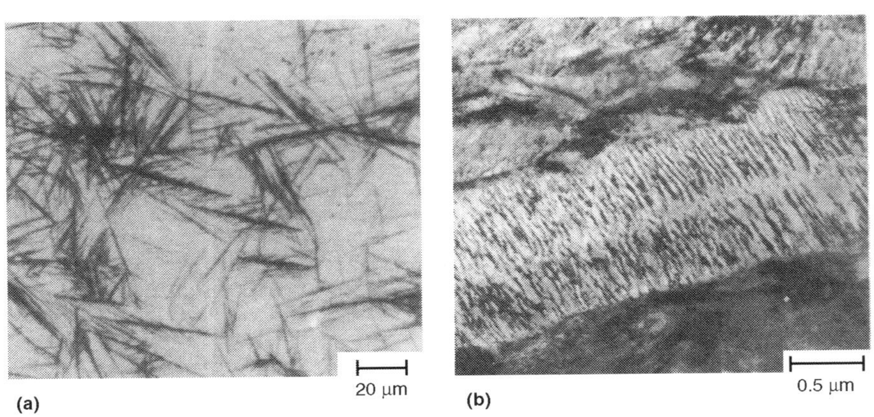

Fig. 15.36 Lower bainite with midribs in a 1.1% C steel transformed at 190 °C (374 °F) for 5 h. (a) Light micrograph. (b) Transmission electron micrograph. Courtesy of H. Okamoto, Tottori University. Source: Ref 15.19

with midrib in a 1.1% C steel transformed at 190 °C (370 °F). The midrib is an isothermally formed thin plate of martensite that provides the interface at which the two-phase carbide-ferrite lower bainitic structure forms.

Bainite Formation Mechanisms. The fact that the classical bainites consist of ferrite and nonlamellar distributions of cementite attests to the need for carbon diffusion during some stage of bainite transformation. However, the relatively low temperatures at which bainites form severely restrict iron atom diffusion. The latter feature of the transformation of austenite to bainite has led to two quite different views of ferrite nucleation in bainite. The first view states that the first-formed ferrite is formed by a diffusionless shear or martensitic transformation. The second view states that the first-formed ferrite nucleates and grows by a ledge-type mechanism where short-range iron atom rearrangement can take place at ledges in the ferrite-austenite interface. There is scientific and experimental support for both sides of the argument about the nucleation and growth mechanisms of bainite.

The empirical B_s equations noted earlier reflect the strong effect of alloying elements on the start of bainitic transformation. Coupled with this characteristic of steels with prominent bainite transformations is the presence of a pronounced bay or region of very sluggish transformation in TTT diagrams. These regions correspond to the temperature ranges that show the marked separation of the transformation curves for pearlite and bainite in Fig. 15.25. An example of such a bay is shown in the isothermal TTT diagram for 4340 steel (Fig. 15.37).

Such bays correlate with the presence of substitutional alloying elements that may partition to or from ferrite and concentrate at austenite-ferrite interfaces, creating a solute drag or significant restraining force on the formation of bainitic ferrite. As noted relative to Fig. 15.26, the isothermal transformation of austenite to bainite may be severely retarded. This phenomenon is referred to as stasis and is also discussed in terms of atom partitioning and solute drag at austenite-ferrite interfaces.

The distribution of very fine carbides in plates of lower bainite suggests that a ferrite crystal has initially formed, perhaps by a martensitic mechanism, and as a consequence of the supersaturation of the ferrite with carbon, fine carbides precipitate within the ferrite. Another explanation for the formation of lower bainite is that a unit of lower bainite forms by a four-step process: (1) precipitation of a nearly carbide-free ferrite spine; (2) sympathetic nucleation of secondary plates of ferrite, usually on only one side of and at an angle of approximately 55 to 60° to the initiating spine; (3) precipitation of carbides in austenite at α:γ boundaries, forming gaps between adjacent secondary (ferrite) plates; and (4) an annealing process in which the gaps are filled in with further growth of ferrite and additional carbide precipitation.

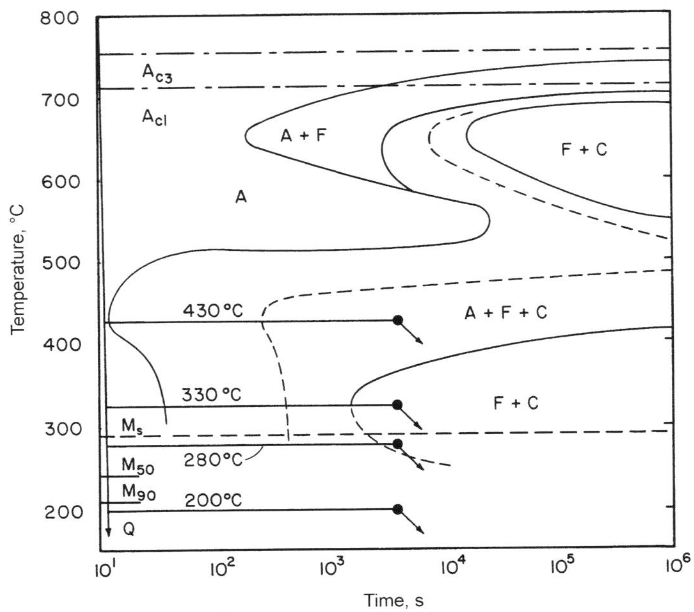

Fig. 15.37 Isothermal transformation diagram for 4340 steel and isothermal heat treatments applied to produce various microstructures for fracture evaluation. Source: Ref 15.24 as published in Ref 15.19

Mechanical Behavior of Ferrite-Carbide Bainites. Steels largely transformed to ferrite-carbide bainitic microstructures develop a wide range of strengths and ductilities. Ultimate tensile strengths of high-carbon lower bainitic microstructures may reach 1400 MPa (200 ksi), and hardness may reach 55 HRC or higher. The strengths are derived from relatively fine ferrite crystal structures, high dislocation densities within the ferritic crystals, and fine dispersions of cementite. The lower the temperature of bainite formation, the finer the carbide dispersions, and the higher the hardness and strength. Lower bainite microstructures compete well with low-temperature tempered martensites in strength and fracture resistance. Often, low-alloy steels are subjected to isothermal holds to form bainite, instead of quenching to martensite, in order to reduce the stresses that produce quench cracking. In addition, bainitic steels do not incur the additional expense of tempering.

The type of bainite affects fracture characteristics. Specimens with upper bainitic microstructures have low toughness and ductility compared with specimens with lower bainitic microstructures, and upper bainites have higher ductile-to-brittle transition temperatures. These observations were confirmed in a study of 4340 steel isothermally transformed at various temperatures, as shown in Fig. 15.37. Specimens quenched in oil and tempered at 200 °C (390 °F) had tempered martensite microstructures with hardness of 52 HRC; those held at 200 °C (390 °F) also transformed to tempered martensite with hardness 52 HRC; those held at 280 and 330 °C (540 and 630 °F) transformed largely to lower bainite with hardness of 50 and 44 HRC, respectively; and those transformed at 430 °C (810 °F) transformed largely to upper bainite with hardness of 32 HRC.

The results of room-temperature instrumented Charpy V-notch testing of the 4340 specimens are shown in Fig. 15.38. Instrumented impact testing measures both initiation and propagation energies. The fracture energy of the specimens with the upper bainitic microstructures was significantly lower than those with tempered martensite or lower bainite. When fracture was initiated in the upper bainite, the propagation energy dropped to zero. Fractography of the upper bainitic specimens showed, except at initiation at the notch root, that the fracture surface consisted entirely of cleavage fracture (Fig. 15.39b), a result attributed to the coarse interlath carbides and common cleavage plane of the parallel ferrite crystal in packets of upper bainite. In contrast, the fracture surfaces of the specimens transformed to tempered martensite consisted of ductile microvoid coalescence (Fig. 15.39a). Although microstructures with lower strength and hardness typically show better ductility and fracture resistance than microstructures with higher hardness, the behavior of upper bainite, with its lower hardness compared with other microstructures in 4340 steel, contradicts this general rule. A study of the fracture behavior of 4150 steel isothermally transformed to lower and upper bainite confirms the strong susceptibility of upper bainite to cleavage fracture despite its lower hardness and strength relative to lower bainitic microstructures.

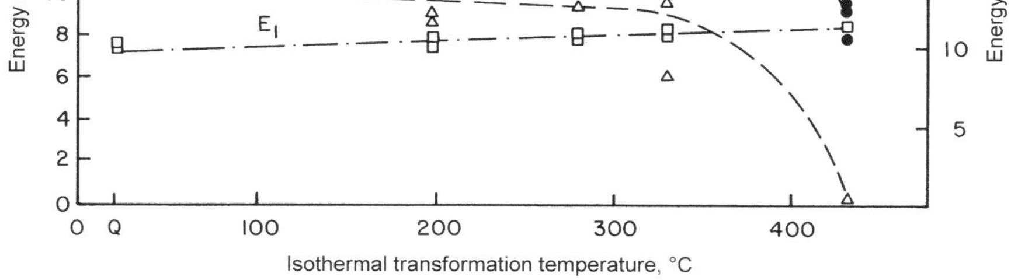

Fig. 15.38 Impact energy absorbed as a function of isothermal transformation temperature for specimens of 4340 steel. E_0, total energy absorbed; E_1, fracture initiation energy; E_2, fracture propagation energy. Source: Ref 15.24 as published in Ref 15.19

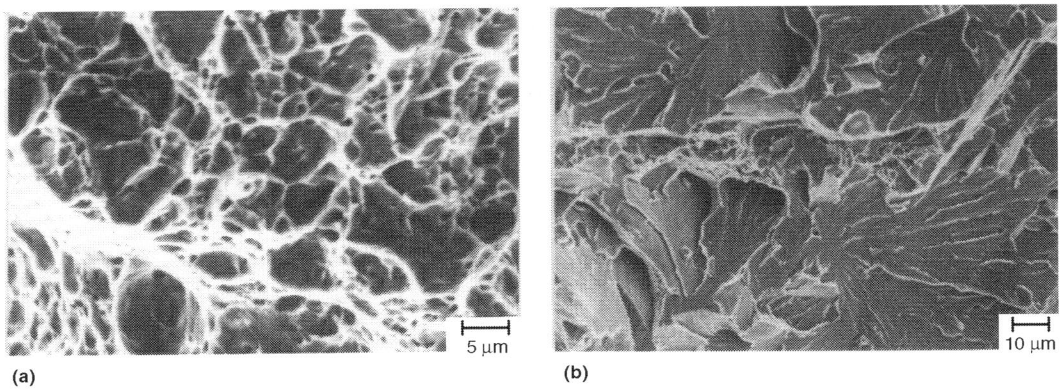

Fig. 15.39 Fracture morphologies of fracture surfaces of 4340 steel Charpy V-notch specimens heat treated as: (a) oil quenched and tempered at 200 °C (390 °F), and (b) isothermally transformed at 430 °C (810 °F). Source: Ref 15.24 as published in Ref 15.19

ACKNOWLEDGMENTS

The contents of this chapter came from *Elements of Metallurgy and Engineering Alloys* by F.C. Campbell, ASM International, 2008; "Martensitic Structures," in *Metallography and Microstructures,* Vol 9, *ASM Handbook*, ASM International, 2004; "Ferrous Martensite," by R.M. Deacon in *Metallography and Microstructures,* Vol 9, *ASM Handbook*, ASM International, 2004; "Nonferrous Martensite," by F.C. Gift, Jr. in *Metallography and Microstructures,* Vol 9, *ASM Handbook*, ASM International, 2004; and "Bainite: An Intermediate Temperature Transformation Product of Austenite," by G. Krauss in *Steels: Processing, Structure, and Performance*, ASM International, 2005.

REFERENCES

15.1 W.F. Smith, *Principles of Materials Science and Engineering*, McGraw-Hill, 1986
15.2 F.C. Campbell, *Elements of Metallurgy and Engineering Alloys*, ASM International, 2008
15.3 Martensitic Structures, *Metallography and Microstructures,* Vol 9, *ASM Handbook*, ASM International, 2004, p 165–178
15.4 B.L. Bramfitt, Effects of Composition, Processing, and Structure on Properties of Irons and Steels, *Materials Selection and Design*, Vol 20, *ASM Handbook,* ASM International, 1997
15.5 G. Krauss, Microstructures, Processing, and Properties of Steels, *Properties and Selections: Irons, Steels, and High-Performance Alloys*, Vol 1, *ASM Handbook,* ASM International, 1990
15.6 D. Aliya and S. Lampman, Physical Metallurgy Concepts in Interpretation of Microstructures, *Metallography and Microstructures*, Vol 9, *ASM Handbook,* ASM International, 2004

15.7 A.K. Sinha, *Ferrous Physical Metallurgy,* Butterworth Publishers, 1989
15.8 G. Krauss, *Principles of Heat Treatment of Steel,* American Society for Metals, 1985
15.9 R.M. Deacon, Ferrous Martensite, *Metallography and Microstructures,* Vol 9, *ASM Handbook*, ASM International, 2004, p 165–170
15.10 A.R. Marder, Structure-Property Relationships in Ferrous Transformation Products, *Phase Transformation of Ferrous Alloys,* TMS, 1984, p 11–41
15.11 B.A. Lindsley and A.R. Marder, The Morphology and Coarsening Kinetics of Spheroidized Fe-C Binary Alloys, *Acta Mater.,* Vol 46, 1997, p 341–351
15.12 H. Scherngell and A.C. Kneissl, *Acta Mater.,* Vol 50, 2002, p 328
15.13 F.C. Gift, Jr., Nonferrous Martensite, *Metallography and Microstructures,* Vol 9, *ASM Handbook*, ASM International, 2004, p 172–175
15.14 J. Dutkiewicz, T. Czeppe, and J. Morgiel, *Mater. Sci. Eng.,* Vol A273–275, 1999, p 706
15.15 D. Schryvers and D. Holland-Moritz, *Mater. Sci. Eng.,* Vol A273–275, 1999, p 699
15.16 D.E. Hodgson, M.H. Wu, and R.J. Biermann, Shape Memory Alloys, *Properties and Selection: Nonferrous Alloys and Special-Purpose Materials*, Vol 2, *ASM Handbook,* ASM International, 1990
15.17 C. Zener, Kinetics of the Decomposition of Austenite, *Transactions AIME,* Vol 167, 1946, p 550–595
15.18 H.K.D.H. Bhadeshia, *Bainite in Steels,* Book No. 504, The Institute of Materials, London, 1992
15.19 G. Krauss, Chapter 6: Bainite: An Intermediate Temperature Transformation Product of Austenite, *Steels: Processing, Structure, and Performance*, ASM International, 2005
15.20 H.I. Aaronson, W.T. Reynolds, Jr., G.J. Shiflet, and G. Spanos, Bainite Viewed Three Different Ways, *Metall. Trans. A,* Vol 21A, 1990, p 1343–1380
15.21 R.F. Hehemann and A.R. Troiano, The Bainite Transformation, *Met. Prog.,* 1956, Vol 70 (No. 2), p 97–104
15.22 F.B. Pickering, The Structure and Properties of Bainite in Steels, *Transformation and Hardenability in Steels,* Climax Molybdenum Company of Michigan, Ann Arbor, MI, 1977, p 109–132
15.23 R.F. Hehemann, Ferrous and Nonferrous Bainitic Structures, *Metallography, Structures and Phase Diagrams,* Vol 8, *Metals Handbook,* 8th ed., American Society for Metals, 1973, p 194–196
15.24 G. Baozhu and G. Krauss, The Effect of Low-Temperature Isothermal Heat Treatments on the Fracture of 4340 Steel, *J. Heat Treating,* Vol 4 (No. 4), 1986, p 365–372

CHAPTER 16

Nonequilibrium Reactions—Precipitation Hardening

SOLID-STATE PRECIPITATION REACTIONS are of great importance in engineering alloys. Phase diagram configurations that give rise to precipitation reactions are shown in Fig. 16.1. The reaction occurs when the initial phase composition (e.g., α_0, β_0, or I_0) transforms into a two-phase product that includes a new phase, or precipitate. The precipitate phase may differ in crystal structure, composition, and/or degree of long-range order from that of the initial single-phase (parent) phase and the resultant product matrix. The matrix retains the same crystal structure as the initial one-phase parent but with a different equilibrium composition (α, β, or I) and usually a different lattice parameter than the parent phase. This general type of phase change is different from reactions at the invariant points of phase transformation (e.g., a eutectic or peritectic), where any change in temperature or composition results in complete transformation of the parent into a matrix with a different crystal structure.

16.1 Precipitation Hardening

Precipitation hardening is used extensively to strengthen aluminum alloys, magnesium alloys, nickel-base superalloys, beryllium-copper alloys, and precipitation-hardening (PH) stainless steels. Precipitation hardening is a three-step process in which the alloy is:

1. Heated to a high enough temperature to take a significant amount of an alloying element into solid solution.
2. Rapidly cooled (quenched) to room temperature, trapping the alloying elements in solution.

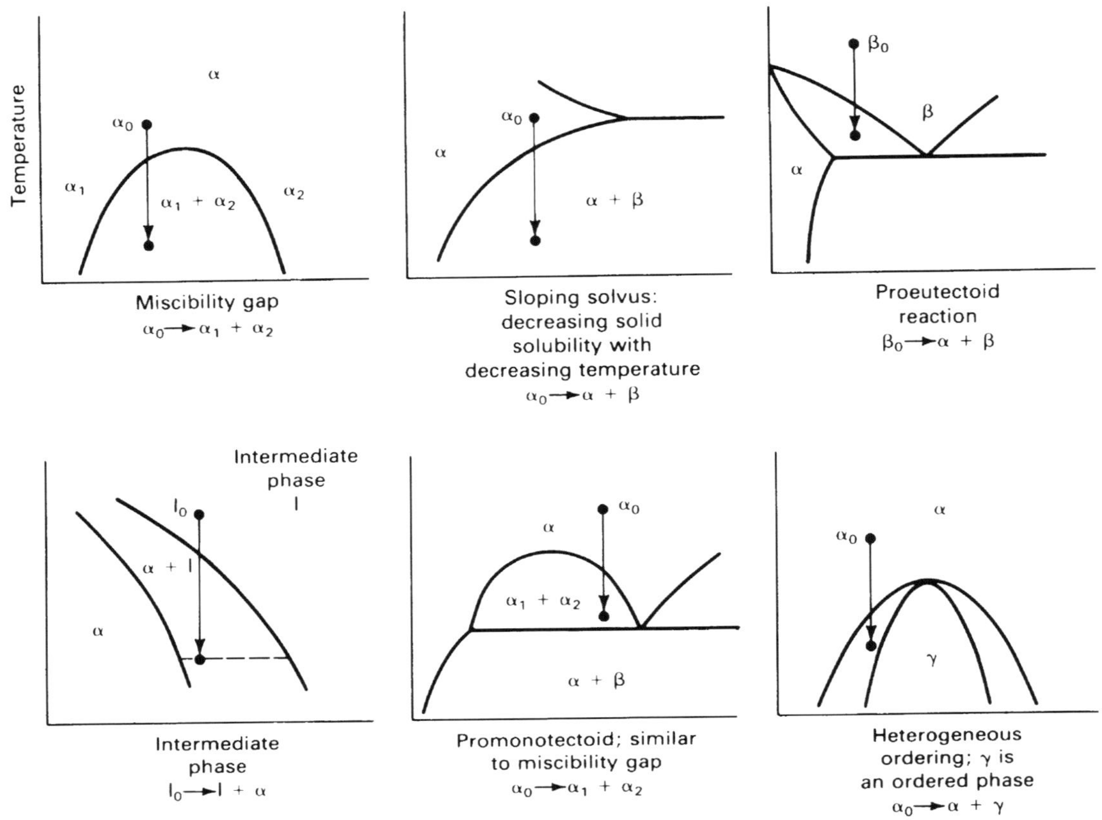

Fig. 16.1 Equilibrium phase diagrams illustrating various conditions for precipitation of a second phase. In all cases, the matrix of the two-phase product has the same crystal structure as the initial one-phase parent, but with a different equilibrium composition (α, β, or I). Source: Ref 16.1 as published in Ref 16.2

3. Reheated to an intermediate temperature so that the host metal rejects the alloying element in the form of fine precipitates that create matrix strains in the lattice. These fine precipitate particles act as barriers to the motion of dislocations and provide resistance to slip, thereby increasing the strength and hardness.

There are two requirements for precipitation hardening: (1) the process must result in an extremely fine precipitate dispersed in the matrix, and (2) there must be a degree of lattice matching between the precipitate particles and the matrix (i.e., the precipitate must be coherent).

Particle hardening is a form of particle, or dispersion, hardening where extremely small particles are dispersed throughout the matrix. When a dislocation encounters a fine particle, it must either cut through the particle or bow (loop) around it, as shown schematically in Fig. 16.2.

Particles are usually classified as deformable or nondeformable, meaning that the dislocation is able to cut through it (deformable) or the particle is so strong that the dislocation cannot cut through (nondeformable). For effective particle strengthening (Fig. 16.3), the matrix should be soft and ductile, while the particles should be hard and discontinuous. A ductile matrix is better in resisting catastrophic crack propagation. Smaller and more numerous particles are more effective at interfering with dislocation motion than larger and more widely spaced particles. Preferably, the particles should

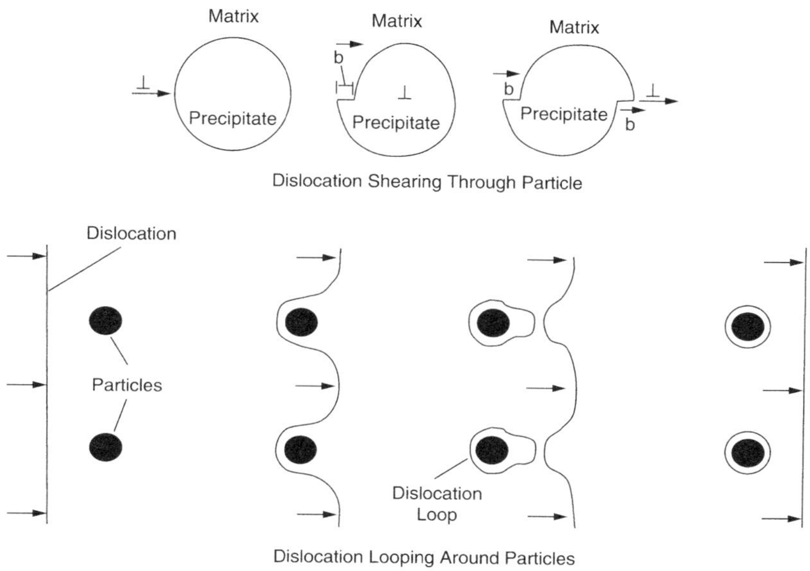

Fig. 16.2 Particle strengthening. Source: Ref 16.3

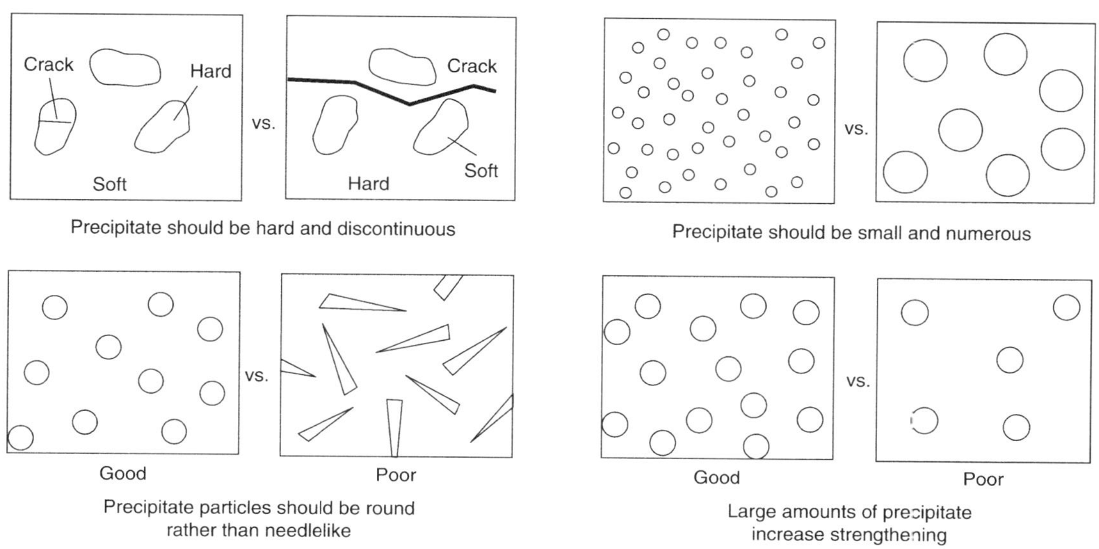

Fig. 16.3 Particle hardening considerations. Source: Ref 16.4 as published in Ref 16.3

be spherical rather than needlelike to prevent stress-concentration effects. Finally, larger amounts of particles increase strengthening.

16.2 Theory of Precipitation Hardening

Precipitation hardening is also known as age hardening or aging, to indicate that the resulting strength increase develops with time. It should be noted that sometimes the term *age hardening* is used to denote alloys that harden on aging at room temperature, while the term *precipitation hardening* is used to denote alloys that must be heated above room temperature for hardening to occur at an appreciable rate. However, in both cases, the hardening mechanism is the precipitation of extremely fine particles, which impedes dislocation movement. A portion of a phase diagram for an aluminum alloy system that has the characteristics required for precipitation hardening is shown in Fig. 16.4. Note that the solvent metal at the left-hand edge of the diagram can absorb much more of the solute metal at elevated temperature than it can at room temperature. When the alloy is heated to the solution heat treating temperature and held for a sufficient length of time, the solvent metal absorbs some of solute metal. Then, when it is rapidly cooled to room temperature, atoms of the solute

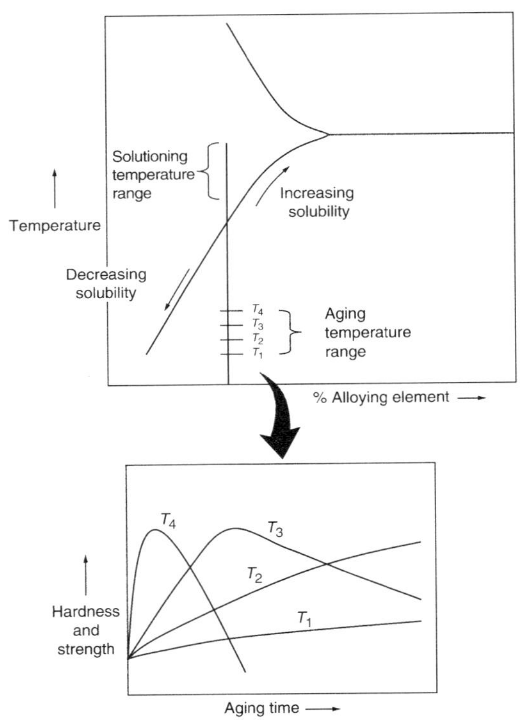

Fig. 16.4 Typical precipitation-hardening heat treatment for an aluminum alloy. Source: Ref 16.5 as published in Ref 16.3

metal are trapped as a supersaturated solid solution in the solvent metal. On reheating to an intermediate aging temperature, the supersaturated solution precipitates very fine particles that act as barriers to dislocation movement. Note the effects of different aging temperatures shown in Fig. 16.4. If the metal is aged at too low a temperature (T_1), the precipitation process will be incomplete and the desired strength will not be achieved, a condition known as underaging. On the other hand, aging at too high a temperature (T_4) also results in lower-than-desired strength because the precipitate particles coarsen, and the alloy is now said to be overaged. Commercial heat treatments are closer to T_2 and T_3, in which the optimal strength can be obtained in a reasonable aging time. The alloy used in this example is one that requires artificial or elevated-temperature aging. Some alloys will age satisfactorily at room temperature, a process called natural aging.

Alloys that harden by precipitation hardening do so by forming coherent precipitates within the matrix. The first step in the aging process is the congregation of solute atoms in the matrix lattice. These solute-rich regions are called clusters and are the embryos for nucleation. Solute atoms then diffuse to the clusters from the surrounding matrix and convert some of them to nuclei of a new phase. During the early phases of precipitation, the equilibrium phase does not immediately form, but an intermediate crystal structure related to it grows in close contact with the solid solution. As long as there tends to be atomic matching, or coherency, between the transition phase and the matrix, the transition phase will create a local strain field within the matrix. The initial precipitate particles are often not spherical but can be platelike or rodlike in shape. The combination of a fine precipitate size and the localized strain fields is an effective barrier to dislocation movement.

For effective precipitation hardening, either a coherent or semicoherent interface must be present. The lattice distortion produced by a coherent precipitate, which impedes dislocation motion, is illustrated in Fig. 16.5. Once the alloy is overaged, an incoherent interface develops that is accompanied by softening of the alloy. The peak strength is usually attained when the alloy is aged to the point where both particle cutting and particle bowing (looping) of dislocations contribute to the alloy strength (Fig. 16.6). The eventual formation of the equilibrium phase is always energetically favorable. If the aluminum alloy is aged at room temperature, equilibrium may never be achieved. However, when the alloy is aged at elevated temperature, either longer aging times or higher temperatures will result in the formation of the final equilibrium precipitate. Because the equilibrium precipitate is larger and more widely spaced, it no longer imposes a strain field on the matrix and is not nearly as effective in blocking dislocation motion, and the strength properties decrease. The equilibrium phase usually nucleates separately from a transition phase(s) and competes with the less stable transition phase for solute atoms, eventually causing the transition phase to redissolve as its particles get smaller. As the volume fraction of equilibrium phase grows, the sizes of the precipitate particles increase, their numbers decrease, and the distance between particles increases, all allowing easier passage of dislocations. It is important to remember that precipitation-hardened alloys are metastable; equilibrium is always lurking around the corner. All precipitation-hardened alloys will eventually soften if heated to high enough temperatures or if they are exposed to somewhat lower temperatures for long periods of time. This being said, some are extremely stable at elevated temperatures, such as the ordered $Ni_3(Al,Ti)$ precipitate in precipitation-hardened nickel-base superalloys that are used in the hottest portions of jet engines.

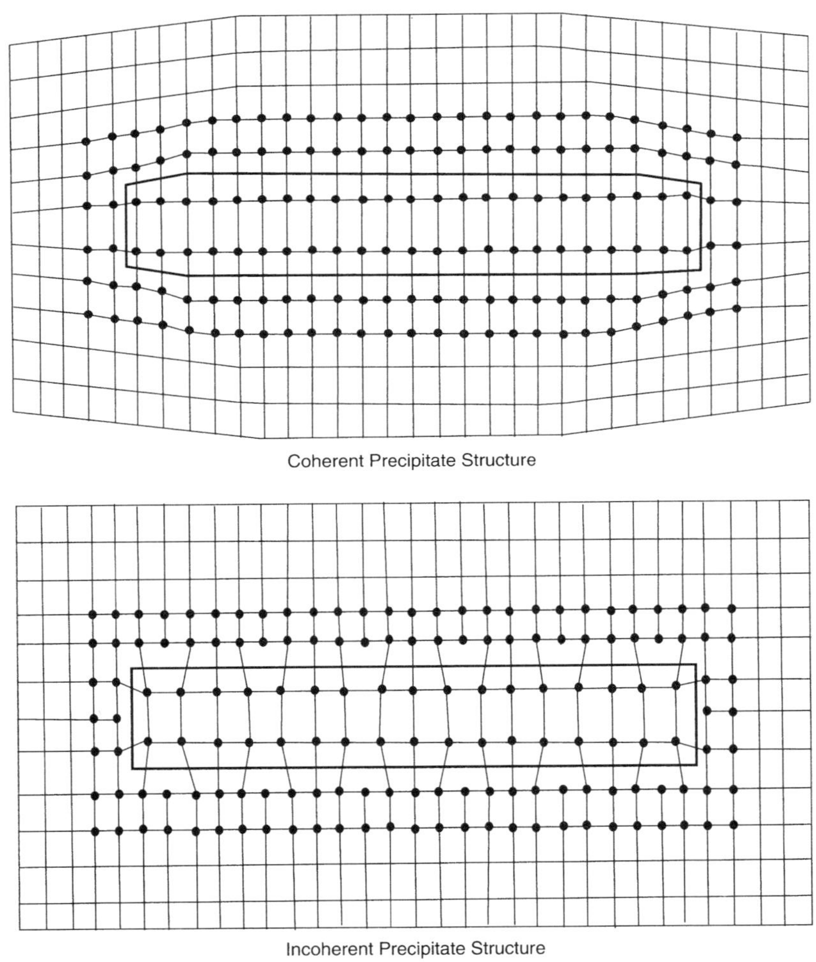

Fig. 16.5 Coherent and incoherent precipitates. Source: Ref 16.3

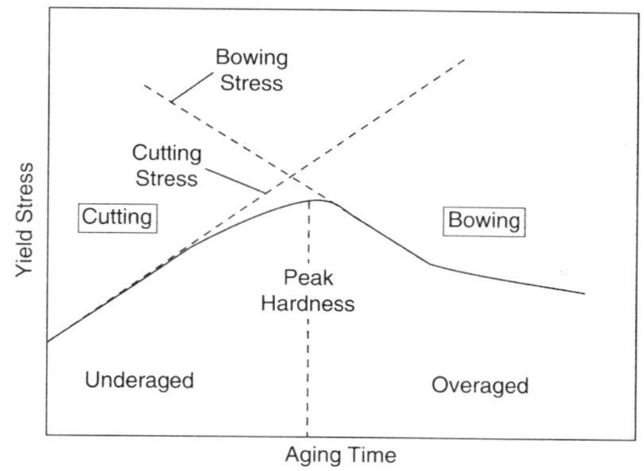

Fig. 16.6 Relative contributions of particle cutting and dislocation bowing. Source: Ref 16.3

Coarsening of the particles occurs because the microstructure of a two-phase alloy is not stable unless the interfacial energy is at a minimum. A lower density of larger particles has less total interfacial energy than a higher density of small particles, which provides the driving force for particle coarsening that occurs by diffusion of solute atoms. Thus, the rate of coarsening increases with temperature. Ostwald ripening is the mechanism by which smaller precipitates dissolve, and the solute is redistributed to larger, stable precipitates. Smaller particles have a higher free energy due to increased pressure as a result of their high surface curvature. Reversion occurs when an alloy containing a coherent precipitate, or an intermediate semicoherent precipitate, is heated above its solvus temperature, allowing the particles to redissolve into the matrix.

Precipitate formation is not always uniform. Remember that grain boundaries are high-energy sites, and precipitates often form along the grain boundaries. Some alloys will form discontinuous precipitates at the grain boundaries, in which lamellae of the second phase are interspersed with the solute-depleted matrix. Small additions of nickel or cobalt are used in beryllium-copper alloys to minimize this effect, because it adversely affects mechanical properties.

A number of alloy systems that can be precipitation hardened are given in Table 16.1. Aluminum alloys are one of the most important series of alloys that can be precipitation hardened, including the 2*xxx* (aluminum-copper), 6*xxx* (aluminum-magnesium-silicon), 7*xxx* (aluminum-zinc), and some of the 8*xxx* (aluminum-lithium) alloys. Some of the copper alloys, in particular beryllium-copper, can be precipitation hardened. The iron- and nickel-base superalloys are another particularly important class of precipitation-hardening alloys. In the nickel-base superalloys, the precipitate $Ni_3(Al,Ti)$ has very little lattice misfit with the nickel matrix ($< 2\%$), which produces very low strain energies (~10 to 30 mJ/m^2) and provides resistance to overaging for prolonged periods at high temperatures.

Table 16.1 Some common precipitation-hardening systems

Matrix	Solute		Transition structures(a)	Equilibrium precipitate
Al	Cu	(i)	Platelike solute-rich GP [1] zones	θ-$CuAl_2$
		(ii)	Ordered GP [2] zones	
		(iii)	θ″ phase	
		(iv)	θ′ phase	
Al	Mg, Si	(i)	GP zones rich in Mg and Si atoms	β″-Mg_2Si
		(ii)	Ordered zones of β″	
Al	Mg, Cu	(i)	GP zones rich in Mg and Cu atoms	*S*-$CuAl_2Mg$
		(ii)	*S*′ platelets	
Al	Mg, Zn	(i)	Spherical zones rich in Mg and Zn	η-$MgZn_2$
		(ii)	Platelets of η′ phase	
Cu	Be	(i)	Be-rich regions	γ-CuBe
		(ii)	γ′ spherical GP zones	
Cu	Co	(i)	Spherical GP zones	β-Co
Ni	Al, Ti	(i)	γ′ cubes	γ-Ni_3(AlTi)
Fe	C	(i)	α′ martensite	Fe_3C
		(ii)	α″ martensite	
		(iii)	ε carbide	
Fe	N	(i)	α′ nitrogen martensite	Fe_4N
		(ii)	α″ nitrogen martensite	

(a) GP, Guinier-Preston. Source: Ref 16.6 as published in Ref 16.3

Nucleation and Growth. Nucleation, growth, and coarsening are important in determining the resultant microstructure of the precipitates and associated mechanical properties. During nucleation, not only is the type of precipitate that forms important, but also the distribution of the precipitates. Distribution of precipitates influences mechanical strength by affecting dislocation motion.

Nucleation can occur either homogenously (uniformly and nonpreferentially) or heterogeneously (preferentially) at specific sites such as grain boundaries or dislocations. Most precipitates involve or require the presence of preferential sites for heterogeneous nucleation, but Guinier-Preston (GP) zones and other fully coherent precipitates (such as Ni_3Al in nickel-base superalloys) nucleate homogeneously. Coherent precipitation occurs when continuity is maintained between the crystal lattice of the precipitate and the lattice of the matrix.

Typical heterogeneous nucleation sites include crystal defects such as grain boundaries, grain corners, vacancies, or dislocations. Heterogeneous nucleation occurs because the elimination of defects and high-energy surfaces (by the nucleation of a new phase) acts to reduce the overall free energy of the system. The rate of heterogeneous nucleation is thus influenced by the density of these irregularities. The free-energy relationships associated with homogenous and heterogeneous nucleation can be described as:

$$\Delta G_{\text{hom}} = -V(\Delta G_{\text{v}} - \Delta G_{\text{s}}) + A\gamma$$

$$\Delta G_{\text{het}} = -V(\Delta G_{\text{v}} - \Delta G_{\text{s}}) + A\gamma - \Delta G_{\text{d}}$$

where:

ΔG_{hom} is the total free-energy change for homogeneous nucleation
ΔG_{het} is the total free-energy change for heterogeneous nucleation
V is the volume of transformed phase
ΔG_v is the volume free energy of transformed phase
ΔG_s is the volume misfit strain energy of transformed phase
$A\gamma$ is the surface area and surface energy term of transformed phase, assuming isotropic
ΔG_d is the free energy resulting from destruction of defect

Typical values for surface energy are summarized in Table 16.2, and various interfaces are shown in Fig. 16.7. Faceted interfaces are often coherent, while nonfaceted interfaces are semicoherent or incoherent. As

Table 16.2 Surface energies for different types of interfaces

Type of interface	Surface energy
Coherent	$\gamma_{coherent} = \gamma_{chemical} \leq 200$ mJ/m^2
Semicoherent	$\gamma_{semicoherent} = \gamma_{chemical} + \gamma_{structural} \approx 200$ to 500mJ/m^2
Incoherent	$\gamma_{incoherent} \approx 500$ to 1000 mJ/m^2

Source: Ref 16.2

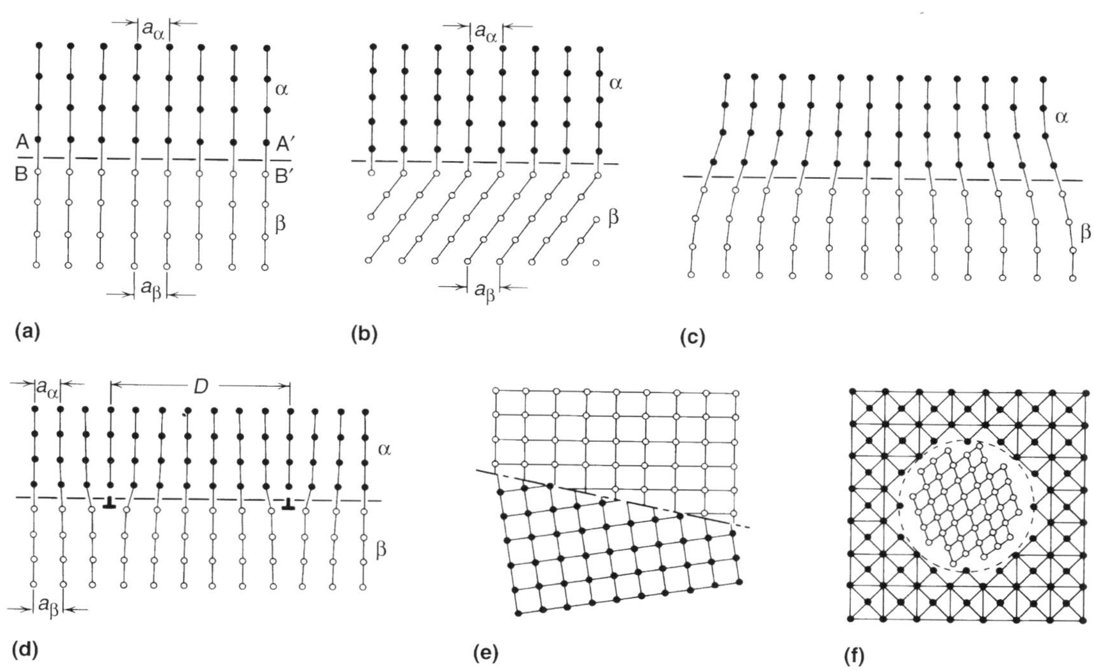

Fig. 16.7 Different types of interfaces. (a) and (b) Fully coherent. (c) and (d) Semicoherent showing lattice strain and the presence of dislocations. (e) and (f) Incoherent. Source: Ref 16.1 as published in Ref 16.2

long as there is a sufficient density of heterogeneous nucleation sites, homogeneous nucleation will not be favored.

A coherent interface (Fig. 16.7a, b) is characterized by atomic matching at the boundary and a continuity of lattice planes, although a small mismatch between the crystal lattices can lead to coherency strains (Fig. 16.7c). Coherent interfaces have a relatively low interfacial energy that typically ranges from 50 to 200 ergs/cm^2 (0.05 to 0.2 J/m^2).

An incoherent interface (Fig. 16.7e, f) is an interphase boundary that results when the matrix and precipitate have very different crystal structures and little or no atomic matching can occur across the interface. The boundary is essentially a high-angle grain boundary characterized by a relatively high interfacial surface energy (~500 to 1000 ergs/cm^2, or 0.5 to 1.0 J/m^2).

Semicoherent interfaces (Fig. 16.7d) represent an intermediate case in which it becomes energetically favorable to partially relax the coherency strains, which would develop if perfect matching occurred across the boundary by introducing an array of misfit dislocations. These interfaces, which are characterized by regions of good fit punctuated by dislocations that accommodate some of the disregistry, have interfacial surface energies from 200 to 500 ergs/cm^2 (0.2 to 0.5 J/m^2).

Dislocations act as nucleation sites only for semicoherent precipitates. Precipitates forming at dislocations is shown in Fig. 16.8. The formation of semicoherent precipitates usually results in the generation of dislocations as a result of the lattice mismatch. The generation of a dislocation maintains coherency by relaxing the strains that develop because of the difference in the lattice parameter at the interface. Vacancies play several roles in the nucleation of precipitates. Vacancies allow for appreciable diffusion at temperatures where diffusion is not expected. They also act to relieve local strains, allowing for the nucleation of coherent precipitates. Vacancy concentration is dependent on temperature, so it is essential for high quenching rates to not only maintain a supersaturated solid solution but to retain a significant number of vacancies.

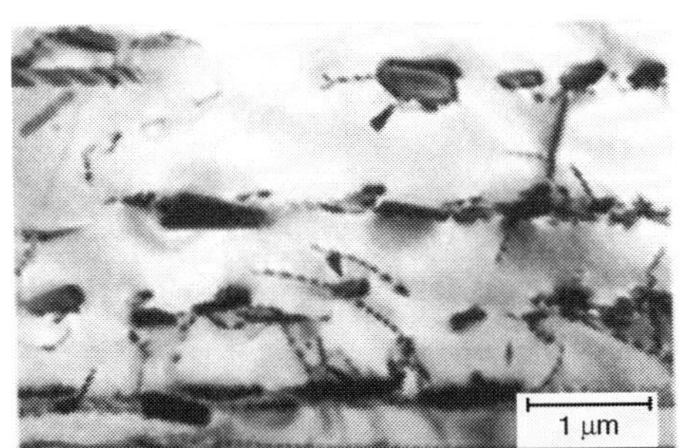

Fig. 16.8 Transmission electron microscopy bright field micrograph showing Ti_5Si_3 precipitates at dislocations in a $Ti_{52}Al_{48}$-$3Si_2Cr$ alloy. Source: Ref 16.7 as published in Ref 16.2

Coarsening of the particles occurs because the microstructure of a two-phase alloy is not stable unless the interfacial energy is at a minimum. A lower density of larger particles has less total interfacial energy than a high density of small particles, which provides the driving force for particle coarsening. Particle coarsening is driven by diffusion of solute atoms; thus, the rate of coarsening increases with temperature. In some cases, coarsening rates are determined by interface control.

Ostwald ripening is the mechanism where the smaller precipitates dissolve and the solute is redistributed to the larger stable precipitates. The higher solubility of the smaller particles in the matrix is termed the capillary effect and can be seen on a free-energy diagram (Fig. 16.9a). Smaller particles have a higher free energy due to increased pressure because of high surface curvature. The point of common tangent (Fig. 16.9a) therefore occurs at a higher solute concentration.

Growth of precipitates occurs by either movement of the incoherent interfaces or a ledge mechanism where coherency is maintained while thickening due to diffusion. Reversion occurs when an alloy containing GP zones or intermediate semicoherent phase is heated above their respective solvus temperatures and they redissolve into the matrix.

Precipitation Sequences. In many precipitation systems and in virtually all effective commercial age-hardening alloys, the supersaturated matrix transforms along a multistage reaction path, producing one or more metastable transition precipitates before the appearance of the equilibrium

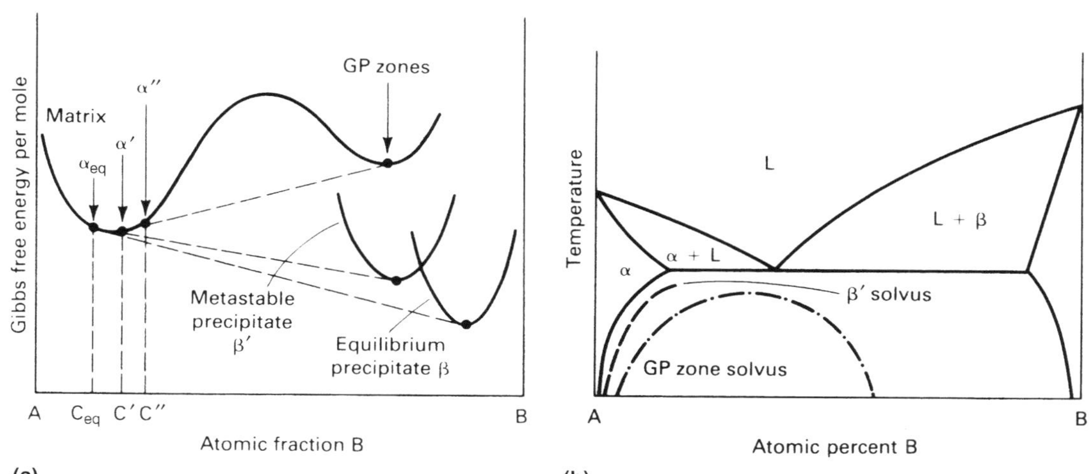

Fig. 16.9 (a) Gibbs free-energy composition diagram and (b) locus of solvus curves of metastable and stable equilibrium phases in a precipitation sequence. (a) The points of common tangency show the relationship between compositions of the matrix phase (C″, C′, and C_{eq}) and the various forms of precipitate phases at a given temperature. From this common tangent construction, it can be see that for the small Guinier-Preston (GP) zones, there is a higher solubility in the matrix. (b) Hypothetical phase diagram showing the locus of metastable and stable solvus curves. Source: Ref 16.1 as published in Ref 16.2

phase. The approach to equilibrium is controlled by the activation (nucleation) barriers separating the initial state from the states of lower free energy. The transition precipitates are often crystallographically similar to the matrix, allowing the formation of a low-energy coherent interface during the nucleation process.

Often, the precipitation sequence begins with the nucleation of small, fully coherent phases known as Guinier and Preston zones (discovered independently by Guinier and Preston from x-ray diffraction studies). Guinier-Preston zones are solute-rich clusters resulting from phase separation or precipitation within a metastable miscibility gap in the alloy system. They may form by homogeneous nucleation and grow at small undercoolings or by spinodal decomposition at large undercoolings or supersaturations.

The GP zones are the first to nucleate because of their small size and coherency with the matrix. The interfacial energy term is extremely low, providing a low barrier to nucleation, although the driving force for nucleation may not be as high as for the final phase to form. The GP zones typically take the shape of small spherical particles or disk-shaped particles (Fig. 16.10) that are approximately two atomic layers thick and several nanometers in diameter, aligned perpendicular to the elastically soft direction in the matrix material crystal structure. The GP precipitates generally grow into more stable transition phases and eventually an equilibrium phase.

The phases that nucleate and grow from the GP zones are termed transition phases. They have an intermediate crystal structure between the matrix and equilibrium phase. This minimizes the strain contribution to energy between the precipitate and the matrix, making it more favorable

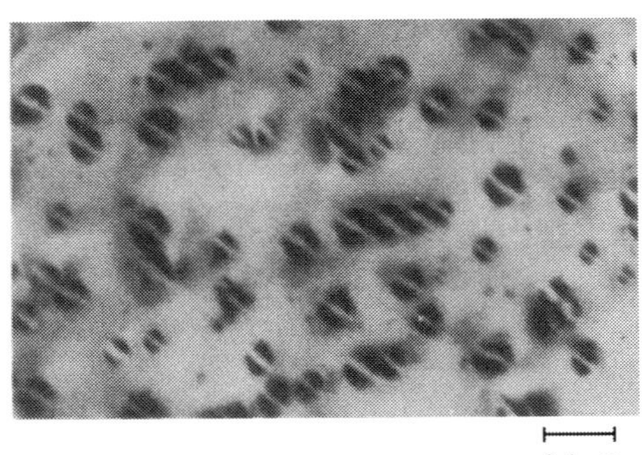

Fig. 16.10 Coherent transition precipitates revealed by strain contrast (dark field) in transmission electron microscopy. The specimen is a Cu-3.1Co alloy aged 24 h at 650 °C (1200 °F). The precipitate is a metastable face-centered cubic (fcc) phase of virtually pure cobalt in the fcc matrix. The particles are essentially spherical, and the "lobe" contrast is characteristic of an embedded "misfitting sphere." This strain contrast reveals the particles indirectly through their coherency strain fields. Original magnification: 70,000×. Source: Ref 16.2

in the nucleation sequence than the equilibrium phases, which is incompatible with the matrix and has high interfacial energy. A typical reaction sequence for aluminum-copper systems is shown in Fig. 16.11 and can be written as:

$$\alpha_0 \rightarrow \alpha_1 + \text{GPZ} \rightarrow \alpha_2 + \theta'' \rightarrow \alpha_3 + \theta' \rightarrow \alpha_{eq} + \theta$$

where θ' and θ'' are transition precipitates and θ is the equilibrium precipitate. Composition of each phase and the matrix can be determined by the common tangent method applied to Fig. 16.12(a). As each new precipitate forms, the matrix, α, becomes more and more depleted in copper. The GP zones and θ'' precipitates are resolved in transmission electron microscopy (TEM) because of the lattice coherency strains. Each step results in the previously precipitated phase being replaced with the new, more stable phase. Figure 16.12(b) outlines the step reductions in total free energy for reactions in the precipitation sequence. The size of the step reduction is the activation energy for a transformation.

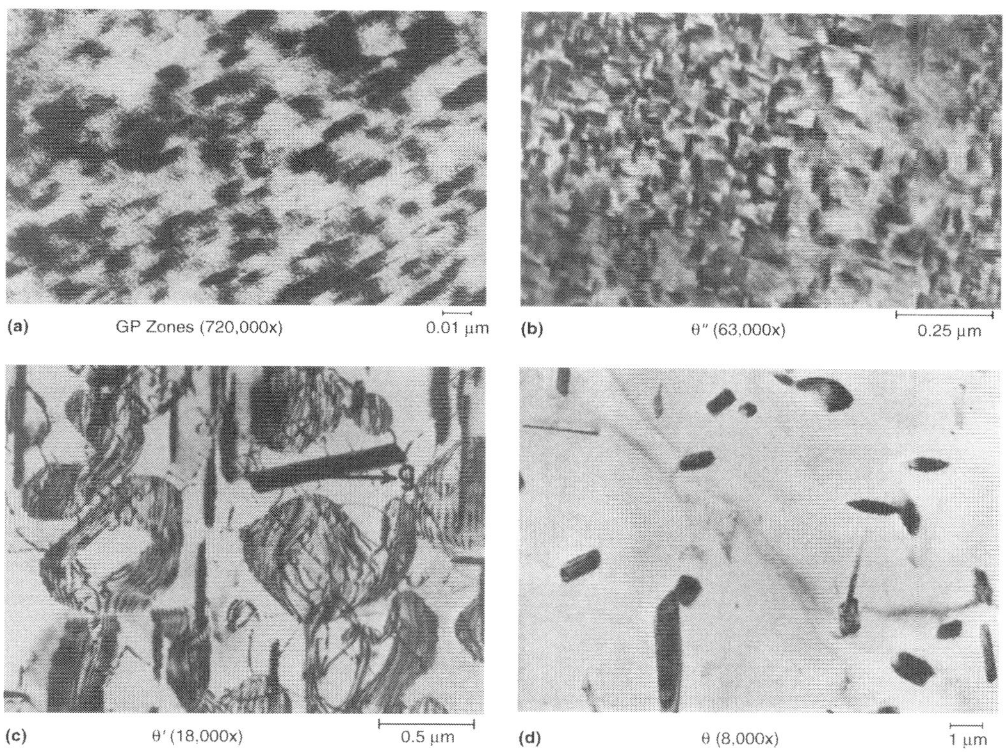

Fig. 16.11 Transmission electron micrographs of precipitation sequence in aluminum-copper alloys. (a) Guinier-Preston zones at 720,000×. (b) θ'' at 63,000×. (c) θ' at 18,000×. (d) θ at 8000×. Source: Ref 16.8 as published in Ref 16.2

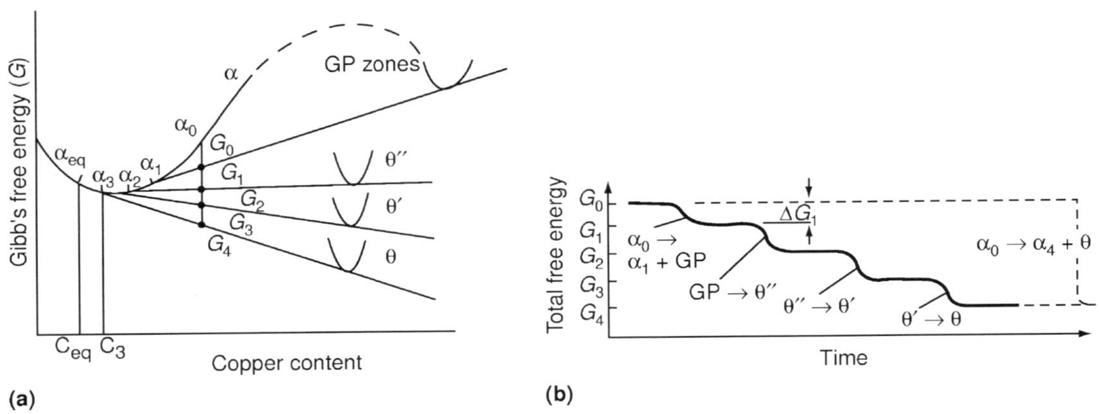

Fig. 16.12 Free-energy plots of precipitation sequence in aluminum-copper alloys. (a) Free-energy curve with common tangent points for phase compositions in the matrix. (b) Step reductions in the free energy as the transformation proceeds. C_{eq} and C_3, copper content of α_{eq} and α_3 phases; ΔG_1, activation energy for $\alpha_0 \rightarrow \alpha_1$ + GP; GP, Guinier-Preston. Source: Ref 16.9 as published in Ref 16.2

16.3 Precipitation Hardening of Aluminum Alloys

The process of strengthening by precipitation hardening plays a critical role in high-strength aluminum alloys. Precipitation hardening consists of three steps: (1) solution heat treating, (2) rapid quenching, and then (3) aging at room or elevated temperature.

In solution heat treating, the alloy is heated to a temperature that is high enough to put the soluble alloying elements in solution. After holding at the solution treating temperature for long enough for diffusion of solute atoms into the solvent matrix to occur, it is quenched to a lower temperature (e.g., room temperature) to keep the alloying elements trapped in solution. During aging, the alloying elements trapped in solution precipitate to form a uniform distribution of very fine particles. Some aluminum alloys will harden after a few days at room temperature, a process called natural aging, while others are artificially aged by heating to an intermediate temperature.

Consider the aluminum-copper system shown in Fig. 16.13. The equilibrium solid solubility of copper in aluminum increases from approximately 0.20% at 250 °C (480 °F) to a maximum of 5.65% at the eutectic melting temperature of 550 °C (1018 °F). It is even lower than 0.20% at temperatures below 250 °C (480 °F). For aluminum-copper alloys containing from 0.2 to 5.6% Cu, two equilibrium solid states are possible. At temperatures above the solvus curve, copper is completely soluble, and when the alloy is held at such temperatures for sufficient time to permit diffusion, copper will be taken completely into the α solid solution. At temperatures below the solvus, the equilibrium state consists of two solid phases—solid solution α plus the equilibrium intermetallic compound θ ($CuAl_2$).

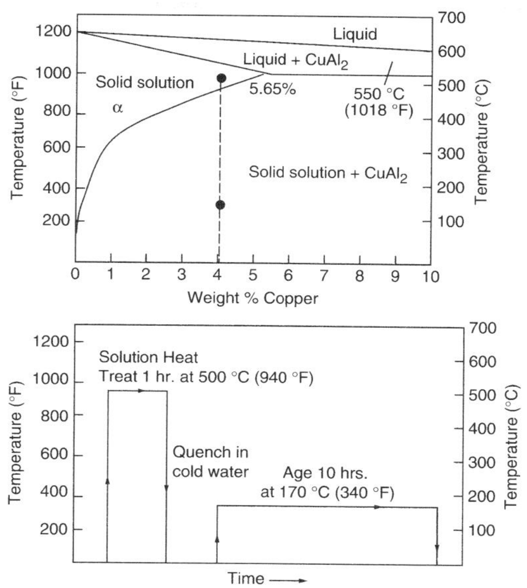

Fig. 16.13 Precipitation hardening of an aluminum-copper alloy. Source: Ref 16.3

If an alloy of aluminum containing 4% Cu is heated to 500 °C (940 °F) and held for 1 h, the copper will go into solution in the aluminum. After solution heat treating, the alloy is quenched in cold water to room temperature to keep the copper in solution. The alloy is then artificially aged at 170 °C (340 °F) for 10 h. During the aging process, very fine particles of aluminum-copper are precipitated and the strength and hardness increase dramatically. Precipitation heat treatments generally are low-temperature, long-term processes. Aging temperatures for aluminum range from 115 to 190 °C (240 to 375 °F) with times between 5 and 48 h.

On quenching, the copper is trapped as a supersaturated solution in the a matrix. There is a strong driving force to precipitate the copper as the equilibrium precipitate θ ($CuAl_2$). However, in aluminum alloys, precipitation occurs by one or more metastable transition precipitates appearing before the appearance of the final equilibrium phase, with each successive stage lowering the free energy of the system. The transition precipitates are often crystallographically similar to the matrix, allowing the formation of a low-energy coherent interface during the nucleation process. Often, the precipitation sequence begins with the nucleation of small, fully coherent GP zones.

The GP zones are the first to nucleate because of their small size and coherency with the matrix. The interfacial energy term is extremely low, providing a low barrier to nucleation. The GP zones are extremely fine, with sizes in the range of tens of angstroms. The exact shape, size, and distribution of the GP zones depends on the specific alloy and on the thermal and mechanical history of the product. The GP zones typically take the shape of small spherical particles or disk-shaped particles that are approximately two atomic layers thick and 10 nm in diameter, with spacings on the order of 10 nm. Their formation requires movements of atoms over very short distances and their densities can approach 1017 to 1018 cm^{-3}. As previously discussed, the GP precipitates then grow into more stable transition phases and eventually the equilibrium phase.

The transition phases that form in aluminum-copper alloys are shown in Fig. 16.14. The transition phases have a crystal structure intermediate between the matrix and the equilibrium phase. This minimizes the strain energy between the precipitate and the matrix, making it more favorable in the nucleation sequence than the equilibrium phase, which is incompatible with the matrix and has a high interfacial energy. A typical reaction sequence for aluminum-copper systems is:

$$\text{Supersaturated solid solution} \rightarrow \text{Clustering} \rightarrow \text{GP zones} \rightarrow \theta'' \rightarrow \theta' \rightarrow \theta$$

where θ'' and θ' are transition precipitates and θ is the final equilibrium precipitate. While the GP zones are totally coherent with the matrix, the much larger transition precipitates are only semicoherent. As each new precipitate forms, the matrix, α, becomes more and more depleted in copper. The GP zones and θ'' precipitates, also known as GP II zones, can be resolved by a transmission electron microscope because of the lattice coherency strains, as shown in Fig. 16.11. The zones themselves are too small to be resolved but the resulting strain fields can be resolved. Each step results in the previously precipitated phase being replaced with a new more stable phase with a lower free energy.

During heating, the GP zones develop an intermediate precipitate, θ'', which has a tetragonal structure that forms as plates and maintains coherency with the matrix and further increases the strain in the matrix, providing peak strength levels. On still further heating, θ'' is replaced by a second intermediate precipitate, θ', which is not coherent with the matrix, and the strength starts to decrease; the alloy is now termed overaged. However, in the highest strength condition, both θ'' and θ' are generally present. Both the precipitate particles themselves and the strains they produce in the lattice structure inhibit dislocation motion, and thus both contribute to strengthening. To maximize strengthening, aging is typically carried out at temperatures between those where precipitation of θ'' and θ' occurs, because this spacing and lattice strain are ideal for hindering dislocation motion. Further heating of the alloy causes θ' to transform to the equilibrium precipitate θ, which is stoichiometric $CuAl_2$.

The solvus lines for the GP zones and the transition phases can be shown as lines on the phase diagram (Fig. 16.15). This series of lines defines the upper temperature limit of the various transition phases for different con-

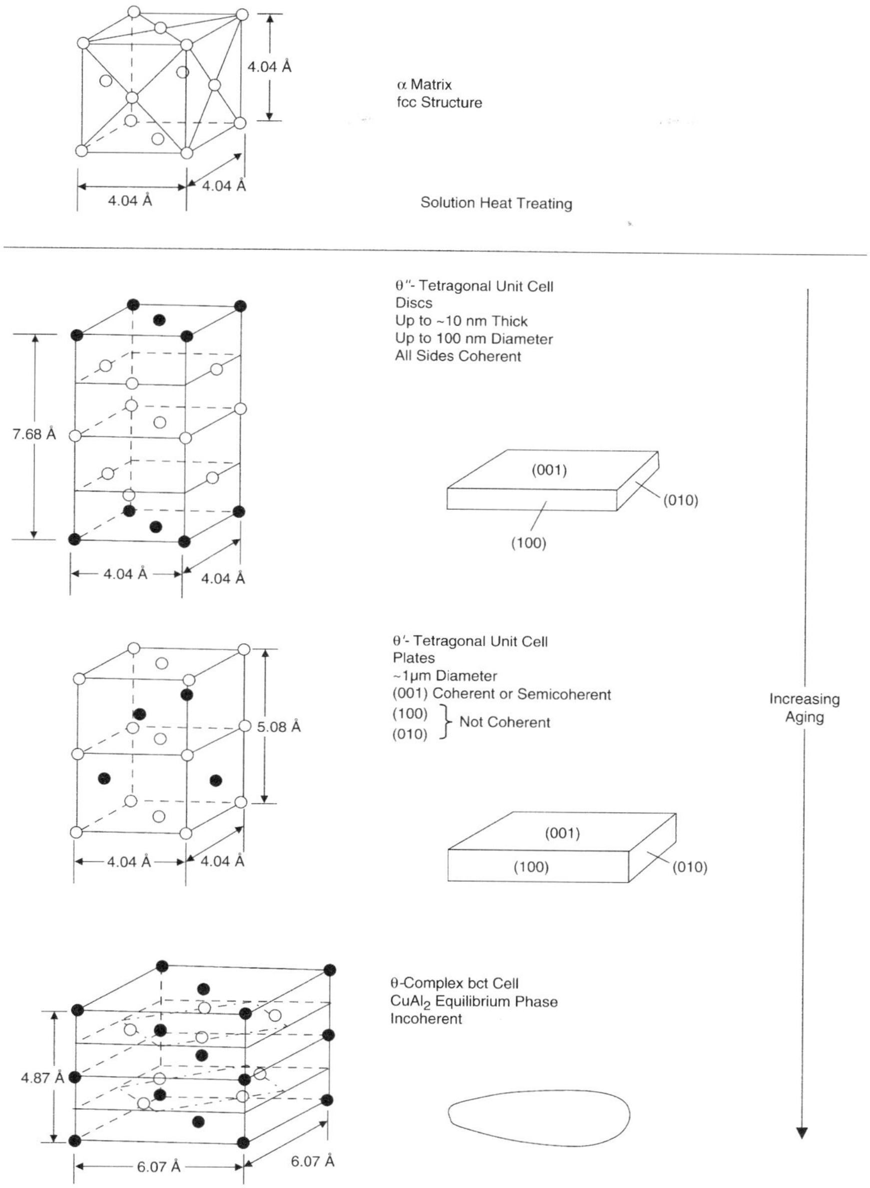

Fig. 16.14 Aluminum-copper precipitation sequence. fcc, face-centered cubic; bct, body-centered tetragonal. Source: Ref 16.8 as published in Ref 16.3

centrations. For example, if aging is carried out above the θ″ solvus but below the θ′ solvus, then the first precipitate to form will be θ′.

Guinier-Preston zones will normally develop on aging at room temperature. The fact that this will happen at room temperature is somewhat surprising and can be attributed to a high vacancy concentration. When the

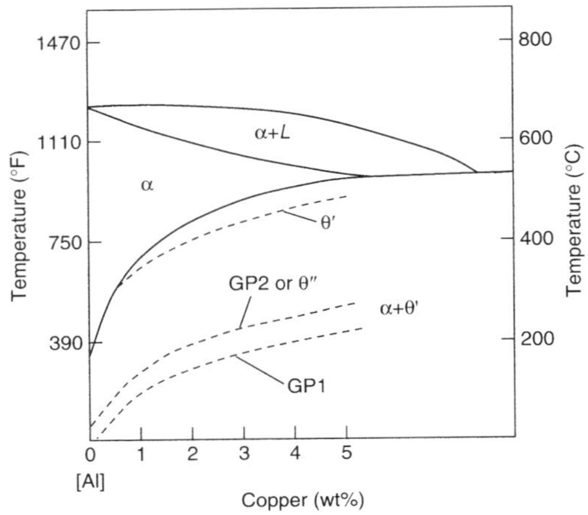

Fig. 16.15 Aluminum-copper binary diagram with GP_I, θ″, and θ′ solvus lines. Source: Ref 16.8 as published in Ref 16.3

alloy is heated to the solution heat treating temperature, the equilibrium number of vacancies increases as the temperature increases. On quenching, the vacancies become trapped in solution. These excess vacancies are then available to help accelerate the nucleation and growth process.

A series of aging curves for an Al-4% Cu alloy is shown in Fig. 16.16. Reactions carried out beyond maximum strengthening are overaged, because the beneficial effects of precipitation strengthening are lost as the precipitates grow larger in size and spacing. Both the underaged and overaged conditions have lower strengths and hardness levels than the peak aged condition.

In aluminum alloys, precipitate-free zones (PFZs) can occur adjacent to the grain boundaries (Fig. 16.17). The grain boundaries themselves and the interior of the grains contain precipitate particles, but there is a zone adjacent to the boundaries with very few particles. These differences in chemical composition can set up galvanic effects that lead to intergranular corrosion. There are two plausible explanations for PFZs. The first is due to vacancy migration. Grain boundaries are a major sink for vacancies that migrate to the grain boundaries and deplete the areas adjacent to the boundaries of vacancies. The lack of vacancies in these depleted areas inhibits the nucleation and growth of precipitate particles, even though the matrix in these areas contains sufficient solute. The second explanation is that precipitation occurs directly on the higher-energy grain boundaries and the adjacent areas become depleted of solute. There is direct experimental evidence for both of these mechanisms. Special heat treatments are used to minimize the formation of PFZs. Precipitate-free zones can be eliminated by a two-stage heat treatment where nucleation is induced homogeneously at a low temperature, and the precipitates are then allowed

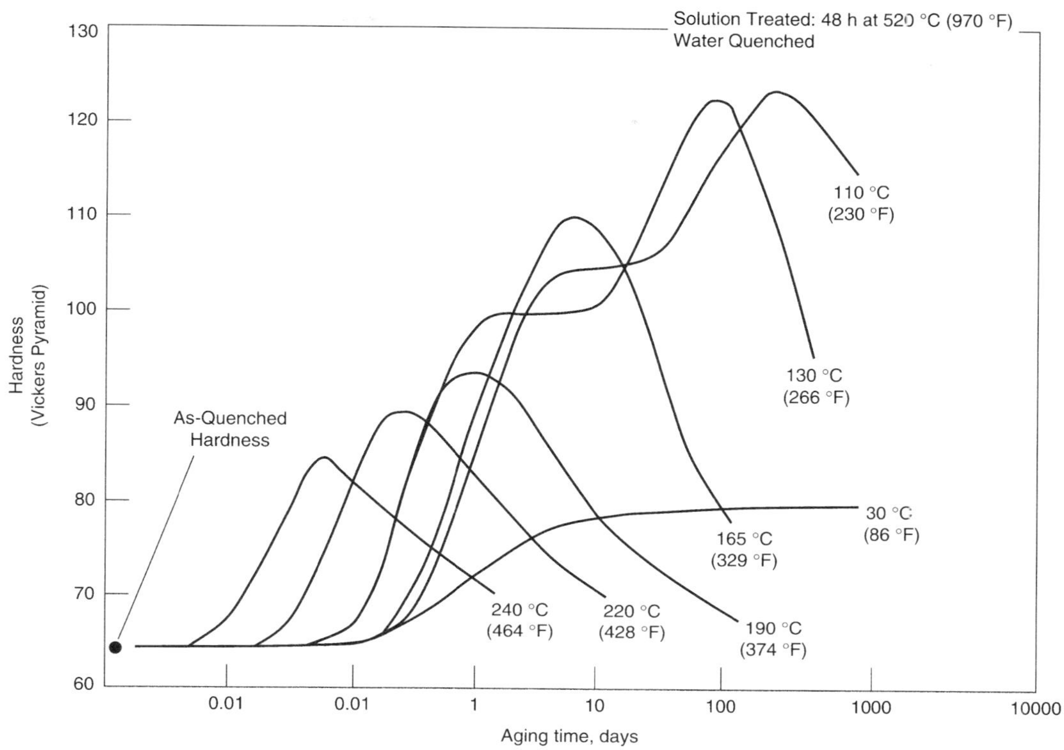

Fig. 16.16 Aging curves for Al-4% Cu alloy. Source: Ref 16.3

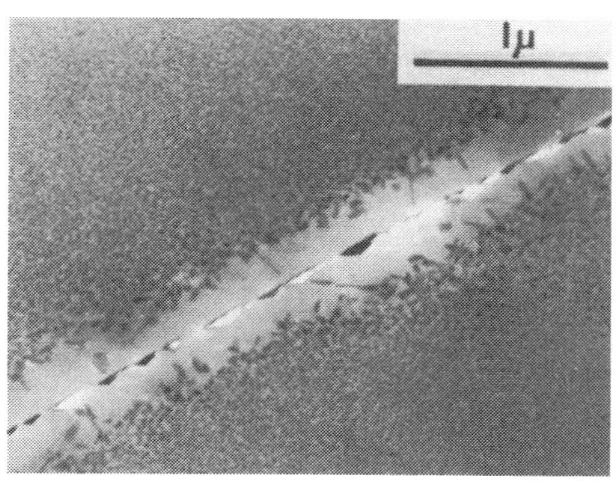

Fig. 16.17 Precipitate-free zone (PFZ) in an Al-Zn-Mg alloy. Source: Ref 16.10 as published in Ref 16.3

to grow during the second, higher-temperature aging treatment. In addition, higher solution treating temperatures and faster quenching rates also reduce the PFZ widths. The beneficial effect of a faster quenching rate is shown in Fig. 16.18.

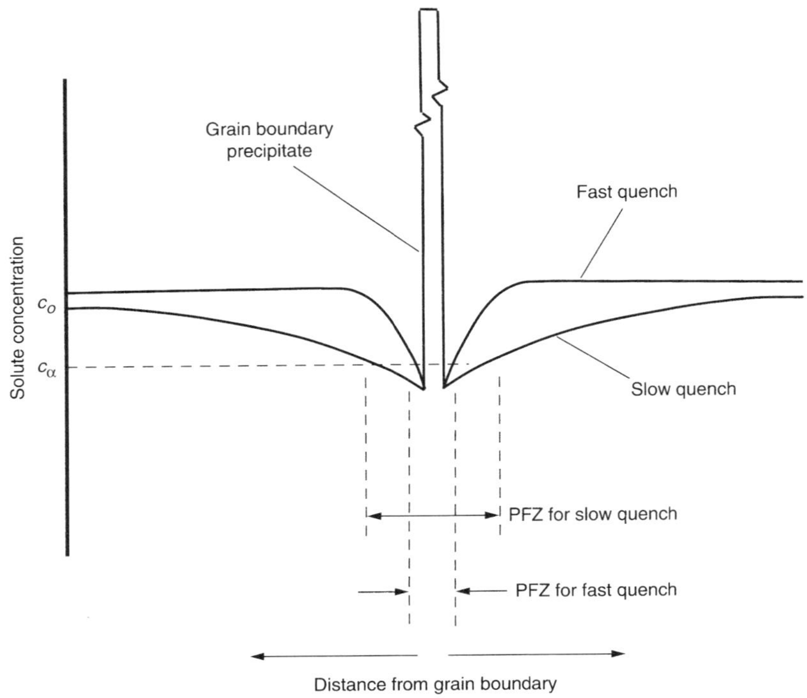

Fig. 16.18 Precipitate-free zone (PFZ) width as function of quench rate. Source: Ref 16.8 as published in Ref 16.3

16.4 Precipitation Hardening of Nickel-Base Superalloys

Another class of important alloys strengthened by precipitation hardening is the nickel-base superalloys. Superalloys are heat-resistant alloys of nickel, iron-nickel, and cobalt that frequently operate at temperatures exceeding 540 °C (1000 °F). Superalloys are the primary materials used in the hot portions of jet turbine engines, such as the blades, vanes, and combustion chambers, constituting over 50% of the engine weight. In general, the nickel-base alloys are used for the highest temperature applications, followed by the cobalt-base alloys and then the iron-nickel alloys.

Nickel-base superalloys are strengthened by a combination of solid-solution hardening, precipitation hardening, and the presence of carbides at the grain boundaries. The face-centered cubic (fcc) nickel matrix, which is designated as austenite (γ), contains a large percentage of solid-solution elements such as iron, chromium, cobalt, molybdenum, tungsten, titanium, and aluminum. Aluminum and titanium, in addition to being potent solid-solution hardeners, are also precipitation strengtheners. At temperatures above 0.6 T_m, which is in the temperature range for diffusion-controlled creep, the slowly diffusing elements molybdenum and tungsten are beneficial in reducing high-temperature creep.

The most important precipitate in nickel- and iron-nickel-base superalloys is γ' fcc-ordered $Ni_3(Al,Ti)$ in the form of either Ni_3Al or Ni_3Ti. The γ' phase is precipitated by precipitation-hardening heat treatments—solution heat treating followed by aging. The γ' precipitate is an A_3B type compound where A is composed of the relatively electronegative elements nickel, cobalt, and iron, and B is composed of the electropositive elements aluminum, titanium, or niobium. Typically, in the nickel-base alloys, γ' is of the form $Ni_3(Al,Ti)$, but if cobalt is added, it can substitute for some nickel as $(Ni,Co)_3(Al,Ti)$. The precipitate γ' has only approximately a 0.1% mismatch with the γ matrix; therefore, γ' precipitates homogeneously with a low surface energy and has extraordinary long-term stability. The coherency between γ' and γ is maintained to high temperatures and has a very slow coarsening rate, so that the alloy overages extremely slowly, even as high as 0.7 T_m, where T_m is the absolute melting temperature. Because the degree of order in $Ni_3(Al,Ti)$ increases with temperature, alloys with a high volume of γ' actually exhibit an increase in strength as the temperature is increased up to approximately 700 °C (1300 °F). The γ/γ' mismatch determines the γ' precipitate morphology, with small mismatches (~ 0.05%) producing spherical precipitates and larger mismatches producing cubical precipitates, as shown in Fig. 16.19.

Wrought precipitation-strengthened nickel-base superalloys are solution heat treated and then aged to produce the desired properties. Solution-treating temperatures range from approximately 980 to 1230 °C (1800 to

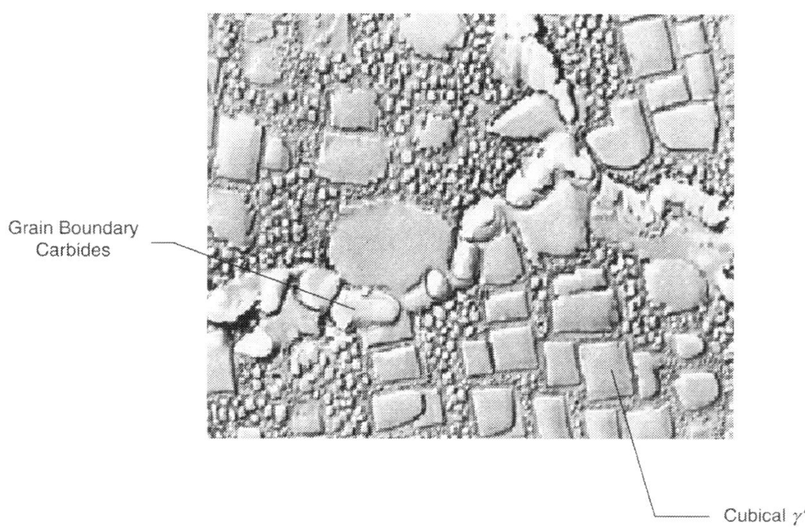

Fig. 16.19 Microstructure of precipitation-hardened nickel-base superalloy. Source: Ref 16.3

2250 °F), or even up to 1315 °C (2400 °F) for some single-crystal alloys. Long exposure times at solution-treatment temperatures can result in partial dissolution of primary carbides with subsequent grain growth. The solution treatment may be above or below the γ' solvus, depending on the desired microstructure and application. A higher temperature is used to develop coarser grain sizes for creep- and stress-rupture-critical applications, while a lower solution temperature will produce a finer grain size for enhanced tensile and fatigue properties. To retain the supersaturated solution obtained during solution treating, the part is rapidly cooled to room temperature using either gas cooling or water or oil quenching.

Aging treatments are used to strengthen precipitation-strengthened alloys by precipitating one or more phases (γ' or γ''). Aging treatments vary from as low as 620 °C (1150 °F) to as high as 1040 °C (1900 °F). Double-aging treatments are also used to produce different sizes and distributions of precipitates. A principal reason for double-aging treatments, in addition to γ' and γ'' control, is the need to precipitate or control grain-boundary carbide morphology. Aging heat treatments usually range from 870 to 980 °C (1600 to 1800 °F), with times of approximately 4 to 32 h. Either single or multiple aging treatments are then used to precipitate γ'. Like the solution temperature, the aging temperatures and times are selected depending on the intended application. Higher aging temperatures will produce coarse γ' precipitates desirable for creep and stress rupture applications, while lower aging temperatures produce finer γ' precipitates for applications requiring strength and fatigue resistance. In general, lower solution-treating temperatures produce better strength, while higher solution-treating temperatures provide better creep and stress rupture properties.

ACKNOWLEDGMENTS

The material for this chapter came from "Structures by Precipitation from Solid Solution," by M. Epler in *Metallography and Microstructures,* Vol 9, *ASM Handbook*, ASM International, 2004, and *Elements of Metallurgy and Engineering Alloys* by F.C. Campbell, ASM International, 2008.

REFERENCES

16.1 W.A. Soffa, Structures Resulting from Precipitation from Solid Solution, *Metallography and Microstructures,* Vol 9, *ASM Handbook,* ASM International, 1985, p 646–650
16.2 M. Epler, Structures by Precipitation from Solid Solution, *Metallography and Microstructures,* Vol 9, *ASM Handbook*, ASM International, 2004, p 134–139
16.3 F.C. Campbell, *Elements of Metallurgy and Engineering Alloys*, ASM International, 2008
16.4 D.R. Askeland, *The Science and Engineering of Materials*, 2nd ed., PWS-Kent Publishing Co., 1989

16.5 F.C. Campbell, *Manufacturing Technology for Aerospace Structural Materials*, Elsevier Scientific, 2006
16.6 R.E. Smallman and R.J. Bishop, *Modern Physical Metallurgy and Materials Engineering,* Butterworth Heinemann, 1999
16.7 S. Celotto and T.J. Bastow, Study of Precipitation in Aged Binary Mg-Al and Ternary Mg-Al-Zn Alloys Using 72AlNMR Spectroscopy, *Acta Mater.,* Vol 49, 2001, p 41–51
16.8 D.A. Porter and K.E. Easterling, *Phase Transformations in Metals and Alloys,* 2nd ed., Chapman and Hall, 1996
16.9 W.H. Van Geertruyden, W.Z. Misiolek, G.G. Lea, and R.M. Kelly, Thermal Cycle Simulation of 6*xxx* Aluminum Alloy Extrusion, *Proc. Seventh International Extrusion Technology Seminar,* Aluminum Extruders Council, 2000
16.10 W.T. Becker and S. Lampman, Fracture Appearance and Mechanisms of Deformation and Fracture, *Failure Analysis and Prevention,* Vol 11, *ASM Handbook,* ASM International, 2002

APPENDIX A

Review of Metallic Structure

THE WORD *METAL,* derived from the Greek *metallon,* is believed to have originated as a verb meaning to seek, search after, or inquire about. Today, a metal is defined as any element that tends to lose electrons from the outer shells of its atoms. The resulting positive ions are held together in crystalline structure by the cloud of these free electrons in what is known as the metallic bond. The metallic bond yields three physical characteristics typical of solid metals: (1) metals are good conductors of electricity, (2) metals are good conductors of heat, and (3) metals have a lustrous appearance. In addition, most metals are malleable, ductile, and generally denser than other elemental substances. Those elements that do not display the characteristics of the metallic elements are called nonmetals. However, there are some elements that behave as metals under some circumstances and as nonmetals under different circumstances. These are now called semimetals, but have also been called metalloids, meaning like metals. The boundaries separating the regions in the periodic table covered by the different classes of elements are not distinct, except that nonmetals never form positive ions. A simplified periodic table is shown in Fig. A.1, highlighting the elements that are currently considered to be metals.

A.1 Periodic Table

In the periodic table, it is the number of electrons in the outer shell that affects the properties of the elements the most. Those elements that have the same number of electrons in their outermost electron shells, and therefore have similar chemical behavior, are placed in columns. For example, lithium, sodium, and potassium each have a single electron in their outer shells and are chemically very similar. They all oxidize very rapidly and react vigorously with water, liberating hydrogen and forming soluble hydroxides. They are physically very similar, being soft light metals with

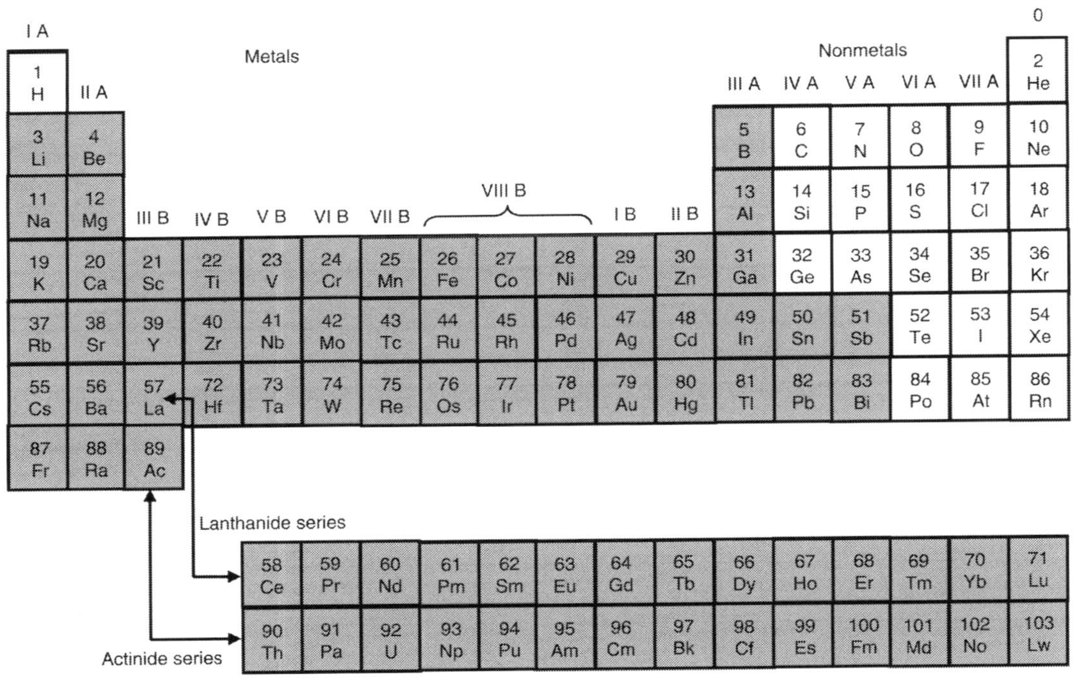

Fig. A.1 Periodic table of the elements. Source: Ref A.1

a somewhat silver color. At the other end of the periodic table, the gases fluorine and chlorine, with seven electrons in their outer shells, also have similar chemical properties. Both are gases with strong nonmetallic properties. At the far right side of the periodic table, the noble gases helium, neon, and argon all contain eight electrons in their outer shells. Because this fills the shell, these gases are nonreactive, or inert, under normal circumstances. Therefore, the chemical interaction between elements is governed by the number of electrons present in the outer shell. When the outer shell is filled, the atom has no further tendency to combine or react with other atoms.

Metallic properties depend on both the nature of their constituent atoms and the way in which they are assembled. Assemblies of atoms can be gases, liquids, or solids. When they are in the solid state, metals are normally arranged in a crystalline structure. The crystalline nature of metals is responsible for their ultimate engineering usefulness, and the crystalline arrangement strongly influences their processing. Although metals can exist as single crystals, they are more commonly polycrystalline solids with crystalline grains of repeating atomic packing sequences. Periodic crystalline order is the equilibrium structure of all solid metals. Crystalline structures are a dominant factor in determining mechanical properties, and crystal structures also play an important role in the magnetic, electrical, and thermal properties. The greatest bonding energy occurs when the atoms are closely packed, and the atoms in a crystalline structure tend to pack as densely as possible. In addition, total metallic bonding energy is increased when each atom has the greatest possible number of nearest-neighbor atoms. However, due to a shared bonding arrangement in some metals that is partially metallic and partially covalent, some metals do not crystallize into these close-packed structures. Covalent tendencies appear as one moves closer to the nonmetals on the periodic table. As one moves rightward across the periodic table, progressively greater numbers of metals have looser-packed structures. Most metals bordering the nonmetals possess more complex structures with lower packing densities, because covalent bonding plays a large role in determining their crystal structures. The directionality of covalent bonding dictates fewer nearest neighbors than exist in densely packed metallic crystals. For metals near the nonmetals on the right side of the periodic table, where electronegativities are high, covalency becomes a major part of the bonding.

Properties important to the engineer are strongly influenced by crystal structure. One of the most important properties related to crystal structure is ductility. Densely packed structures usually allow motion on one or more slip planes, permitting the metal to deform plastically without fracturing. Ductility is vital for easy formability and for fracture toughness, two properties that give metals a great advantage over ceramic materials for many engineering uses.

A.2 Bonding in Solids

Bonding in solids may be classified as either primary or secondary bonding. Methods of primary bonding include the metallic, ionic, and covalent bonds. Secondary bonds are much weaker bonding mechanisms that are only predominant when one of the primary bonding mechanisms is absent. When two atoms are brought close to each other, there will be a repulsion between the negatively charged electrons of each atom. The repulsion force increases rapidly as the distance of separation decreases. However, when the separation is large, there is attraction between the positive nucleus charge and the negative charge of the electrons. At some equilibrium distance, the attractive and repulsive forces balance each other and the net force is zero. At this equilibrium distance, the potential energy is at a minimum, as shown in Fig. A.2. The magnitude of this energy is known as the bond energy, usually expressed in kJ/mol. Primary bond energies range from 100 to 1,000 kJ/mol, while the much weaker secondary bonds are on the order of only 1 to 60 kJ/mol. The equilibrium distance, a_o, is the bond length. Strong primary bonds have large forces of attraction with bond lengths of 1 to 2 Å, while the weaker secondary bonds have larger bond lengths of 2 to 5 Å. While it is convenient to discuss the four major types of bonding separately, it should be recognized that although metallic bonding may be predominate, other types of bonding, in particular covalent bonding, may also be present. A comparison of the some of the properties of the different bond types is given in Table A.1.

Metallic bonding occurs when each of the atoms of the metal contributes its valence electrons to the formation of an electron cloud that surrounds the positively charged metal ions, as illustrated in Fig. A.3. Hence, the valence electrons are shared by all of the atoms. In this bond, the positively charged ions repel each other uniformly, so they arrange themselves into a regular pattern that is held together by the negatively charged electron cloud. Because the negative electron cloud surrounds each of the positive ions that make up the orderly three-dimensional crystal structure, strong electronic attraction holds the metal together. A characteristic of metallic bonding is the fact that every positive ion is equivalent. Ideally, a

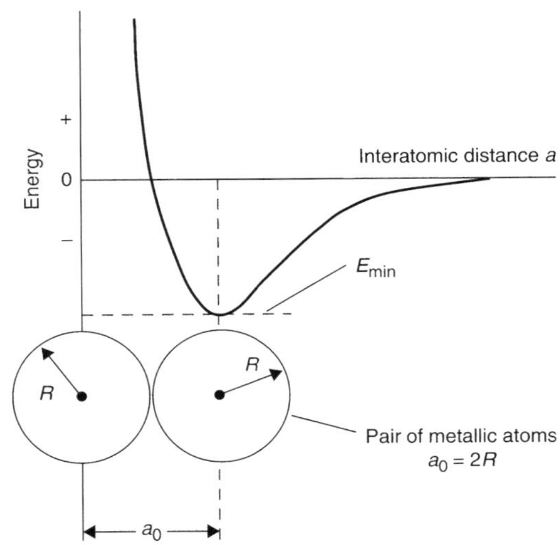

Fig. A.2 Bond energy in metallic bond. Source: Ref A.1

Table A.1 General characteristics of bond types

Properties	Metallic bond	Covalent bond	Ionic bond	Secondary bonds
Examples	Copper, nickel, iron	Diamond, carborundum	NaCl, $CaCl_2$	Wax, argon
Mechanical	Weaker than ionic or covalent bond. Tough and ductile. Nondirectional	Very hard and brittle. Fails by cleavage. Strongly directional	Hardness increases with ionic charge. Fails by cleavage. Nondirectional	Weak and soft. Can be plastically deformed
Thermal	Moderately high melting points. Good conductors of heat	Very high melting points. Thermal insulators	Fairly high melting points. Thermal insulators	Low melting points
Electrical	Conductors	Insulators	Insulators	Insulators
Optical	Opaque and reflecting	Transparent or opaque. High refractive index	Transparent. Colored by ions	Transparent

Source: Ref A.1 as published in Ref A.2

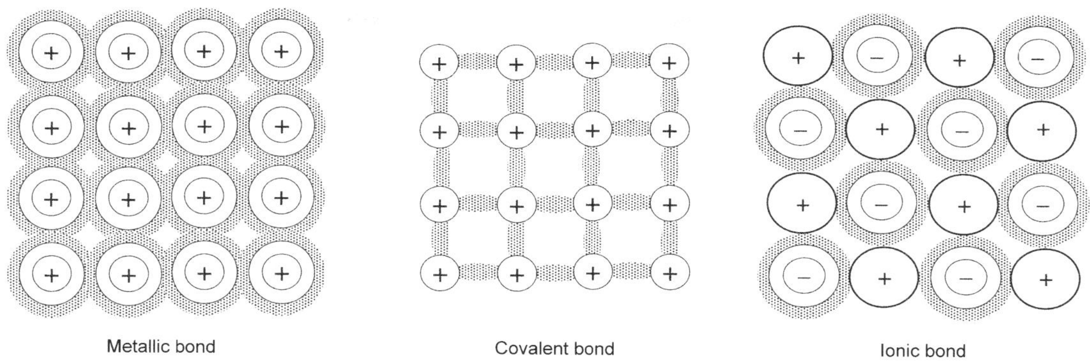

Fig. A.3 Primary bonding mechanisms. Source: Ref A.1

symmetrical ion is produced when a valence electron is removed from the metal atom. As a result of this ion symmetry, metals tend to form highly symmetrical, close-packed crystal structures. They also have a large number of nearest-neighboring atoms (usually 8 to 12), which helps to explain their high densities and high elastic stiffness.

Because the valence electrons are no longer attached to specific positive ions and they are free to travel among the positive ions, metals exhibit high electrical and thermal conductivity. The opaque luster of metals is due to the reflection of light by the free electrons. A light wave striking the surface causes the free electrons to vibrate and absorb all the energy of the wave and prevent transmission. The vibrating electrons then re-emit, or reflect, the wave from the surface. The ability of metals to undergo significant amounts of plastic deformation is also due to the metallic bond. Under the action of an applied shearing force, layers of the positive ion cores can slide over each other and re-establish their bonds without drastically altering their relationship with the electron cloud. The ability to alloy, or mix several metals together in the liquid state, is one of the keys to the flexibility of metals. In the liquid state, solubility is often complete, while in the solid state, solubility is generally much more restricted. This change in solubility with temperature forms the basis for heat treatments that can vary the strength and ductility over quite wide ranges.

In general, the fewer the valence electrons and more loosely they are held, the more metallic is the bonding. Metals such as copper and silver, which have few valence electrons, are very good conductors of electricity and heat, because their few valence electrons are highly mobile. As the number of valence electrons increases and the tightness with which they are held to the nucleus increases, the valence electrons became localized and the bond becomes more covalent. The transition metals, such as iron and nickel, have incomplete *d*-shells and exhibit some covalent bonding, which helps explain their relatively high melting points. Tin is interesting in that it has two crystalline forms, one that is mostly metallic and ductile and another that is mostly covalent and very brittle. Intermetallic compounds can also be formed between two metals in which the bonding is partly metallic and partly ionic. As the electronegativity difference between the two metals increases, the bonding becomes more ionic in nature. For example, aluminum and vanadium both have an electronegativity of 1.5 and the difference is 0, so the compound Al_3V is primarily metallic. On the other hand, aluminum and lithium (electronegativity of 1.0) have an electronegativity difference of 0.5; thus, when they form the compound AlLi, the bond is a combination of metallic and ionic.

Ionic bonding, also shown in Fig. A.3, is a result of electrical attraction between alternately placed positive and negative ions. In the ionic bond, the electrons are shared by an electropositive ion (cation) and an electronegative ion (anion). The electropositive ion gives up its valence electrons, while the electronegative ion captures them to produce ions having full electron orbitals or suborbitals. As a consequence, there are no free electrons available to conduct electricity. In ionically bonded solids such as salts, there are very few slip systems along which atoms can move. This is a consequence of the electrically charged nature of the ions. For slip in some directions, ions of like charge must be brought into close proximity to each other, and because of electrostatic repulsion, this mode of slip is very restricted. This is not a problem in metals, because all atoms are electronically neutral. No electrical conduction of the kind found in metals is possible in ionic crystals, but weak ionic conduction occurs as a result of the motion of the individual ions. When subjected to stresses, ionic crystals tend to cleave, or break, along certain planes of atoms rather than deform in a ductile fashion as metals do.

Ionic bonds form between electropositive metals and electronegative nonmetals. The further apart the two are on the periodic table, the more likely they are to form ionic bonds. For example, sodium (Na) is on the far left side of the periodic table in Group I, while chlorine (Cl) is on the far left side in Group VII. Sodium and chlorine combine to form common table salt (NaCl). As shown in Fig. A.4, the sodium atom gives up its outer valence electron, which is transferred to the outer electron shell of the chlorine atom. Because the outer shell of chlorine now contains eight electrons, similar to the noble gases, it is an extremely stable configuration. In terms of symbols, the sodium ion is written as Na^+, and the chlorine ion is written as Cl^-. When the two combine to form an ionic bond, the compound (NaCl) is neutral because the charges balance. Because the positively charged cation can attract multiple negatively charged anions, the ionic bond is nondirectional.

Covalent Bonding. Many elements that have three or more valence electrons are bound into crystal structures by forces arising from the

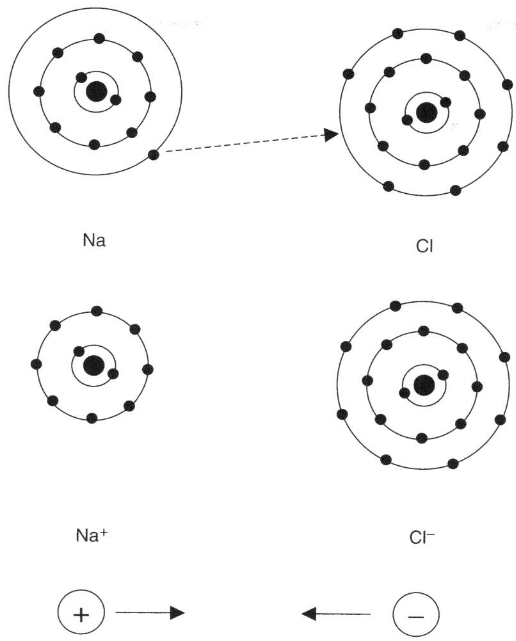

Fig. A.4 Ionic bonding in NaCl. Source: Ref A.1

sharing of electrons. The nature of this covalent bonding is shown schematically in Fig. A.3. To complete the octet of electrons needed for atomic stability, electrons must be shared with 8-*N* neighboring atoms, where *N* is the number of valence electrons in the given element. High hardness and low electrical conductivity are general characteristics of solids of this type. In covalently bonded ceramics, the bonding between atoms is specific and directional, involving the exchange of electron charge between pairs of atoms. Thus, when covalent crystals are stressed to a sufficient extent, they exhibit brittle fracture due to a separation of electron pair bonds, without subsequent reformation. It should also be noted that ceramics are rarely either all ionically or covalently bonded; they usually consist of a mix of the two types of bonds. For example, silicon nitride (Si_3N_4) consists of approximately 70% covalent bonds and 30% ionic bonds.

Covalent bonds also form between electropositive elements and electronegative elements. However, the separation on the periodic table is not great enough to result in electron transfer as in the ionic bond. Instead, the valence electrons are shared between the two elements. For example, a molecule of methane (CH_4), shown in Fig. A.5, is held together by covalent bonds. Note that hydrogen, in Group I on the periodic table, and carbon in Group IV, are much closer together than sodium and chlorine that form ionic bonds. In a molecule of methane gas, four hydrogen atoms are combined with one carbon atom. The carbon atom has four electrons in its outer shell, and these are combined with four more electrons, one from each of the four hydrogen atoms, to give a completed stable outer shell of eight electrons held together by covalent bonds. Each shared electron passes from an orbital controlled by one nucleus into an orbital shared by two nuclei. Covalent bonds, because they do not ionize, will not conduct electricity and are nonconductors. Covalent bonds form the basis for many organic compounds, including long-chain polymer molecules. As the molecule size increases, the bond strength of the material also increases. Likewise, the strength of long-chain molecules also increases with increases in chain length.

Secondary Bonding. Secondary, or van der Waals, bonding is weak in comparison to the primary metallic, ionic, and covalent bonds. Bond energies are typically on the order of only 10 kJ/mol (0.1 eV/atom). Although secondary bonding exists between virtually all atoms or molecules, its presence is usually obscured if any of the three primary bonding types is present. While van der Waals forces only play a minor role in metals, they are an important source of bonding for the inert gases that have stable electron structures, some molecular compounds such as water, and thermoplastic polymers where the main chains are covalently bonded but are held to other main chains by secondary bonding. Van der Waals bonding between two dipoles is illustrated in Fig. A.6.

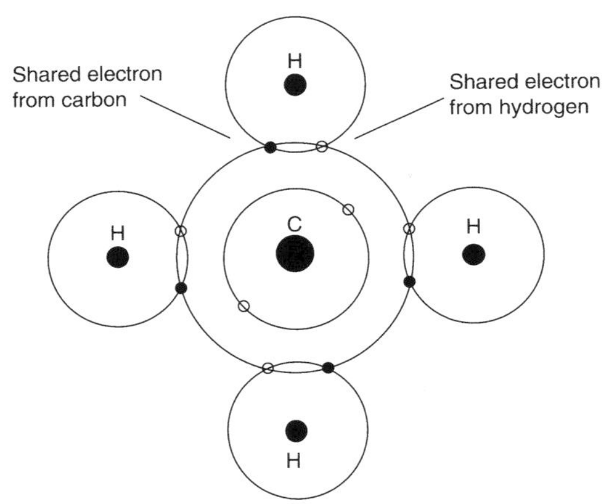

Fig. A.5 Covalent bonding in methane. Source: Ref A.1

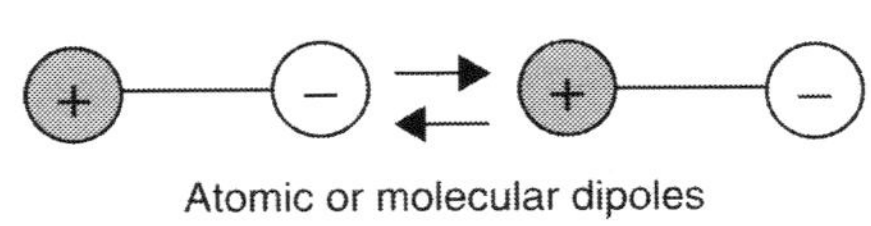

Fig. A.6 Van der Waals bonding between two dipoles. Source: Ref A.1

A.3 Crystalline Structure

When a substance freezes on cooling from the liquid state, it forms a solid that is either an amorphous or a crystalline structure. An amorphous structure is essentially a random structure. Although there may be what is known as short-range order, in which small groups of atoms are arranged in an orderly manner, it does not contain long-range order, in which all of the atoms are arranged in an orderly manner. Typical amorphous materials include glasses and almost all organic compounds. However, metals, under normal freezing conditions, normally form long-range, orderly crystalline structures. Except for glasses, almost all ceramic materials also form crystalline structures. Therefore, metals and ceramics are, in general, crystalline, while glasses and polymers are mostly amorphous.

Space Lattices and Crystal Systems. A crystalline structure consists of atoms, or molecules, arranged in a pattern that is repetitive in three dimensions. The arrangement of the atoms or molecules in the interior of a crystal is called its crystalline structure. A distribution of points (or atoms) in three dimensions is said to form a space lattice if every point has identical surroundings, as shown in Fig. A.7. The intersections of the lines, called lattice points, represent locations in space with the same kind of atom or group of atoms of identical composition, arrangement, and orientation. The geometry of a space lattice is completely specified by the lattice constants a, b, and c and the interaxial angles α, β, and γ. The unit cell of a crystal is the smallest pattern of arrangement that can be contained in a parallelepiped, the edges of which from the a, b, and c axes of the crystal.

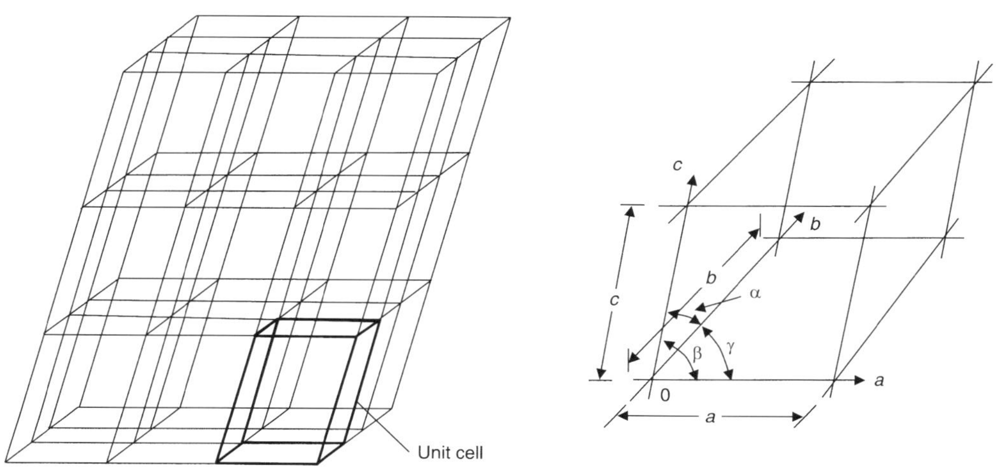

Fig. A.7 Space lattice and unit cell. Source: Ref A.1

When discussing crystal structure, it is usually assumed that the space lattice continues to infinity in all directions. In terms of a typical crystal (or grain) of, for example, iron that is 0.2 cm^3 in size, this may appear to be a preposterous assumption, but when it is realized that there are 10^{18} iron atoms in such a grain, the approximation to infinity seems much more plausible.

All crystal systems can be grouped into one of seven basic systems, as defined in Table A.2, which can be arranged in 14 different ways, called Bravais lattices, as shown in Fig. A.8. Almost all structural metals crystallize into one of three crystalline patterns: face-centered cubic (fcc), hexagonal close-packed (hcp), or body-centered cubic (bcc), illustrated in Fig. A.9. It should be noted that the unit cell edge lengths and axial angles are unique for each crystalline substance. The unique edge lengths are called lattice parameters. Axial angles other than 90° or 120° can also change slightly with changes in composition. When the edges of the unit cell are not equal in all three directions, all unequal lengths must be stated to completely define the crystal. The same is true if all axial angles are not equal.

Face-Centered Cubic System. The fcc system is shown in Fig. A.10. As the name implies, in addition to the corner atoms, there is an atom centrally located on each face. Because each of the atoms located on the faces belongs to two unit cells and the eight corner atoms each belong to eight unit cells, the number of atoms belonging to a unit cell is four. The atomic packing factor (the volume of atoms belonging to the unit cell divided by the volume of the unit cell) is 0.74 for the fcc structure. This is the densest packing that can be obtained. The fcc structure is the most efficient, with 12 nearest atom neighbors (also referred to as the coordination number, CN), that is, the fcc structure has a CN = 12. As shown in Fig. A.11, the stacking sequence for the fcc structure is *ABCABC*. The fcc structure is found in many important metals such as aluminum, copper, and nickel.

Hexagonal Close-Packed System. The atoms in the hcp structure (Fig. A.12) are also packed along close-packed planes. It should also be noted that both the fcc and hcp structures are what is known as close-packed

Table A.2 Seven crystal systems

Crystal system	Edge lengths	Interaxial angles
Triclinic	$a \neq b \neq c$	$\alpha \neq \beta \neq \gamma \neq 90°$
Monoclinic	$a \neq b \neq c$	$\alpha = \gamma = 90° \neq \beta$
Orthorhombic	$a \neq b \neq c$	$\alpha = \beta = \gamma = 90°$
Tetragonal	$a = b \neq c$	$\alpha = \beta = \gamma = 90°$
Hexagonal	$a = b \neq c$	$\alpha = \beta = 90°, \gamma = 120°$
Rhombohedral	$a = b = c$	$\alpha = \beta = \gamma \neq 90°$
Cubic	$a = b = c$	$\alpha = \beta = \gamma = 90°$

Source: Ref A.1

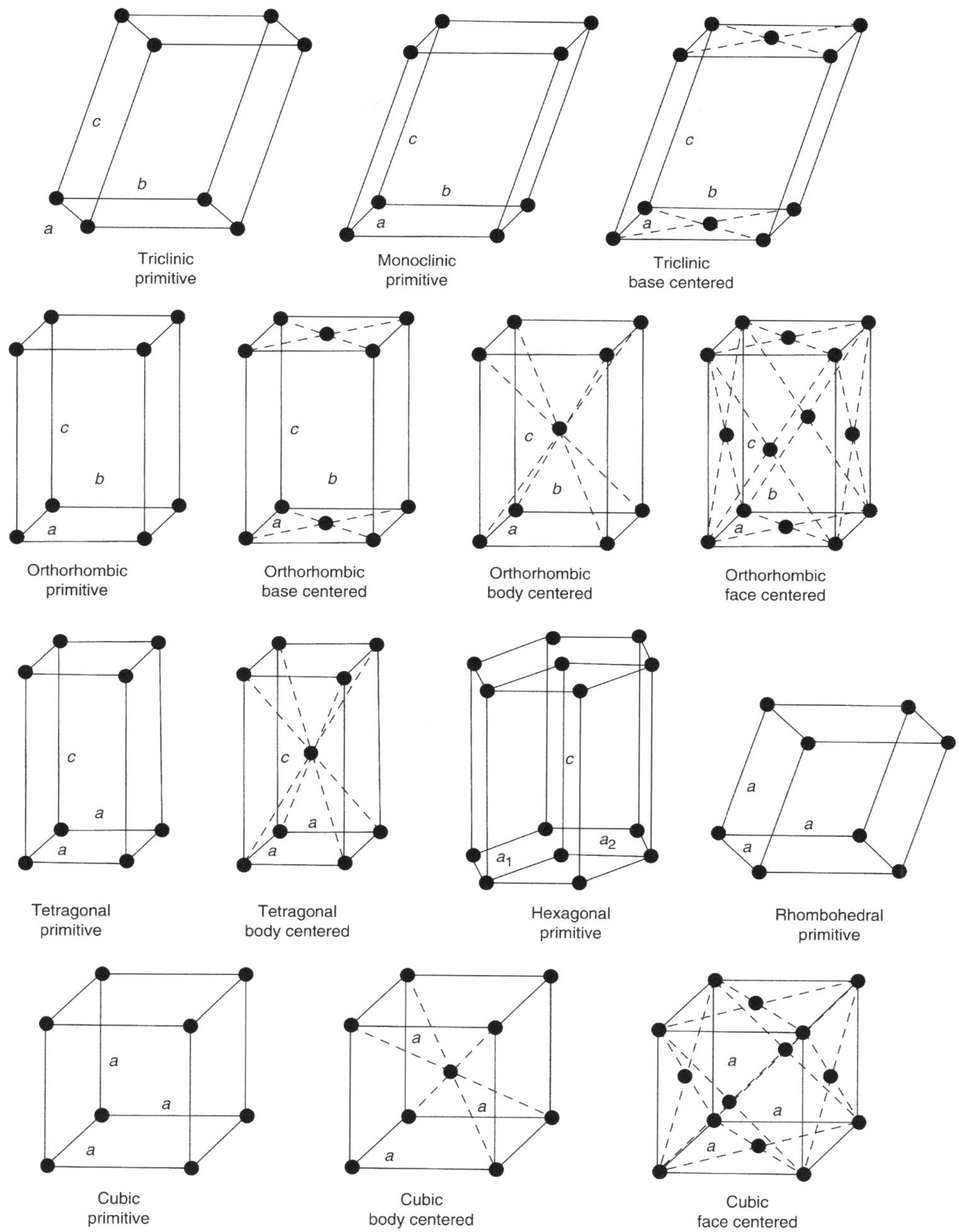

Fig. A.8 The fourteen Bravais lattices. Source: Ref A.1

structures with crystallographic planes having the same arrangement of atoms, however, the order of stacking the planes is different. Atoms in the hcp planes (called the basal planes) have the same arrangement as those in the fcc close-packed planes. However, in the hcp structure, these planes repeat every other layer to give a stacking sequence of ...*ABA*... . The

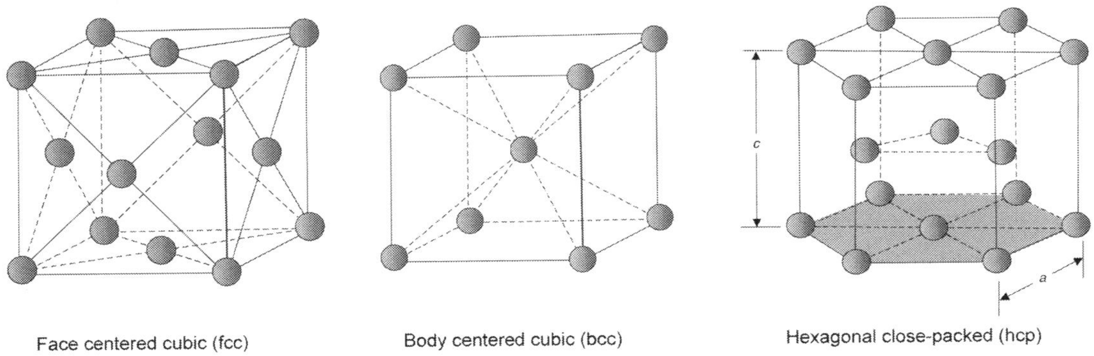

Fig. A.9 Common metallic crystalline structures. Source: Ref A.1

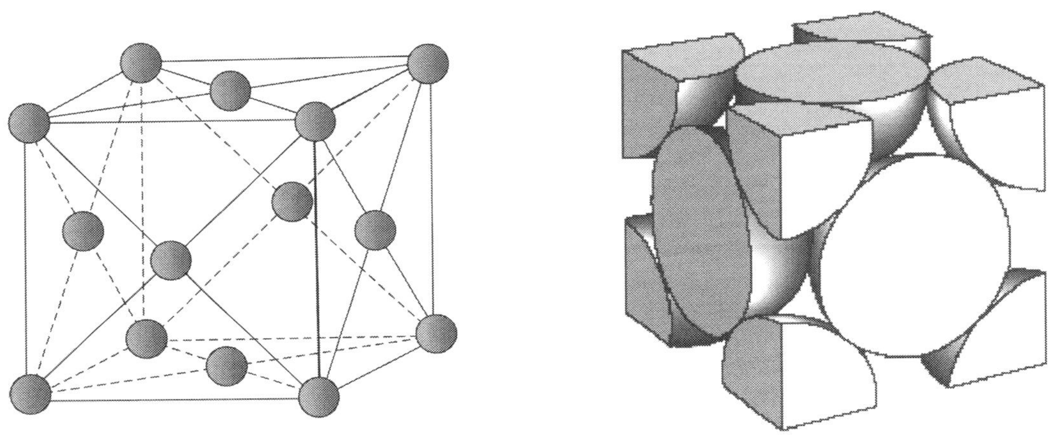

Fig. A.10 Face-centered cubic (fcc) structure. Source: Ref A.1

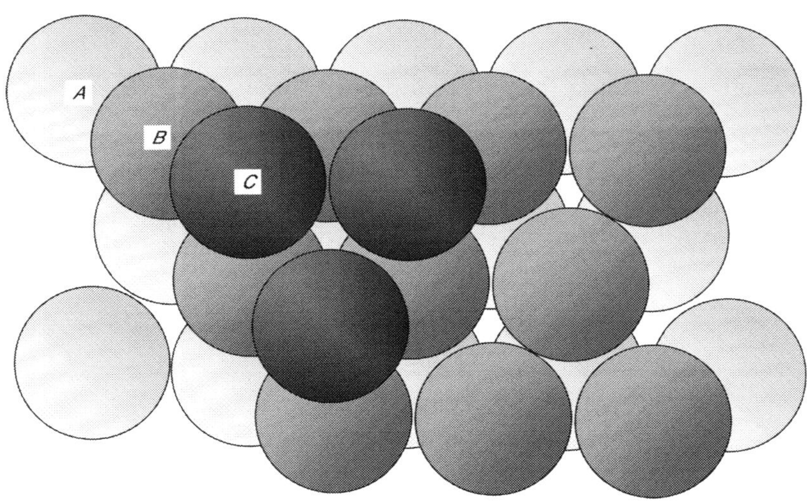

Fig. A.11 Close packing of planes. Source: Ref A.1

number of atoms belonging to the hcp unit cell is six and the atomic packing factor is 0.74. Note that this is the same packing factor as was obtained for the fcc structure. Also, the CN obtained for the hcp structure, 12, is the same as that for the fcc structure. A basic rule of crystallography is that if the CNs of two different unit cells are the same, then they will both have the same packing factors.

Two lattice parameters, c and a, also shown in Fig. A.12, are needed to completely describe the hcp unit cell. In an ideal hcp structure, the ratio of the lattice constants c/a is 1.633. In this ideal packing arrangement, the layer between the two basal planes in the center of the structure is located

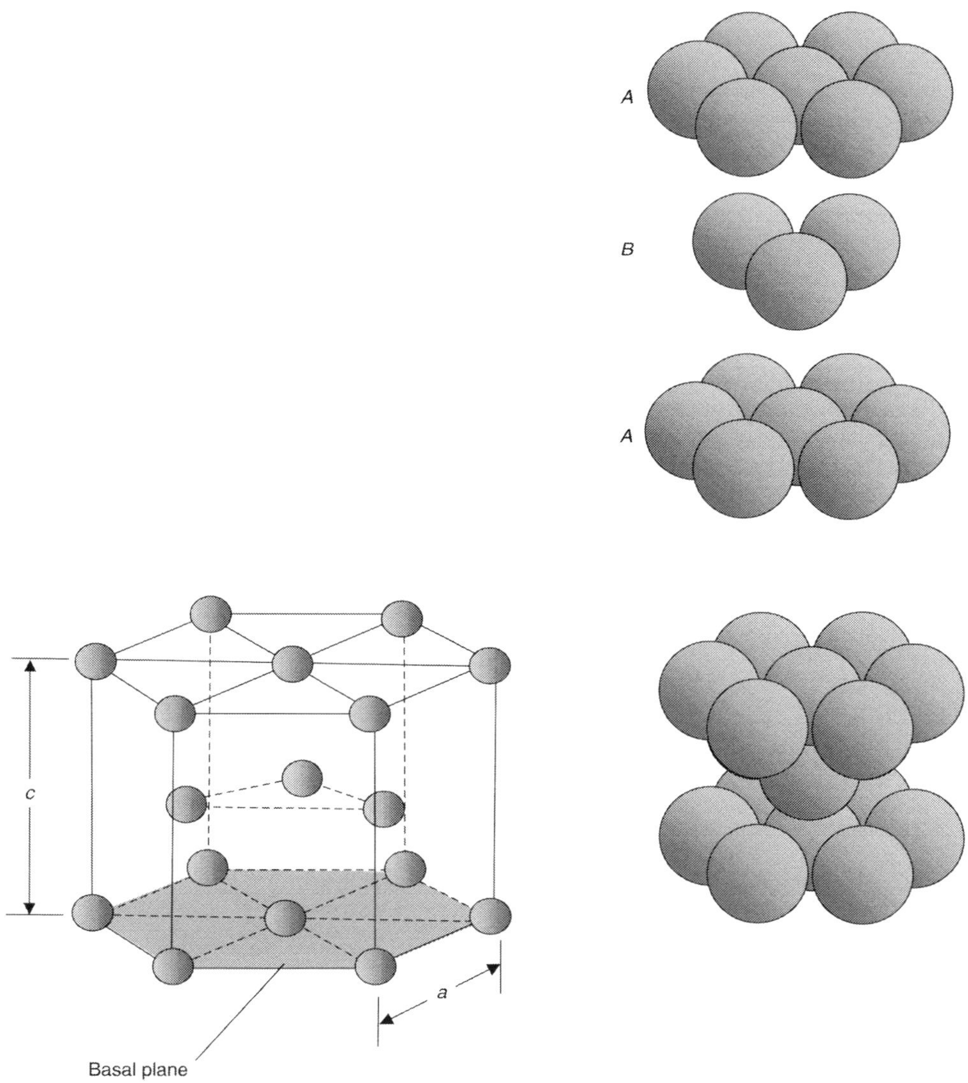

Fig. A.12 Hexagonal close-packed (hcp) structure. Source: Ref A.1

close to the atoms on the upper and lower basal planes. Therefore, any atom in the lattice is in contact with 12 neighboring atoms, and therefore CN = 12. It should be noted that there is often some deviation from the ideal ratio of $c/a = 1.633$. If the ratio is less than 1.633, it means that the atoms are compressed in the c-axis direction, and if the ratio is greater than 1.633, the atoms are elongated along the c-axis. In these situations, the hcp structure can no longer be viewed as truly being close packed and should be described as just being hexagonal. However, structures deviating from the ideal packing are still normally described as being hcp. For example, beryllium is described as having an hcp structure, but its c/a ratio of 1.57 is unusually low and causes some distortion in the lattice. This distortion and the unusually high elastic modulus of beryllium (3×10^5 MPa, or 42×10^3 ksi) result from a covalent component in its bonding. Contributions from covalent bonding are also present in the hcp metals zinc and cadmium, with c/a ratios greater than 1.85. This lowers their packing density to approximately 65%, considerably less than the 74% of the ideal hcp structure.

Body-Centered Cubic System. The body-centered cubic (bcc) system is shown in Fig. A.13. The bcc system is similar to the simple cubic system except that it has an additional atom located in the center of the structure. Because the center atom belongs completely to the unit cell in question, the number of atoms belonging to the bcc unit cell is two. The CN for the bcc structure is eight, because the full center atom is in contact with eight neighboring atoms located at the corner points of the lattice. The atomic packing factor for the bcc structure is 0.68, which is less than that of the fcc and hcp structures. Because the packing is less efficient in the bcc structure, the closest-packed planes are less densely packed (Fig. A.14).

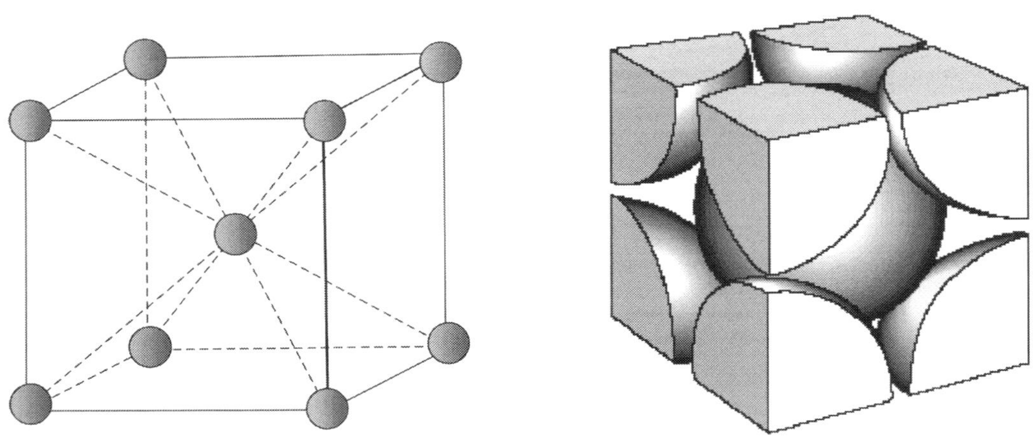

Fig. A.13 Body-centered cubic (bcc) structure. Source: Ref A.1

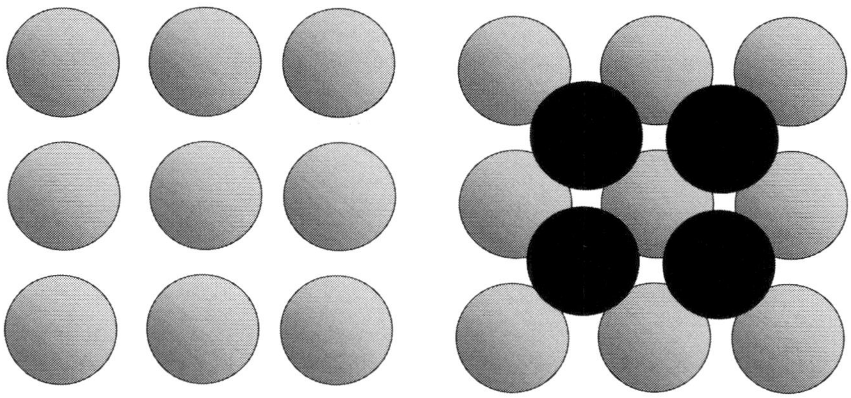

Fig. A.14 Loose packing in bcc structure. Source: Ref A.1

Even though the bcc crystal is not a densely packed structure, it is the equilibrium structure of 15 metallic elements at room temperature, including many of the important transition elements. This is attributable to two factors: (1) Even though each atom has only eight nearest neighbors, the six second-nearest neighbor atoms are closer in the bcc structure than in the fcc structure. Calculations indicate that these second-nearest neighbor bonds make a significant contribution to the total bonding energy of bcc metals; and (2) in addition, the greater entropy of the less densely packed bcc structure gives it a stability advantage over the more tightly packed fcc structure at high temperatures. As a consequence, some metals that have the close-packed structures at room temperature transition to bcc structures at higher temperatures.

A.4 Crystalline System Calculations

The atomic packing factor (APF) and coordination number (CN) of the important crystalline structures can be calculated from their geometries.

Cubic Systems. The cubic crystal systems are regular cubes with a lattice parameter such that $a = b = c$ and $\alpha = \beta = \gamma = 90°$. Therefore, only one lattice parameter (a) is required to define the cubic lattice. Within the cubic family of systems, there are three important variations: (1) the simple or primitive cubic, in which there are atoms only at the corner points of the lattice, (2) the body-centered cubic (bcc) structure, which has an additional atom located at the center of the structure, and (3) the face-centered cubic (fcc) structure, which has an extra atom located on each of the six faces. The bcc and fcc structures are extremely important in metallurgy, with about 90% of industrially important metals crystallizing into one of these two structures.

Simple Cubic System. Two types of graphical depictions of a simple cubic system are shown in Fig. A.15, namely, the point model and the hard ball model. Although the hard ball model, with the outermost electron shells in contact, is the more realistic of the two, the point model is more commonly used because it is easier to visualize all of the atoms.

The relationship between the atomic radius, r, and the lattice constant, a, is shown in Fig. A.15 and is equal to:

$$a = 2r \qquad \text{(Eq A.1)}$$

Another important value is the number of atoms, N, belonging to a unit cell. Examination of the hard ball model of Fig. A.15 reveals that each atom located at the corner point of the unit cell belongs to eight unit cells. Therefore, only 1/8 of each atom can be considered to belong to one unit cell. Because there are eight corners, or atomic positions, in the unit cell, the number of atoms belonging to the unit cell is:

$$N = 8\left(\frac{1}{8}\right) = 1 \qquad \text{(Eq A.2)}$$

Thus, in a simple cubic crystal system, one full atom belongs to each unit cell.

In metals, the atoms tend to pack or occupy positions as close as possible to each other and to form the most dense lattice structure possible. The APF, which indicates the part of the volume of the unit cell that is actually occupied by atoms, is used to describe the denseness of the unit cell.

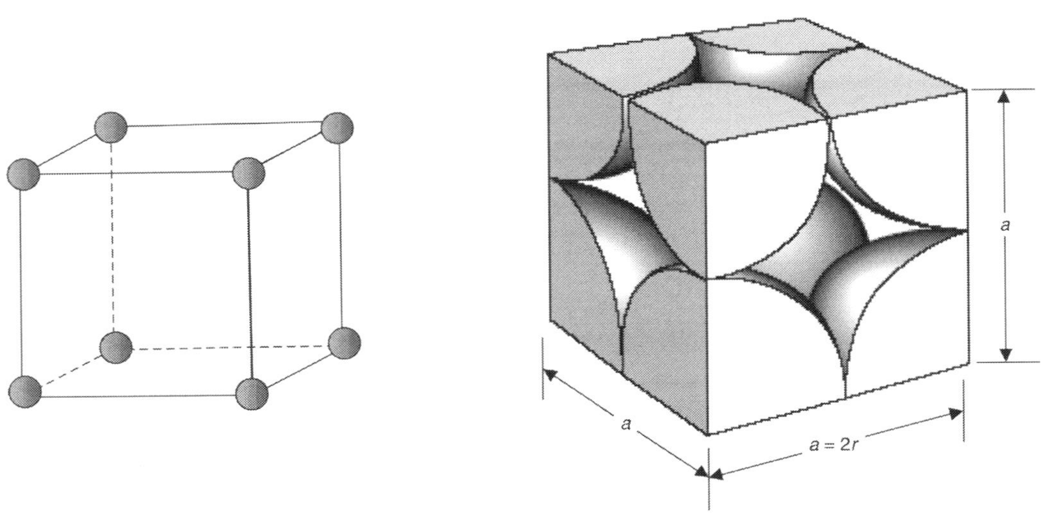

Fig. A.15 Simple cubic structure. Source: Ref A.1

$$\text{APF} = \frac{V_a}{V_c} = \frac{\text{Volume of atoms belonging to unit cell}}{\text{Volume of unit cell}} \quad \text{(Eq A.3)}$$

The volume of atoms belonging to the simple unit cell can be calculated as:

$$V_a = N\frac{4\pi r^3}{3} \quad \text{(Eq A.4)}$$

Likewise, the volume of the simple cubic system is:

$$V_c = a^3 = 8r^3 \quad \text{(Eq A.5)}$$

Therefore, the APF is:

$$\text{APF} = \frac{N\frac{4\pi r^3}{3}}{8r^3} = \frac{N\pi}{6} \quad \text{(Eq A.6)}$$

Because $N = 1$ for the simple cubic system, the APF reduces to:

$$\text{APF} = \frac{\pi}{6} = 0.52 \quad \text{(Eq A.7)}$$

This means that only approximately half of the simple cubic system is occupied with atoms. Because metals tend to pack as closely as possible during crystallization, this rather loose packing explains why very few metals crystallize into the simple cubic system.

Another parameter used to define crystal systems is the CN, or number of nearest neighboring atoms each atom in the structure possesses. In the hard ball model, it is the number of atoms that touch a given atom. Referring again to Fig. A.15, it can be seen that any atom is in contact with six neighboring atoms, thus CN = 6.

Body-Centered Cubic System. The body-centered cubic (bcc) system is shown in Fig. A.16. The bcc system is similar to the simple cubic system except that it has an additional atom located in the center of the structure. Because the center atom belongs completely to the unit cell in question, the number of atoms belonging to the bcc unit cell is:

$$N = 8 \times \frac{1}{8} + 1 = 2 \quad \text{(Eq A.8)}$$

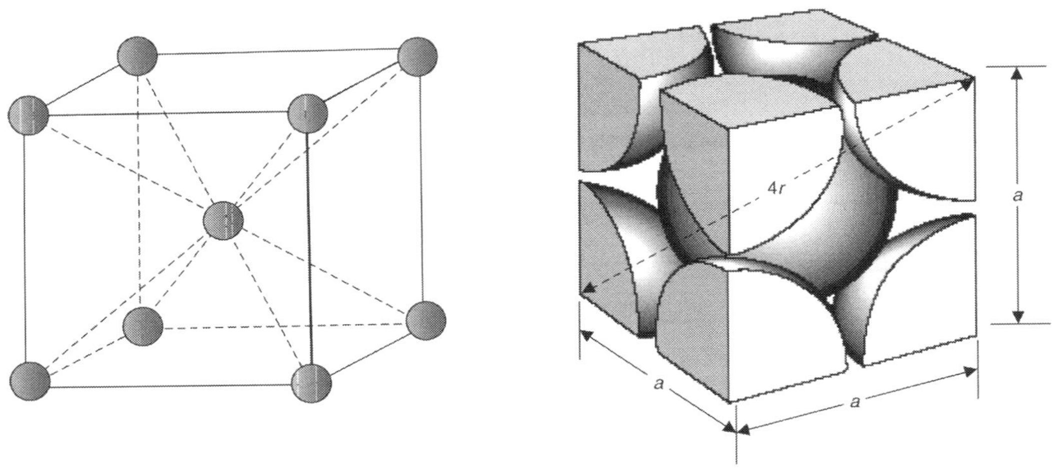

Fig. A.16 Body-centered cubic structure. Source: Ref A.1

The relationship between the lattice parameter, a, and the atomic radius, r, is:

$$a = \frac{4r}{\sqrt{3}} \qquad \text{(Eq A.9)}$$

The APF can then be calculated to be:

$$\text{APF} = \frac{\frac{4N\pi r^3}{3}}{a^3} = \frac{\frac{4N\pi r^3}{3}}{\frac{64r^3}{3\sqrt{3}}} \qquad \text{(Eq A.10)}$$

$$\text{APF} = \frac{\pi\sqrt{3}}{8} = 0.68 \qquad \text{(Eq A.11)}$$

Because the APF for the bcc structure is significantly higher (0.68) than that for the simple cubic structure (0.52), the bcc structure is significantly denser and many important metals, α-iron for example, crystallize into the bcc structure.

For the bcc structure, CN = 8, because the full center atom is in contact with eight neighboring atoms located at the corner points of the lattice.

Face-Centered Cubic System. The face-centered cubic (fcc) system is shown in Fig. A.17. As the name implies, in addition to the corner atoms,

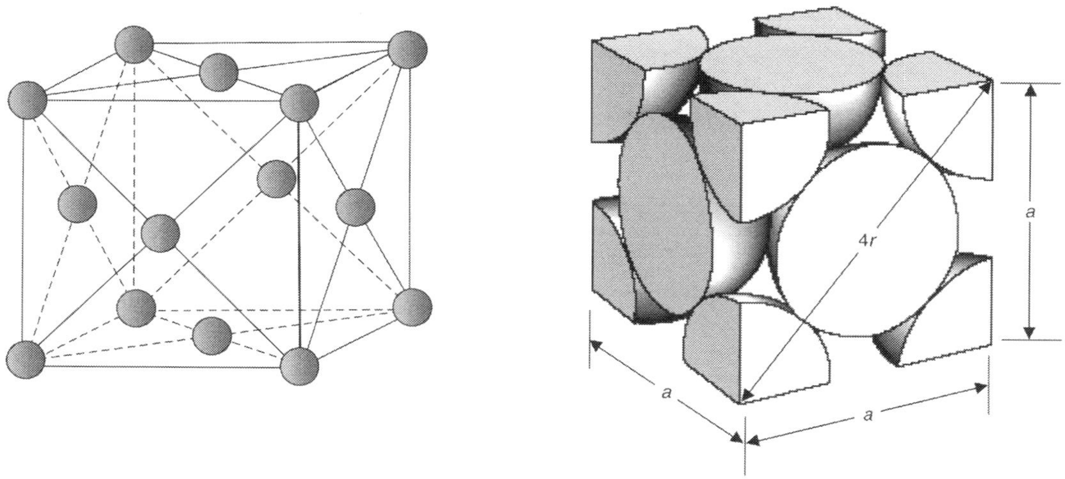

Fig. A.17 Face-centered cubic structure. Source: Ref A.1

there is an atom centrally located on each face. Because each of the atoms located on the faces belong to two unit cells and the eight corner atoms each belong to eight unit cells, the number of atoms belonging to a unit cell is:

$$N = 8 \times \frac{1}{8} + 6 \times \frac{1}{2} = 4 \qquad \text{(Eq A.12)}$$

Because, in the fcc structure, atoms are in contact along the face diagonals, the relationship between the lattice parameter and atomic radius is:

$$a = \frac{4r}{\sqrt{2}} \qquad \text{(Eq A.13)}$$

The APF can then be calculated as:

$$\text{APF} = \frac{\frac{4N\pi r^3}{3}}{a^3} = \frac{\frac{4N\pi r^3}{3}}{\frac{64r^3}{2\sqrt{2}}} \qquad \text{(Eq A.14)}$$

Substituting $N = 2$ into the equation, the APF becomes:

$$\text{APF} = \frac{\pi\sqrt{2}}{6} = 0.74 \qquad \text{(Eq A.15)}$$

Because 74% of the fcc lattice is occupied by atoms, this is an even denser packing than that for the bcc structure. It occurs in many important metals such as aluminum, copper, and nickel.

The fcc structure has a CN = 12. With 12 nearest atom neighbors, the fcc structure is the most efficient of the cubic structures.

Hexagonal System. The hexagonal close-packed (hcp) system is shown in Fig. A.18. The lattice parameters are:

$$a_1 = a_2 = a_3 \neq c$$
$$\alpha_1 = \alpha_2 = \alpha_3 = 120°;\ \gamma = 90°$$

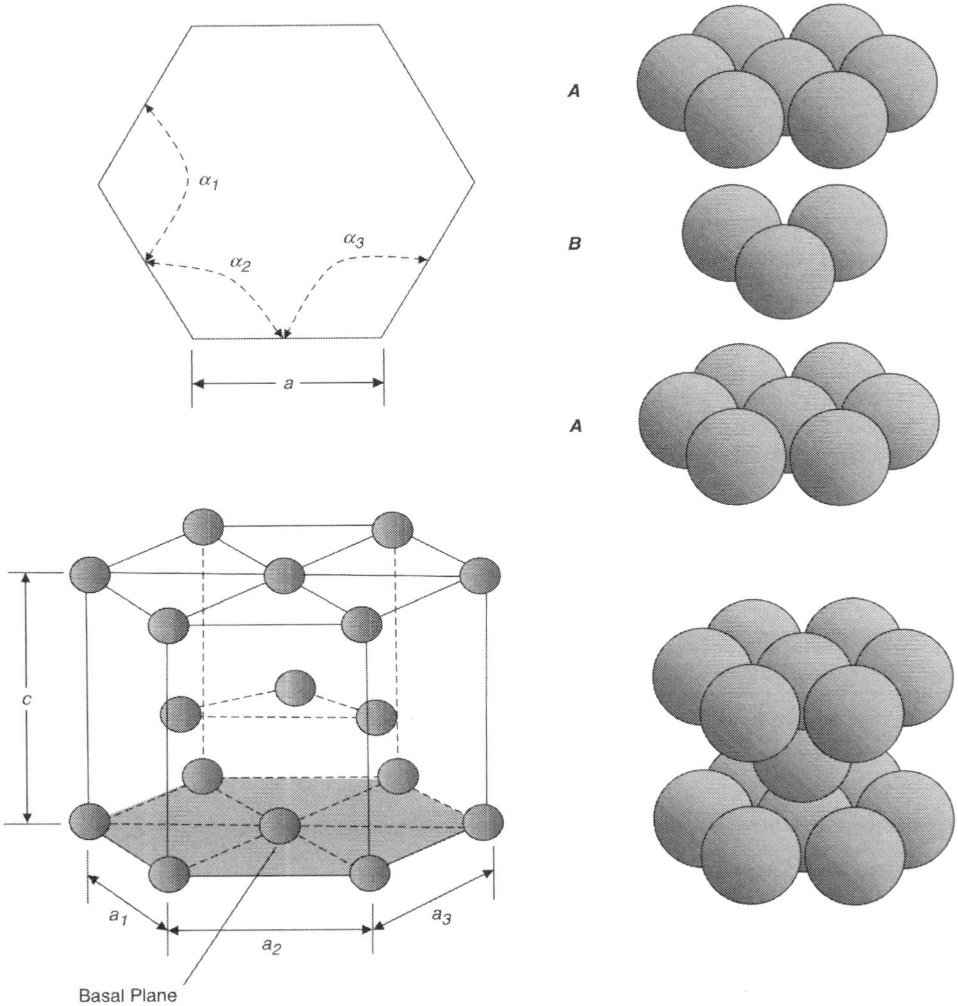

Fig. A.18 Hexagonal closed-packed structure. Source: Ref A.1

To determine the number of atoms belonging to a unit cell, note that atoms at the corner points belong to the six neighboring lattices and they contribute 1/6 of an atom to the unit cell. Atoms at the face centers belong to two adjacent lattices and contribute 1/2 of an atom to the unit cell. Finally, the three atoms in the interior all belong to the unit cell. The number of atoms belonging to the unit cell is therefore:

$$N = 12 \times \frac{1}{6} + 2 \times \frac{1}{2} + 3 \times 1 = 6 \qquad \text{(Eq A.16)}$$

In an ideal hcp structure, the ratio of the lattice constants c/a is 1.633. In this ideal packing arrangement, the layer between the two basal planes in the center of the structure is located close to the atoms on the upper and lower basal planes. Therefore, any atom in the lattice is in contact with 12 neighboring atoms and therefore CN = 12. It should be noted that there is often some deviation from the ideal ratio of $c/a = 1.633$. If the ratio is less than 1.633, it means that the atoms are compressed in the c-axis direction, and if the ratio is greater than 1.633, the atoms are elongated along the c-axis.

The relationships between the lattice parameters, a and c, and the atomic radius, r, for an ideal hcp structure ($c/a = 1.633$) are:

$$a = 2r$$
$$c = 1.6333 \times 2r = 3.266r \qquad \text{(Eq A.17)}$$

To determine the atomic packing factor, first determine the volume of the hexagonal lattice using Fig. A.19. The area of the base hexagon is:

$$A = 3a^2 \sin 60° = \frac{3\sqrt{3}}{2} a^2 \qquad \text{(Eq A.18)}$$

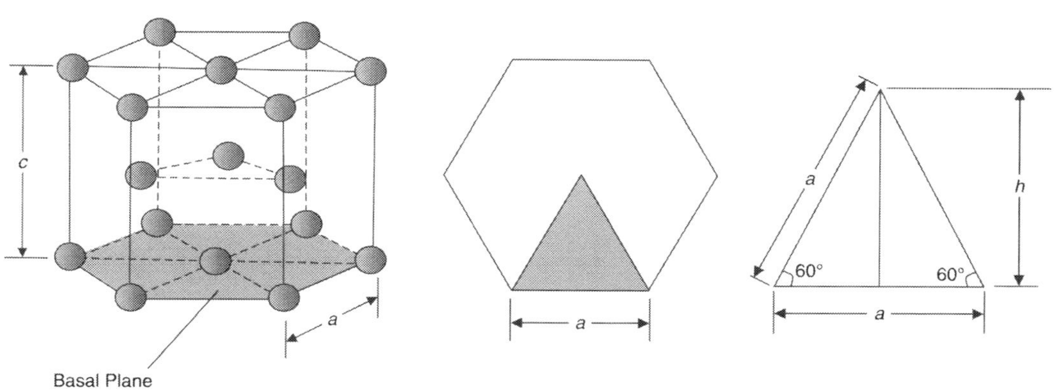

Fig. A.19 Calculation of volume of hexagonal lattice. Source: Ref A.1

The volume of the lattice is:

$$V_c = \frac{3\sqrt{3}}{2}1.633a^3 \qquad \text{(Eq A.19)}$$

The APF is then:

$$\text{APF} = \frac{V_a}{V_c} = \frac{\frac{4N\pi r^3}{3}}{\frac{3\sqrt{3}}{2}1.633a^3} \qquad \text{(Eq A.20)}$$

Substituting $a = 2r$ in Eq A.20 gives:

$$\text{APF} = \frac{N\pi}{(1.633)9\sqrt{3}} \qquad \text{(Eq A.21)}$$

Finally, substituting the number of atoms belonging to the unit cell, $N = 6$, gives:

$$\text{APF} = 0.74 \qquad \text{(Eq A.22)}$$

Note that this is the same value as was obtained for the fcc structure. Also, the CN obtained for the fcc structure, 12, is the same as that for the hcp structure. A basic rule of crystallography is that if the CNs of two different unit cells are the same, then they will both have the same packing factors. It should also be noted that both the fcc and hcp structures are what is known as close-packed structures with crystallographic planes having the same arrangement of atoms; however, the order of stacking the planes is different.

A.5 Slip Systems

Plastic deformation takes place by sliding (slip) of close-packed planes over one another. A reason for this slip plane preference is that the separation between close-packed planes is greater than for other crystal planes, and this makes their relative displacement easier. Furthermore, the slip direction is in a close-packed direction. The combination of planes and directions on which slip takes place constitutes the slip systems of the material. In polycrystalline materials, a certain number of slip systems must be available in order for the material to be capable of plastic deformation. Other things being equal, the greater the number of slip systems, the greater the capacity for deformation. Face-centered cubic metals have

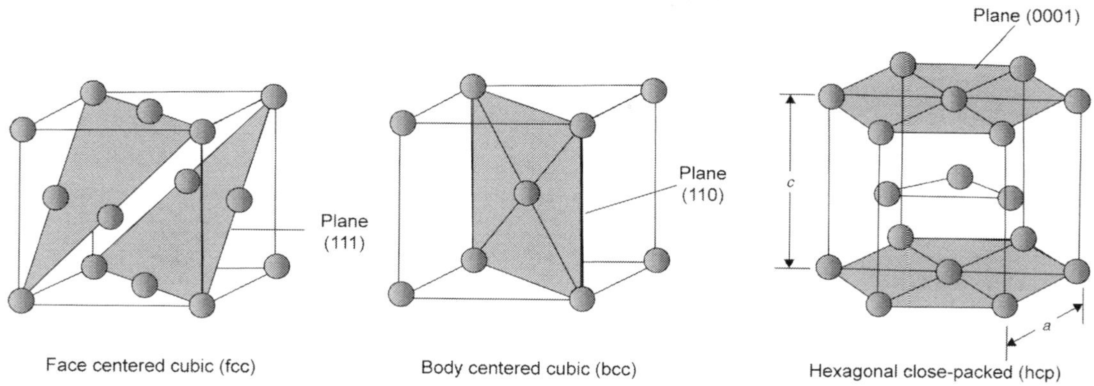

Fig. A.20 Close-packed planes. Source: Ref A.1

a large number of slip systems (12) and are capable of moderate to extensive plastic deformation even at temperatures approaching absolute zero. A number of close-packed planes for the fcc, bcc, and hcp structures are illustrated in Fig. A.20.

Materials having the bcc structure also often display 12 slip systems, although this number comes about differently than it does for the fcc lattice. A closest-packed bcc plane is defined by a unit cell edge and face diagonal. There are only two close-packed directions (the cube diagonals) in the closest-packed bcc plane, but there are six nonparallel planes of this type. Over certain temperature ranges, some bcc metals display slip on other than close-packed planes, although the slip direction remains a close-packed one. Thus, bcc metals have the requisite number of slip systems to allow for plastic deformation. However, bcc metals often become brittle at low temperatures as a result of the strong temperature sensitivity of their yield strength, which causes them to fracture prior to undergoing significant plastic deformation.

Depending on their c/a ratio, polycrystalline hcp metals may or may not have the necessary number of slip systems to allow for appreciable plastic deformation. The ideal hcp structure has only three slip systems, because there is only one nonparallel close-packed plane in it (the basal plane, which contains three nonparallel close-packed directions). However, three slip systems are insufficient to permit polycrystalline plastic deformation, and so hcp polycrystals for which slip is restricted to the basal plane are not malleable. When c/a is less than the ideal ratio, basal planes become less widely separated, and other planes compete with them for slip activity. In these instances, the number of slip systems increases, and material ductility is beneficially affected. In addition, polycrystalline hcp metals can also deform by a mechanism called twinning.

A.6 Crystallographic Planes and Directions

This section will give a short introduction to identifying crystalline planes and crystalline directions.

Miller Indices for Cubic Systems. Special planes and directions within metal crystal structures play an important part in plastic deformation, hardening reactions, and other aspects of metal behavior. Crystallographic planes are defined by what are called Miller indices. Consider the general plane shown in Fig. A.21. The Miller indices for a given plane can be determined as follows:

1. The plane should be displaced, if necessary, with a parallel displacement to a position where it does not pass through the origin of the coordinate system. This is permissible because (1) the coordinate system chosen is an arbitrary one that could be chosen anywhere on the vast expanse of a space lattice, and (2) as we shall see, parallel planes are equivalent.
2. The a, b, and c intercepts of the plane with the x, y, and z coordinates are determined.
3. The reciprocals of these numbers are determined. The reciprocals are denoted by the letters h, k, l, respectively:

$$\frac{1}{a} \rightarrow h; \frac{1}{b} \rightarrow k; \frac{1}{c} \rightarrow l$$

4. The reciprocals normally result in fractions. Apply either multiplication or division to determine the smallest set of h, k, and l values that yield integer numbers. These numbers, when enclosed in brackets such as $(h\ k\ l)$, give the Miller indices for the plane.

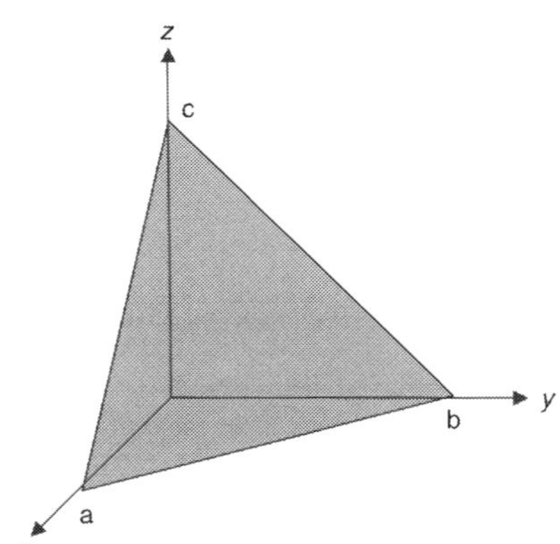

Fig. A.21 Crystallographic plane indices for cubic systems. Source: Ref A.1

Several examples can be used to best show the calculation method. The Miller indices for the plane shown in Fig. A.22(a) would be:

1. Intercepts $x = 1, y = 1, z = 1$
2. Reciprocals $\frac{1}{x} = 1, \frac{1}{y} = 1, \frac{1}{z} = 1$
3. No fractions to clear
4. (111)

The Miller indices for the plane shown in Fig. A.22(b) would be:

1. The plane never intercepts the z-axis, so $x = 1, y = 2, z = \infty$.
2. Reciprocals $\frac{1}{x} = 1, \frac{1}{y} = \frac{1}{2}, \frac{1}{z} = \frac{1}{\infty} = 0$
3. Clear fractions $\frac{1}{x} = 2, \frac{1}{y} = 1, \frac{1}{z} = \frac{1}{\infty} = 0$
4. (210)

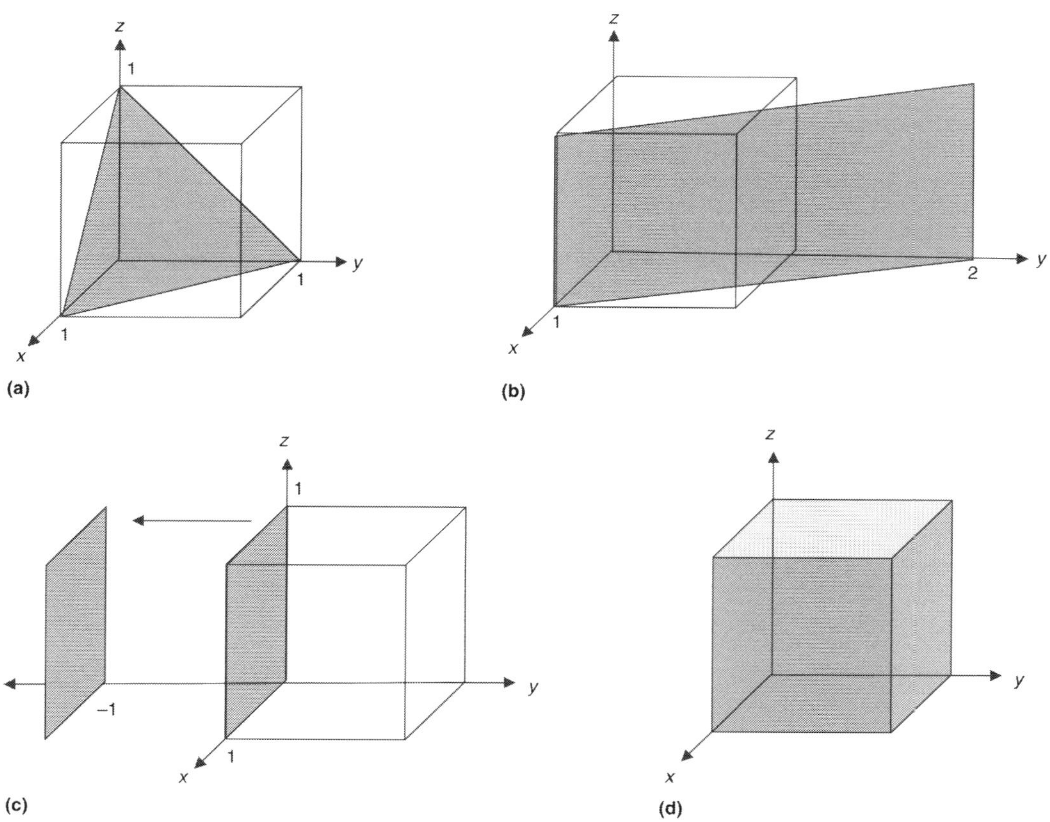

Fig. A.22 Examples of Miller indices for planes. Source: Ref A.1

The Miller indices for the plane shown in Fig. A.22(c) would be:

1. Because the plane passes through the origin, the plane must be moved. If it is moved one lattice parameter in the y-direction, the $x = \infty$, $y = -1$, $z = \infty$.
2. Reciprocals $\frac{1}{x} = 0, \frac{1}{y} = -1, \frac{1}{z} = 0$
3. No fractions to clear
4. $(0\bar{1}0)$ Note that the negative intercept of –1 for the y-axis has a bar placed over the top of it (i.e., $\bar{1}$).

Finally, examine Fig. A.22(d) in which all of the side faces of the cube are indicated with planes. Remembering that the space lattice, for all purposes, is infinite in all three coordinate directions, and that our designation of the coordinate system, x, y, and z, is also arbitrary, it can be concluded that all of the face planes in the cube shown have the same atomic arrangement. These are known as crystallographic equivalent planes. A little analysis will show that all of these planes consist of the same three numbers, namely 1,0,0. Crystallographic equivalent planes are known as family planes and are denoted by putting the integers in braces. In this case, the notation {100} represents all side faces of the cube collectively as a family plane.

Miller-Bravais Indices for Hexagonal Crystal Systems. The hexagonal crystal systems use a slightly different procedure for specifying the crystalline planes. Besides the h, k, and l indices, there is a fourth index, i, that is used for hexagonal systems. In the Miller-Bravais system, the three axes in the basal plane are denoted by a_1, a_2, and a_3 indices, while l is used to denote the intercept with the z-axis. Thus, a plane will have the designation ($h\ k\ i\ l$). As mentioned earlier, there are two different lattice parameters. The lattice parameter a is measured along the three axes of the basal plane, while the c-axis is measured in the direction of the z-axis perpendicular to the basal plane. Thus, $a_1 = a_2 = a_3 = a \neq c$.

The Miller indices for the basal plane shown in Fig. A.23(a) would be:

1. Intercepts $a_1 = a_2 = a_3 = \infty$; c = 1
2. Reciprocals $\frac{1}{a_1} = \frac{1}{a_2} = \frac{1}{a_3} = 0; \frac{1}{c} = 1$
3. No fractions to clear
4. (0001)

The Miller indices for the basal plane shown in Fig. A.23(b) would be:

1. Intercepts $a_1 = 1, a_2 = 1, a_3 = -\frac{1}{2}; c = 1$

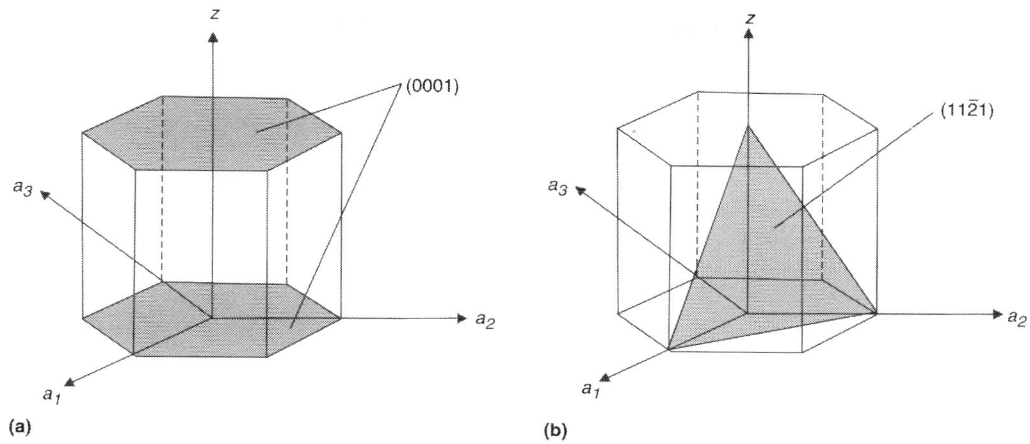

Fig. A.23 Examples of Miller-Bravais indices for hexagonal planes. Source: Ref A.1

2. Reciprocals $\frac{1}{a_1} = 1, \frac{1}{a_2} = 1, = \frac{1}{a_3} = -2; \frac{1}{c} = 1$
3. No fractions to clear
4. $(11\bar{2}1)$

Because the four axes representation contains redundancy, the equation $h + k = -i$ is valid for the indices. In other words, the third index in the basal plane can be calculated from the other two.

Crystallographic Directions in Cubic Crystal Structures. A direction in a crystal is indicated by bracketed indices, for example, [110], and in the cubic system, the direction is always perpendicular to the plane having the same indices (Fig. A.24). More generally, a line in a given direction, such as [110], can be constructed in the following manner. Draw a line from the origin through the point having the coordinates $x = 1, y = 1, z = 0$, in terms of axial lengths. This line and all lines parallel to it are then in the given direction. Note that reciprocals are not involved in obtaining the Miller indices of directions. A family of equivalent directions, such as [100], [$\bar{1}$00], [010], [0$\bar{1}$0], [001], and [00$\bar{1}$], is designated as {100}.

Crystallographic Directions in Hexagonal Crystal Structures. The direction system for the Miller-Bravais system for hexagonal crystal structures is extremely cumbersome and confusing. An explanation is attempted here, and then its use will be avoided whenever possible. Miller-Bravais indices of direction are also expressed in terms of four digits. Like the indices for planes, the third digit must always equal the negative sum of the first two digits (i.e., $h + k = -i$). Thus, if the first two digits are 2 and −1, then the third digit must be −1; that is, (2) + (−1) = 1 → −1. The method for finding the direction of the a_1 axis is shown in Fig. A.25. This axis has the same direction as the vector sum of the three vectors, one of them of

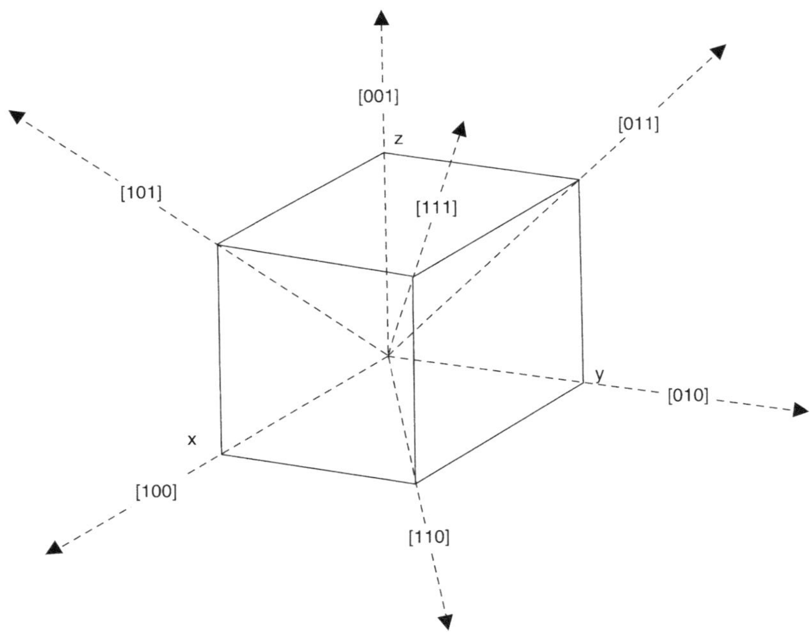

Fig. A.24 Important directions in a cubic cell. Source: Ref A.1

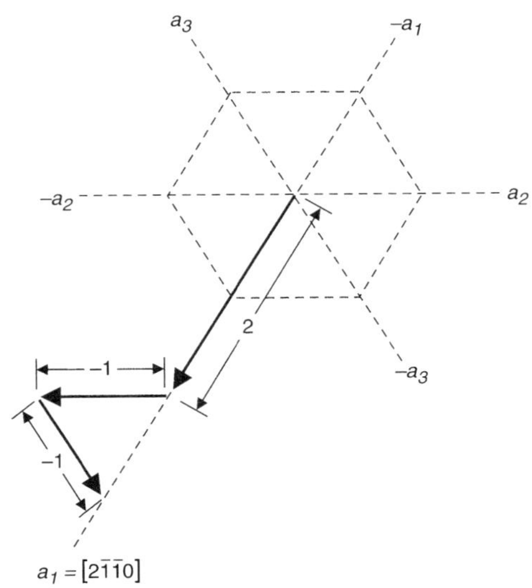

Fig. A.25 Example of Miller-Bravais directional indices for hexagonal planes. Source: Ref A.3 as published in Ref A.1

length +2 along the a_1 axis, another of length –1 along the a_2 axis, and the third of length –1 along the a_3 axis. This yields direction indices of [$2\bar{1}\bar{1}0$]. This cumbersome method is required to satisfy the relationship $h + k = -i$. The corresponding indices of the a_2 and a_3 axes are [$\bar{1}2\bar{1}0$] and [$\bar{1}\bar{1}20$], respectively.

A.7 X-ray Diffraction for Determining Crystalline Structure

X-rays are a form of electromagnetic radiation that has high energies and short wavelengths. Because the wavelengths of x-rays are approximately the same as the atomic spacings on metallic lattices, a number of x-ray techniques have been developed over the years to determine crystalline structure. Diffraction occurs when a wave encounters a series of regularly spaced obstacles that (1) are capable of scattering the wave, and (2) have spacings that are comparable in magnitude to the wavelength. A diffracted beam is one in which a large number of scattered waves mutually reinforce one another.

When a beam of x-rays with a wavelength λ strikes at set of crystalline planes at some arbitrary angle, there will usually be no reflected beam because the rays reflected from the various crystal planes must travel paths of different lengths. In other words, although the incident rays are in phase, the reflected ways are out of phase and thus cancel one another. However, if each ray is out of phase with the preceding one by exactly one wavelength, or any whole integer wavelength (n = 1, 2, 3,...), then the reflected beam will consist of rays that are in phase again. The angle at which reflection occurs is known as the Bragg angle, θ.

Two parallel planes of atoms identified as A-A' and B-B', which have the same h, k, l Miller indices separated by the interplanar spacing d_{hkl}, are shown in Fig. A.26. Two in-phase rays, labeled 1 and 2, having a wavelength λ strike the surface at an angle θ. On impact, rays 1 and 2 are scattered by atoms P and Q. Constructive interference of the scattered rays 1′ and 2′ occurs at an angle to the planes, if the path length difference between 1-P-1 and 2-Q-2 (i.e., $\overline{SQ}+\overline{QT}$) is equal to a whole number, n, of wavelengths. This condition for diffraction is:

$$n\lambda = \overline{SQ}+\overline{QT}$$
$$n\lambda = d_{hkl}\sin\theta + d_{hkl}\sin\theta$$
$$n\lambda = 2d_{hkl}\sin\theta \qquad \text{(Eq A.23)}$$

Equation A.23 is known as Bragg's law; also, n is the order of reflection, which may be any integer (1, 2, 3, ...) consistent with sin θ not exceeding unity. Bragg's law is a simple expression relating the x-ray wavelength and interatomic spacing to the angle of the diffracted beam. If Bragg's law is

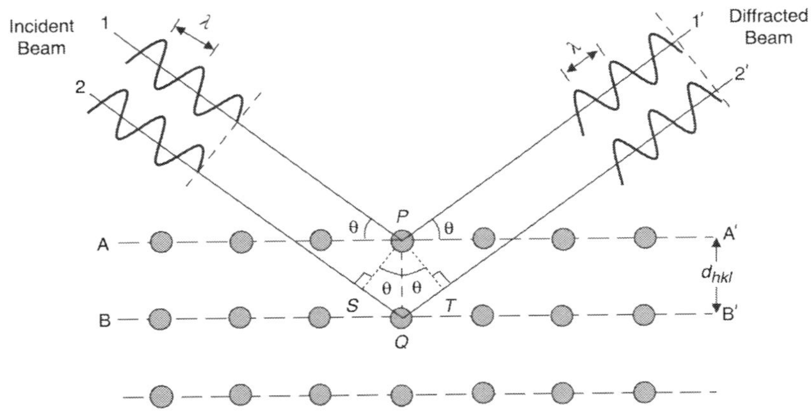

Fig. A.26 Diffraction of x-rays by planes of atoms. Source: Ref A.4 as published in Ref A.1

not satisfied, then the reflected rays will be out of phase and cancel each other. Although for a given metallic substance, n and d_{hkl} will ordinarily have only a few different values that will satisfy Bragg's law, the values for λ and θ can be varied continuously over a wide range.

The magnitude of the distance between two adjacent and parallel planes of atoms (i.e., the interplanar spacing d_{hkl}) is a function of the Miller indices (h, k, and l) as well as the lattice parameter(s). For example, for crystal structures that have cubic symmetry:

$$d_{hkl} = \frac{a}{\sqrt{h^2 + k^2 + l^2}} \qquad \text{(Eq A.24)}$$

in which a is the lattice parameter. There are also relationships similar to this one for the cubic system for the other crystalline systems.

There are a number of x-ray techniques that have been developed for studying crystalline structures. The Laue method uses a narrow collimated beam of white x-rays that strike a stationary single crystal. It is not suitable for the determination of lattice parameters but is used for the determination of crystal orientation and the detection of lattice imperfections. The rotating single-crystal method uses a constant wavelength beam while the diffracting crystal is rotated. This method is used mostly for determining the structure of single crystals. The Debye-Scherrer powder method uses an x-ray beam of constant wavelength and a specimen consisting of thousands of tiny crystals. Because there are a large number of powder particles with many different orientations, the diffracted beam produces a cone of radiation. Different reflection cones are recorded on a film strip that allows the diffraction angles to be measured. The distance d_{hkl} between the reflecting atomic planes is then calculated using Bragg's law.

In addition to the photographic methods, there are diffractometer methods that measure the intensity of the beam in counts per second diffracted from the specimen over a range of angles. A powdered or polycrystalline specimen consisting of many fine and randomly oriented particles is exposed to monochromatic x-radiation. Each powder particle (or grain) is a crystal, and having a large number of them with random orientations ensures that some particles are properly oriented such that every possible set of crystallographic planes will be available for diffraction. In a typical diffractometer (Fig. A.27), a specimen "S" is supported so that it rotates about the axis labeled "O." The monochromatic x-ray beam is generated at point "X," and the intensities of diffracted beams are detected with a counter labeled "C." The counter is mounted on a movable carriage that may also be rotated about the O-axis, with its angular position in terms of 2θ marked on a graduated scale. The carriage and specimen are mechanically coupled such that a rotation of the specimen through θ is accompanied by a 2θ rotation of the counter, which assures that the incident and reflection angles are maintained equal to one another. Collimators are incorporated within the beam path to produce a well-defined and focused beam. A filter provides a near-monochromatic beam. As the counter moves at constant angular velocity, a recorder automatically plots the diffracted beam intensity monitored by the counter as a function of 2θ, where 2θ is the diffraction angle and measured experimentally. A diffraction pattern for a powdered specimen of lead is shown in Fig. A.28. The high-intensity peaks result when the Bragg diffraction condition is satisfied by some set of crystallographic planes. One of the primary uses of x-ray diffractometry is for the determination of crystal structure. The unit cell size and geometry may be resolved from the angular positions of the diffraction peaks, while the arrangement of atoms within the unit cell is associated with the relative intensities of these peaks.

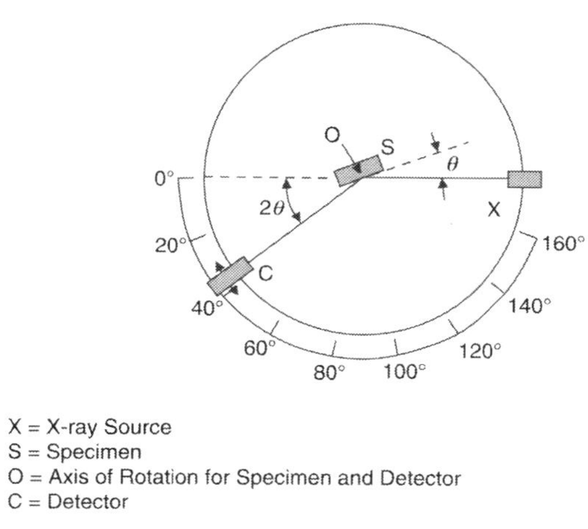

Fig. A.27 Schematic of x-ray diffractometer. Source: Ref A.4 as published in Ref A.1

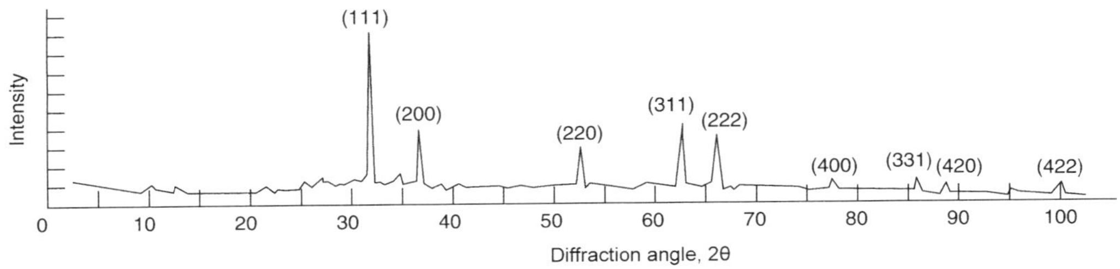

Fig. A.28 Typical diffraction pattern. Source: Ref A.4 as published in Ref A.1

A.8 Crystalline Imperfections

In a perfect crystalline structure, there is an orderly repetition of the lattice in every direction in space. However, real crystals are not perfect—they always contain a considerable number of imperfections, or defects, that affect their physical, chemical, mechanical, and electronic properties. It should be noted that defects do not necessarily have adverse effects on the properties of materials. They play an important role in processes such as deformation, annealing, precipitation, diffusion, and sintering.

All defects and imperfections can be conveniently classified under four main divisions: point defects, line defects, planar defects, and volume defects. Point defects are inherent to the equilibrium state and thus determined by temperature, pressure, and composition. However, the presence and concentration of the other defects depends on the way the metal was originally formed and subsequently processed.

Point Defects. A point defect is an irregularity in the lattice associated with a missing atom (vacancy), an extra atom (interstitial), or an impurity atom (substitutional). Due to their small size, point defects generally produce only very local distortions in the crystalline lattice. However, their presence can be significant, for example, in aiding diffusion in the crystalline lattice. Vacancies are the simplest defect. A vacancy is simply an atom missing from the crystalline lattice, as illustrated in Fig. A.29. Vacancies are created during solidification due to imperfect packing. They also occur during processing at elevated temperatures. In an otherwise completely regular lattice, the atoms are constantly being displaced from their ideal locations by thermal vibrations. The frequency of vibration is almost independent of temperature, but the amplitude increases with increasing temperature. For copper, the amplitude near room temperature is approximately one-half its value near the melting point and approximately twice its value near absolute zero. As the temperature is increased, the lattice vibrations become larger and atoms have a tendency to jump out of their normal positions, leaving a vacant lattice site behind.

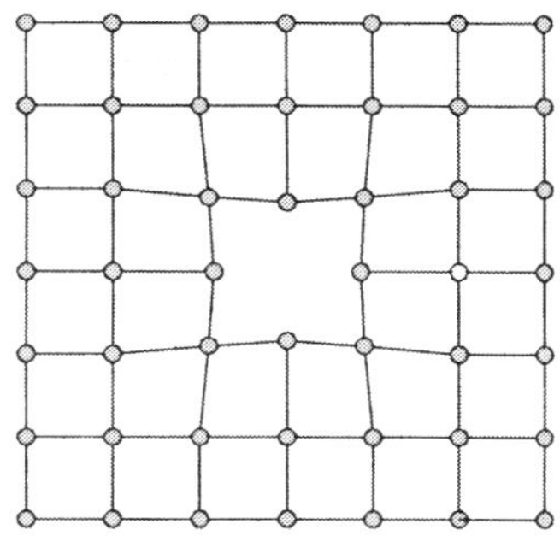

Fig. A.29 Vacancy point defect. Source: Ref A.5 as published in Ref A.1

The number of vacancies increases exponentially with temperature according to:

$$n_v = Ne^{-E_v/kT} \quad \text{(Eq A.25)}$$

where n_v is the number of vacancies at temperature T, N is the total number of lattice sites, E_v is energy necessary to form a vacancy, K is Boltzmann's constant (1.38×10^{-24} J/K), and T is absolute temperature in degrees Kelvin.

While the number of vacancies would be zero at absolute zero, it is on the order of 10^{-3} for metals near their melting point. According to Eq A.25, at any temperature above absolute zero, the equilibrium condition for a metal will contain vacancies; that is, the presence of vacancies is a condition of equilibrium.

Vacancies affect the properties of the metal. Density slightly decreases as the number of vacancies increases. The electrical resistance also decreases as the number of vacancies increases. Vacancies enhance atomic diffusion. Vacancy diffusion is the movement of a vacancy through the lattice, thereby assisting the diffusion of atoms. The number of dislocations is reduced when the vacancies diffuse to grain boundaries or surfaces, which act as sinks. If a metal is heated to a high temperature, the number of vacancies increases. If it is then suddenly quenched to room temperature, the vacancies are trapped in the lattice because they do not have time to diffuse out.

Vacancies can form by several mechanisms. In the Frenkel mechanism (Fig. A.30), an atom is displaced from its normal lattice position into an interstitial site. However, this requires quite a bit of energy–the energy to form a vacancy and the energy to form an interstitial. Therefore, the probability is quite low. A more realistic, and lower-energy method, is the Schottky mechanism (Fig. A.31), in which vacancies originate at free surfaces and move by diffusion into the crystal interior.

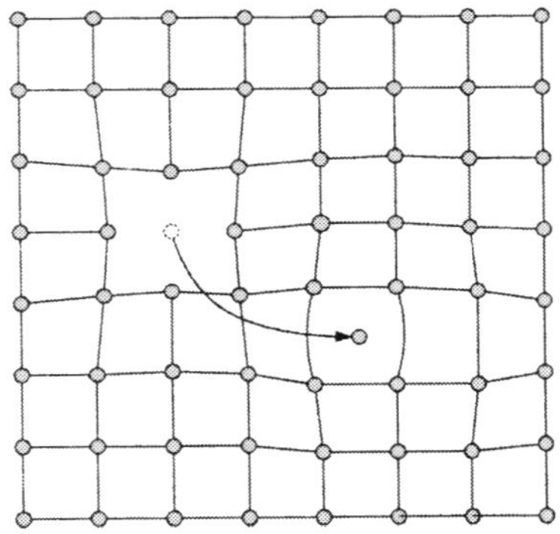

Fig. A.30 Frenkel mechanism. Simultaneous formation of vacancy and interstitial atom. Source: Ref A.5 as published in Ref A.1

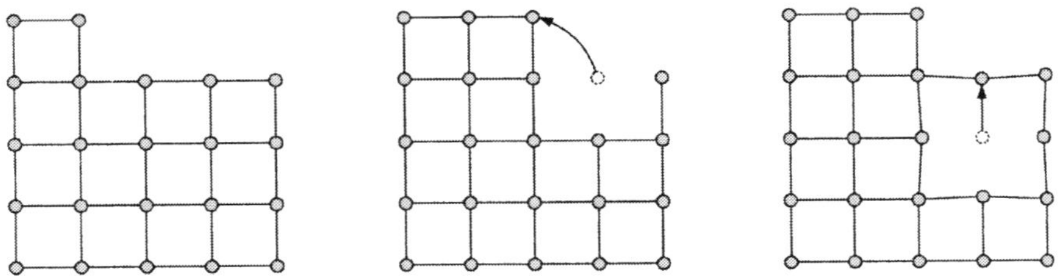

Fig. A.31 Schottky mechanism for vacancy formation. Source: Ref A.5 as published in Ref A.1

Solute atoms of a second metal can be present as impurities or added as intentional alloying elements. These solute atoms can substitute on the crystalline lattice for solvent atoms and form substitutional point defects, or they can be located in the interstitial locations between the atoms of the crystalline lattice to form interstitial defects (Fig. A.32). If the solute impurities are close to the same diameter as the solvent atoms, they will substitute for solvent atoms to form substitutional defects. Small atoms that can fit in between the larger solvent atoms of a crystalline structure are called interstitials. To form interstitial defects, the atomic diameter of the impurity has to be significantly smaller than the solvent atom diameter. Therefore, only atoms with very small diameters, such as carbon, nitrogen, hydrogen, and boron, can form interstitial defects. If the foreign atoms cause harmful or undesirable effects, they are called impurities, while if they are helpful, they are referred to as alloying elements.

Point defects influence solid-state processes such as diffusion, dislocation motion, phase transformations, and electrical conductivity. Point defects typically strengthen a metal and decrease its ductility by impeding the motion of dislocations. Point defects also decrease electrical conductivity, because they interfere with the flow of electrons through the lattice.

Line Defects. One of the most important defects is the line or edge dislocation. The existence of line defects in crystals, called dislocations, provides the mechanism that allows mechanical deformation. A crystalline metal without dislocations, although extremely strong, would also be extremely brittle and practically useless as an engineering material. Thus, dislocations play a central role in the determination of such important properties as strength and ductility. In fact, virtually all mechanical properties of metals are to a significant extent controlled by the behavior of line imperfections.

As shown in Fig. A.33, an edge dislocation can be visualized as resulting from the insertion of an extra half-plane of atoms above (or below) the

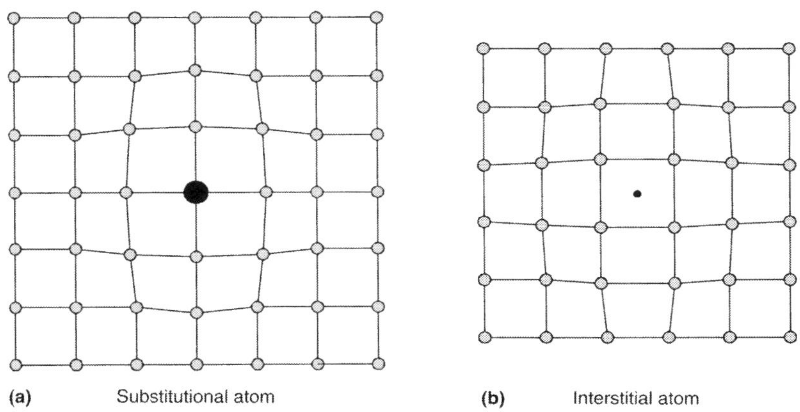

Fig. A.32 Foreign atom point defects. Source: Ref A.5 as published in Ref A.1

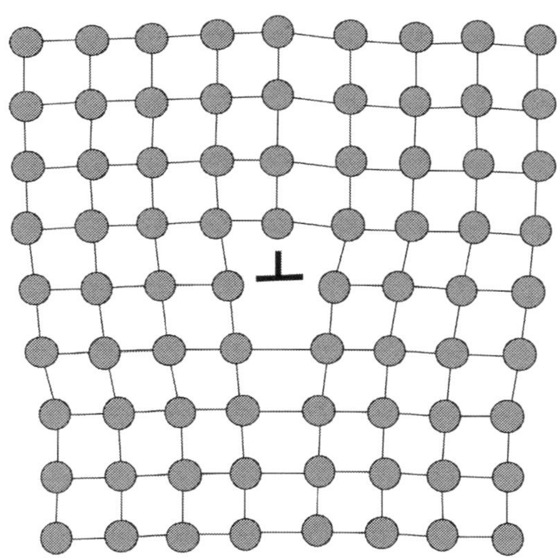

Fig. A.33 Line dislocation. Source: Ref A.1

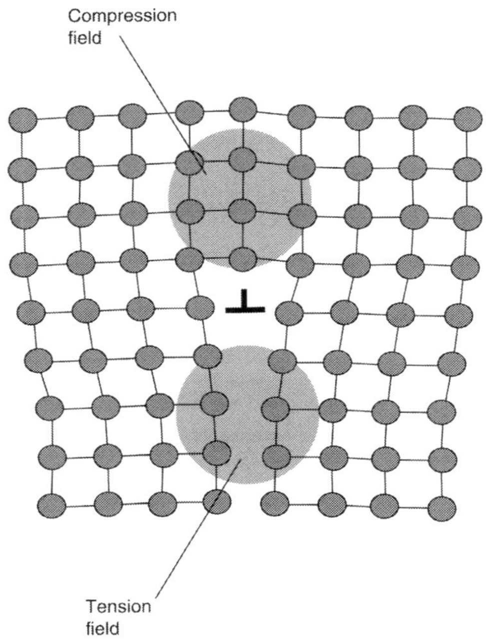

Fig. A.34 Stress field around line dislocation. Source: Ref A.1

dislocation line. By definition, the dislocation shown in Fig. A.33 is a positive dislocation. A negative dislocation has the extra half-plane below the dislocation line. An edge dislocation creates a zone of elastic deformation around the dislocation (Fig. A.34). The lattice below the dislocation is in a state of tension, while above the dislocation there is a compressive stress field. In the lattice below a dislocation, interstitial atoms usually cluster in regions where the tensile stresses help make more room for them.

A quantitative description of dislocations is given by the Burgers vector, b, illustrated in Fig. A.35. This vector is defined using what is called the Burgers circuit, which is an atom-to-atom path that makes a closed loop in a dislocation-free part of the crystal lattice. Now, if the same Burgers circuit is made to encircle a dislocation, the loop does not close. The vector needed to close the loop (the vector from the end of the Burgers circuit to its starting point) is the Burgers vector, b, describing the dislocation. The displacement vector between the two parts of the crystal is denoted by u, and the axis of the dislocation is t. For an edge dislocation, the Burgers vector, b, is perpendicular to the axis of the dislocation, t ($b \perp t$), and parallel to the displacement vector, u ($b \parallel u$).

The other important type of line dislocation is the screw dislocation (Fig. A.36). The term *screw dislocation* is used because of the spiral surface formed by the atomic planes around the screw dislocation line. When a

Burgers circuit is used to determine the Burgers vector of a screw dislocation, the vector is found to be parallel to the dislocation line rather than perpendicular to it, as in the case of an edge dislocation. A screw dislocation is somewhat like a spiral ramp with an imperfection line running down its axis. In a screw dislocation, the Burgers vector, b, is parallel to both the axis of the dislocation, t, and the displacement vector, u; that is, $(b \parallel t \parallel u)$.

An important characteristic of a dislocation is that it cannot end inside the crystal; it must end at a surface such as a grain boundary or at a surface of the crystal. It is possible for a dislocation to change its character inside the crystal, as shown in the mixed dislocation for an aluminum alloy in Fig. A.37. Here, an edge dislocation is converted to a screw dislocation; in this

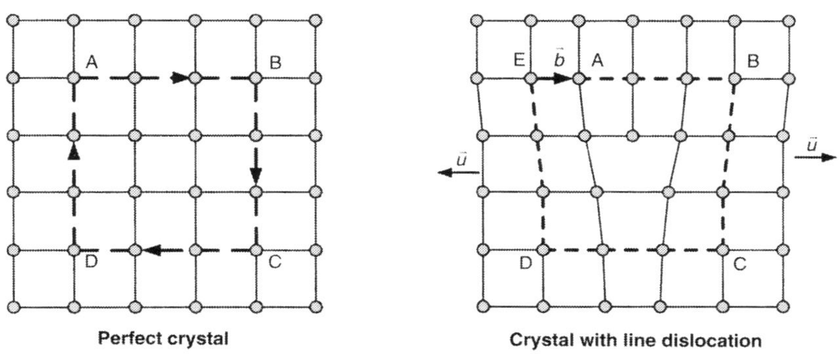

Fig. A.35 Burgers circuit and vector for line dislocation. Source: Ref A.5 as published in Ref A.1

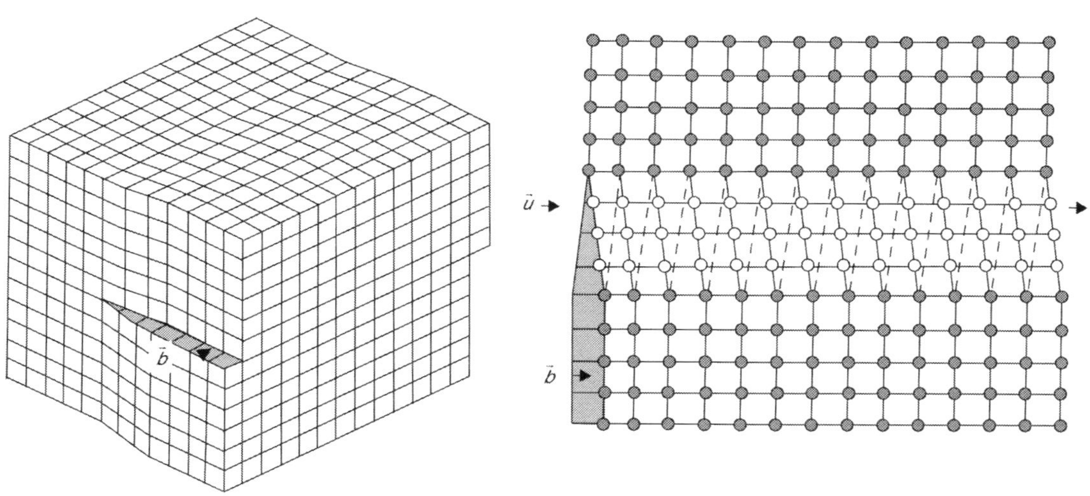

Fig. A.36 Screw dislocation. Source: Ref A.1

case, the screw dislocation is shown ending at the surface of the crystal. Dislocations will also form closed loops within a crystal, changing from an edge to a screw and then back to another edge and finally back to a screw to enclose the loop. The material within the loop is visualized as having slipped on the specified slip plane relative to the material around it.

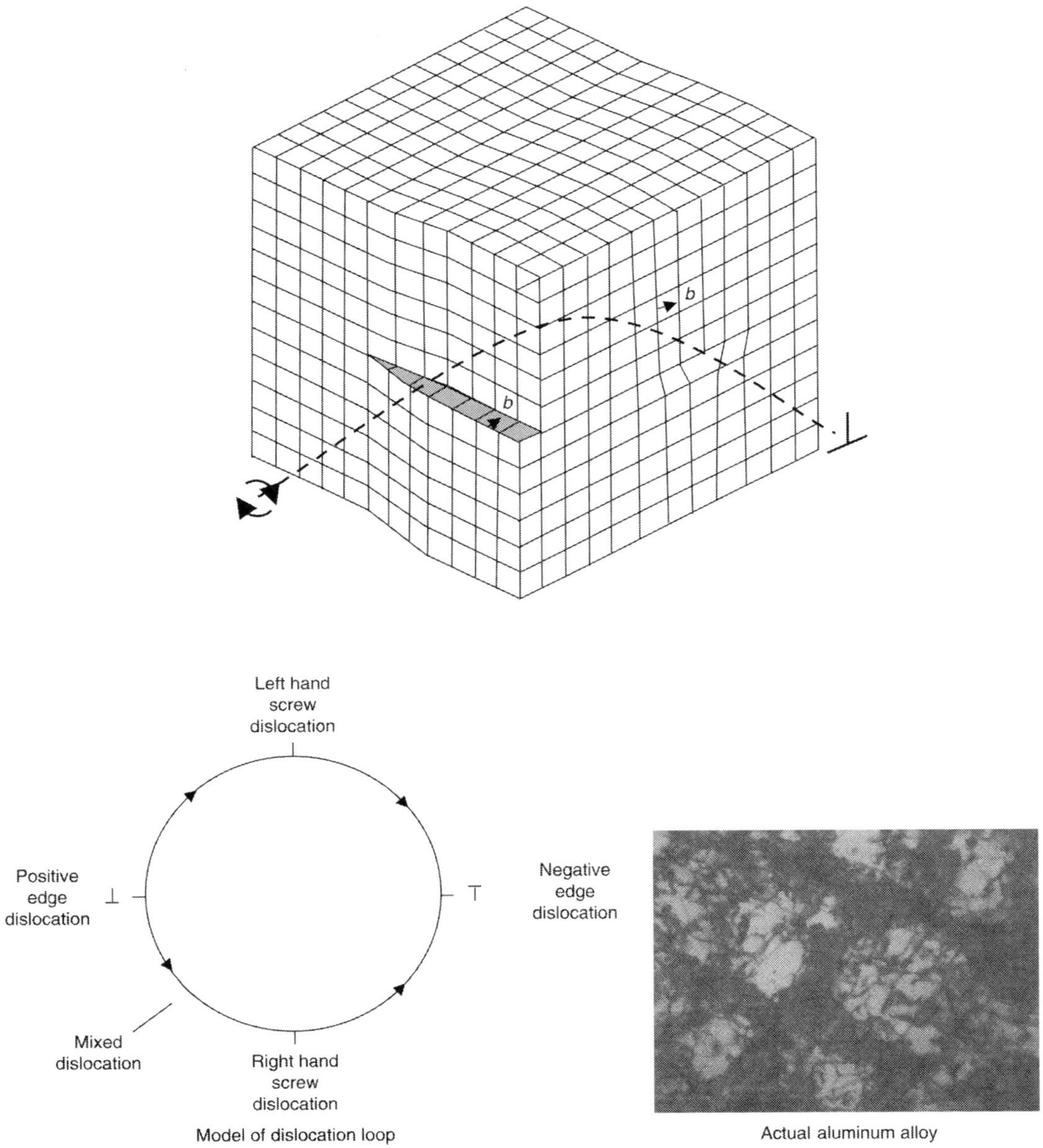

Fig. A.37 Combination screw and line dislocations. Source: Ref A.1

A.9 Plastic Deformation

When a mechanical shear load is applied to a metal, it deforms under the applied stress, as shown schematically in Fig. A.38. If the load is small, the bonds between the atoms will be stretched but will return to their normal lattice positions when the load is removed; this is elastic deformation. However, if a ductile metal is stretched beyond the elastic capability of the bonds, when the load is removed it will not return to its original shape. It is then said to have undergone plastic deformation. On the other hand, if the material is brittle rather than ductile, when the elastic stretching of the bonds is exceeded, it immediately fails with little or no evidence of plastic deformation. This is the type of failure that is normally encountered in covalently and ionically bonded solids, such as glasses and crystalline ceramics. The question becomes: Why do metals exhibit moderate-to-large amounts of plastic deformation, while other materials, such as glasses and ceramics, exhibit almost no ductility and fail in a brittle manner?

The nondirectional metallic bond allows metals to deform by shear as illustrated in Fig. A.39. For the atoms in the upper plane to slide over those in the lower plane, strong interatomic forces must be overcome by the applied shear stress. When the atoms in the upper plane have been displaced by one-half of their transit distance, the crystal energy is at a maximum and then falls once they reach their new equilibrium positions. The shear stress required to cause slip is initially zero when the two planes

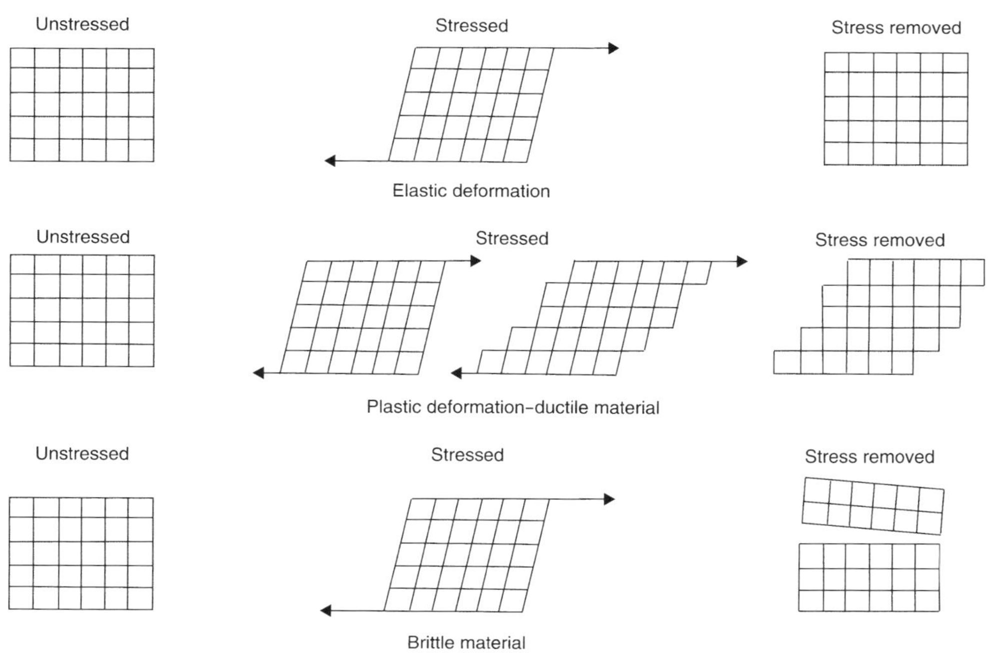

Fig. A.38 Material behavior under stress. Source: Ref A.1

are in coincidence and when the atoms of the top plane are midway between those of the bottom plane, because this is a symmetry position. Between these positions, each atom is attracted to its nearest atom of the other row, so that the shearing stress is a periodic function of the displacement. This shear, or slip, takes the path of least resistance and thus occurs along the close-packed planes in close-packed directions. Atomic bonds are broken and then re-established as the metallic ions move pass one another. This is much more difficult for covalently and ionically bonded solids. In the covalent bond, the bonds between two atoms are well established and do not want to be broken. Remember that covalent bonds are both strong and highly directional, while metallic ions share their valence electrons, allowing freer movement through the electron cloud. The problem with ionic bonds is that one ion is positive and the other is negative. During any type of shear mechanism, when two positively charged ions, or two negatively charged ions, approach each other too closely, a strong electrical repulsive force will develop between the two and resist plastic movement.

During tensile testing of a single crystal, shown schematically in Fig. A.40, an applied stress will be reached when the shear stress, resolved onto a slip plane in a slip direction, attains a critical value so that dislocations on that slip plane slip or glide. If the normal, n, of the slip plane lies at an

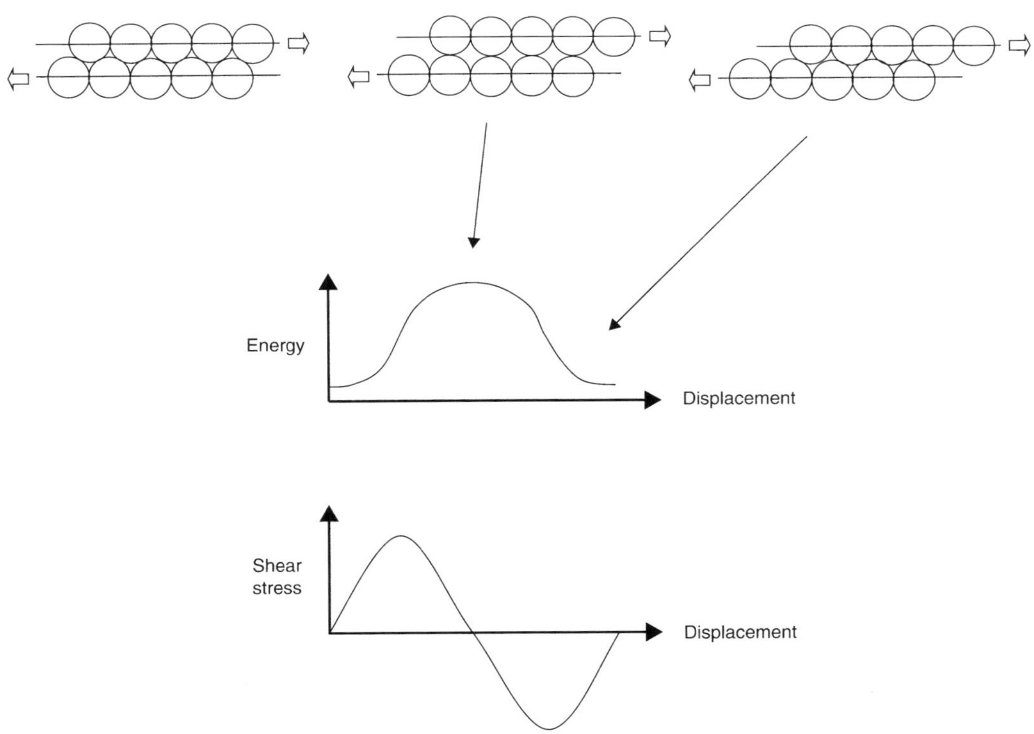

Fig. A.39 Planar slip. Source: Ref A.1

angle, ϕ, to the tensile axis, its area will be $A/\cos\phi$. Similarly, if the slip plane lies at an angle, λ, to the tensile axis, the component of the axial force, P, acting on the slip direction will be $P\cos\lambda$. The resolved shear stress, τ_r, is then given by:

$$\tau_r = \frac{P\cos\lambda}{(A/\cos\phi)} = \sigma\cos\phi\cos\lambda \qquad \text{(Eq A.26)}$$

For a given metal, the value of τ at which slip occurs is usually found to be a constant, known as the critical resolved shear stress, τ_c. This relationship is known as Schmid's law, and the quantity $\cos\phi\cos\lambda$ is called the Schmid factor. Because the shear stress at which slip occurs is the yield stress, σ_y, it follows that:

$$\tau_c = \sigma_y \cos\phi\cos\lambda \qquad \text{(Eq A.27)}$$

However, most metals used in industry are polycrystalline, not single crystals. Under an applied axial load, the Schmid factor will by different for each grain. For randomly oriented grains, the average value of the Schmid factor is ~1/3, which is referred to as the Taylor factor. It then follows that the yield strength should have a value of approximately $3\tau_c$.

Dislocations and Plastic Flow. From our knowledge of the metallic bond, it is possible to derive a theoretical value for the stress required to

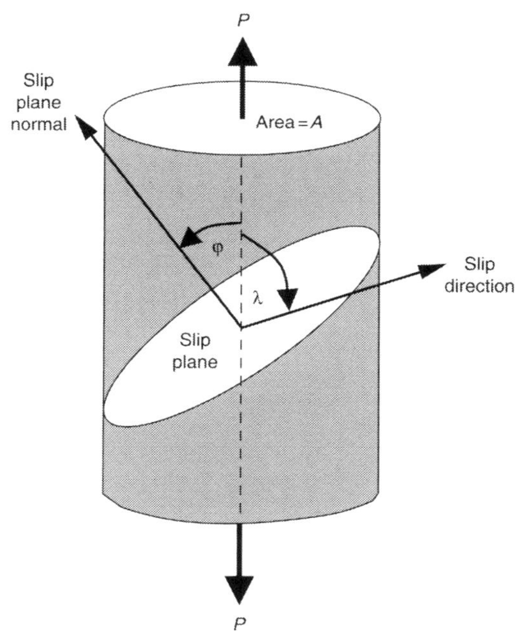

Fig. A.40 Tensile test of single crystal. Source: Ref A.1

produce slip by the simultaneous movement of atoms along a plane in a metallic crystal. However, the strength actually obtained experimentally on single crystals is only approximately one-thousandth (1/1000) of the theoretical value, assuming simultaneous slip by all atoms on the plane. Obviously, slip does not occur by the simple simultaneous block movement of one layer of atoms sliding over another, as previously shown in Fig. A.39. Nor does such a simple interpretation of slip explain work hardening that takes place during mechanical deformation. Earlier theories that sought to explain slip by the simultaneous gliding of a complete block of atoms over another have now been discarded, and the modern concept is that slip occurs by the step-by-step movement of dislocations through the crystal.

When force is applied such that it shears the upper portion of the crystal to the right, as shown in Fig. A.41, the plane of atoms above the dislocation can easily establish bonds with the lower plane of atoms to its right, with the result that the dislocation moves one lattice spacing at a time. Note that only single bonds are being broken at any one time, rather than the whole row, as shown in Fig. A.39. The atomic distribution is again similar to the initial configuration and so the slipping of atom planes can be repeated. The movement is much like that of advancing a carpet along a floor by using a wrinkle that is easily propagated down its length. This stress required to cause plastic deformation is orders of magnitude less when dislocations are present than in a dislocation-free, perfect crystalline structure. If a large number of dislocations move in succession along the same slip plane, the accumulated deformation becomes visible, resulting in macroscopic plastic deformation. Slip can take place by both edge and screw dislocations, as shown in Fig. A.42. Note that although the mechanisms are different, the unit slip produced by both is the same.

Dislocations do not move with the same degree of ease on all crystallographic planes nor in all crystallographic directions. Ordinarily, there are preferred planes, and in these planes, there are specific directions along which dislocation motion can occur. These planes are called slip planes, and the direction of movement is known as the slip direction. The combination of a slip plane and a slip direction forms a slip system. For a

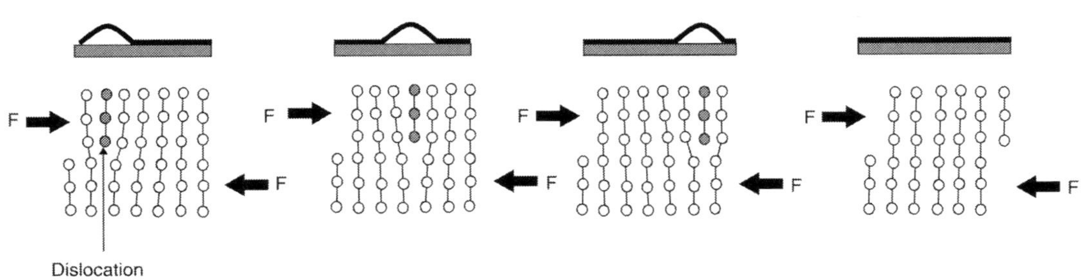

Fig. A.41 Line dislocation movement. Source: Ref A.1

particular crystal structure, the slip plane is that plane having the most dense atomic packing; that is, it has the greatest planar density. The slip direction corresponds to the direction, in this plane, that is most closely packed with atoms, that is, has the highest linear density. Because plastic deformation takes place by slip, or sliding, on the close-packed planes, the greater the number of slip systems available, the greater the capacity for plastic deformation. The major slip systems for the common metallic crystalline systems are summarized in Table A.3.

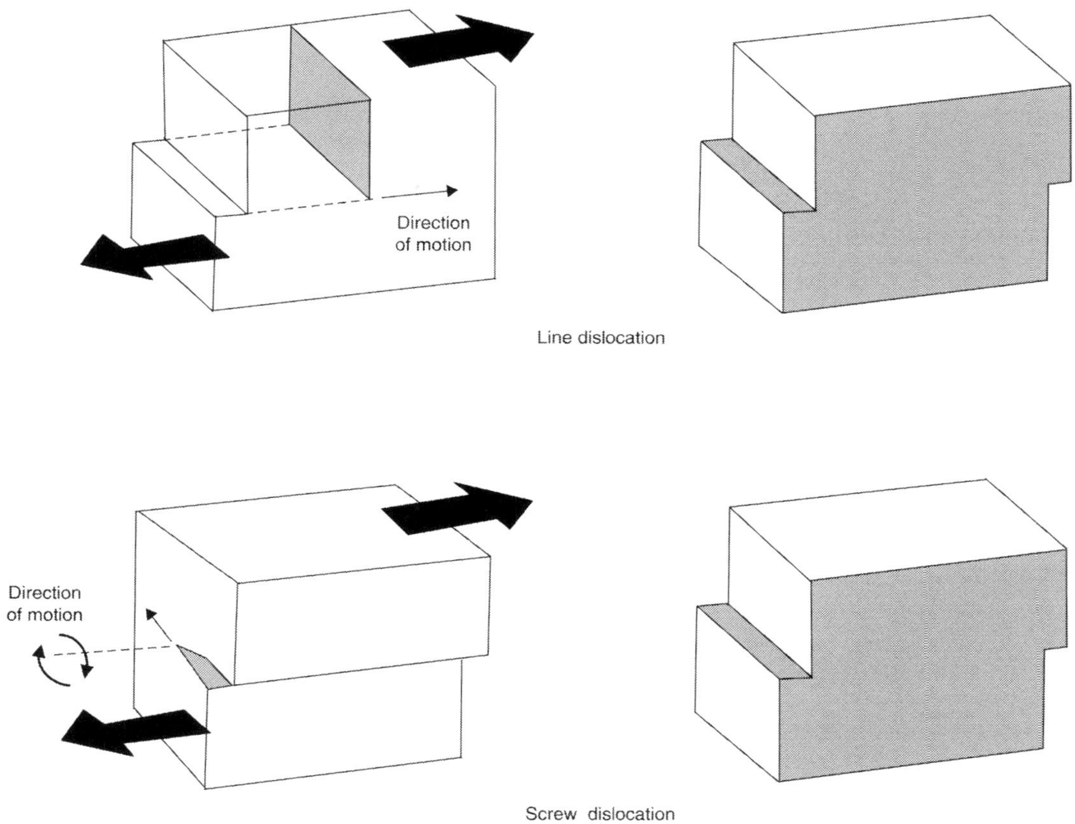

Fig. A.42 Displacements caused by line and screw dislocations. Source: Ref A.1

Table A.3 Major slip systems for common crystal systems

Crystal system(a)	Slip planes		Slip directions		No. of slip systems
Notation	Notation	Number	Notation	Number	Number
bcc	{110}	6	<111>	2	12
fcc	{111}	4	<110>	3	12
hcp	(0001)	1	$[11\bar{2}0]$	3	3

(a) bcc, body-centered cubic; fcc, face-centered cubic; hcp, hexagonal close-packed. Source: Ref A.5 as published in Ref A.1

Face-centered cubic (fcc) metals have a large number of slip systems (12) and are therefore capable of moderate to extensive plastic deformation. Although body-centered cubic (bcc) systems often have up to 12 slip systems, some of them, like steel, exhibit a ductile-to-brittle transition as the temperature is lowered due to the strong temperature sensitivity of their yield strength, which causes them to fracture prior to reaching their full potential of plastic deformation. In general, the number of slip systems available for hexagonal close-packed (hcp) metals is less than that for either the fcc or bcc metals, and their plastic deformation is much more restricted. The hcp structure normally has only three to six slip systems, only one-fourth to one-half the available slip systems in fcc and bcc structures. Therefore, metals with the hcp structure have poor to only moderate room-temperature ductility. Thus, the hcp metals, such as alloys of magnesium, beryllium, and titanium, often require heating to elevated temperatures, where slip becomes much easier, prior to forming operations.

Dislocations can have two basic types of movement: glide and climb. Glide, or slip, is the type of dislocation movement that has been discussed thus far. It occurs in the plane containing both the dislocation line and the Burgers vector. During each glide step, a single row of atoms changes position with a closest-neighbor atom, and the passage of the dislocation displaces the upper part of the grain with respect to the lower part of the grain. The simultaneous glide of many identical dislocations under an applied stress is known as slip and is the typical mechanism of plastic deformation in metals. The second type of dislocation motion is known as climb. As illustrated in Fig. A.43, climb is directly dependent on vacancies. For an edge dislocation to climb, vacancies must be either created or destroyed. If vacancies are not present in large quantities, climb cannot occur because it is dependent on diffusion. Although glide can occur at

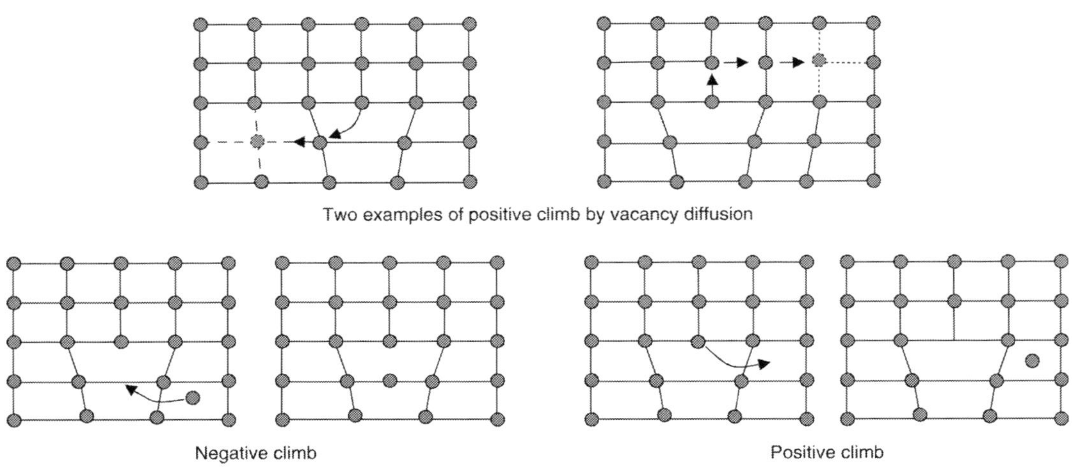

Fig. A.43 Dislocation climb associated with interstitial atoms. Source: Ref A.1

all temperatures, climb is practically nonexistent at temperatures below approximately 0.4 T_m, where T_m is the absolute melting point. However, climb becomes an important deformation mechanism when the metal is subjected to stresses at temperatures exceeding approximately 0.4 T_m.

The dislocation density for crystals is approximately 10^8 cm^{-2}, corresponding to an average distance between dislocations of a few thousand atoms. If each dislocation produces only one unit of slip, this relatively small number of dislocations could not produce large-scale plastic deformation. Thus, a large number of dislocations must be present to produce macroscopic slip. The Frank-Read spiral mechanism (Fig. A.44) explains how dislocations can multiply and increase their effectiveness a thousandfold. If a dislocation line becomes immobilized and is pinned at its ends, it will tend to bow out under the influence of an applied shear stress. Eventually, the loop becomes circular and then starts closing in on itself at the ends. This allows the formation of a new loop that again bows out under the influence of a shear stress. This process is repeated over and over again, each time generating a new dislocation loop.

Dislocations are influenced by the presence of other dislocations and interact with each other, as shown for a number of different interactions in Fig. A.45. Dislocations with Burgers vectors of the same sign will repel each other, while dislocations of opposite signs will attract each other and, if they meet, annihilate each other. If the two dislocations of opposite signs are not on the same slip plane, they will merge to form a row of vacancies. These types of interactions occur because they reduce the internal energy of the system. When a dislocation becomes pinned by an obstacle

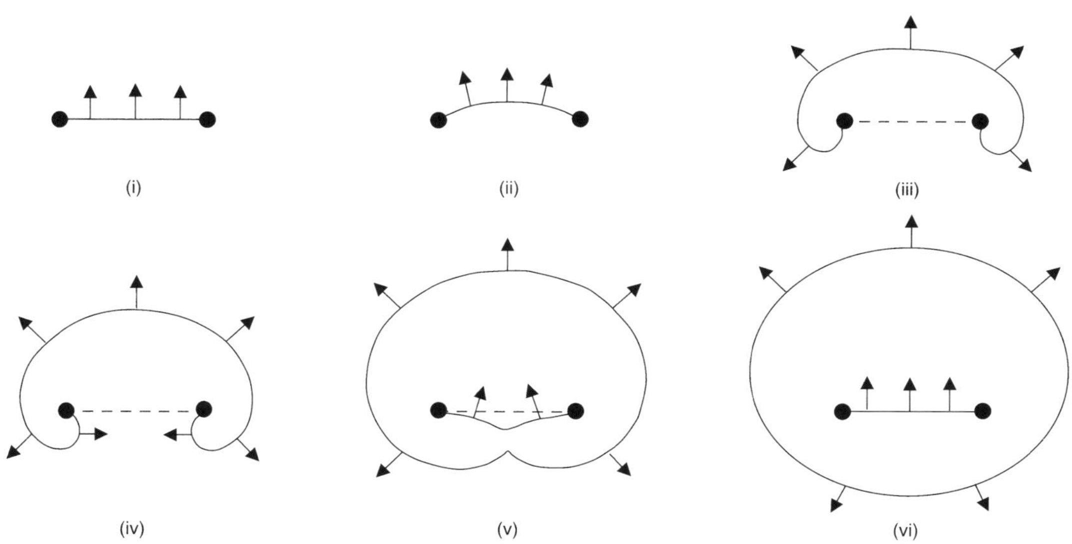

Fig. A.44 Frank-Read mechanism for dislocation multiplication. Source: Ref A.1

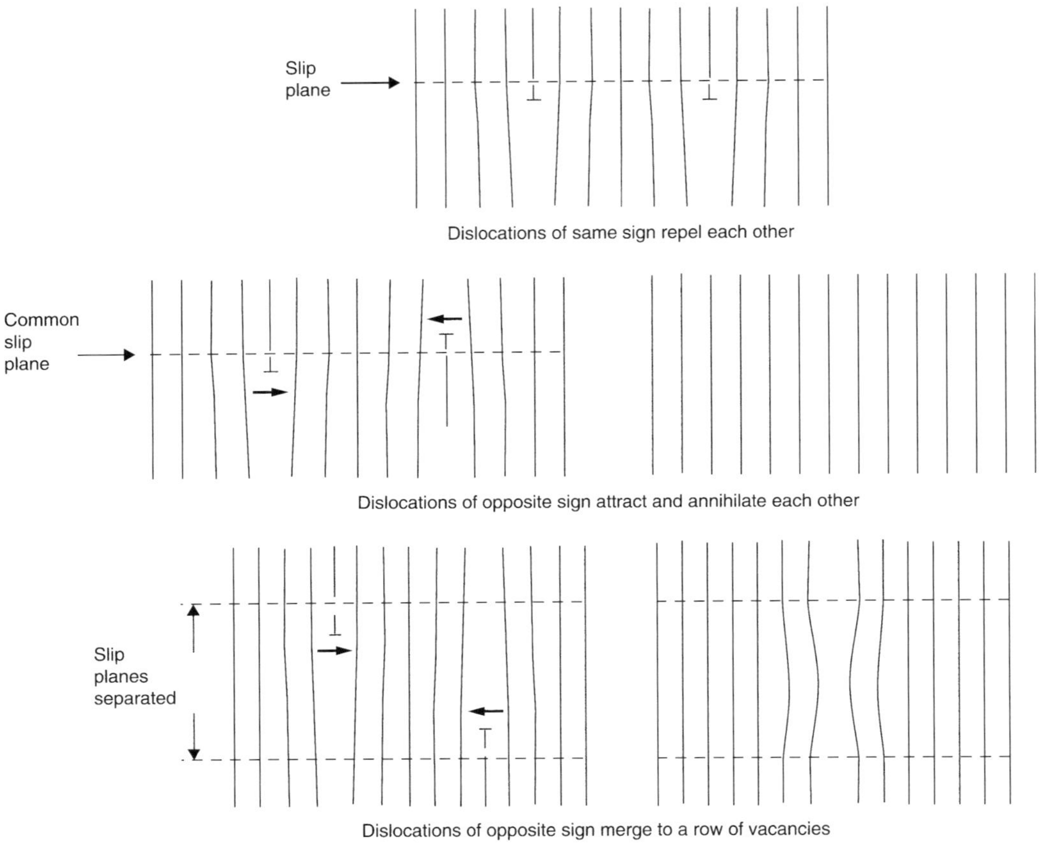

Fig. A.45 Examples of dislocation interactions. Source: Ref A.1

and is immobile, it is termed a sessile dislocation. A dislocation that is not impeded and can move through the lattice is called a glissile dislocation.

If dislocations could only move by gliding on a single slip plane, they would soon be impeded by obstacles and their motion would be restrained. However, screw dislocations can bypass obstacles on their slip plane by cross slipping onto an alternate plane (Fig. A.46). While line dislocations cannot cross slip, they can often convert themselves into screw dislocations while they cross slip. Vacancy diffusion also contributes significantly to high-temperature creep.

Work Hardening. While slip is required to facilitate plastic deformation, and therefore allow a metal to be formed into useful shapes, strengthening metals requires increasing the number of barriers to slip and reducing the ability to plastically deform. Increasing the interference to slip, and increasing the strength, can be accomplished by methods such as plastic deformation. As a metal is plastically deformed, new dislocations

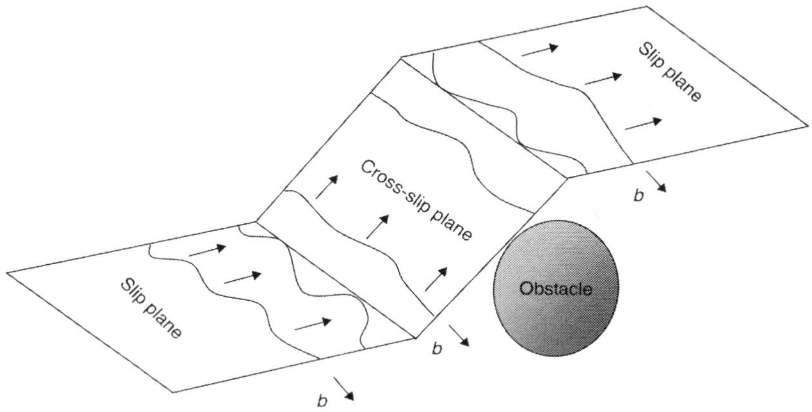

Fig. A.46 Cross slip of screw dislocation. Source: Ref A.1

are created, so that the dislocation density becomes higher and higher. In addition to multiplying, the dislocations become entangled and impede each others' motion. The result is increasing resistance to plastic deformation with increasing dislocation density. The number of dislocations is defined by the dislocation density, ρ, which is the length of dislocations per unit volume of material. Therefore, the units of ρ are cm/cm^3 or cm^{-2}. The dislocation density of an annealed metal usually varies between approximately 10^6 to 10^7 cm^{-2}, while that for a cold worked metal may run as high as 10^8 to 10^{11} cm^{-2}.

This continual increase in resistance to plastic deformation is known as work hardening, cold working, or strain hardening. Work hardening results in a simultaneous increase in strength and a decrease in ductility. Because the work-hardened condition increases the stored energy in the metal and is thermodynamically unstable, the deformed metal will try to return to a state of lower energy. This generally cannot be accomplished at room temperature. Elevated temperatures, in the range of ½ to ¾ of the absolute melting point, are necessary to allow mechanisms, such as diffusion, to restore the lower-energy state. The process of heating a work-hardened metal to restore its original strength and ductility is called annealing. Metals undergoing forming operations often require intermediate anneals to restore enough ductility to continue the forming operation. Approximately 5% of the energy of deformation is retained internally as dislocations when a metal is plastically deformed, while the rest is dissipated as heat.

A.10 Surface or Planar Defects

Surface, or planar, defects occur whenever the crystalline structure of a metal is discontinuous across a plane. Surface defects extend in two directions over a relatively large surface with a thickness of only one or two lattice parameters. Grain boundaries and phase boundaries are independent of crystal structure, while coherent phase boundaries, twin boundaries, and stacking faults all depend on the crystalline structure.

A.10.1 Grain Boundaries

The most important surface defect is the grain boundary. Most metals are polycrystalline and consist of many small crystallites called grains. The interfaces between these grains are called grain boundaries and are only

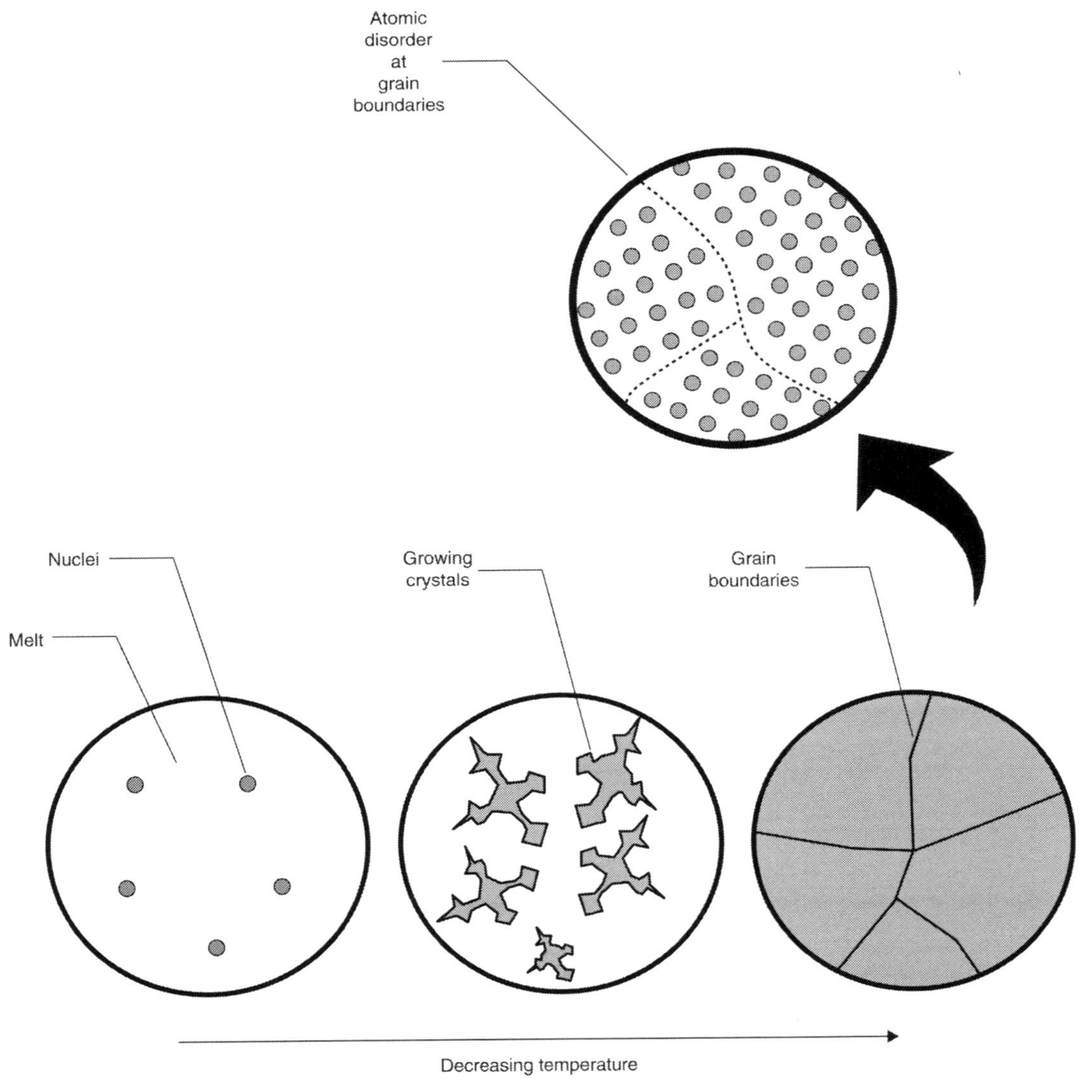

Fig. A.47 Solidification sequence for a metal. Source: Ref A.1

one or two atoms thick, because the system wants to reduce the free energy as much as possible. Atoms within the grain boundaries are highly strained and distorted; therefore, grain boundaries are high-energy sites. The average diameter of the individual grains within a polycrystalline metal defines the metal grain size. Grain boundaries are a result of the solidification process and occur as a result of the misorientation of the grains as they are frozen into position (Fig. A.47).

Small-angle grain boundaries occur when the misorientation between grains is small, usually less than 5°. These small-angle misorientations can be represented by a row of somewhat parallel edge dislocations, as shown for the low-angle tilt boundary in Fig. A.48. The regions between the dislocations consist of an almost perfect fit and have low strain, while regions at the dislocation cores have poor fit and are high-strain regions. The regions surrounded by low-angle grain boundaries are called subgrains or subcrystals and are essentially free of dislocations. The spacing between dislocations, D, is:

$$D = \frac{b}{\sin\theta} \cong \frac{b}{\theta} \qquad \text{(Eq A.28)}$$

where θ is the angular misorientation across the boundary. A gross grid of two screw dislocations can also form a low-angle grain boundary, in this instance it is called a low-angle twist boundary.

If the misorientation is greater than approximately 10 to 15°, then high-angle grain boundaries will form, as previously illustrated in Fig. A.48. Because high-angle grain boundaries result in a less ordered arrangement of the atoms with large areas of misfit and a relatively more open structure,

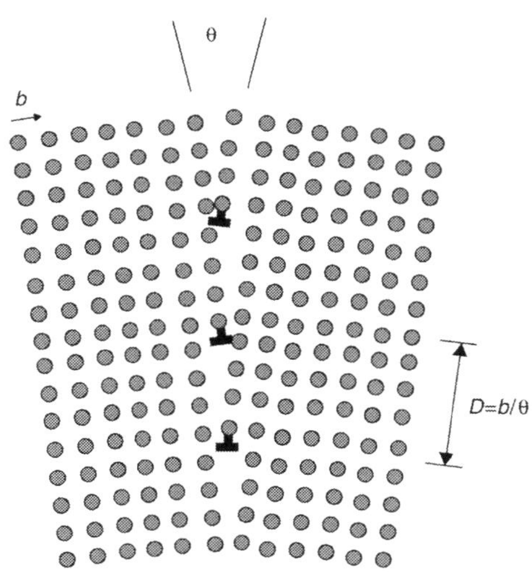

Fig. A.48 Low-angle tilt boundary. Source: Ref A.1

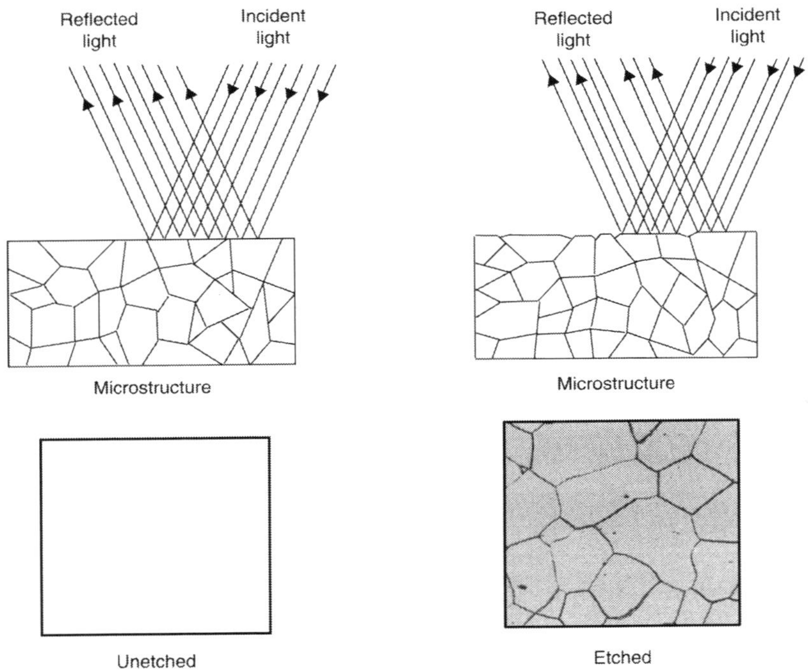

Fig. A.49 Metallography of unetched and etched samples. Source: Ref A.1

the atoms along the grain boundaries have a higher energy than the atoms within the grain interiors. Thus, grain boundaries are regions with many irregularly placed atoms, dislocations, and voids. Compared with high-angle boundaries, low-angle boundaries have less severe defects, obstruct plastic flow less, and are less susceptible to chemical attack and segregation of alloying constituents. In general, mixed types of grain-boundary defects are common. All grain boundaries are sinks into which vacancies and dislocations can disappear.

Grain boundaries, as well as other microstructural features, are often observed by polishing a metal surface, lightly etching it with an acid, and then examining it with a light microscope at magnification. Grain boundaries become visible when the polished surface is etched with the proper acid, creating a microscopically uneven surface that reflects the light slightly differently (Fig. A.49) than the unetched surface.

A grain boundary tends to minimize its area in order to reduce the internal energy of the system. The driving force for this energy reduction is surface tension, which can be reduced by straightening of the irregular-shaped boundaries. If a grain has less than six boundaries, then each boundary will be concave inward and will be unstable. On the other hand, if a grain has more than six sides, the boundaries will be planar and stable. At high temperatures, for example, during annealing operations at $T > 0.5$

T_m, there is an exponential increase in the mobility of the atoms. Grains with six or more boundaries will tend to grow, while grains with less than six sides will shrink and be consumed by larger grains. Atoms in the shrinking grains will migrate across the boundary interface to join the larger growing grains. The presence of second-phase particles helps to pin the grain boundaries and impede grain growth. At equilibrium (Fig. A.50), all three grain boundaries will have the same surface tension, γ, and all three will have angles $\theta = 120°$. If all of the boundaries meet at 120°, then the shape of the grain will fill all of the available area and is called a tetrakaidecahedron which contains 14 faces, 36 edges, and 24 corners. A stack of six tetrakaidecahedra is shown in Fig. A.51.

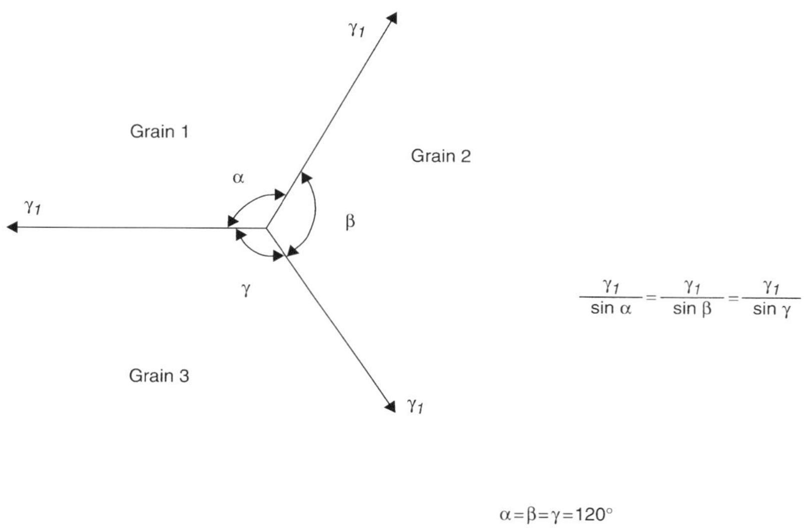

Fig. A.50 Grain boundaries in equilibrium. Source: Ref A.6 as published in Ref A.1

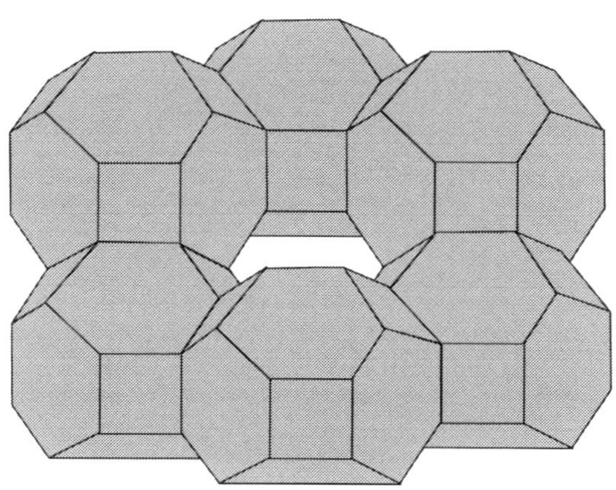

Fig. A.51 Stack of tetrakaidecahedra. Source: Ref A.6 as published in Ref A.1

Grain boundaries are preferential regions for accumulation and segregation of many types of impurities. Weakening or embrittlement can also occur by preferential phase precipitation or absorption of environmental species, such as hydrogen or oxygen, in the grain boundaries. At room temperature, the grain boundaries are usually stronger than the grain interiors, and failure usually occurs through the grain themselves (transgranular). However, at high temperatures, the grain boundaries typically become the weak link, and failure occurs through the grain boundaries (intergranular). Thus, a coarse grain structure is desirable for high-temperature applications, while fine grains and finely divided phase regions are preferred for most room- and low-temperature applications.

Polycrystalline Metals. In the single-crystal tensile stress where the critical resolved shear stress was determined (Fig. A.40), the slip planes were not restricted by the presence of other grains, and slip occurs as in the left-hand portion of Fig. A.52. However, in polycrystalline metals, the orientation of the slip planes in adjoining grains is seldom aligned, and the slip plane must change direction when traveling from one grain to

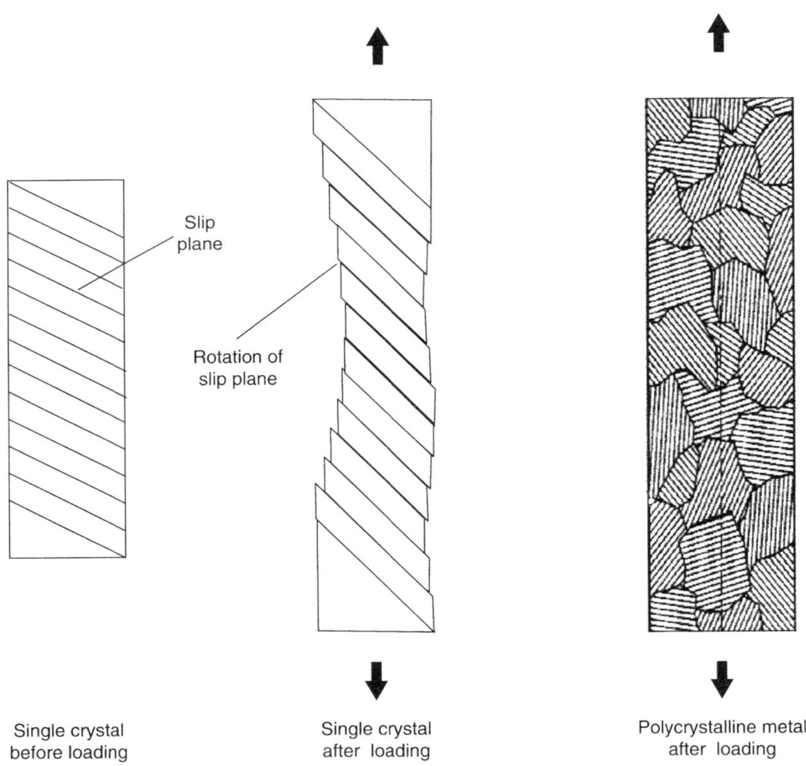

Fig. A.52 Tension loading of single and polycrystalline metals. Source: Ref A.1

another. Reducing the grain size produces more changes in direction of the slip path and also lengthens it, making slip more difficult; therefore, grain boundaries are effective obstacles to slip. In addition, dislocations cannot cross the high-energy grain boundaries; instead they are blocked and pile up at the boundaries (Fig. A.53). Decreasing the grain size is effective in both increasing strength and also increasing ductility, and, as such, is one of the most effective strengthening mechanisms. Fracture resistance also generally improves with reductions in grain size, because the cracks formed during deformation, which are the precursors to those causing fracture, are limited in size to the grain diameter. The yield strength of many metals and their alloys has been found to vary with grain size according to the Hall-Petch relationship:

$$\sigma_y = \sigma_0 + k_y d^{-1/2} \quad \text{(Eq A.29)}$$

where k_y is the Hall-Petch coefficient, a material constant; d is the grain diameter; and σ_0 is the yield strength of an imaginary polycrystalline metal having an infinite grain size.

The Hall-Petch relationships for a number of metals are shown in Fig. A.54. The value of the Hall-Petch coefficient varies widely for different metals, and grain size refinement is more efficient for some metals than others. For example, grain size refinement significantly increases the yield strength of low-carbon steels (up to 275 MPa, or 40 ksi), while it provides only approximately a 60 MPa (9 ksi) increase for a typical aluminum alloy.

Because the grain size of a metal or alloy has important effects on the structural properties, a number of methods have been developed to measure the grain size of a sample. In all methods, some form of microexamination is used in which a small sample is mounted, polished, and then etched to reveal the grain structure. The most direct method is then to count the

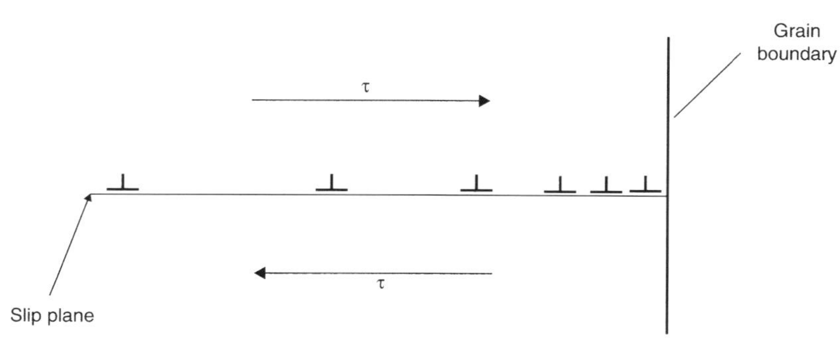

Fig. A.53 Dislocation pile-up at grain boundary. Source: Ref A.1

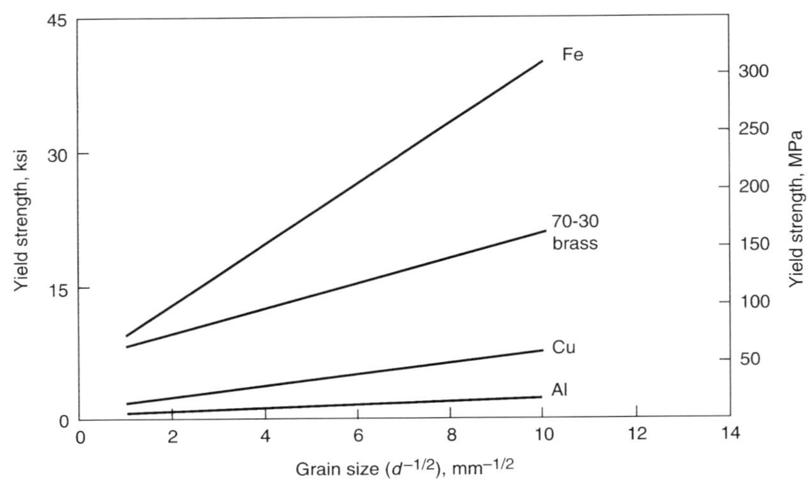

Fig. A.54 Hall-Petch relationship. Source: Ref A.1

number of grains present in a known area of the sample so the grain size can be expressed as the number of grains/area.

ASTM International has developed standard procedures for determining average grain size. The ASTM grain size number provides a convenient method for communicating grain sizes. For materials with a uniform grain size distribution, the ASTM grain size number is derived from the number of grains/in.^{2} when counted at a magnification of 100×. The ASTM index, N, is given by:

$$n = 2^{(N-1)} \quad \text{(Eq A.30)}$$

where n is the number of grains/in.^{2} at 100× magnification. To obtain the number of grains per square millimeter at 1×, multiply n by 15.50. This can be rewritten as:

$$\log n = (N - 1) \log 2 \quad \text{(Eq A.31)}$$

or

$$N = \frac{\log n}{0.3010} + 1 \quad \text{(Eq A.32)}$$

A listing of ASTM grain size numbers and the corresponding grain size is given in Table A.4. Note that larger ASTM grain size numbers indicate more grains per unit area and finer, or smaller, grain sizes.

A.10.2 Phase Boundaries

While a grain boundary is an interface between grains of the same composition and same crystalline structure (α/α interface) with different orientations, a phase boundary is one between two different phases (α/β interface) that can have different crystalline structures and/or different compositions. In two-phase alloys, such as copper-zinc brass alloys containing more than 40% Zn, second phases, such as the one shown in Fig. A.55, can form due to the limited solid solubility of zinc in copper.

There are three different types of crystalline interfaces that can develop between two phases (Fig. A.56): coherent, semicoherent, and incoherent. A fully coherent phase boundary (Fig. A.56a, b) occurs when there is perfect atomic matching and the two lattices are continuous across the interface. The interfacial plane will have the same atomic configuration in both planes. Because there is perfect matching at the interface, the

Table A.4 ASTM grain size, $n = 2^{(N-1)}$

Grain size no.	Average number of grains/unit area: No./in.² at 100×	Average number of grains/unit area: No./mm² at 1×
$N = 1$	$n = 1.00$	$n = 15.50$
2	2.00	31.00
3	4.00	62.00
4	8.00	124.00
5	16.00	496.00
6	32.00	992.00
7	64.00	1984.0
8	128.00	3968.0
9	256.00	3968.0
10	512.00	7936.0

Source: Ref A.7 as published in Ref A.1

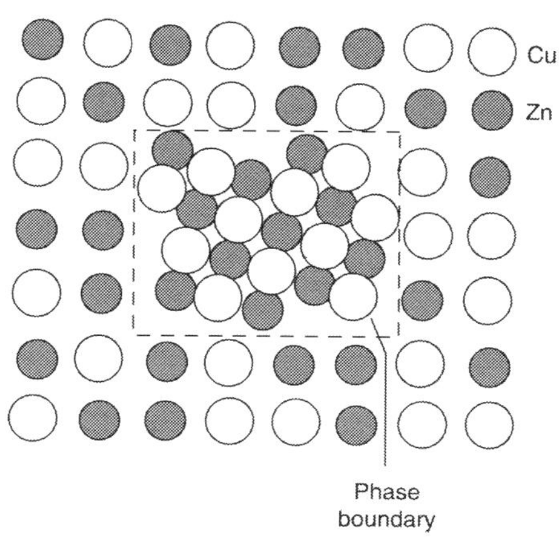

Fig. A.55 Phase boundary in copper-zinc system. Source: Ref A.8 as published in Ref A.1

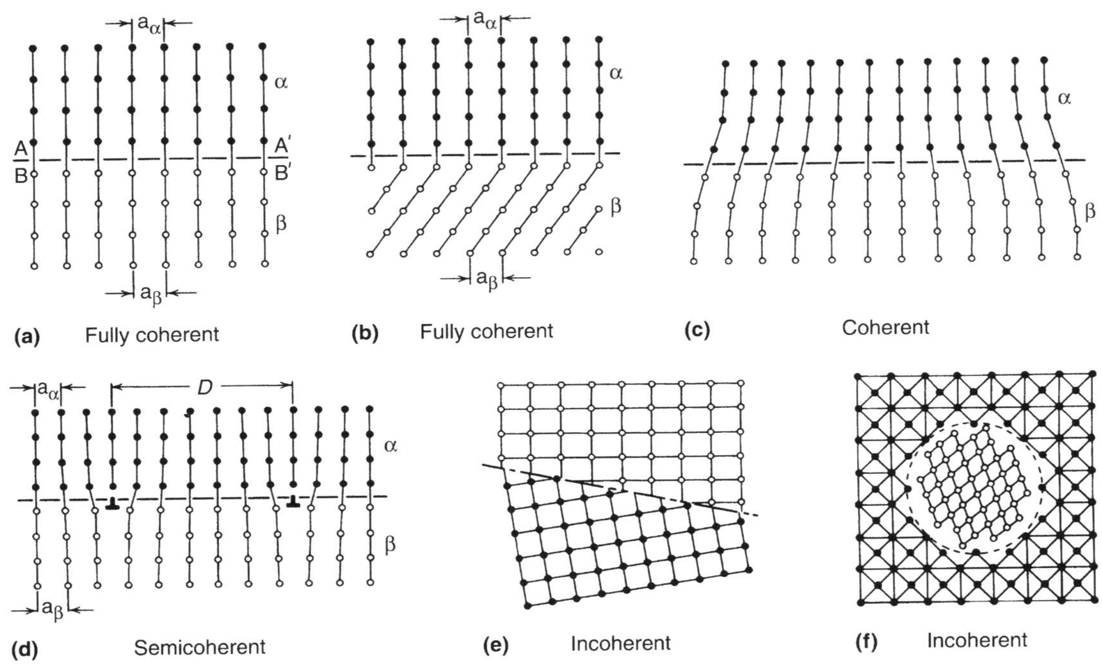

Fig. A.56 Phase boundaries. Source: Ref A.9 as published in Ref A.1

interfacial energy is low, typically up to approximately 200 mJ/m^2. When the distances between atoms at the interface are not identical (Fig. A.56c), coherency strains start to develop; however, because there is still perfect atomic matching, it is still a coherent phase boundary, only the interfacial energy will be higher than one with no distortion. When the mismatch becomes sufficiently large, dislocations form to accommodate the growing disregistry. The result is called a semicoherent interface (Fig. A.56d) that has an interfacial energy of 200 to 500 mJ/m^2. Finally, an incoherent interface (Fig. A.56e, f) is an interphase boundary that results when the matrix and precipitate have very different crystal structures, and little or no atomic matching can occur across the interface. The interfacial energy is even greater, reaching values between 500 and 1000 mJ/m^2. An incoherent boundary is essentially the same as a high-angle grain boundary.

In many instances, second phases have a tendency to form at the grain boundaries. This occurs because they reduce their interfacial energy by occupying a grain boundary; that is, by occupying a grain boundary, part of the interfacial energy is eliminated and the total energy of the system is reduced. Consider the case where two grains of α phase meet with one grain of β phase, as shown in Fig. A.57. The surface energy, γ, will be in equilibrium if:

$$\gamma_{\alpha\alpha} = 2\gamma_{\alpha\beta} \cos\frac{\theta}{2} \quad \text{(Eq A.33)}$$

The angle, θ, is called the dihedral angle at the α-to-β interface. If $\gamma_{\alpha\beta} > \frac{1}{2}\gamma_{\alpha\alpha}$, θ will have a finite value; if $\gamma_{\alpha\beta} = \gamma_{\alpha\alpha}$, $\theta = 120°$; and if $\gamma_{\alpha\beta} > \gamma_{\alpha\alpha}$, $\theta > 120°$. However, if $\gamma_{\alpha\alpha} > 2\gamma_{\alpha\beta}$, Eq A.30 cannot be satisfied and no equilibrium will exist. Instead, the β phase will wet the grain boundary and spread out as a thin grain-boundary film. In this case, if the β phase is brittle, or has a low melting point, the mechanical properties of the alloy will be impaired even though the α matrix is strong and tough. This potentially disastrous condition, known as grain-boundary embrittlement, is shown in Fig. A.58.

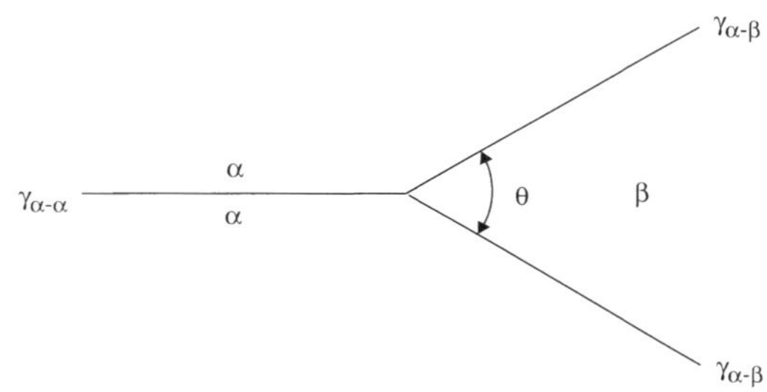

Fig. A.57 Dihedral angle, θ, between two interfaces of differing phases. Source: Ref A.6 as published in Ref A.1

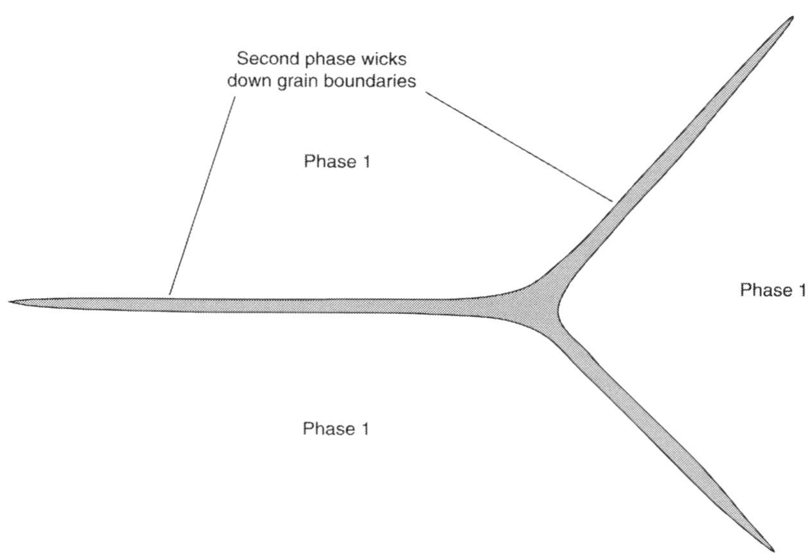

Fig. A.58 Grain-boundary embrittlement. Source: Ref A.2 as published in Ref A.1

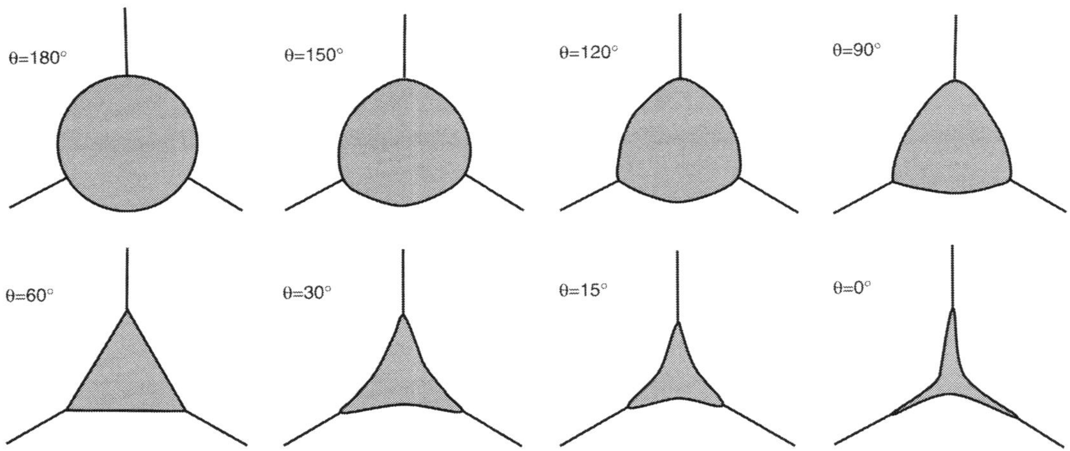

Fig. A.59 Effects of dihedral angle on second-phase shape. Source: Ref A.6 as published in Ref A.1

When the second phase is located at the juncture of three grains, it can form different shapes (Fig. A.59), depending again on the dihedral angle, θ. At small angles, such as θ = 0°, the second phase can penetrate the grain boundaries and possibly affect alloy properties, while at the other extreme (θ = 180°), it will form round particles that should not inhibit alloy performance. Second-phase particles of lead are often added to alloys to improve machinability by forming cleaner chips. Because they form round particles, such as in the case where θ = 180°, they do not adversely affect strength or ductility. There are even some instances where grain-boundary wetting is desirable, such as during liquid phase sintering of carbide cutting tools. Here, a cobalt matrix wets the tungsten carbide particles and binds them together during sintering.

A.10.3 Twinning

Twinning is another mechanism that causes plastic deformation, although it is not nearly as important as dislocation movement. Mechanical twinning is the coordinated movement of large numbers of atoms that deforms a portion of the crystal by an abrupt shearing motion. Atoms on each side of the twinning plane, or habit plane, form a mirror image with those on the other side of the plane (Fig. A.60). Shear stresses along the twin plane cause atoms to move a distance that is proportional to the distance from the twin plane. However, atom motion with respect to one's nearest neighbors is less than one atomic spacing. Twins occur in pairs, such that the change in orientation of the atoms introduced by one twin is restored by the second twin. Twinning occurs on a definite crystallographic plane and in a specific direction that depends on the crystalline structure. Twins

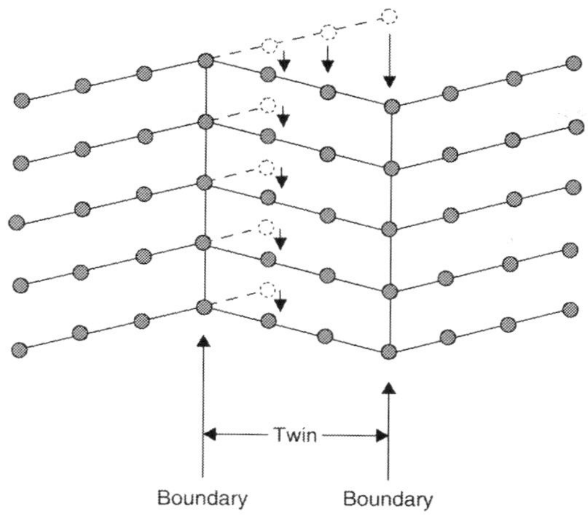

Fig. A.60 Deformation by twinning. Source: Ref A.1

can occur as a result of plastic deformation (deformation twins) or during annealing (annealing twins). Mechanical twinning occurs in bcc and hcp metals, while annealing twins are fairly common in fcc metals.

Mechanical twinning increases the strength because it subdivides the crystal, thereby increasing the number of barriers to dislocation movement. Twinning is not a dominant deformation mode in metals with multiple slip systems, such as fcc structures. Mechanical twinning occurs in metals that have bcc and hcp crystalline structures at low temperatures and at high rates of loading, conditions in which the normal slip process is restricted due to few operable slip systems. The amount of bulk plastic deformation in twinning is small compared to slip. The real importance of twinning is that crystallographic planes are reoriented so that additional slip can take place.

Unlike slip, the shear movements in twinning are only a fraction of the interatomic spacing, and the shear is uniformly distributed over volume rather than localized on a number of distinct planes. Also, there is a difference in orientation of the atoms in the twinned region compared to the untwinned region that constitutes a phase boundary. Twins form suddenly, at a rate approaching the speed of sound, and can produce audible sounds such as "tin cry." Because the amount of atom movement during twinning is small, the resulting plastic deformation is also small.

A comparison of the slip and twinning mechanisms is shown in Fig. A.61. The differences between the two deformation mechanisms include:

- *Orientation*. In slip, the orientation above and below the slip plane is the same after slip, while in twinning there is an orientation change across the twin plane.

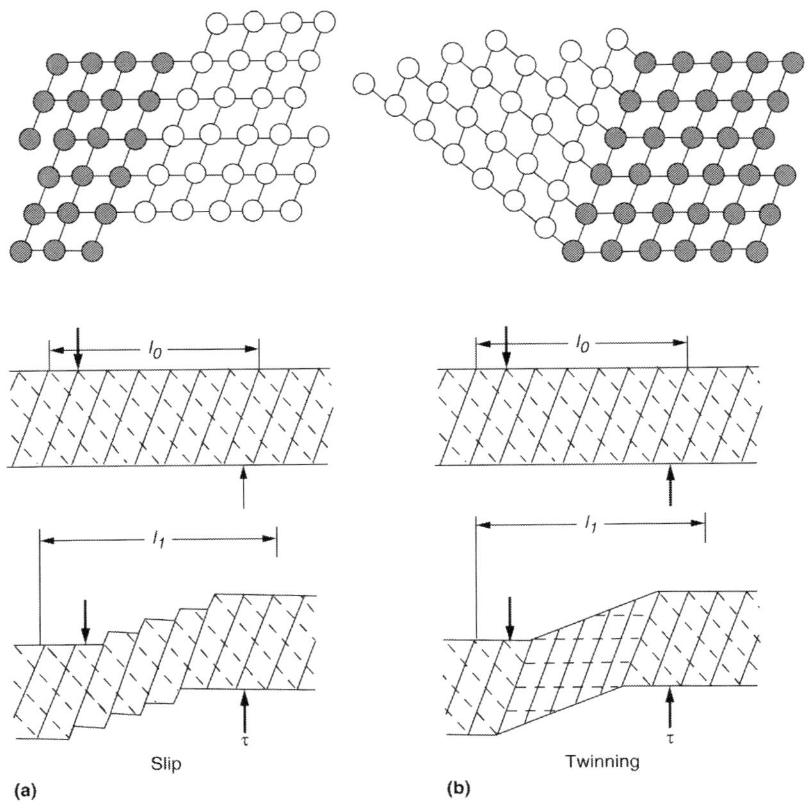

Fig. A.61 Comparison of slip and twinning deformation mechanisms occurring over a length, l, under a shear stress, τ. Source: Ref A.5 as published in Ref A.1

- *Mirror image*. Atoms in the twinned portion of the lattice form a mirror image with the untwinned portion. No such relationship exists in slip.
- *Deformation*. In slip, the deformation is nonhomogeneous because it is concentrated in bands, while the metal adjacent to the bands is largely undeformed. In twinning, the deformation is homogeneous because all of the atoms move cooperatively at the same time.
- *Stress*. In slip, a lower stress is required to initiate it, while a higher stress is required to keep it propagating. In twinning, a high stress is required to initiate, but a very low stress is required for propagation. The shear stress required for twinning is usually higher than that required for slip.

Twin boundaries generally are very flat, appearing as straight lines in micrographs, and are two-dimensional defects of lower energy than high-angle grain boundaries. Therefore, twin boundaries are less effective as sources and sinks of other defects and are less active in deformation and corrosion than ordinary grain boundaries.

Another type of deformation similar to twinning is kink band formation. Kink band formation usually occurs in hcp metals under compressive loading. It occurs when the applied stress is nearly perpendicular to the principal slip plane, normally the basal plane in hcp metals. In this mechanism, the metal shears by the formation of dislocation arrays that produce buckling along the slip plane direction that shears the metal several degrees away from its previous position. As opposed to twinning, the atomic positions do not form a mirror image after shear displacement.

A.10.4 Stacking Faults

The passage of a total dislocation through a crystalline lattice leaves the perfection of the lattice undisturbed. Each atom is shifted from one normal position in the lattice to an adjacent normal position. However, the energy of the system can sometimes be lowered if a total dislocation splits into a partial dislocation (Fig. A.62). It takes less energy if the total dislocation splits into two partial dislocations that can move in a zigzag path through the valley between atoms, rather than having to climb over an atom. Instead of an atom moving directly from its lattice position to a new position, indicated by the tip of the arrow of the Burgers vector, it can move first to an intermediate vacant site and then again to the final site. Thus, two short jumps are made instead of one longer one, which requires less energy. However, the passage of a partial dislocation leaves behind a planar region of crystalline imperfection.

The planar imperfection produced by the passage of a partial dislocation is called a stacking fault, as illustrated in Fig. A.63. In a fcc structure, the stacking sequence changes from the normal *ABCABC* to *ABAB*, which is the stacking sequence for the hcp structure. Passage of the second partial dislocation restores the normal *ABCABC* stacking sequence. These partial dislocations are often referred to as Schottky partials. The two partial dis-

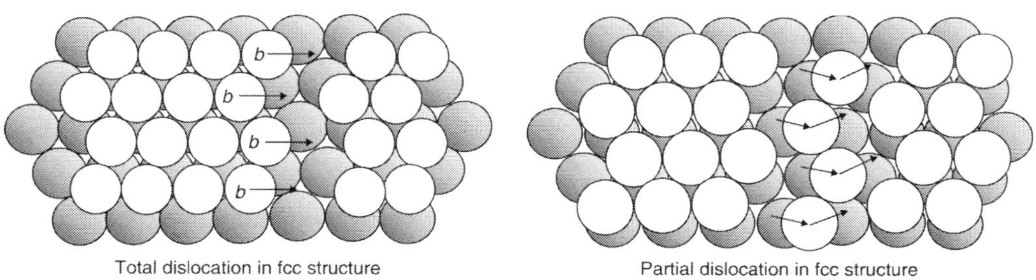

Fig. A.62 Concept of partial dislocation. Source: Ref A.3 as published in Ref A.1

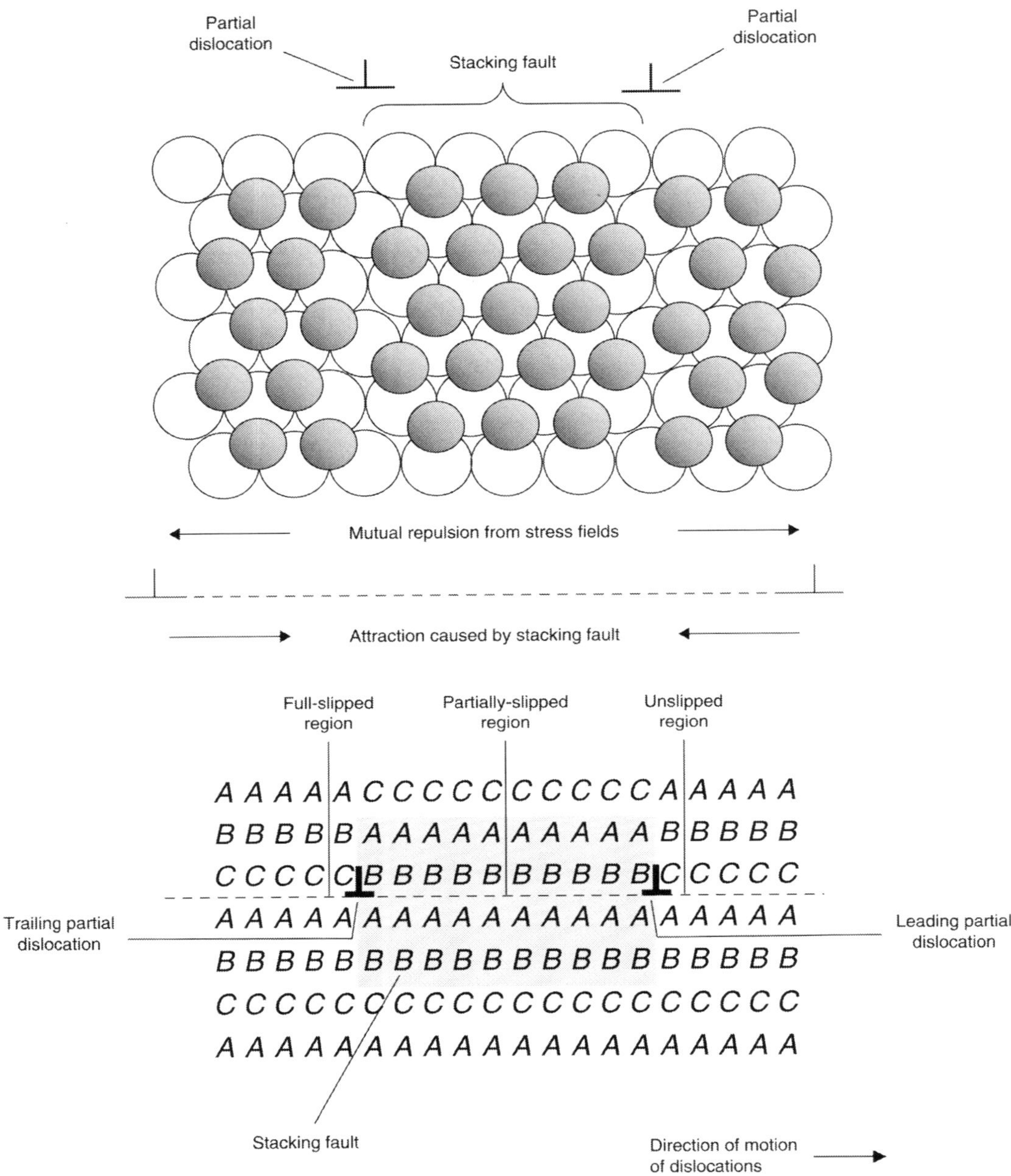

Fig. A.63 Stacking fault and extended dislocation. Source: Ref A.10 and A.3 as published in Ref A.1

locations that are separated by the faulted area are known as an extended dislocation.

The total energy of a perfect lattice is lower than one with a stacking fault. Thus, a stacking fault has an energy associated with it. The difference in energy between a perfect lattice and one with a stacking fault is known as the stacking fault energy (SFE). Equilibrium occurs between the repulsive energy of the two partials and the surface energy of the fault. The larger the separation between the partial dislocations, the smaller is the repulsive force between them. On the other hand, the surface energy associated with the stacking fault increases with the distance between the two partial dislocations. In general, if the separation between the partial dislocations is small, the metal is said to have a high stacking fault energy. If the separation is large, the metal would have a low stacking fault energy. For example, the separation in aluminum (high SFE) is on the order of an atomic spacing, while that of copper (low SFE) is approximately 12 atomic spacings.

Stacking fault energy plays a role in determining deformation textures in fcc and hcp metals. Stacking faults also influence plastic deformation characteristics. Metals with wide stacking faults (low SFE) strain harden more rapidly and twin more readily during annealing than those with narrow stacking faults (high SFE). Some representative SFEs are given in Table A.5.

A.11 Volume Defects

Volume defects, such as porosity and microcracks, almost always reduce strength and fracture resistance. The reductions can be quite substantial, even when the defects constitute only several volume percent. Shrinkage during solidification can result in microporosity; that is, porosity having diameters on the order of micrometers. In metals, porosity is much more likely to be found in castings than in wrought products. The extensive plastic deformation during the production of wrought metals is usually sufficient to heal or close microporosity.

Powder metallurgy products also frequently contain porosity. Powder metallurgy products are usually produced by blending metal powders, pressing them into a shape, and then sintering them at temperatures just below the melting point. Porosity in powder metallurgy products can be reduced if pressure is used during the sintering process by either hot pressing in a press or hot isostatic pressing under gas pressure.

Table A.5 Approximate stacking fault energies (SFEs)

Metal	SFE, mJ/m^2
Brass	<10
Austenitic Stainless Steel	<10
Silver	20–25
Gold	50–75
Copper	80–90
Nickel	130–200
Aluminum	200–250

Source: Ref A.1

ACKNOWLEDGMENT

The contents of this chapter came from *Elements of Metallurgy and Engineering Alloys* by F.C. Campbell, ASM International, 2008.

REFERENCES

A.1 F.C. Campbell, *Elements of Metallurgy and Engineering Alloys*, ASM International, 2008

A.2 V. Singh, *Physical Metallurgy*, Standard Publishers Distributors, 1999

A.3 R.E. Reed-Hill and R. Abbaschian, *Physical Metallurgy Principles*, 3rd ed., PWS Publishing Company, 1991

A.4 W.D. Callister, *Fundamentals of Materials Science and Engineering*, 5th ed., John Wiley & Sons, Inc., 2001

A.5 M. Tisza, *Physical Metallurgy for Engineers*, ASM International, 2001

A.6 R.M. Brick, A.W. Pense, and R.B. Gordon, *Structure and Properties of Engineering Materials*, 4th ed., McGraw-Hill Book Company, 1977

A.7 "Determining Average Grain Size," E112-96, *Annual Book of ASTM Standards,* ASTM International, 1999, p 237–259

A.8 D.R. Askeland, *The Science and Engineering of Materials*, 2nd ed., PWS-Kent Publishing Co., 1989

A.9 M. Epler, Structures by Precipitation from Solid Solution, *Metallography and Microstructures*, Vol 9, *ASM Handbook*, ASM International, 2004

A.10 A.G. Guy, *Elements of Physical Metallurgy*, 2nd ed., Addison-Wesley Publishing Company, 1959

SELECTED REFERENCES

- M.F. Ashby and D.R.H. Jones, *Engineering Materials 1—An Introduction to Their Properties, and Applications*, 2nd ed., Butterworth Heinemann, 1996
- H. Baker, The Chemical Elements, *Metals Handbook Desk Edition*, 2nd ed., ASM International, 1998
- H. Baker, Introduction to Alloy Phase Diagrams, *Alloy Phase Diagrams*, Vol 3, *ASM Handbook*, ASM International, 1992, reprinted in *Desk Handbook: Phase Diagrams for Binary Alloys*, 2nd ed., H. Okamoto, Ed., ASM International, 2010
- T.H. Courtney, Fundamental Structure-Property Relationships, *Engineering Materials, Materials Selection and Design*, Vol 20, *ASM Handbook*, ASM International, 1997
- A.M. Russell and K.L. Lee, *Structure-Property Relationships in Nonferrous Alloys*, Wiley-Interscience, 2005

- D.R. Sadoway, "The Imperfect Solid State," Lecture notes, Introduction to Solid-State Chemistry, Department of Materials Science and Engineering, Massachusetts Institute of Technology, Fall 2006
- W.F. Smith, *Principles of Materials Science and Engineering*, McGraw-Hill, 1986

APPENDIX B

Fundamentals of Solidification

ALMOST ALL METALS AND ALLOYS are produced from liquids by solidification. Sometimes, the liquid metal is poured in a mold with a shape that is close to the desired final shape, a process called casting. The cast part then requires little or no final machining prior to use. On the other hand, simple castings called ingots are initially cast and then shaped by combinations of hot and cold working into wrought product forms, such as plate, sheet, or bar stock. For both castings and wrought products, the solidification process has a major influence on both the microstructure and mechanical properties of the final product.

B.1 The Liquid State

The liquid state is intermediate between a gas and a crystalline solid. Liquids possess neither the long-range order of solids nor the complete disorder of gases. Liquids do, however, possess short-range order. In fact, the structure of liquid metals is very close to the solid that freezes from it. For close-packed metals with the face-centered cubic (fcc) and hexagonal close-packed (hcp) crystalline structures, melting causes a volume decrease of only 3.5 to 6% and even less in the more loosely packed body-centered cubic (bcc) metals, which usually exhibit a volume decrease of only 1 to 3% on freezing. Some elements with even looser atomic packing, such as silicon, actually expand on freezing. In addition, the latent heat of fusion is only approximately 3 to 4% of the corresponding latent heat of vaporization. The small amount of volume change and the relatively small latent heats of fusion are evidence that the atomic bonding in liquids and solids is very similar.

X-ray diffraction studies of liquid metals indicate that the atoms in liquids are arranged in an orderly manner over short distances, but lack long-range order. This is probably due to the presence of increased amounts of defects such as vacancies, interstitials, and dislocations. This large population of defects is believed to be responsible for the much higher rates of diffusion in liquid metals. The higher diffusion rates leads to a picture of liquid metals in which the structure is continuously changing. Atoms form clusters with a high degree of short-range order that may exist for a moment and then fall apart to appear again in another location. As the liquid is cooled towards its freezing point, the degree of short-range order increases and the ordered clusters become larger and more stable.

B.2 Solidification Interfaces

The solidifying solid-liquid interface can exhibit one of three types of interfacial growth in the liquid: planar, cellular, or dendritic. As shown in Fig. B.1, the type of growth is controlled by the manner in which heat is removed from the system. When the liquid ahead of the solid-liquid interface has a positive temperature gradient, heat is removed from the liquid by conduction through the growing solid. Because the temperature gradient is linear and uniform perpendicular to the interface, a smooth interface is maintained and the growth is planar into the liquid (Fig. B.1a). When there is a temperature inversion and the temperature decreases ahead of the solid-liquid interface, then either cellular or dendritic growth will occur (Fig. B.1b, c). The difference between the two is a matter of degree; small undercoolings will tend to produce cellular growth, while large undercoolings will tend to produce dendritic growth. In pure metals, undercooling can result from thermal supercooling, while in alloys, it can result from a combination of thermal and constitutional supercooling.

Thermal supercooling. Cellular growth, as illustrated in Fig. B.2, occurs when the advancing planar solid-liquid interface becomes unstable, and a small spike appears on the interface that then grows into a cellular-type structure. The planar surface becomes unstable because any part of the interface that grows ahead of the remainder enters a region in the liquid that is at a lower temperature. The initial spikes that form remain isolated at first, because as they grow by solidification, they release their latent heat of fusion into the adjacent liquid, causing a localized increase in the temperature. Consequently, parallel spikes of almost equal spacing advance into the liquid.

Dendritic growth is a further manifestation of cellular growth in which the spikes develop side protrusions. At still higher undercooling and higher growth velocities, the cells grow into rapidly advancing projections, sometimes of complex geometry. Their treelike forms (Fig. B.3) have given them the name dendrites, after the Greek word *dendros* for tree. The secondary arms of dendrites develop perpendicular to the primary arms because, as the primary arm solidifies and gives off its latent heat of reaction, the

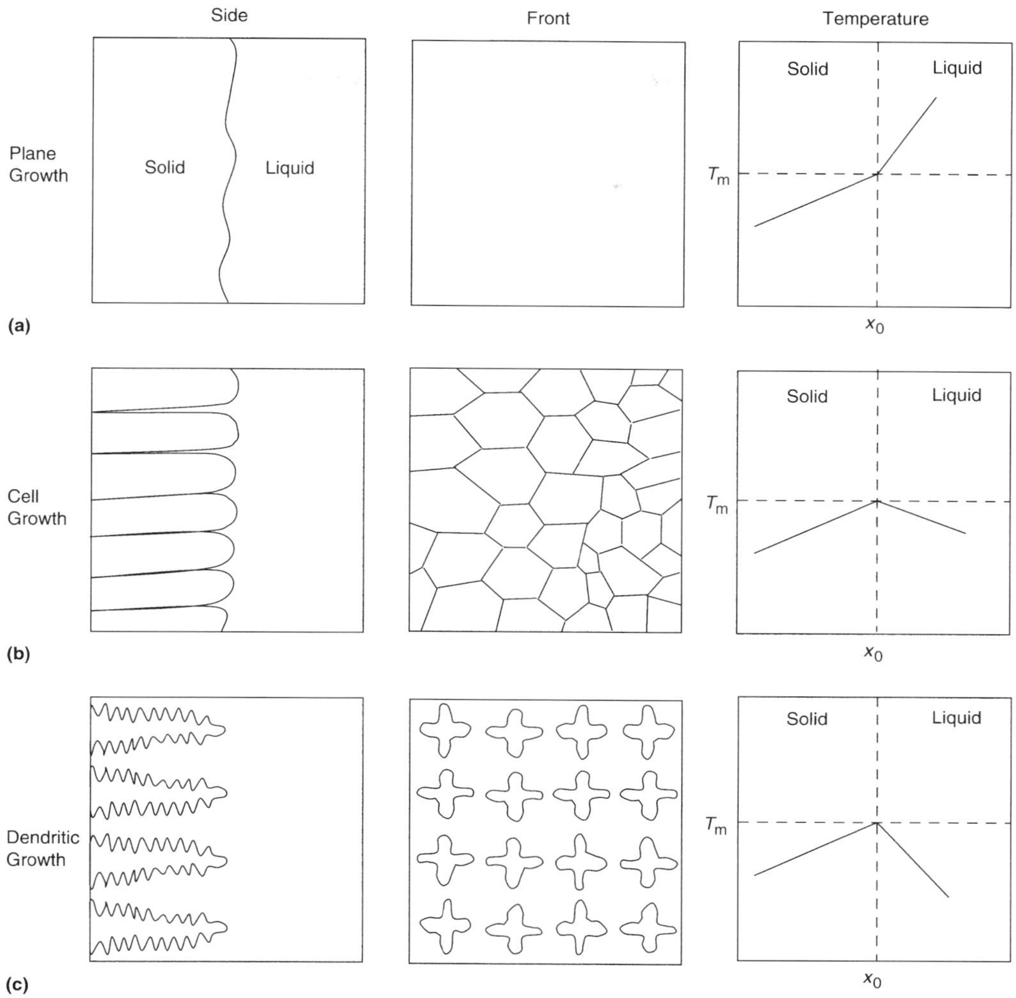

Fig. B.1 Effects of undercooling on solidification structures. Source: Ref B.1 as published in Ref B.2

temperature immediately adjacent to the primary arm increases. This creates another temperature inversion in the liquid between the primary arms, so secondary arms shoot out in that direction. A similar argument can be made for the formation of the tertiary arms. The spacing of the secondary arms is proportional to the rate at which heat is removed from the casting during solidification, with faster cooling rates producing smaller dendrite arm spacings. Dendrites start as long thin crystals that grow into the liquid and thicken. The change in interface morphology of a succinonitrile-4% acetone solution with increasing solidification velocity (Fig. B.4) demonstrates the evolution of a dendritic structure.

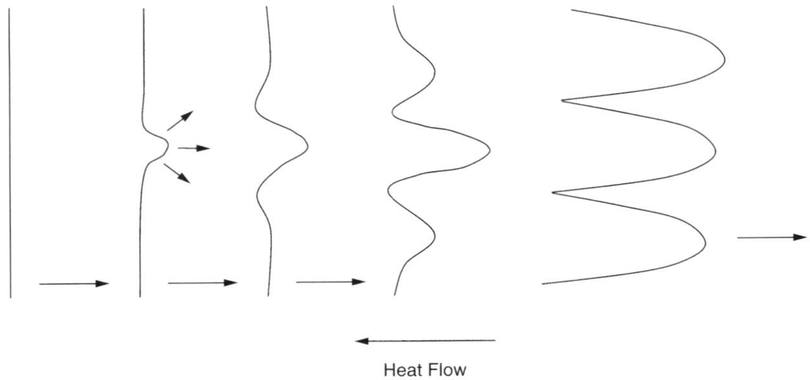

Fig. B.2 Transition from planar to cellular growth. Source: Ref B.2

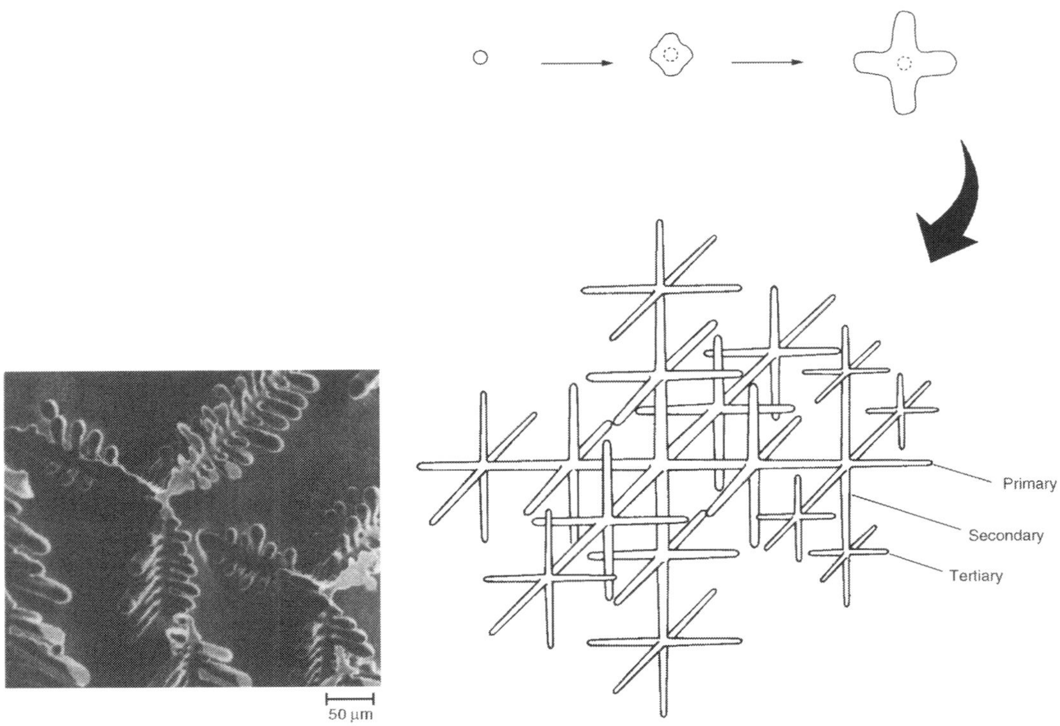

Fig. B.3 Dendrite formation. Source: Ref B.3 as published in Ref B.2

Constitutional Supercooling. While dendrites can form to a limited extent in pure metals due to temperature inversions, they are more prevalent in alloys because of the additional undercooling due to constitutional supercooling. Constitutional supercooling arises because of the

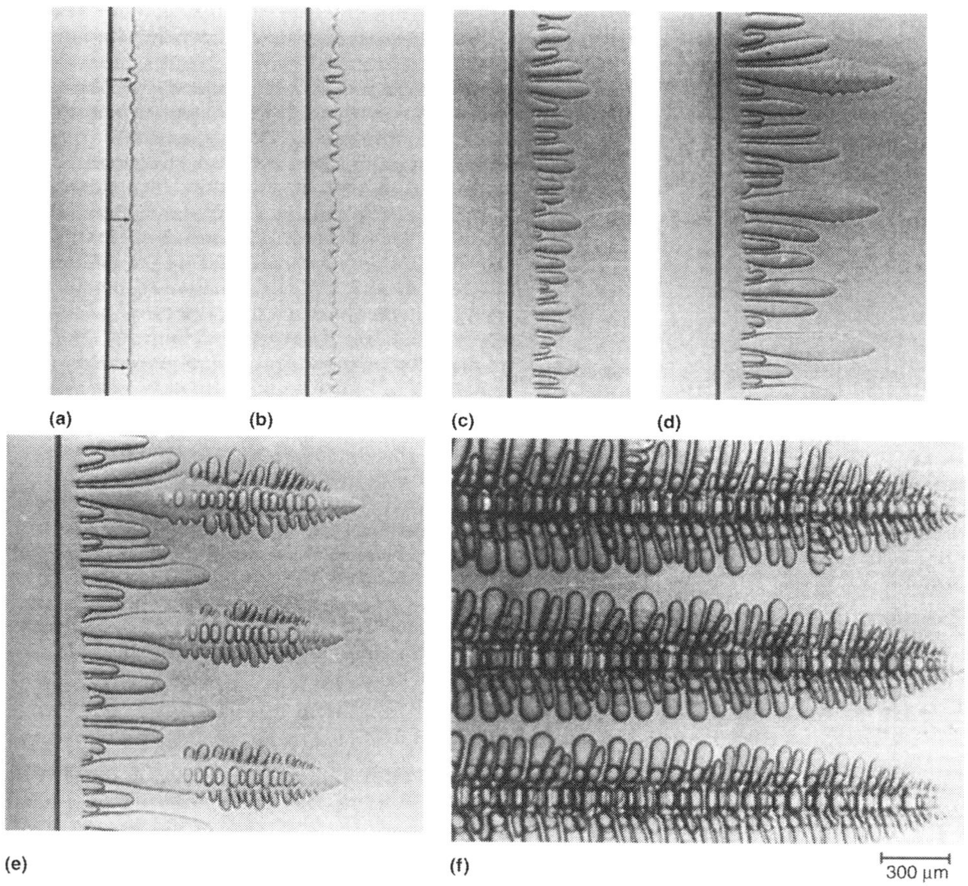

Fig. B.4 Dendrite formation in succiononitrile-4% acetone solution. Source: Ref B.3 as published in Ref B.2

segregation of alloying elements ahead of the solid-liquid front. The extra concentration of alloying elements reduces the melting point of the liquid. If this reduction is sufficient to reduce the melting point to below the actual temperature at that point, then the liquid is said to be locally constitutionally supercooled; that is, it is effectively undercooled because of a change in the constitution of the liquid.

B.3 Solidification Structures

A metal cast into a mold can have up to two or three distinct zones: a chill zone, a zone containing columnar grains, and a center-equiaxed grain zone (Fig. B.5). It should be noted that all three zones do not always occur. For example, pure metals can exhibit a chill zone and a columnar zone but do not contain a center-equiaxed zone.

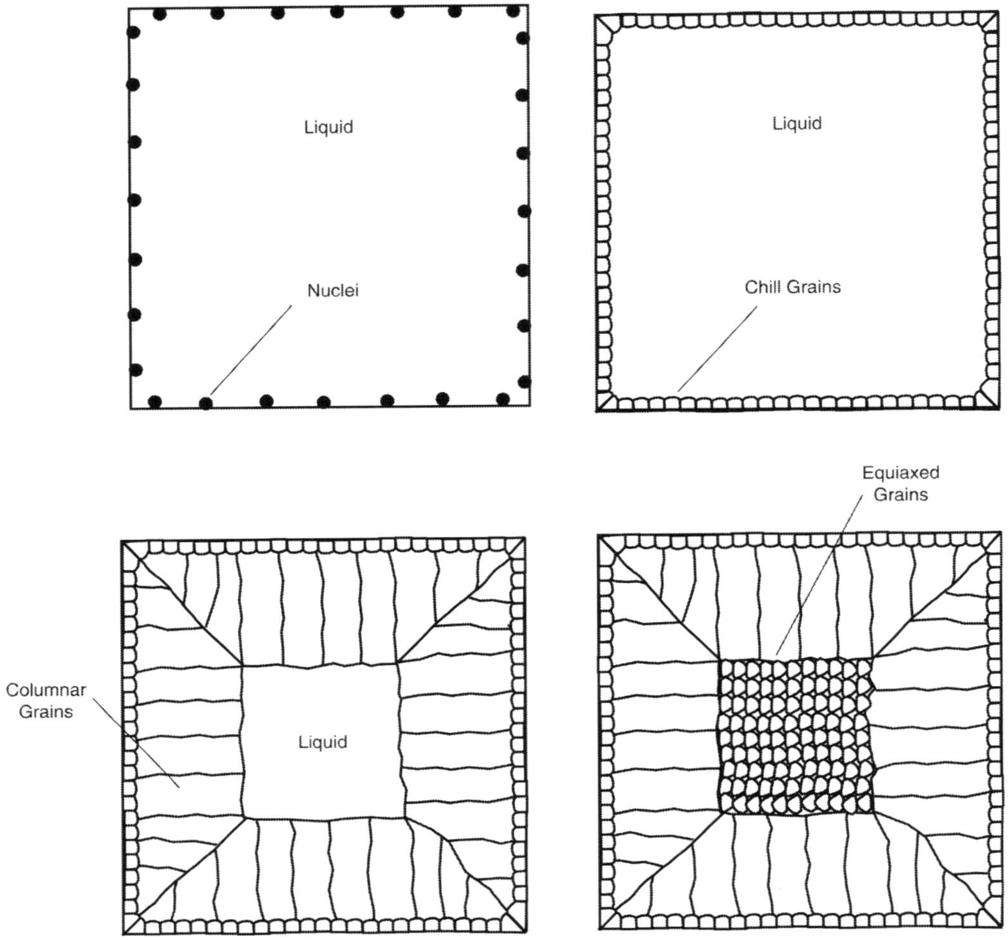

Fig. B.5 Freezing sequence for an alloy casting. Source: Ref B.3 as published in Ref B.2

Chill Zone. Because the mold is cooler than the metal, nucleation will occur over the interior surface of the mold. Because the mold wall is much cooler than the liquid metal, the nucleation rate is high and thus the average grain size is small. Each nucleation event produces an individual crystal, or grain, that grows dendritically in a direction roughly perpendicular to the mold wall until it impinges on other grains. As a large amount of latent heat of fusion is released from the solidifying grains, and as the superheat of the liquid is dissipated, the rate of growth decreases. The chill zone grains are oriented randomly with respect to the mold; that is, the major axis of each grain is randomly oriented.

Columnar Zone. Inside of the chill zone, there are a series of columnar, or column-shaped, grains that are oriented almost parallel to the heat flow direction. Because each metal grows more favorably in one principal crystallographic direction, only those grains favorably oriented with their growth direction most perpendicular to the mold wall will grow into the center of the casting. The axes of the columnar grains are parallel to the direction of heat flow and they grow along specific crystalline planes. As they grow, the more perfectly oriented grains grow ahead of the less perfectly oriented grains and crowd them out (Fig. B.6). The fact that the grains most favorably oriented to the mold wall grow the fastest means that the final shape of the grains in a pure metal casting is columnar, with the parallel columns growing progressively from the mold wall into the center of the casting. The final columnar structure results from parallel growth of different colonies of dendrites and the gradual lateral growth, or thickening, between them.

Center-Equiaxed Zone. A third region at the center of some alloy castings consists of smaller grains that are randomly oriented and nearly equiaxed. As freezing progresses, the thermal gradient decreases, and this causes the dendrites to become very long. Breakdown of columnar growth may occur as a result of fracturing of the very long dendrite grains by convection currents in the melt. These broken arms (Fig. B.7) can then serve as nuclei for new grains. Another possibility is that new grains nucleate as a result of a low thermal gradient and segregation that is occurring during freezing. At low casting temperatures, the entire the casting may solidify with an equiaxed structure.

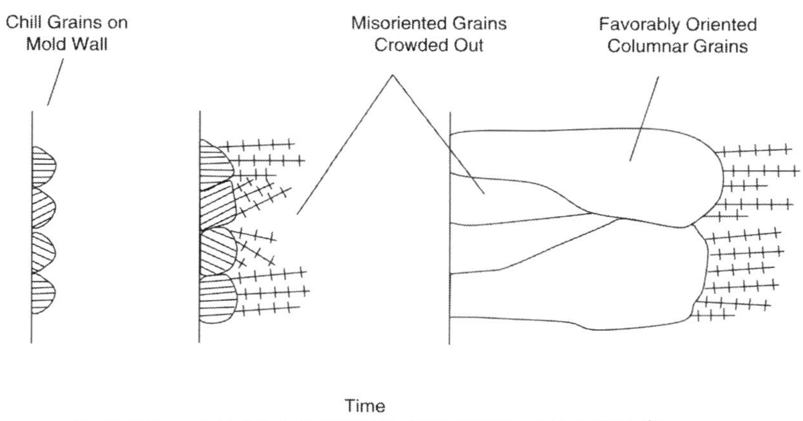

Fig. B.6 Dendritic growth from mold wall. Source: Ref B.1 as published in Ref B.2

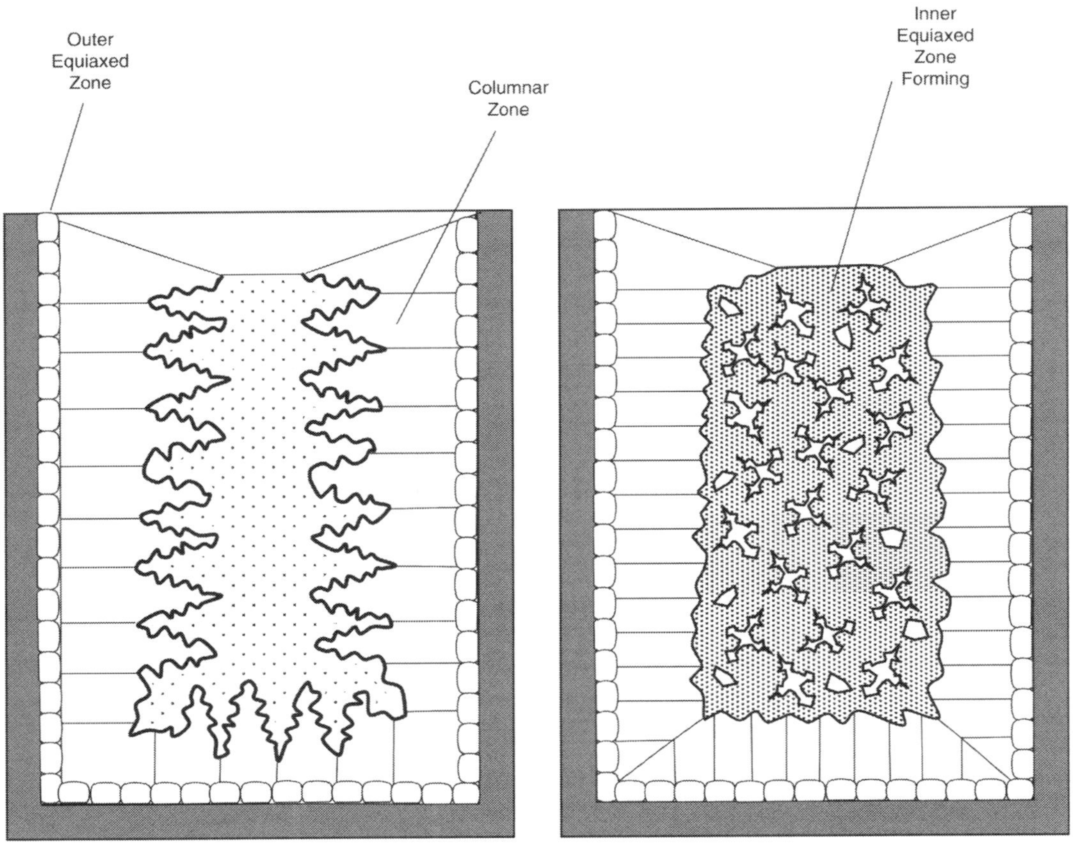

Fig. B.7 Formation of equiaxed zone in alloy casting. Source: Ref B.4 as published in Ref B.2

The amount of the final cast structure that is columnar or equiaxed depends on the alloy composition and on the thermal gradient at the liquid-solid interface during solidification. A thermal gradient is most easily controlled by controlling the rate of heat extraction from the casting, or the cooling rate. Alloys that have a wide spread between the liquidus and the solidus temperatures solidify with a mostly equiaxed grain structure at normal cooling rates, whereas alloys with small differences in solidus and liquidus temperatures solidify with a mostly columnar structure. High cooling rates encourage columnar solidification because they establish high thermal gradients at the liquid-solid interface. Low thermal gradients encourage equiaxed solidification.

B.4 Segregation

Segregation may be defined as any departure from a uniform distribution of the chemical elements in the alloy. Because of the way in which alloys partition on freezing, it follows that all castings are segregated to different degrees. Alloys can exhibit several types of segregation, and more than one type can be present in any given casting.

Normal Segregation. Normal segregation is the result of rejection of the alloy solute at an advancing interface because it is more soluble in the liquid than the solid. As freezing progresses, there is a buildup of the solute in the liquid that freezes last, such as at the center of the casting. Such long range variations in composition are called macrosegregation. Normal segregation frequently occurs when the direction of growth is inward, as in columnar growth.

An examination of the relevant portion of a phase diagram (Fig. B.8) can be used to show the effects of segregation during freezing. The original melt of composition C_o starts to freeze at the liquidus temperature, T_L. The first solid to appear has the composition $k\ C_o$, where k is known as the partition coefficient. The partition coefficient usually has a value less than 1. For example, for $k = 0.5$, the first solid has only 50% of the concentration of alloying element compared to the original melt; the first metal to appear is therefore usually rather pure. In general, k defines how the solute alloy partitions between the solid and liquid phases. Thus:

$$k = C_S/C_L \qquad \text{(Eq B.1)}$$

where C_S is the solute fraction by weight in solid and C_L is the solute fraction by weight in liquid.

For solidus and liquidus lines that are straight, k is constant for all compositions. However, even where they are curved, the relative matching of the curvatures often means that k is still reasonably constant over wide ranges of composition. When k is close to 1, the close spacing of the liquidus and solidus lines indicates little tendency toward segregation. When k is small, then the wide spacing of the liquidus and solidus lines indicates a strongly partitioning alloying element and a tendency to segregate on solidification.

On forming the solid that contains only $k\ C_o$ amount of alloying element, the alloying element remaining in the liquid has to be rejected ahead of the advancing front. Thus, although the liquid ahead of the front was initially of composition C_o, after an advance of approximately a millimeter or so, the composition ahead of the front builds up to a peak value of C_o/k.

Gravity Segregation. Gravity segregation is another type of macrosegregation that occurs due to differences in density in the liquid melt. Denser constituents tend to sink toward the bottom, while lighter ones float toward the top.

Microsegregation. As the dendrite grows into the melt, and as secondary arms spread from the main dendrite stem, solute is rejected, effectively

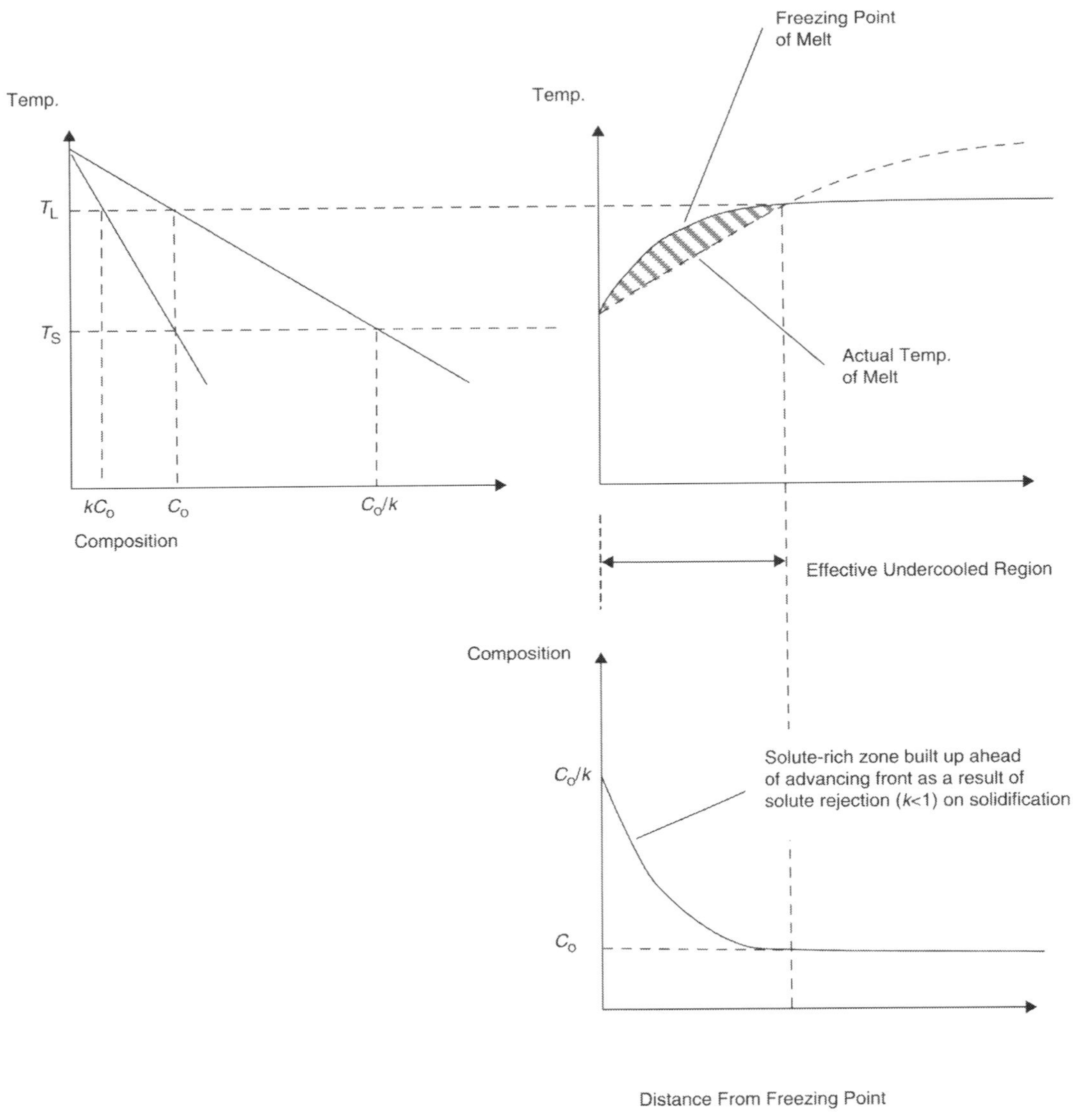

Fig. B.8 Constitutional supercooling. Source: Ref B.1 as published in Ref B.2

being pushed aside to concentrate in the tiny regions enclosed by the secondary dendrite arms. This gradation of composition from the inside to the outside of the dendrite is called coring, because on etching a polished section of such dendrites, the progressive change in composition appears as onionlike layers around a central core.

Inverse Segregation. Inverse segregation occurs when the solute content at, or near, the surface is higher than it is at the center—exactly opposite of what is expected. Under certain conditions, channels form between the dendrites that are growing toward the center of the casting, forming a path in which the last-to-freeze solute-enriched liquid can flow back from the center to the outside. This reverse flow path could be caused by differential contraction between the mold and the ingot creating a suction effect that pulls the solute-enriched liquid toward the surface. Another possible mechanism is that a large amount of gas evolving at the end of solidification pushes the solute-enriched liquid through the dendritic channels toward the surface.

It should be recognized that this discussion on segregation is rather cursory. An examination of the segregation patterns in the aluminum-killed steel ingot of Fig. B.9 shows the complexity of segregation that can be observed in real castings. While ingot homogenization treatments can be effective in smoothing out microsegregation, they are not very effective

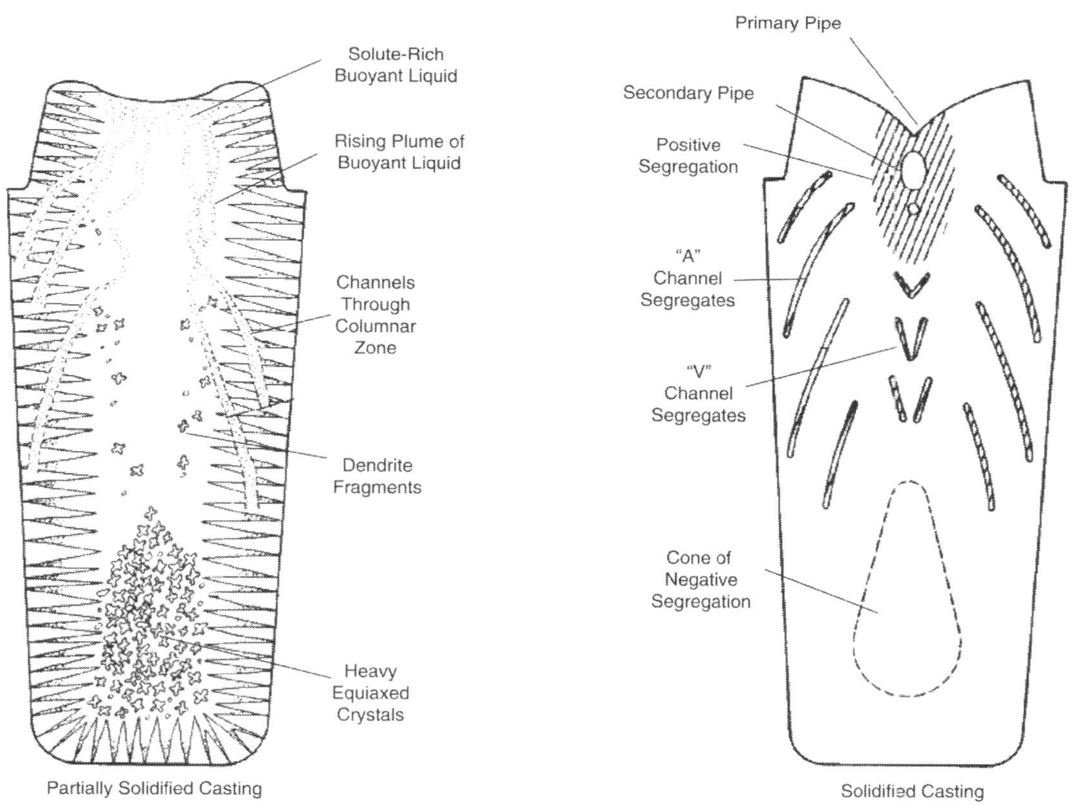

Fig. B.9 Segregation in killed steel ingot. Source: Ref B.1 as published in Ref B.2

for macrosegregation. Because macrosegregation occurs over distances ranging from 1 to 100 cm (0.4 to 40 in.), the diffusion lengths are just too long and the times required are prohibitive. Segregation is one of the prime reasons that wrought products are often preferred over castings. The severe plastic deformation that occurs during hot working will heal internal porosity and help break up and eliminate segregated areas.

B.5 Grain Refinement and Secondary Dendrite Arm Spacing

Rapid solidification rates are usually desirable because they tend to produce finer grain sizes and finer dendrites. Grain size control for castings is important because fine grain sizes result in higher strengths and greater ductility. Because the size of porosity in a casting often corresponds somewhat with grain size, finer porosity also goes with finer grain sizes. In addition, shrinkage and hot cracking are more prevalent in castings with a coarse grain size. Grain size is a function of pouring temperature, solidification rate, and the presence or absence of a grain-refining agent. Low pouring temperatures, faster solidification rates, and grain refiners, such as titanium and boron in aluminum, all produce finer grain sizes.

The solidification rate also affects the size of the dendrites that form during freezing. The dendrite size is normally measured as the distance between the secondary dendrite arms, referred to as the secondary dendrite arm spacing (SDAS). As shown in Fig. B.10, faster solidification rates produce smaller SDAS. The SDAS is related to the solidification time, t, by:

$$\text{SDAS} = kt^n \qquad \text{(Eq B.2)}$$

where k and n are constants that depend on the composition of the metal. Small secondary dendrite arm spacings are associated with higher strengths and improved ductility (Fig. B.11). Homogenization heat treatments are also dependent on the time, t, required to diffuse a solute over a given average distance, x. If the coefficient of diffusion in the solid is D, then the relationship between time and distance is approximately:

$$x = \sqrt{Dt} \qquad \text{(Eq B.3)}$$

Thus, finer SDAS results in shorter homogenization times, or better homogenization in similar times.

B.6 Porosity and Shrinkage

Porosity is the presence of pores in the casting. These pores may be connected to the surface, or they may be wholly internal. While mild porosity

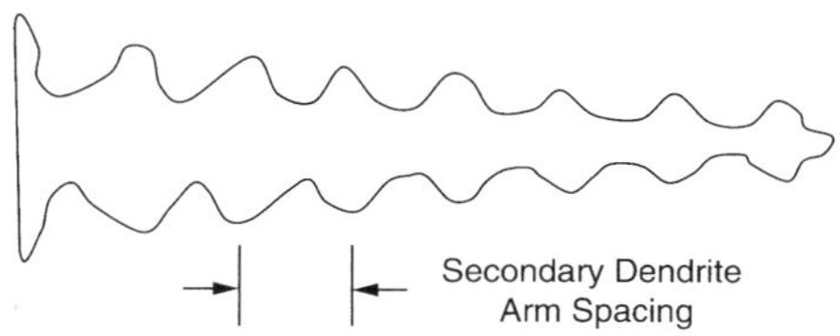

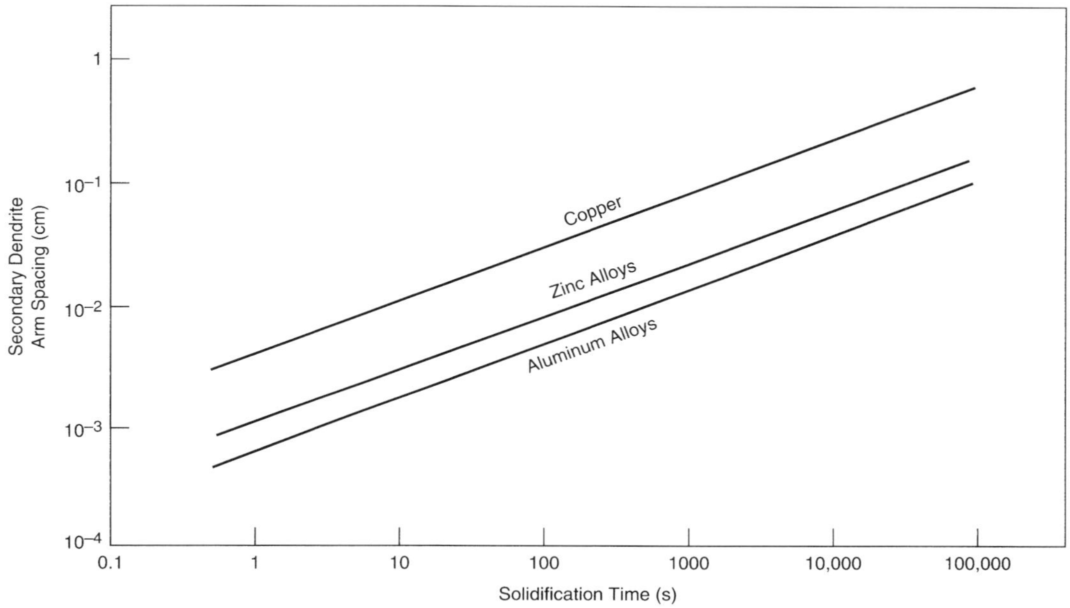

Fig. B.10 Effect of solidification time on secondary dendrite arm spacing. Source: Ref B.5 as published in Ref B.2

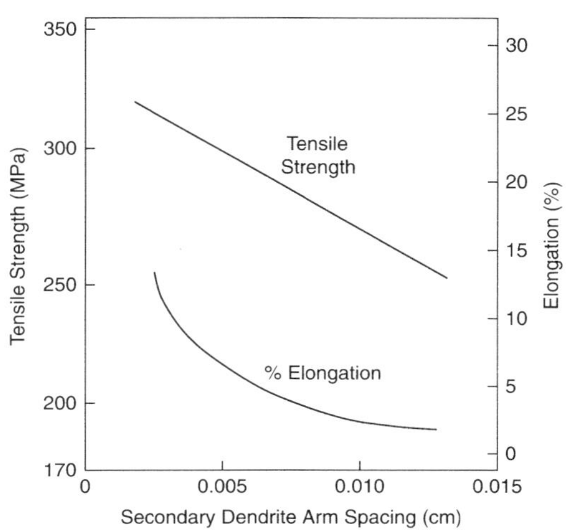

Fig. B.11 Effect of secondary dendrite arm spacing on properties of aluminum casting alloy. Source: Ref B.5 as published in Ref B.2

does not normally degrade static properties, such as tensile strength, it can severely impact dynamic properties, such as fatigue strength. However, a reduction in static properties can occur if the porosity is extensive enough to significantly reduce the cross-sectional area or if it has sharp corners that produce stress concentrations. If the surfaces of small porosity in ingots are not oxidized, they will generally be welded shut during hot working such as rolling or forging. Small porosity in finished castings must either be accepted or healed by processes such as hot isostatic pressing, where the metal is subjected to significant heat and pressure. Porosity that is open to the surface in finished castings is sometimes ground out and weld repaired. Both gas evolution and metal shrinkage during casting can produce porosity.

Gas porosity is caused by the absorption of gases in the liquid metal prior to casting. The most troublesome gas is hydrogen, which usually comes from either moisture in the air or from burning fossil fuels for heating and melting. Moisture and liquid metal can react to form molecular hydrogen according to:

$$xM + yH_2O \rightarrow M_xO_y + yH_2 \qquad \text{(Eq B.4)}$$

where M is the metal, and x and y are stoichiometric coefficients balancing the equation. At the high temperatures in the liquid melt, molecular hydrogen decomposes into atomic hydrogen that dissolves in the metal. The solubility of hydrogen in liquid metals increases sharply with temperature, as shown in Fig. B.12 for nickel and iron. If the hydrogen is not removed prior to solidification, the large decrease in solubility during freezing results in the formation of porosity. One problem with gas porosity is the formation of blisters during subsequent processing such as heat treating or annealing. The equilibrium gas concentration in a liquid metal can be determined by Sievert's law:

$$G = k\sqrt{p_g} \qquad \text{(Eq B.5)}$$

where G is the equilibrium gas solubility, k is the equilibrium constant, and p_g is the partial pressure of gas at the metal surface.

Hydrogen is often removed by vacuum degassing prior to casting. For example, premium-quality steels are commonly vacuum degassed. In addition, some metals, such as titanium, are so reactive that all melting and casting operations must be done under vacuum. In open-air casting operations, hydrogen is removed from the melt by using degassing fluxes. The removal of hydrogen is a mechanical, not chemical, process in which the hydrogen attaches itself to the fluxing gas.

The presence of nonmetallic inclusions has an effect on gas porosity formation. It has been shown that nucleation of pores is difficult in the

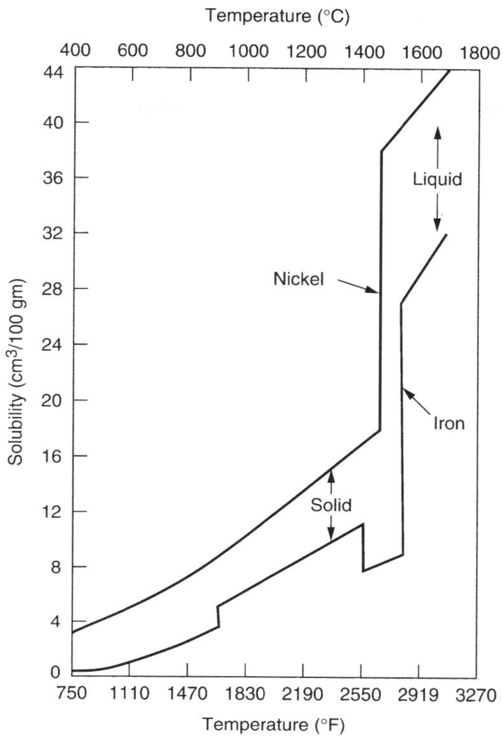

Fig. B.12 Maximum solubility of hydrogen in nickel and iron. Source: Ref B.6 as published in Ref B.2

absence of some sort of substrate, such as a nonmetallic inclusion, a grain refiner, or a second-phase particle. Investigations have shown that castings free of inclusions have less porosity than castings that contain inclusions.

As a metal casting cools and solidifies, the volume decreases, contributing to three types of shrinkage: liquid shrinkage, solidification shrinkage, and solid shrinkage. The shrinkage occurring in the liquid metal itself does not cause casting problems. The extra liquid metal required to compensate for this small reduction in volume is provided by a slight extension to the pouring time or by a slight fall in feeder level.

Solidification shrinkage occurs at the freezing point when the metal transitions from a liquid to a more highly ordered crystalline solid. The volume change on melting for a number of metals is given in Table B.1. Solidification shrinkage is usually compensated for by designing the mold with reservoirs for liquid metal, called risers, which allow liquid metal to flow in and fill the areas being vacated by the solidifying metal. Shrinkage porosity often forms in areas that the liquid metal from the risers cannot reach. For example, it is difficult to effectively feed metal into the interdendritic areas where shrinkage is occurring. Because this type of

Table B.1 Volume changes on melting

Metal	Crystal Structure(a)	Volume change on melting, %
Al	fcc	6.0
Au	fcc	5.1
Cu	fcc	4.2
Zn	hcp	4.2
Ag	fcc	3.8
Pb	fcc	3.5
Fe	bcc	3.4
Sn	bct	2.8
Nb	bcc	0.9
Sb	Rhombohedral	−0.95
Si	Diamond cubic	−12

(a) fcc, face-centered cubic; hcp, hexagonal close-packed; bcc, body-centered cubic; bct, body-centered tetragonal. Source: Ref B.2

porosity occurs late in solidification, especially in alloys with wide freezing ranges and a large mushy zone, it is particularly difficult to eliminate. A comparison of shrinkage porosity and gas porosity is shown in Fig. B.13. Shrinkage porosity tends to follow the directions of the dendritic arms and forms along the grain boundaries, while gas porosity occurs throughout the matrix.

Solid shrinkage occurs after the casting has solidified and is cooling down to room temperature. As cooling progresses, and the casting attempts to contract, it is constrained to some extent by the mold. This constraint always leads to the casting being somewhat larger than would be expected from free contraction alone. This can lead to difficulties in accurately designing the size of the pattern, because the contraction allowance is not easy to quantify. Mold constraint during cooling can also lead to problems such as hot tearing or cracking of the casting.

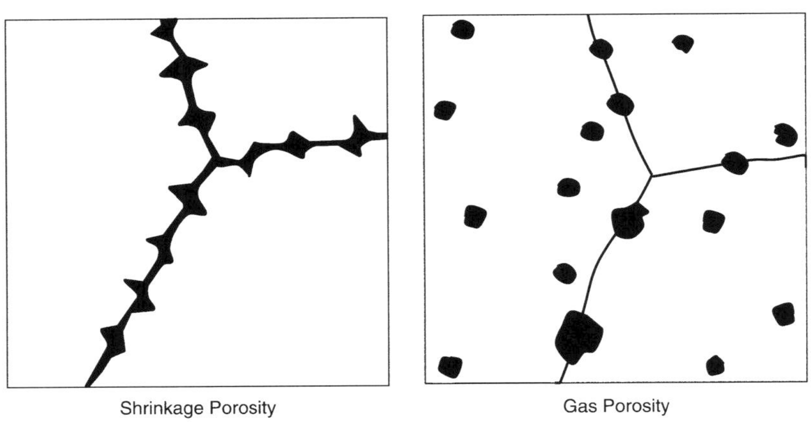

Fig. B.13 Porosity in cast metals. Source: Ref B.7 as published in Ref B.2

ACKNOWLEDGMENT

The material in this chapter came from *Elements of Metallurgy and Engineering Alloys* by F.C. Campbell, ASM International, 2008.

REFERENCES

B.1 J. Campbell, *Castings,* Butterworth Heinemann, 1991

B.2 F.C. Campbell, *Elements of Metallurgy and Engineering Alloys,* ASM International, 2008

B.3 D.M. Stefanescu and R. Ruxanda, Fundamentals of Solidification, *Metallography and Microstructure,* Vol 9, *ASM Handbook,* ASM International, 2004

B.4 M.F. Ashby and D.R.H. Jones, *Engineering Materials 2—An Introduction to Microstructures, Processing and Design,* 2nd ed., Butterworth Heinemann, 1998

B.5 D.R. Askeland, *The Science and Engineering of Materials,* 2nd ed., PWS-Kent Publishing Co., 1989

B.6 A.G. Guy and J.J. Hren, *Elements of Physical Metallurgy,* 3rd ed., Addison-Wesley Publishing Company, 1974

B.7 R.A. Higgins, *Engineering Metallurgy—Applied Physical Metallurgy,* 6th ed., Arnold, 1993

Index

D

E

F

G

H

I

J

K

L

M

Q

R

S

T

U

V

W

X

Z

FSC
www.fsc.org
MIX
Papier | Fördert gute Waldnutzung
FSC® C083411

Kolibri 360 GmbH
Ferdinand-Jühlke-Straße 7
99095 Erfurt, Deutschland
produktsicherheit@kolibri360.de